Green Energy and Technology

Climate change, environmental impact and the limited natural resources urge scientific research and novel technical solutions. The monograph series Green Energy and Technology serves as a publishing platform for scientific and technological approaches to "green"—i.e. environmentally friendly and sustainable—technologies. While a focus lies on energy and power supply, it also covers "green" solutions in industrial engineering and engineering design. Green Energy and Technology addresses researchers, advanced students, technical consultants as well as decision makers in industries and politics. Hence, the level of presentation spans from instructional to highly technical.

More information about this series at http://www.springer.com/series/8059

Akshai K. Runchal · Ashwani K. Gupta
Abhijit Kushari · Ashoke De
Suresh K. Aggarwal
Editors

Energy for Propulsion

A Sustainable Technologies Approach

Springer

Editors
Akshai K. Runchal
The CFD Innovators
Analytic & Computational Research, Inc. (ACRi)
Los Angeles, CA
USA

Ashwani K. Gupta
Department of Mechanical Engineering
University of Maryland
College Park, MD
USA

Abhijit Kushari
Department of Aerospace Engineering
Indian Institute of Technology Kanpur
Kanpur, Uttar Pradesh
India

Ashoke De
Department of Aerospace Engineering
Indian Institute of Technology Kanpur
Kanpur, Uttar Pradesh
India

Suresh K. Aggarwal
Department of Mechanical and Industrial Engineering
University of Illinois at Chicago
Chicago, IL
USA

ISSN 1865-3529 ISSN 1865-3537 (electronic)
Green Energy and Technology
ISBN 978-981-10-7472-1 ISBN 978-981-10-7473-8 (eBook)
https://doi.org/10.1007/978-981-10-7473-8

Library of Congress Control Number: 2018941990

Printed on acid-free paper

This Springer imprint is published by the registered company Springer Nature Singapore Pte Ltd.
The registered company address is: 152 Beach Road, #21-01/04 Gateway East, Singapore 189721, Singapore

Preface

Energy, power, and propulsion, the prime movers of economy and society, pose challenging problems as one seeks cleaner and efficient conversion with due consideration to sustainability. The exergy of contemporary power and propulsion systems is only about a third to half of that theoretically possible so that the scientists and engineers continue to strive for higher efficiency and improved performance. The main focus of the practicing engineers has been to design reliable, efficient, and cost-effective systems with low footprint on the environment. To date, fossil fuels continue to play a pivotal role in all sectors of energy use. With the global increase in CO_2 and other harmful emissions to the environment, one must seek cost-effective measures to control emissions without sacrificing the quality of life or productivity. These environmental concerns, and potential depletion of fossil fuels, have led to increasing global effort to exploit renewable and alternative energy resources.

Presently, there are no real substitutes for fossil fuels but renewable and non-conventional energy sources, such as wind, solar, biomass, algae, geothermal, hydro, and wave, are beginning to have some impact. Per EIA renewables currently contribute over 13% of the total worldwide energy use. This contribution is likely to rise steadily in view of the increasing efforts on R&D to expand the available choices and to improve the conversion and utilization efficiencies. One major advantage of the renewables is that they are mostly environmentally friendly with near-zero carbon emissions. However, continuous and ready availability at all places is a major issue with some of the renewables. Their large-scale use is rather limited. Harnessing the alternative energy resources and using efficient conversion methodologies are pivotal to the future of our society. Several nations have formulated plans to enhance commercial development of these resources through technological advances with due considerations to cost-effectiveness and minimal environmental impact. As of today, renewables offer favorable technical performance compared to fossil fuels, although cost, availability, and robustness could vary at a specific location. However, for a number of reasons, currently, they play only a supportive role in the global energy use. The present book is an effort to explore these technologies in a balanced perspective with due focus on the fossil fuels as well as the emerging renewable resources.

In a short span of two decades, we have seen remarkable advances in the search of alternative fuels and alternative energy conversion systems. In addition, we have seen remarkable performance improvements in fossil fuel power and propulsion systems. The analytical tools have benefitted from advances in computational sciences, with more powerful computing systems and significant improvements in our abilities to make use of evolving computer architecture with multiple processors, and parallel and distributed computing. From its meager beginnings in the 1960s with the advent of the electronic computer, computational sciences have revolutionized our theoretical analysis capabilities, and computational fluid dynamics (CFD) tools have now become ubiquitous in academia and industry. Concurrently, new non-invasive diagnostic techniques are becoming firmly established with the development of advanced instrumentation that provides spatial and temporally resolved measurements of complex flows. Also, in parallel, advances in sensors and active control technology have further helped to improve system efficiencies, reduce emission levels, prolong equipment life, and mitigate catastrophic failures. In spite of these major advances in our predictive and diagnostic capabilities for designing more efficient and cleaner energy systems, challenges remain due to growing energy demand and environmental concerns, as the world GDP is expected to double and energy consumption to increase by some 50% over the next 25 years.

This monograph is the result of deliberations at an "International Workshop on Energy, Propulsion and Environment" held at the Indian Institute of Technology (IIT), Kanpur, India, during March 8–11, 2017. The workshop was sponsored by leading research organizations and universities in the USA and India. Many world-renowned scientists and researchers from the USA, UK, India, Thailand, Malaysia, and Saudi Arabia presented their latest research findings and shared their ideas in formal talks and exchange forums. This research monograph brings together the latest research and wealth of knowledge presented by these eminent scientists at this workshop. A common theme of the monograph is energy, power, and propulsion. It is divided into sections that deal with various aspects of this theme. The chapters on conventional and renewable fuels focus on more efficient and environmentally friendly techniques and systems for power production. Dynamics and stability of flames in combustion systems is a major concern. Some of the chapters present novel ideas on these topics. Finally, the gap in our knowledge and understanding of physical processes that underlie computational simulations and analysis are covered in chapters dealing with turbulence, its interaction with chemical kinetics and the complexities of two-phase processes.

Presently, there are no silver bullets for energy and environment sustainability. Collective efforts between academia, industry, and governments are needed to achieve these goals. We hope this book will contribute to further R&D efforts worldwide to develop cutting-edge and novel technologies in renewable alternatives, conversion efficiency, and environmental sustainability.

We would like to express our gratitude to the authors for submitting their work in a timely manner and adhering to tight deadlines. We acknowledge the support received from various agencies and organizations for the successful conduct of the workshop. These included the National Science Foundation, USA; Office of Naval Research Global, Singapore (with special thanks to Dr. R. Kolar); TEQIP program of MHRD; and Bagell House and Springer. We also thank our industry partners who participated at the workshop that further contributed to its success.

Los Angeles, USA — Akshai K. Runchal
College Park, USA — Ashwani K. Gupta
Kanpur, India — Abhijit Kushari
Kanpur, India — Ashoke De
Chicago, USA — Suresh K. Aggarwal

Contents

About the Editors

Dr. Akshai K. Runchal is the Founder and Director of CFD Virtual Reality Institute and ACRI group of companies. His expertise is in computational fluid dynamics (CFD). He has consulted widely on flow, heat and mass transfer, combustion, environmental impact, hazardous and nuclear waste, groundwater, and decision analysis to over 200 clients in 20 countries. He is the principal author of PORFLOW, TIDAL, ANSWER, and RADM simulation models that are used worldwide. He received his Ph.D. from Imperial College London, UK, and Bachelor of Engineering from PEC. He has taught in the USA, India, and UK. He has authored seven books and over 200 technical publications. He has received many honors and awards, and given many invited contributions at conferences.

Dr. Ashwani K. Gupta is a distinguished university professor at the University of Maryland (UMD), USA. He obtained his Ph.D. and also higher doctorate (D.Sc.) from the University of Sheffield, UK. He was awarded D.Sc. from Southampton University, UK; Honorary doctorates from the University of Wisconsin–Milwaukee, USA, King Mongkut's University of Technology North Bangkok, Thailand, and University of Derby, UK. He received the President Kirwan Research Award and College of Engineering Research Award at UMD. He has received many honors and awards from AIAA and ASME, and has authored over 700 papers, three books, and edited 12 books in the areas of combustion, swirl flows, high temperature air combustion (HiTAC), distributed combustion, wastes to energy, acid gas treatment, fuel reforming, and air pollution. He is Honorary Fellow of ASME and Fellow of AIAA, SAE, AAAS, and RAeS (UK).

Dr. Abhijit Kushari is Professor in the Department of Aerospace Engineering, Indian Institute of Technology Kanpur (IIT Kanpur), Kanpur, India. Dr. Kushari received his Ph.D. from Georgia Institute of Technology, Atlanta, Georgia, in 2000, and joined the IIT Kanpur as an assistant professor in 2001. He became a professor in 2014. His research interests are aerospace propulsion, gas turbine engines, turbo-machinery, liquid atomization flow control, and combustion dynamics.

He has authored more than 130 technical papers. He has supervised six Ph.D. and 50 M.Tech theses in aerospace propulsion fields.

Dr. Ashoke De is Associate Professor in the Department of Aerospace Engineering, Indian Institute of Technology Kanpur (IIT Kanpur), Kanpur, India. He is the recipient of IEI Young Engineers Award 2014, DST Young Scientist Award 2015, and P.K. Kelkar Research Fellowship from IIT Kanpur. He is a member of ASME, SIAM, FMFP, ISHMT, and Combustion Institute. He received his MS from IIT Kanpur and Ph.D. from Louisiana State University, Louisiana, USA. He has authored over 80 peer-reviewed papers. His research interests include combustion modeling, hybrid RANS/LES model development, supersonic flows, and fluid–structure interactions (FSI) with a focus on computational mechanics in combustion and turbulent flows.

Dr. Suresh K. Aggarwal received his Ph.D. from Georgia Institute of Technology, Atlanta, Georgia, USA. He was a member of the research staff at Princeton University, New Jersey, USA, and a Senior Research Engineer at CMU. He then joined the University of Illinois at Chicago, USA, and was promoted to Professor in 1995. He served as Director of Graduate Studies and was also Visiting Scientist at ANL; Visiting Professor at Ecole Centrale-Paris, France; and Guest Professor at Jiangsu University, China. His research interests include combustion, multiphase reacting flows, emissions, clean energy, and fire suppression. He has authored over 340 publications and graduated 17 Ph.D. and 46 MS students. He is Fellow of ASME, AAAS, and Associate Fellow of AIAA.

Introduction

Abstract Availability of clean and sustainable energy is of paramount importance in all applications of energy, power, mobility, and propulsion. This research monograph provides state-of-the-art advances in areas of energy, power, and propulsion using fossil and renewable biofuels, their efficiency and performance improvement, and their environmental impact. This book is written by internationally renowned experts from and around the globe on specific topics of current interest to students, researchers, and engineers from academia and industry to provide the latest innovations in cleaner energy utilization for a wide range of devices. Novel developments in the areas of biofuels and products from various feedstock materials are also presented along with thermal management, emission control, and environmental issues. The energy and environment sustainability requires a multipronged approach involving development and utilization of new and renewable fuels, design of fuel-flexible combustion systems, and novel and environmentally friendly technologies for improved fuel use. This monograph will serve as a reference source for practicing engineers, educators, and research professionals.

Keywords Sustainability · Energy · Propulsion · Power · Biofuels
CFD · Alternative energy sources · Environmentally friendly technologies

Meeting global energy demand in a sustainable manner represents one of the major challenges of the twenty-first century. The word "sustainable" implies providing energy security and addressing climate change concerns caused by greenhouse gas (GHG) emissions, and providing energy in a carbon-neutral manner. The global energy demand continues to rise and is expected to increase by 48%, as the population grows from 7 to 9 billion by the year 2040. Most of this growth is expected to occur in developing nations, with non-OECD countries contributing about 71% of the increase. The world GDP (expressed in purchasing power parity) is a key determinant of growth in energy demand and expected to grow by 3.3% per year during this period. Fossil fuels have been the dominant source of global energy and thus of GHG emissions. They will continue to play a similar role in the foreseeable future and are estimated to provide about 78% of energy demand in 2040. However, challenges posed by climate change and growing energy need have led to increasing global effort to develop alternative and renewable energy resources. During the last three decades, we have seen remarkable progress in our analytical and diagnostic capabilities. Major strides have been made in the computational fluid dynamics (CFD) simulations of reacting and non-reacting flows, as well as in non-intrusive diagnostic capabilities. As a consequence, both computational and diagnostic tools are increasingly being used to analyze, test, modify, develop, and improve energy conversion devices, methodologies, and designs in both academia and industry. Rapid advances in computing power, numerical algorithms, physical models, turbulence models, and detailed reaction mechanisms for a variety of fuels, as well as advanced diagnostics, are further accelerating the pace of development of more efficient and less polluting energy and propulsion systems. Significant developments have also occurred in computational techniques and simulations of complex flows in realistic geometries. Examples include shock-capturing techniques for supersonic flows, and various methodologies, based on Reynolds-averaged Navier–Stokes (RANS), large eddy simulation (LES), and direct numerical simulations (DNS), for turbulent reacting flows. Concurrently, new non-invasive diagnostic techniques are becoming firmly established with the development of advanced instrumentation that provides spatial and temporally resolved measurements at high frequency for use in complex flows at microscales. Also, in parallel, advances in sensors and active and passive control technology have further helped to improve system efficiency, reduce emissions, and mitigate catastrophic failures. In spite of these major advances on many fronts, challenges still remain due to growing demands on energy and dangers posed by climate change. Thus, the research and development efforts must continue to grow, so that we seek our quest for even greater efficiency from the existing and new combustion devices, and broadening the overall energy portfolio by increasing the contributions of cleaner and renewable energy sources.

In order to provide a global perspective on the advances and challenges outlined above, we have been organizing international workshops in India on the broad theme of Energy: Power and Propulsion, since 2004. To date, we have organized eight workshops, which have been sponsored by various agencies, universities, and industries in the USA and India. The last workshop, "International Workshop on Energy, Propulsion and Environment", was held at IIT Kanpur, India, during March 8–11, 2017, and was sponsored by NSF, ONR Global, DRDO, and TEQIP organizations in India. It was attended by many world-renowned scientists and researchers from the USA, UK, India, Thailand, Malaysia, and Saudi Arabia, who presented their latest research findings and shared their ideas in formal talks and exchange forums. This research monograph brings together the latest research presented by these scientists at this workshop. A common theme of the monograph is energy, power, and propulsion. The monograph is divided into five sections dealing with various aspects of this theme. The first part contains chapters on combustion and propulsion systems for improved efficiency and reduced environmental footprint using conventional and renewable fuels. Dynamics, stability, and modeling of flames in combustion systems continue to be of major concern. Chapters in Part II present novel ideas on these topics. Chapters in Part III and Part IV provide reviews of current research dealing with various aspects of turbulence, including the modeling of turbulence–chemistry interactions, and simulations of turbulent reacting flows, as well as two-phase flows in solid rocket motors. Part V comprises of review chapters dealing with the production of renewable fuels from various feedstocks and their utilization in transportation and power generation systems. Each chapter is self-contained with a review of latest research and comprehensive references on each topic. This monograph is expected to be of special interest to researchers and engineers as well as graduate students working in the field of energy, biofuels, combustion, power propulsion, and environment. It will also serve as supplementary reading material, in addition to providing cutting-edge research and development technologies on various topics.

Los Angeles, USA — Akshai K. Runchal
College Park, USA — Ashwani K. Gupta
Kanpur, India — Abhijit Kushari
Kanpur, India — Ashoke De
Chicago, USA — Suresh K. Aggarwal

Part I
Combustion and Propulsion Systems

On Lean Direct Injection Research

Xiao Ren, Chih-Jen Sung and Hukam C. Mongia

Abstract This chapter reviews and discusses the recent development in Lean Direct Injection (LDI) combustion technology. An extended definition of LDI is also provided, followed by a broad description of LDI concepts and examples. Recognizing the needs and opportunities to expand the operability range of LDI, fundamental research has been undertaken to elucidate the effects of air swirler vane angle, pressure drop, air swirler rotation direction, and overall equivalence ratio on LDI flow field and flame behavior. Additional investigation was further conducted to understand fundamental differences between representative LDI and airblast injectors. Results of these fundamental studies are discussed to help identify design changes for improving LDI performance.

1 Introduction

NASA and industry have been working continuously since early 1970 on developing energy efficient low-emissions aviation engines [1–3]. The lean direct injection (LDI) concept started from 1990 [4] shows promise for further technology development and demonstration as summarized in Sect. 2. A broader definition of LDI is provided in Sect. 3 so that one can see that LDI technology development has continued since the early 1980s [5] as summarized in Sect. 4. Here, we mention that next-generation LDI has potential for further reducing NO_X by 50% compared to data presented in Sect. 2. A series of fundamental LDI studies have also been conducted, as highlighted in Sect. 5. In particular, high quality diagnostics data on single element LDI mixers have been acquired to illustrate the effects of swirler vane angle and rotating direction on the flow field and combustion characteristics of LDI, as well as to compare and

X. Ren · C.-J. Sung
University of Connecticut, Storrs, CT, USA

H. C. Mongia (✉)
CSTI Associates, LLC, Yardley, PA, USA
e-mail: hmongia43@hotmail.com

A. K. Runchal et al. (eds.), *Energy for Propulsion*, Green Energy and Technology,
https://doi.org/10.1007/978-981-10-7473-8_1

contrast the performance of swirl-venturi LDI with that of conventional airblast injector configuration.

2 LDI Technology Development Status

Based on one of the LDI concepts, multipoint, fuel-injection, multi-burning zone, the first generation of LDI consists of 9 same simplex elements injectors [6], as shown in Fig. 1. Each fuel injector had an axial air swirler for quick mixing of the fuel and air before burning. Various swirl strengths were tested and the NOx emissions were found to decrease as the swirl strength decreases. However, the operability range was also noted to decrease as the swirl strength decreases. Figure 2 demonstrates that the $9 \times 45°$ clockwise (CW) LDI-1 configuration has low NOxEI of 2 g-NOx/kg-fuel at T4 = 2600 °F, while it perhaps would increase to approximately 10 g-NOx/kg-fuel at T4 = 3200 °F.

The second generation of LDI has three variations based on the baseline 9-point swirl-venturi (SV) LDI configuration [7]. The three second generation SV-LDI variations with 13 element injectors are called the flat dome configuration, the 5-recess configuration, and the 9-recess configuration, as shown in Fig. 3. Table 1 also lists the specifications of injectors and swirlers used in the three LDI-2 configurations. LDI-2 configurations have better performance of low power operability than the LDI-1 configurations. Of importance, LDI-2 configurations have low NOx emissions. The 5-Recess configuration has generally slightly lower NOx than the flat dome or 9-recess configurations, as shown in Fig. 4. 5-Recess LDI-2 configuration gives 85%

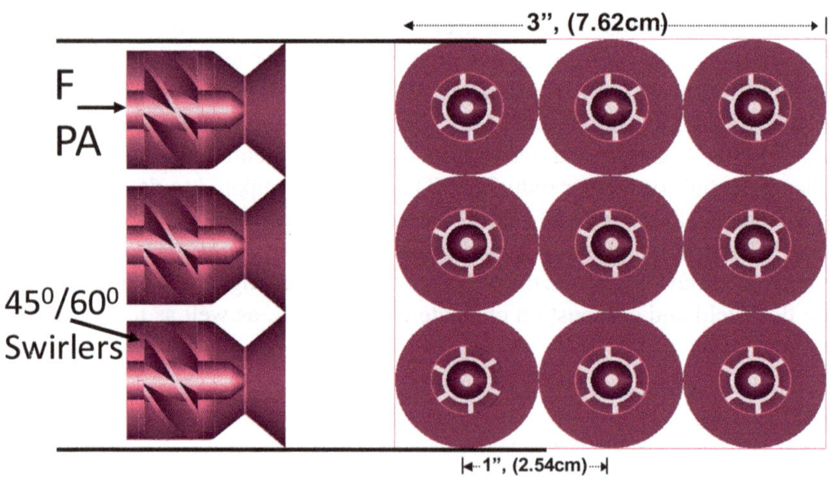

Fig. 1 1st generation 9-element lean direct injection schematic, LDI-1 [6]

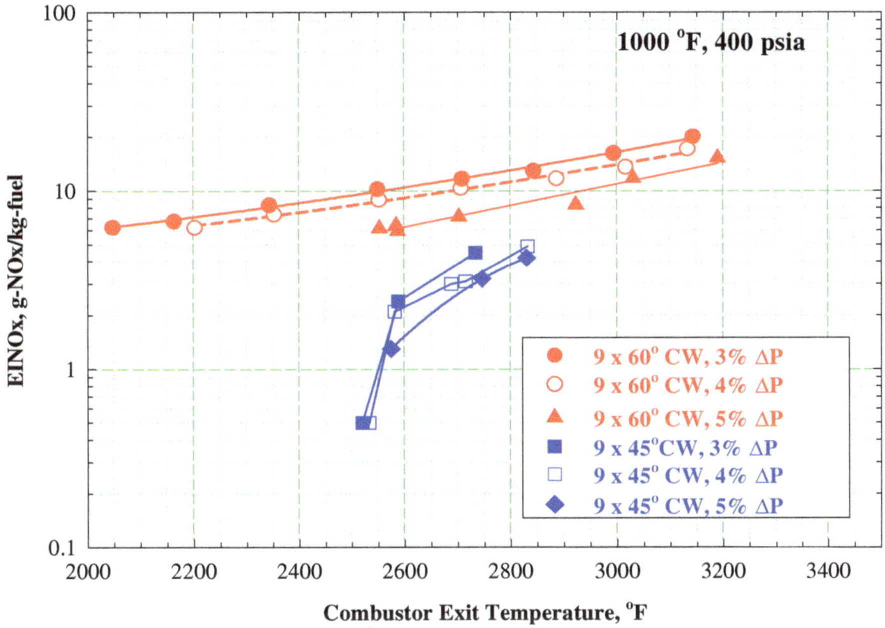

Fig. 2 Typical LDI-1 NOxEI versus combustor exit temperature T4 at combustor inlet temperature T3 and pressure P3 of 1000 °F and 400 psi, respectively [6]

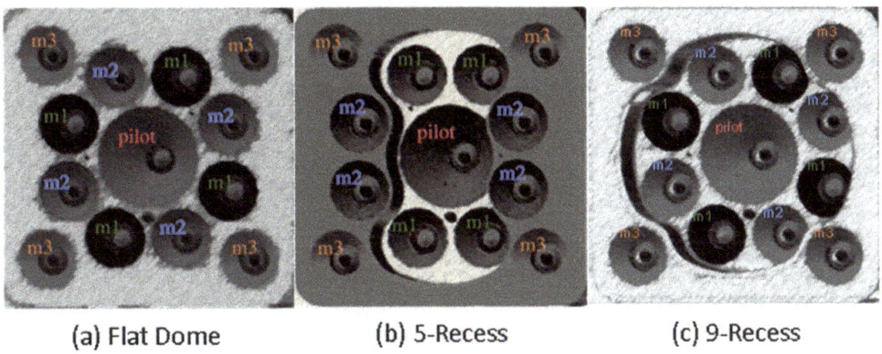

Fig. 3 2nd generation 13-element lean direct injection schematic, LDI-2 [7]

landing takeoff (LTO) NOxEI margin from CAEP/6 standards, which is much better than two premier large engines, Trent1000 and GEnx, as shown in Fig. 5.

Tacina et al. [8] studied combustion dynamics with the three LDI-2 combustor configurations. The dynamic pressure measurements showed that combustion dynamics were typically small, except that all three configurations had high combustion dynamics at low inlet temperatures. It was also found that shifting half of the main-stage fuel flow from the simplex main-1 stage to the airblast main-2 stage,

Table 1 Definitions of injectors and swirlers used in LDI-2 configurations [7]. For each stage, the type of fuel injector and the air swirler angle(s) are given. *OAS* outer air swirler; *IAS* inner air swirler; *CW* clockwise; *CCW* counterclockwise

Configuration	Pilot injector	Pilot swirler	Main1 injector	Main1 swirler	Main 2 injector	Main 2 swirler	Main 3 injector	Main 3 swirler
Flat Dome	Simplex	55°ccw	Simplex	45°ccw	Airblast	IAS:45°cw OAS:45°cw	Airblast	IAS:45°cw OAS:45° cw
5-Recess	Airblast	IAS:57°cw OAS:57°ccw	Simplex	45°cw	Airblast	IAS:45°cw OAS:45°ccw	Airblast	IAS:45°cw OAS:45° ccw
9-Recess	Airblast	IAS:57°cw OAS:57°ccw	Simplex	45°ccw	Airblast	IAS:45°cw OAS:45°cw	Airblast	IAS:45°cw OAS:45° cw

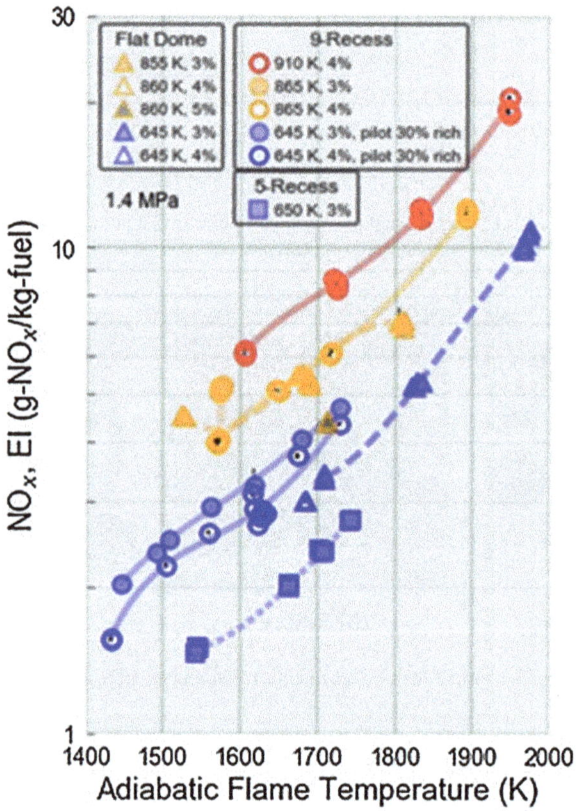

Fig. 4 NOxEI comparison between no-recess, 5- and 9-element recess dome LDI-2 [7]

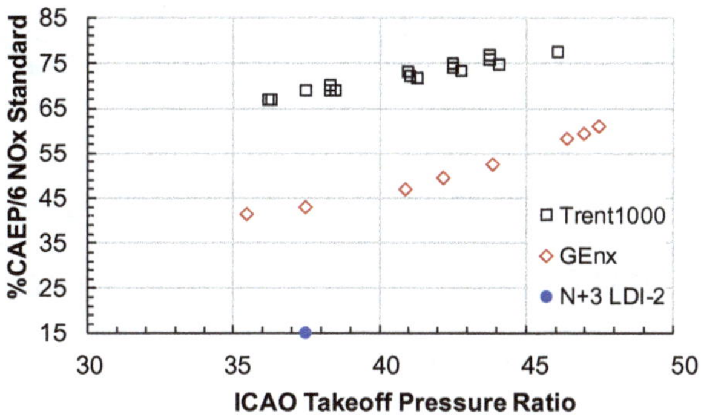

Fig. 5 Comparison of LDI-2 technology potential with two premier large engines' LTO NOx emissions (data taken from ICAO Aircraft Engine Emissions Databank [9])

without changing the overall equivalence ratio, could greatly reduce the combustion dynamics, which could decrease NOx emissions as well.

3 Extended Definition of LDI

Let us provide a descriptive definition of LDI by recognizing its major subcomponents. We can describe LDI in Fig. 1 [6] as single swirler, single venturi, simplex pressure atomizer LDI. Continuing with this descriptive definition, we can call filming airblast nozzle, as shown in Fig. 6a, as twin co- or counter-rotating swirlers with two venturis, without divergence. The latter is called flare by GE combustion engineers. These subcomponents do not change when air-atomizing function of filming airblast nozzle is replaced by pressure atomizing simplex atomizer tip shown as Fig. 6b, known popularly as simplex airblast nozzle. However, specific details of the swirlers and their passages may be very different between filming and simplex airblast nozzles.

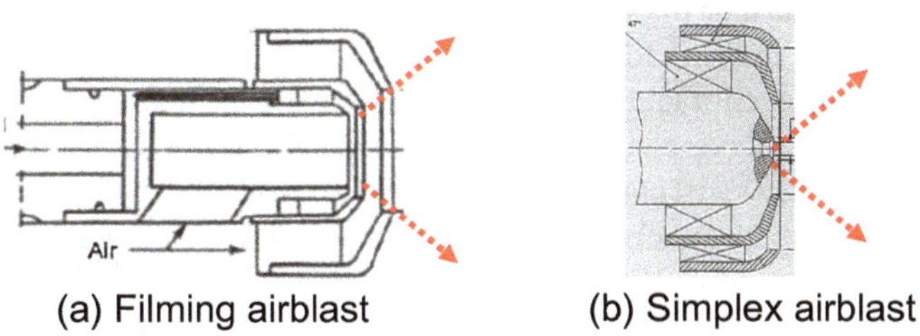

Fig. 6 Conventionally recognized filming and simplex airblast nozzles

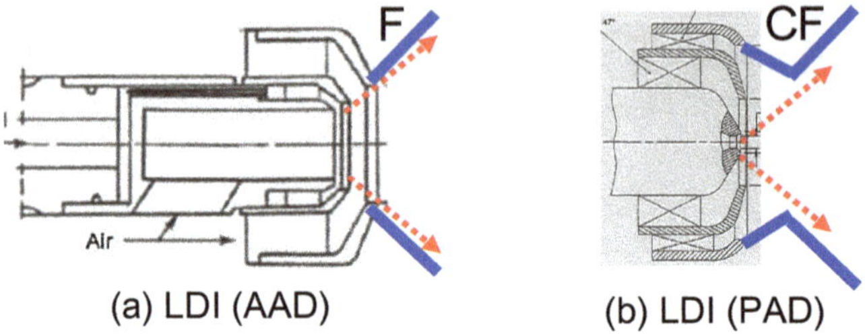

Fig. 7 Descriptive definitions of LDI concepts

As we proceed with this process, bringing in simple or complex flares, configurations of Fig. 6 transform into conventional definition of LDI [10], as shown in Fig. 7. Here we show air-atomizing devices (AAD), pressure atomizing devices (PAD), simple flare (F), and complex flare (CF). Configurations based on schematic shown in Fig. 7a have been included as elements of LDI-2, as shown in Fig. 3 and Table 1.

Therefore, one should not object to simplex atomizer surrounded by appropriately designed co- or counter-rotating swirlers and their converging and diverging passages (cf. Fig. 7b) be classified as LDI. One could define another version of LDI where simplex pressure atomizer is replaced by a dual-orifice atomizer.

4 Examples of LDI Concepts and Entitlement

If we interpret LDI broadly along this line of reasoning, we can fold in several product mixers, as illustrated in Fig. 8, as one of LDI class of mixers. Overall equivalence ratio of the main swirl cup used in the dual-annular combustors (DAC) of the CFM56-5B engine emissions tested in 1995 has primary and secondary venturis as shown in Fig. 8 [11]. In a subsequent engine configuration tested in 1996, it did not have primary venturi. However, their takeoff and climbout NOxEI were comparable. Consequently, for the CFM56-7B design tested in 1997, both primary and secondary venturis were included. However, the desired level of high-power NOx emissions was not achieved, indicating need for further improving the overall design features of swirl cup.

There are several reasons for not continuing further technology development of lean DAC [12]. However, if takeoff NOx emissions were the main reason (see Fig. 9) for not moving forward, one can envision an advanced LDI concept as an extension of current lean DAC main swirl cup. Figure 9 shows takeoff NOxEI of the three CFM56 DAC designs tested respectively in 1994, 1996, and 1997. Also shown are the results for the GE90 DAC tested in 1995 which can be represented by $TONOxEI_{GE90} = 0.0359PR^{2.0028}$, where *PR* is the pressure ratio. Since these values were considered too high for next-generation engine GEnx, a new lean dome technology known popularly as TAPS was used for GEnx giving much lower takeoff NOxEI given by

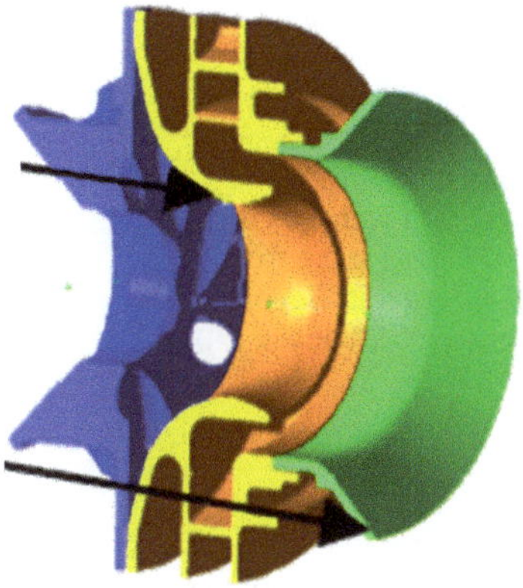

Fig. 8 Main swirl cup of the CFM56-5/7B dual annular combustors classified as LDI [11]

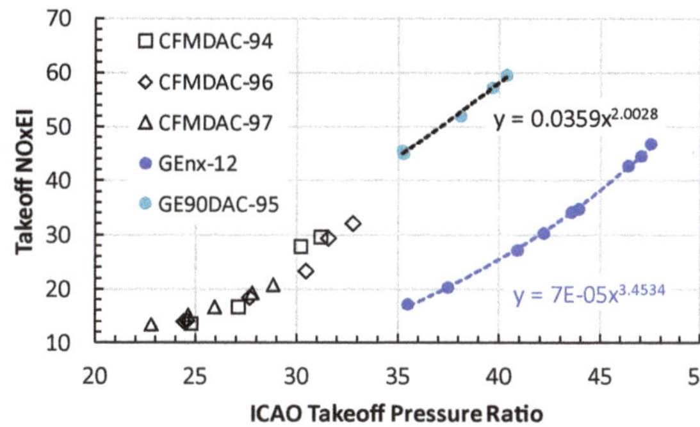

Fig. 9 Takeoff NOxEI versus Takeoff Pressure Ratio of lean DAC's compared with lean dome TAPS (data taken from ICAO Aircraft Engine Emissions Databank [9])

$TONOxEI_{GEnx} = 7 \times 10^{-5} PR^{3.4534}$ giving 50% reduction from the GE90 DAC at PR = 40.

If the CFM56 engine were to increase its maximum takeoff pressure ratio to 35, can we be competitive with GEnx in regard to takeoff NOx, namely 16.0 EI? In other words, can we use lessons learned from the LDI technology development to improve DAC's swirl cup to reduce takeoff NOx by 57%?

People involved in marketing and strategic planning prefer to plot ICAO LTO NOx emissions results in terms of % ICAO standards or alternatively thereof. The resulting data as shown in Fig. 10 are represented well by straight-line trendlines which can be extrapolated with more confidence than power law based correlations for takeoff NOx shown previously in Fig. 9. For the two reasons listed below, the marketing and strategic planning folks would like to have 30% margin for their next product. Looking back at the track record of past NOx stringency rules, it is a fair assumption to make that future NOx stringency might be another 15% stricter. In order for a new engine family to be financially successful, it should offer a reasonable range of rated thrust levels. For example, GEnx rated thrust ranges between 250 and 350 kN. The corresponding range of takeoff pressure ratio is 35 and 48. The resulting LTO NOx values range between 47 and 67% CAEP8 NOx standard; a loss of 20%

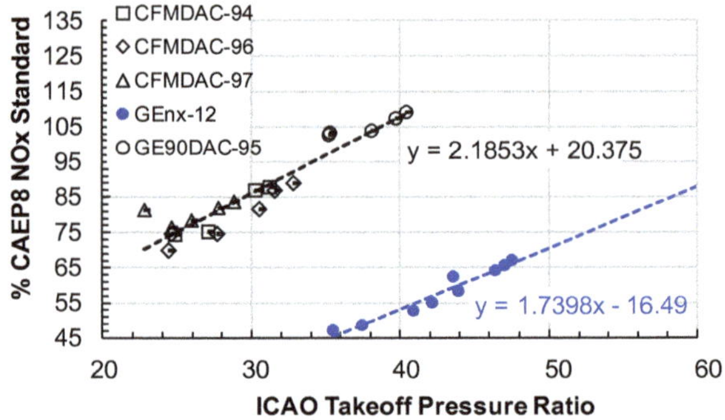

Fig. 10 Landing takeoff NOx (expressed as % CAEP8 NOx Standard) versus Takeoff Pressure Ratio of lean DAC's compared with lean dome TAPS (data taken from ICAO Aircraft Engine Emissions Databank [9])

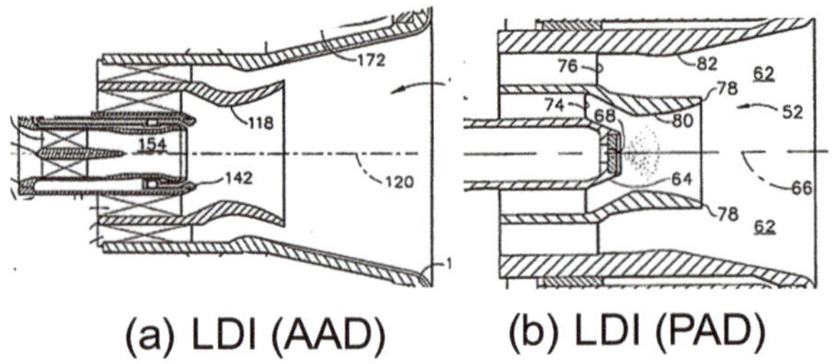

Fig. 11 Pilot part of TAPS [13] extracted to illustrate it as a subset of LDI group defined here

margin from the standard. Therefore, it made sense to develop TAPS technology with target LTO NOx reduction of 50% from the GE90 DAC at PR = 40. As it turned out, 51% NOx reduction at PR = 40 was achieved.

From an academic perspective, GEnx (a replacement of the CF6-80C2) maximum takeoff pressure ratio of 47 is higher than that of GE90 by 7. Therefore, it is reasonable to expect that next generation GE90 should have at least 7 PR higher than that of GE90-115B, namely 49. It is therefore, not surprising to see that GE developed N + 2 low NOx technology for a 55 PR engine; and that advertised pressure ratio of GE9X is 60. The extrapolated value from Fig. 10 at PR = 60 is 88 compared to the corresponding value of GEnx at its maximum PR value is 67 giving 33% margin. Therefore, next generation DAC TAPS for GEnx should have target goal of 33% margin from CAEP8, resulting in only 25% reduction from current technology.

The pilot portion of TAPS has been developed by using LDI concept involving air-atomizing device (AAD) and pressure atomizing device (PAD), as shown in Fig. 11a and 11b, respectively [13]. Three swirlers, prefilming, three venturis, and two flares are involved in forming LDI (AAD) submodule. On the other hand, LDI (PAD) comprises of simplex pressure atomizer, two swirlers, two venturis, and two flares. Later as part of improved performance, simplex pressure atomizer was replaced by a dual orifice pressure atomizer as an extension of LDI (PAD) configuration.

We should not be surprised if in future ultra-low NOx combustion technology development effort, someone investigates a concept based on the configuration sketched in Fig. 12, so called LDI (PAD, AAD).

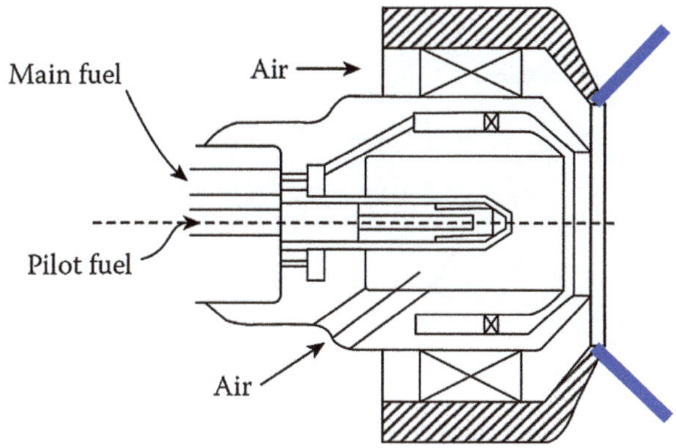

Fig. 12 LDI (PAD, AAD) concept for potential LDI investigations

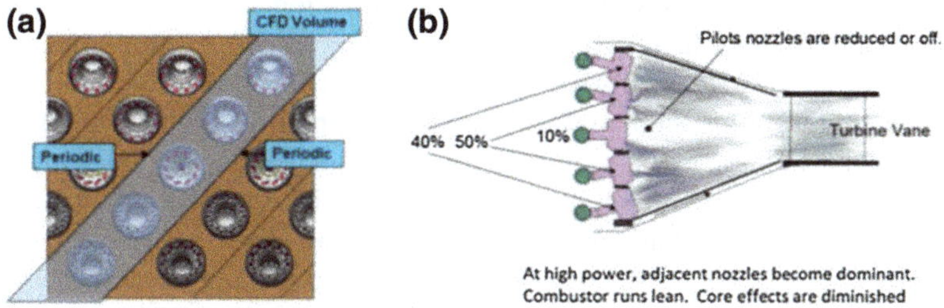

Fig. 13 Goodrich's **a** multi-point LDI modular and **b** multi-zone multi-stage LDI concept [14]

Lee et al. [14] summarized recent LDI works from Goodrich's (currently UTC Aerospace Systems) and Parker's groups. Goodrich's multi-point combustion system used discrete-jet-based airblast fuel nozzles, as shown in Fig. 13a. Near to where fuel is injected, intense mixing turbulence is generated through strong shear layers. This design has rapid breakup and vaporization, which results in more homogeneous and burnable mixture quickly, especially at low power conditions. Multi-zone multi-stage array, shown in Fig. 13b, provides ignition stability with the row of slightly recessed pilot nozzles that are fueled independently.

Parker introduced a multi-staged 3-zone injector module, as shown in Fig. 14. The module consists of miniature mixing cups each fueled by a pressure-swirl nozzle. Multiple fuel stages were used to shift fuel spatially to have low NOx emissions and stable burning. Goodrich and Parker both have obtained good results at the lower pressure flametube testing: demonstrated ignition, flame propagation, lean blowout capabilities, and NOx reduction.

Combustion technology groups have been having healthy dialogue on how far premixing type devices can go in competing against lean premix prevaporized (LPP) combustion system. We will share two good examples in this respect. Ross et al. [15] developed a very impressive premix can combustor for regenerative automotive gas turbine which operated at very high combustor inlet temperature lev-

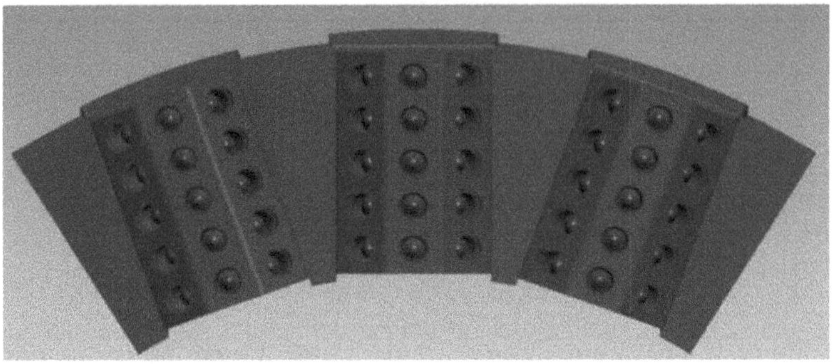

Fig. 14 Parker's conceptual 3-zone module implementation [14]

els ($1674 \leq T3 \leq 1941$ °F) during steady-state operation from idle to maximum power while combustor inlet pressure levels ($19 \leq P3 \leq 71$ psi), overall fuel/air ratios ($0.0032 \leq FAR \leq 0.0132$), and corrected airflow rates ($0.24 \leq W_c \leq 0.37$ lb./s) were small. Their combustion system as shown in Fig. 15 comprises of three fuel insertion devices, namely starter injector, pilot injector, and main fuel injector; the latter comprised of axial swirler (AS) to create a liquid film which was shattered by a radial inflow swirler (RS) which along with axial swirler passage (ASP), variable geometry, and optimized premix/pre-vaporizing (PM/PV) insured no flashback or autoignition for the limited time of operation. It met the desired goal of idle COEI and maximum power NOxEI [16], as shown in Fig. 16.

Sanborn et al. [5] proposed (cf. Fig. 17), designed, and tested proof of feasibility configuration (cf. Fig. 18) of a combustion system which under the expanded definition can be called an LDI. Similar to the AGT100, the AGT101 is a variable geometry combustor that controls simultaneously effective areas of the dome radial swirler (RS) and dilution orifices so that the dome's overall equivalence ratio can be optimized for meeting the emissions requirements for its entire steady operating conditions in addition to controlling residence time within the mixer.

As shown in Fig. 18, fuel nozzle comprises of three swirlers ranging from weak to moderate vane turning angle along with empirically designed passages and resulting three venturis followed by two flares. Discrete liquid jets in crossflow at the two venturi locations were used for two design options, namely flowing into outer and inner swirling streams. Initially, these two streams were controlled by their individual flares, followed by one common flare. Finally, majority portion of the combustion air entered the mixer through high-swirl radial inflow swirler. Details on both AGT100 and AGT101 combustors and their technology development process are provided by Mongia [16].

Feasibility test results of the final configuration are given in Fig. 16 showing potential for this LDI concept to proceed. Unfortunately, this program did not go any further. However, extensive fundamental investigations on this fuel nozzle were conducted by Gutmark and his colleagues [17–23].

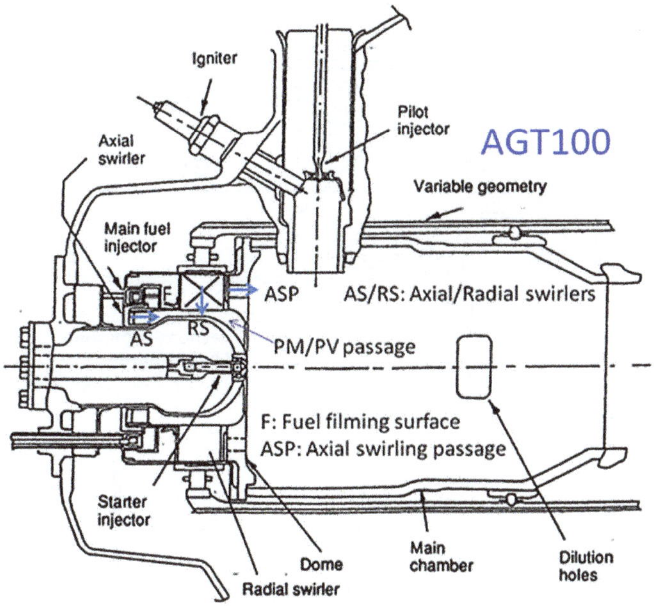

Fig. 15 Lean premix/prevaporized (LP/PV) variable-geometry combustor for regenerative gas turbine application [16]

Fig. 16 LP/PV combustor's CO and NOx characteristics over its operating envelop [16]; shown also are LDI concept described in Fig. 18

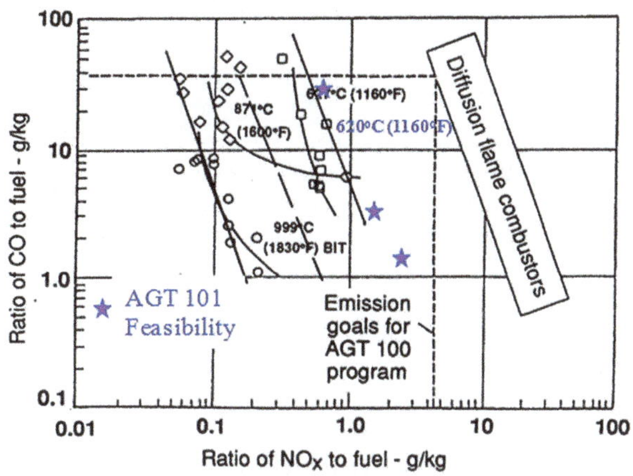

LDI NOx technology entitlement levels should be compared with two data sets, namely accelerating swirl passage premix micromixers reported by Mongia [16] and complex multi-swirlers premixing configuration reported by Mongia [24]. The latter for the LDI test conditions listed in Fig. 2 achieved NOxEI values ranging between 0.3 and 1.0 g-NOx/kg-fuel indicating that LDI technology potential for additional 50% reduction is possible.

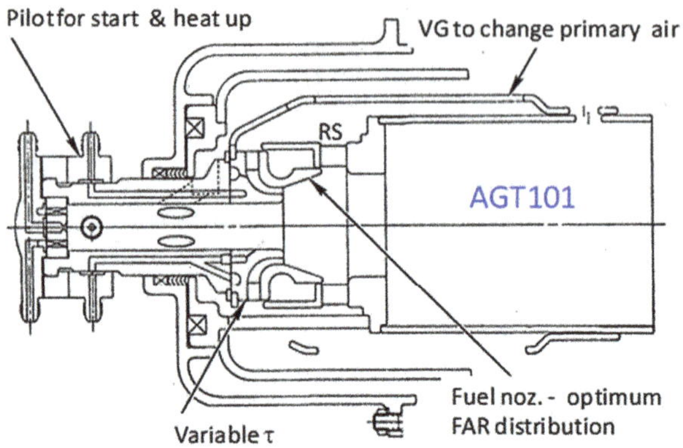

Fig. 17 AGT101 combustion system schematic [16]

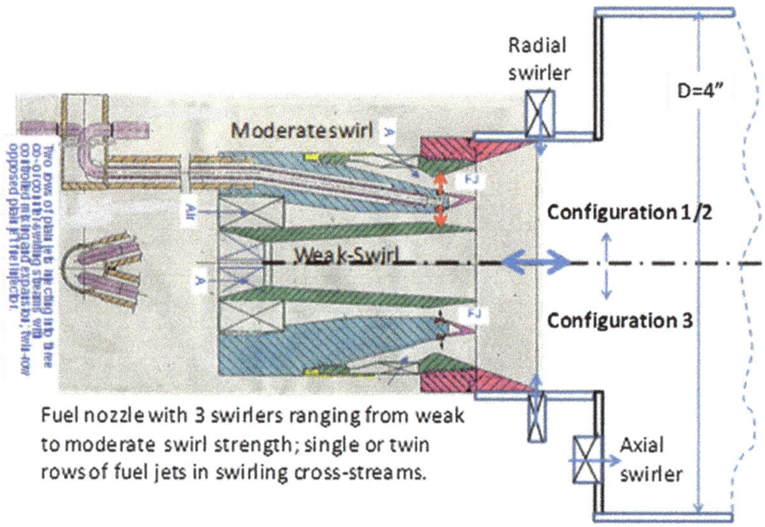

Fig. 18 Schematic details of fuel nozzle and mixer installed in a 4-inch diameter can combustor [16]

5 LDI Related Fundamental Research

The LDI related fundamental research has been conducted using advanced diagnostic techniques to understand fundamental differences between representative LDI and airblast injectors and extend knowhow to propose design changes to the first generation lean dome designs and/or identify changes to generic LDI designs for improving operability.

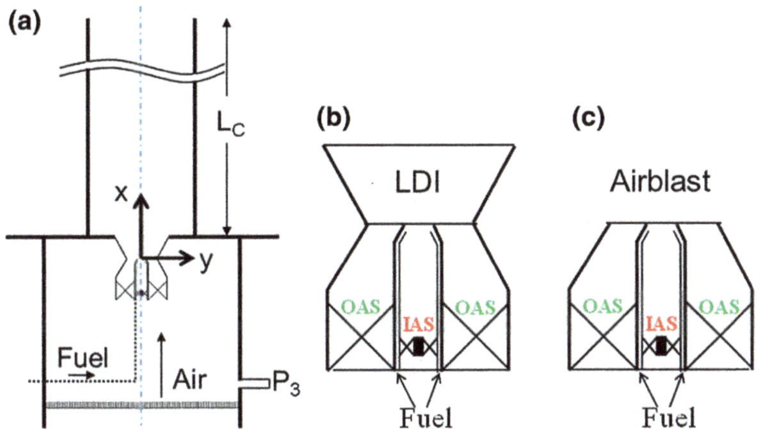

Fig. 19 Schematic of **a** test rig for fundamental experiments, **b** LDI nozzle, and **c** airblast nozzle

5.1 Experimental Specifications

The present experimental rig setup has been developed to investigate single-cup swirling flow/fuel injection systems in an optically-accessible environment under atmospheric conditions, as shown in Fig. 19a. The entrance manifold consists of a 2.5 in × 4 in (length × diameter) cylinder, while the swirler entrance manifold measures 1.5 in × 4 in. Separating the two manifolds is a set of three stacked fine-mesh inserts (40 × 40 openings-per-inch) designed to provide a uniform velocity profile to the entrance of the swirler manifold. Near the midpoint of the swirler manifold, an Omega PX303 pressure transducer with full scale accuracy of 0.25% and 0.01 Torr resolution is included to facilitate pressure measurements upstream of the swirler assembly. Fuel is supplied to the fuel injector through a 1/8 in tube that runs through the axial centerline of the burner assembly (manifolds and swirler assembly). The complete burner assembly is attached to the dump plate such that the exit plane of the venturi is flush with the dump plate surface. The test chamber—consisting of a 2 in × 2 in × 12 in (interior dimensions) section constructed from quartz—is likewise secured to the dump plate, concentric with the swirler assembly. The combustion chamber exit is under an exhaust hood without any area constriction. In addition, the complete flow cell is mounted on a three-axis machine table to facilitate its movement relative to the diagnostic setup. More details of the present test rig can be found in Ren et al. [25].

The single element LDI injector used in this study was provided by Woodward FST and its schematic is shown in Fig. 19b. It consists of a fuel tip with two concentric axial air swirlers, an inner air swirler (IAS) contained within the airblast-type fuel tip and an outer air swirler (OAS), located inside a converging-diverging venturi body. For the present study, helical 45° or 60° OAS vanes are installed with both clockwise (cw) and counter-clockwise (ccw) rotation directions, while the IAS vanes are fixed

Table 2 Specifications of swirlers and venturi used in the present fundamental research

Configuration	Swirler (IAS/OAS)	Venturi
LDI-60°cw	60°ccw/60°cw	With flare
Airblast-60°cw	60°ccw/60°cw	Without flare
LDI-60°ccw	60°ccw/60°ccw	With flare
LDI-45°cw	60°ccw/45°cw	With flare
Airblast-45°cw	60°ccw/45°cw	Without flare
LDI-45°ccw	60°ccw/45°ccw	With flare

at a 60° counter-clockwise rotation. The venturi throat diameter is $D_t = 0.52$ in, and its axial location is denoted as $x = 0$. The LDI flare exit has an angle of 110°, a diameter of $D_f = 1$ in, and a length of $x_f = 0.37$ in. As the combustor length is $L_C = 12$ in, $L_C/D_f = 12$. A representative airblast injector shown in Fig. 19c, by simply removing the flare (the diverging section of the present LDI's venturi), was used to compare and illustrate the differences in fundamental characteristics of LDI and airblast injectors. The six configurations used in our LDI related fundamental studies are summarized in Table 2.

Measurements of non-reacting air flow field and reacting air-methane flow field were carried out using a Dantec Dynamics time-resolved particle image velocimetry (TR-PIV) system, while the combustion zone was characterized via OH* chemiluminescence and direct flame imaging by averaging 15 consecutive images with 0.1 s exposure each. For the present TR-PIV measurements, repetition rates were set to 1–5 kHz for axial vector maps and aluminum oxide particles with 1 μm nominal mean diameter were used. Interrogation areas were set to correspond to approximately 0.6 mm × 0.6 mm (1 mm × 1 mm) areas for non-reacting (reacting) flows, with 50% overlap. Laser pulse delay times were set to approximately follow the "1/4 rule"—frame-to-frame particle movement of approximately ¼ of the interrogation area dimensions—within regions of interest. All velocity data collected were conducted with the laser sheet aligned along the axial centerline. The mean velocity map of capture area was 2 in × 2.5 in (2 in × 4 in) from the venturi exit for non-reacting (reacting) cases. Mean velocity maps were processed from the full 1-second (2-second) collection period at 5 kHz (1 kHz) for non-reacting (reacting) flow measurements, and hence 5000 (2000) pairs of 800 × 800-pixel array images were captured for each non-reacting (reacting) data set.

When presenting our experimental results in the following, the axial and radial distances are in units of the venturi throat diameter, D_t, while all velocity components are normalized by the mean axial velocity based on the venturi throat area, U_t, defined by the following equation.

$$U_t = \frac{\dot{m}}{\rho A_t}$$

Here, $\dot{m}$ is the total mass flow rate, ρ is the mixture density, and A_t is the throat area.

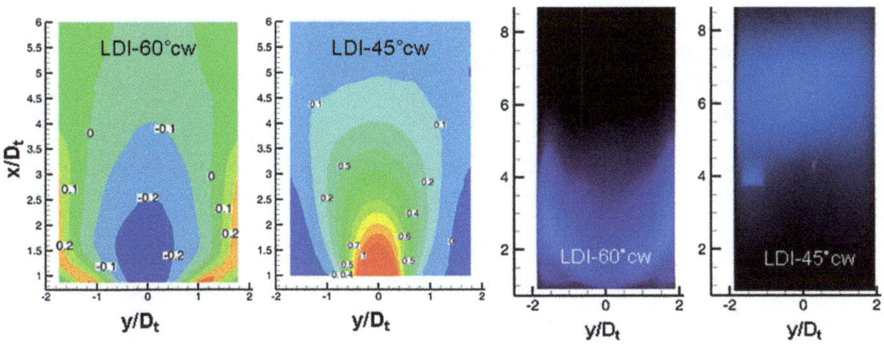

Fig. 20 Effect of OAS vane angle on non-reacting flow field at pressure drop of 3% and flame characteristics at pressure drop of 1%

5.2 Effect of Outer Air Swirler (OAS) Vane Angle

Swirler vane angle plays an important role in influencing the swirling flow strength and determining the type of swirling flow field. Here, two configurations, LDI-60°cw and LDI-45°cw, are used to study the OAS vane angle effect on LDI non-reacting flow field and flame characteristics.

As shown in Fig. 20, the resulting non-reacting flow field in the LDI-60°cw configuration exhibits a center recirculation zone (CRZ) near the venturi exit. On the other hand, no such CRZ is observed in the resulting non-reacting flow field of the LDI-45°cw configuration; instead the velocity at the exit of the venturi is strongly positive and smaller corner recirculation zones (CNRZs) are observed along each side of the combustion chamber. With larger vane angle, the ratio of tangential velocity to axial velocity increases, leading to higher swirl strength and the formation of CRZ.

It is also seen from Fig. 20 that the overall structure of the methane-fueled reacting flow field agrees with that of the non-reacting counterpart. With the same overall equivalence ratio of $\varphi = 0.7$, only the LDI-60°cw configuration exhibits CRZ and the resulting flame resides near the venturi exit, while the combustion zone of the LDI-45°cw configuration is located far downstream due to the positive axial velocities in the center region near the venturi exit. As discussed earlier, Tacina et al. [6] noted that the NOx emissions decrease with decreasing swirl strength. Based on the present TR-PIV results, the lack of CRZ and large positive axial velocities at the venturi exit in the LDI-45°cw configuration would reduce the average residence time in the combustion zone, thereby resulting in reduced NOx emissions.

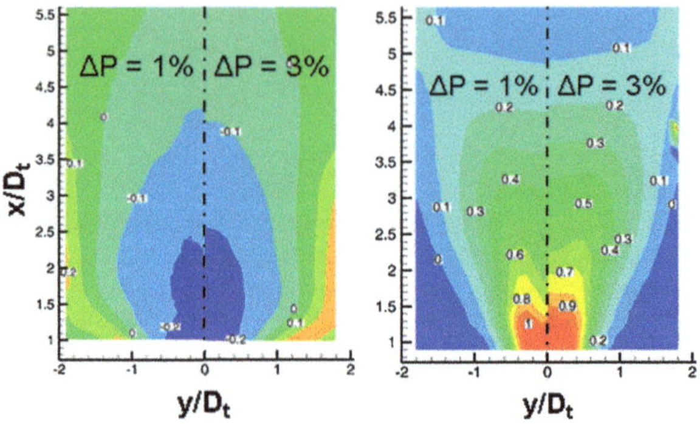

Fig. 21 Effect of pressure drop on non-reacting flow fields of LDI-60°cw (left) and LDI-45°cw (right) configurations

5.3 Effect of Pressure Drop

The effect of pressure drop on non-reacting flow field is studied and shown in Fig. 21 with two configurations of LDI-60°cw and LDI-45°ccw. For both configurations, the normalized non-reacting flow fields generally show a good similarity when varying pressure drop, ΔP, from 1 to 3%. In addition, the overall CRZ dimensions associated with the LDI-60°cw configuration appear to be similar at two different pressure drops. For the LDI-45°cw configuration, the normalized axial velocity fields at ΔP = 1 and 3% are also scaled well. Therefore, Fig. 21 demonstrates that the normalized non-reacting flow field for LDI remains similar under varying pressure drop conditions.

5.4 Effect of Outer Air Swirler (OAS) Rotation Direction

The relative rotation directions between IAS and OAS, co- or counter-rotating, could impact on the air and fuel mixing process, as well as the resulting flow field and the subsequent combustion characteristics. By keeping the IAS vanes at 60°ccw, Fig. 22 compares the non-reacting flow fields of LDI-60°cw, LDI-60°ccw, LDI-45°cw, and LDI-45°ccw at pressure drop of 3%. It is seen from Fig. 22 that the change in OAS rotation direction does not alter the general flow features, CRZ or CNRZ, for both LDI-60° and LDI-45°. However, LDI-60°ccw has a narrower and shorter CRZ than LDI-60°cw, while LDI-45°ccw has a wider region of high positive axial velocity jet flow than LDI-45°cw.

Since the OAS rotation direction also affects the magnitude of axial velocity, Fig. 23 compares the mean axial velocity profiles along the centerline (y = 0) for all four configurations shown in Fig. 22. Comparing the profiles of LDI-60° from x/D_t = 1.5 to 4, it is seen that the reverse flow of LDI-60°cw is stronger than LDI-60°ccw. In addition, Fig. 23 shows that the mean axial velocity near the venturi exit (x/D_t = 1.5–2.5) is higher for LDI-45°cw than LDI-45°ccw. The present TR-PIV results therefore illustrate that when the IAS and OAS are directionally counter-rotating

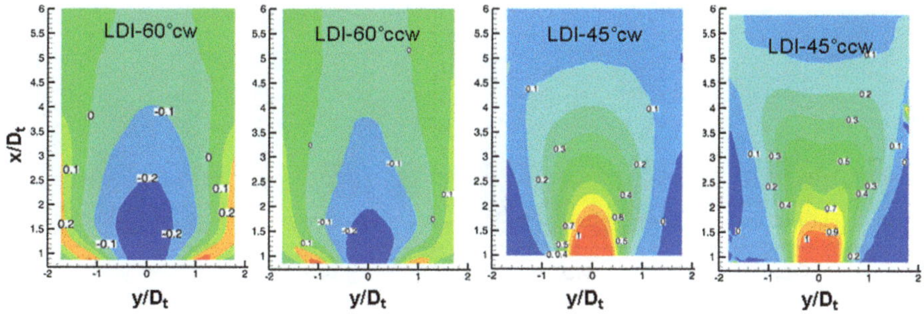

Fig. 22 Mean non-reacting axial velocity contours with different OAS rotation directions at pressure drop of 3%

Fig. 23 Mean non-reacting axial velocity profiles along centerline (y = 0) impacted by OAS rotation direction at pressure drop of 3%

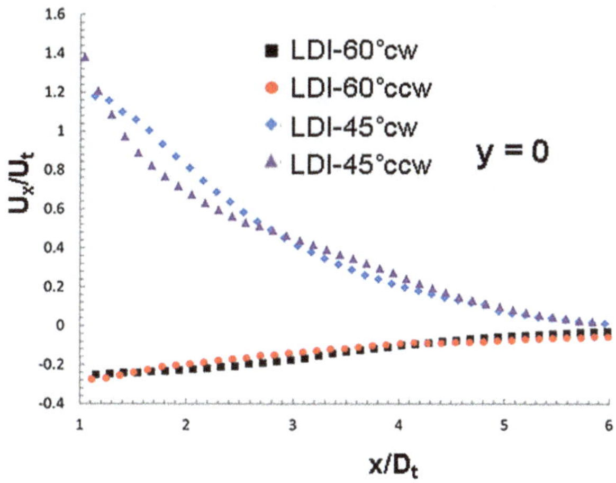

(co-rotating), the resulting swirl strength increases (decreases), which in turn affects the central reverse flow in LDI-60° and the central positive-flow jet in LDI-45°.

5.5 *Effect of Venturi Geometry: LDI Versus Airblast*

The venturi geometries, with and without flare, represent the configurations of LDI and conventional airblast nozzles, respectively. The configurations of LDI-60°cw, Airblast-60°cw, LDI-45°cw, and Airblast-45°cw listed in Table 2 are used to investigate the effect of venturi geometry on non-reacting and reacting flow fields.

At pressure drop of 3%, Figs. 24 and 25 respectively compare the non-reacting flow fields and the methane-fueled reacting flow fields with $\varphi = 0.7$ for the four selected configurations. It is seen that the flare has a prominent impact on flow field. For both non-reacting and reacting flows, removing the flare leads to the change

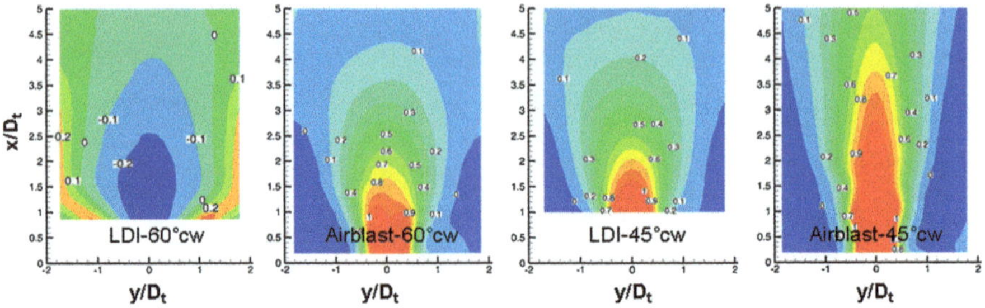

Fig. 24 Mean non-reacting flow field of LDI-60°cw, Airblast-60°cw, LDI-45°cw, and Airblast-45°cw at pressure drop of 3%

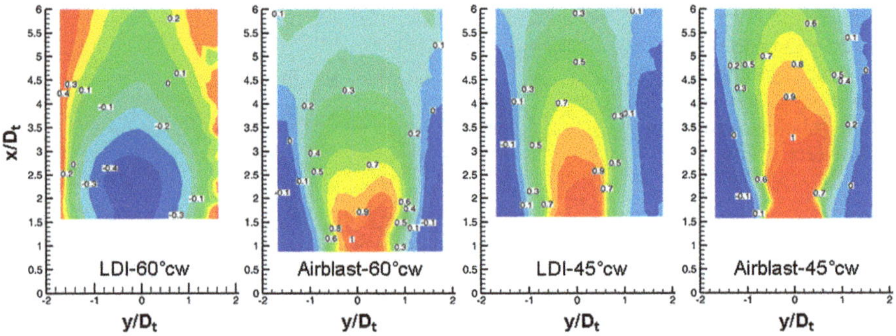

Fig. 25 Mean reacting flow fields of LDI-60°cw, Airblast-60°cw, LDI-45°cw, and Airblast-45°cw with $\varphi = 0.7$ at pressure drop of 3%

in flow field characteristics from the CRZ flow in the LDI-60°cw configuration to the high-velocity positive-flow jet in the Airblast-60°cw configuration. As these two venturi configurations are identical up to the venturi throat and are also operated under identical conditions, the observed difference in flow field structure is caused by the diverging section of the venturi, especially the swirl strength in Airblast-60°cw is much weaker than that in LDI-60°cw. Furthermore, the comparison of LDI-45°cw and Airblast-45°cw in both Figs. 24 and 25 shows that while the high positive-velocity jet structure still remains when removing the flare, the central positive jet flow expands further downstream in the Airblast-45°cw configuration.

Figure 26 further compares the mean axial velocity profiles along the centerline ($y = 0$) for non-reacting and reacting flows of the four selected configurations. It can be observed that only LDI-60°cw exhibits a reversed flow structure downstream of the venturi. In addition, the magnitude of the centerline axial velocity follows the ranking of Airblast-45°cw > LDI-45°cw > Airblast-60°cw, for both non-reacting and reacting flows. These TR-PIV results clearly demonstrate that the removal of the diverging section of the venturi weakens the swirl strength and promotes the positive-velocity jet structure.

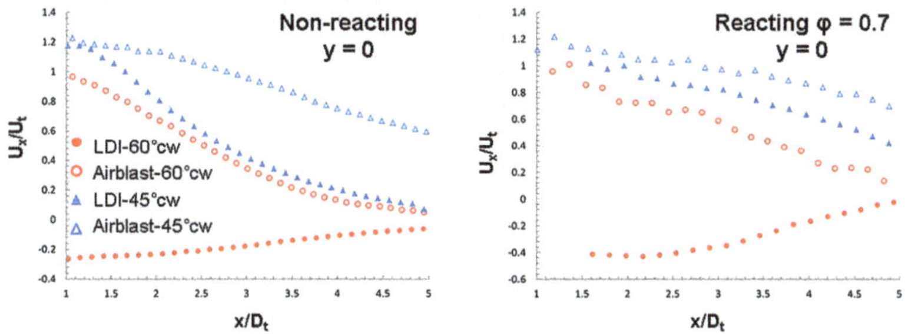

Fig. 26 Mean axial velocity along centerline (y = 0) for non-reacting and reacting cases at pressure drop of 3%

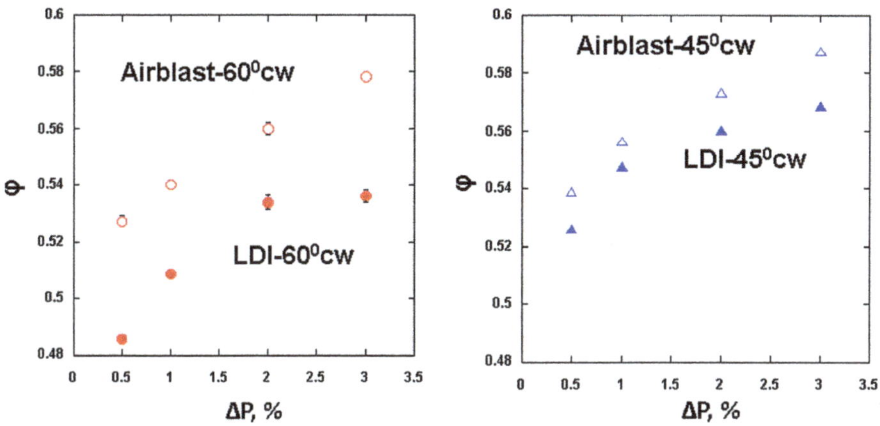

Fig. 27 Impact of venturi geometry on lean blowout limits

The effect of venturi geometry on lean blowout (LBO) limit is also investigated for the four selected configurations, as shown in Fig. 27. LBO is seen to be impacted by the divergent section of the venturi as well. In general, the LDI-type injector has lower LBO limits than the airblast-type one. Comparing LDI-60°cw with Airblast-60°cw, the CRZ structure near the venturi exit associated with the former is beneficial for flame stability. However, this stabilization mechanism also increases average residence time in the combustion zone, resulting in higher NOx emissions, as discussed earlier. Regarding the LBO differences between LDI-45°cw and Airblast-45°cw, the lower limits of the former is due to its jet flow of smaller positive-velocities, as shown in Fig. 26.

5.6 Effect of Overall Equivalence Ratio on Flame Response

Using OH* chemiluminescence and direct flame imaging, the variation of flame characteristics with overall equivalence ratio for the LDI-60°cw configuration at pressure drop of 1% is demonstrated in Fig. 28. It is seen that as overall equivalence ratio is reduced after establishing a vigorously-burning flame anchored at the venturi exit ($\varphi = 0.7$), the flame gradually weakens and extends further downstream until flame lift-off is observed (e.g., $\varphi = 0.55$). It is noted that similar flame structure variation with overall equivalence ratio is also observed for the LDI-60°ccw configuration. For a given pressure drop, the critical equivalence ratio below which the flame loses the anchor point near the venturi exit and moves downstream is determined as the flame lift-off limit. Figure 29 shows the flame lift-off limit as a function of pressure drop for the configurations of LDI-60°cw and LDI-60°ccw. The lower flame lift-off limits of LDI-60°cw is due to its counter-rotating air swirler arrangement that results in a stronger reverse flow in the CRZ and a larger CRZ region than that of LDI-60°ccw, as discussed in Sect. 5.4.

Figure 30 plots and compares the mean flow fields for the four flame macrostructures shown in Fig. 28. In this LDI-60°cw configuration, all four flames exhibit a CRZ structure, while the CRZ length varies with overall equivalence ratio. It is also of interest to note that Flame II ($\varphi = 0.6$) has the widest CRZ at $x = 2D_t$ among the four flames. Figure 31 further shows the mean axial velocity profiles along the centerline and the radial profiles of axial velocity at $x = 2D_t$ for Flames I–IV. It can be observed that Flame II has the strongest reverse flow in the CRZ—the largest magnitude of negative-velocity along the centerline and in the radial direction at $x = 2D_t$.

Figure 32 illustrates the flame structure variation with overall equivalence ratio for the LDI-45°cw configuration through OH* chemiluminescence and direct flame images. Since in this configuration the flow structure is high positive jet flow in nature, the flame is generally longer than that shown in Fig. 28 and extends in both upstream and downstream directions. As the overall equivalence ratio is reduced, the

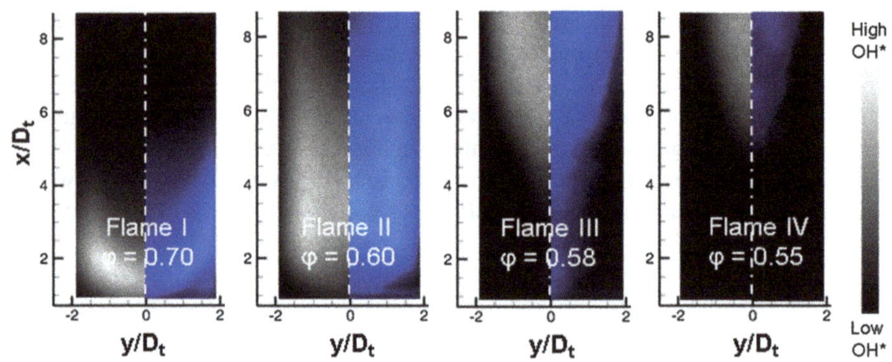

Fig. 28 Flame structure variation of LDI-60°cw with varying overall equivalence ratios at pressure drop of 1%, represented by mean OH* chemiluminescence and direct flame images

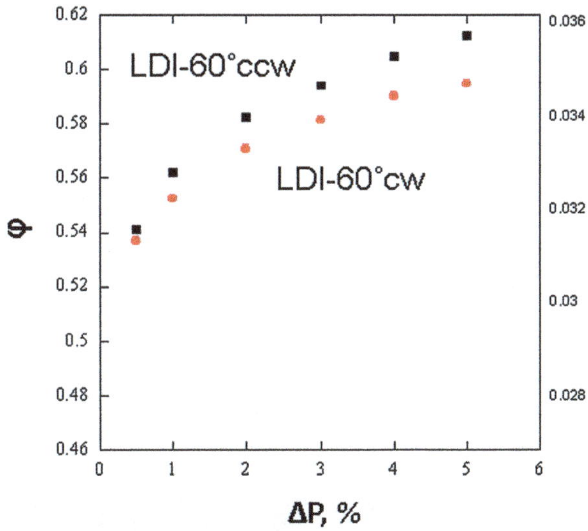

Fig. 29 Lift-off limits for the LDI-60°cw and LDI-60°ccw configurations

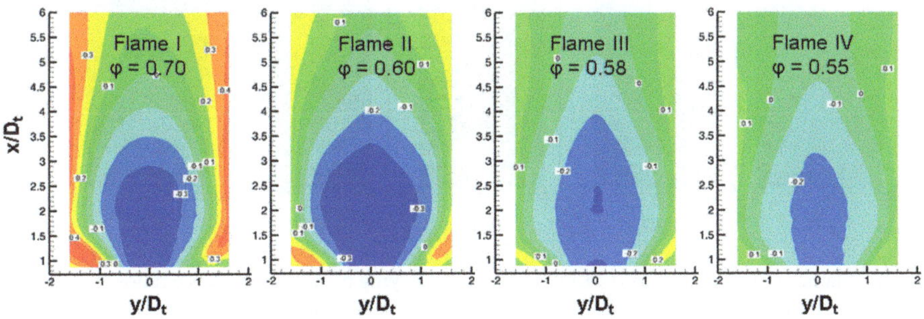

Fig. 30 Mean axial velocity field variation of LDI-60°cw with varying overall equivalence ratios at pressure drop of 1%

flame weakens and gradually moves downstream until it extinguishes when reaching the LBO limit.

6 Concluding Remarks

One of the most promising, robust, and simple lean-front-end combustion concepts is lean direct injection (LDI). Recent technology development and demonstration efforts are summarized and discussed. In addition, extended definition of LDI is elaborated by recognizing its major subcomponents as continued technology development since the early 1980s. Based on the broadly-defined LDI class of mixers, exam-

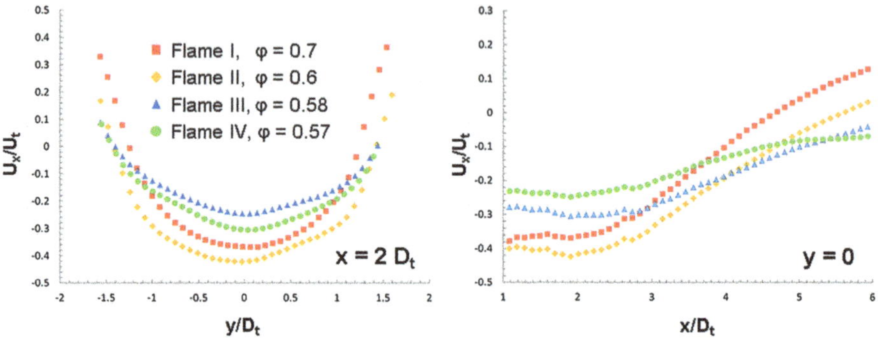

Fig. 31 Radial variations of mean axial velocity at $x = 2D_t$ (left) and mean axial velocity profiles along centerline (right) at varying overall equivalence ratios and pressure drop of 1% for the LDI-60°cw configuration

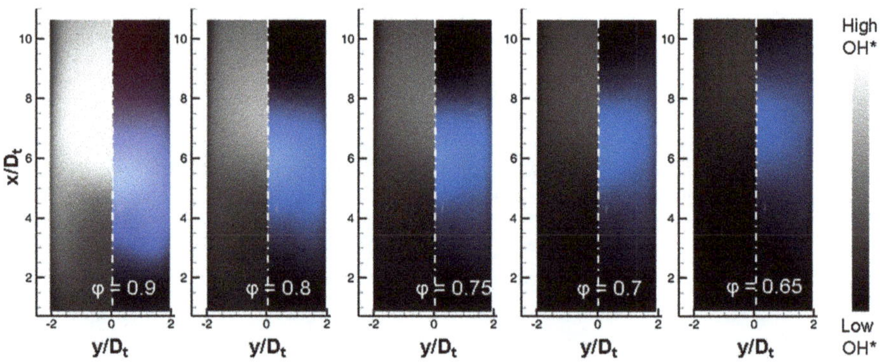

Fig. 32 Flame structure variation of LDI-45°cw with varying overall equivalence ratios at pressure drop of 1%, represented by mean OH* chemiluminescence and direct flame images

ples of LDI concepts and entitlement are provided for developing next-generation, low-emissions LDI technology.

In order to provide insight into technology development of future LDI concepts for lower emissions and expanded operability range, a systematic LDI related fundamental research has been carried out using advanced diagnostic techniques. The present work has investigated and demonstrated the effects of swirler vane angle, varied air swirler rotation directionality, pressure drop, and overall equivalence ratio on the resulting flow field and flame structure/response. It was found that the swirl strength plays an important role in determining the LDI performance. Furthermore, the impact of venturi geometry—either converging-diverging or converging-only—on the flow structure and flame behavior was explored. Without the diverging venturi section, the flow pattern can change dramatically from a center recirculation zone to a jet flow with corner recirculation zones, thereby affecting the average residence time within the combustion zone and the subsequent combustion characteristics. Therefore, some degree of optimization of the venturi shape and its angle of the expansion may improve the venturi's ability to enhance LDI performance.

Acknowledgements The authors would like to thank Drs. Xin Xue, Kyle B. Brady, and Xin Hui for their help in the fundamental single-cup experiments.

References

1. H.C. Mongia, N+3 and N+4 generation aeropropulsion engine combustors part 1: large engines' emissions. ASME Turbo Expo GT2013-94570 (2013)
2. H.C. Mongia, N+3 and N+4 generation aeropropulsion engine combustors part 2: medium size rich-dome engines and lean-domes. ASME Turbo Expo GT2013-94571 (2013)
3. H.C. Mongia, N+3 and N+4 generation aeropropulsion engine combustors part 3: small engine emissions and axial staging combustion technology. ASME Turbo Expo GT2013-94572 (2013)
4. R.R. Tacina, Low NOx potential of gas turbine engines. AIAA 90-0550 (1990)
5. J.W. Sanborn, H.C. Mongia, J.R. Kidwell, Design of a low-emission combustor for an automotive gas turbine. AIAA 1983-0338 (1983)
6. R.R. Tacina, P. Lee, C. Wey, A lean-direct-injection combustor using a 9 point swirl-venturi fuel injector. ISABE 2005-1106 (2005)
7. K.M. Tacina, D.P. Podboy, Z.J. He, P. Lee, B. Dam, H.C. Mongia, A comparison of three second-generation swirl-venturi lean direct injection combustor concepts. AIAA 2016-4891 (2016)
8. K.M. Tacina, P. Lee, C. Chang, Z. He, B. Dam, D. Podboy, An assessment of combustion dynamics in a low-NOx second-generation swirl-venturi lean direct injection combustion concept. ISABE 2015-20249 (2015)
9. ICAO Aircraft Engine Emissions Databank. https://www.easa.europa.eu/easa-and-you/environment/icao-aircraft-engine-emissions-databank
10. K.S. Im, M.C. Lai, R.R. Tacina, A parametric spray study of the swirler/venturi injectors. AIAA 98-3269 (1998)
11. H.C. Mongia, Engineering aspects of complex gas turbine combustion mixers part IV: swirl cup. AIAA 2011-5526 (2011)
12. H.C. Mongia, Engineering aspects of complex gas turbine combustion mixers part III: 30 OPR. AIAA 2011-5525 (2011)
13. H.C. Mongia, Engineering aspects of complex gas turbine combustion mixers part V: 40 OPR. AIAA 2011-5527 (2011)
14. C.M. Lee, C. Chang, J.T. Herbon, S.K. Kramer, NASA project develops next generation low-emissions combustor technologies. AIAA 2013-0540 (2013)
15. P.T. Ross, J.R. Williams, D.N. Anderson, Combustor development for automotive gas turbines. AIAA J. Energy **7**, 429–435 (1983)
16. H.C. Mongia, Engineering aspects of complex gas turbine combustion mixers part II: high T3. AIAA 2011-0106 (2011)
17. C. Fureby, F.F. Grinstein, G. Li, E.J. Gutmark, An experimental and computational study of a multi-swirl gas turbine combustor. Proc. Combust. Inst. **31**, 3107–3114 (2007)
18. F.F. Grinstein, T.R. Young, E.J. Gutmark, G. Li, G. Hsiao, H.C. Mongia, Flow dynamics in a swirl combustor. J. Turbul. **3**(30), 1–19 (2002)
19. E.J. Gutmark, G. Li, F.F. Grinstein, Characterization of multiswirling flow. Eng. Turbul. Model. Exp. **5**, 873–884 (2002)
20. E.J. Gutmark, S. Verfaillie, J. Bonnet, F. Grinstein, Linear stochastic estimation of a swirling jet. AIAA J **44**(3), 457–468 (2006)
21. G. Li, E.J. Gutmark, Effect of exhaust nozzle geometry on combustor flow field and combustion characteristics. Proc. Combust. Inst. **30**, 2893–2901 (2005)

22. T. Yi, E.J. Gutmark, Combustion instabilities and control of a multiswirl atmospheric combustor. ASME J. Eng. Gas Turbines Power **129**(1), 31–37 (2007)
23. X. Zhou, J.B. Jeffries, R.K. Hanson, G. Li, E.J. Gutmark, Wavelength-scanned tunable diode laser temperature measurements in a model gas turbine combustor. AIAA J **45**(2), 420–425 (2007)
24. H.C. Mongia, Recent progress in comprehensive modeling of gas turbine combustion. AIAA 2008-1445 (2008)
25. X. Ren, X. Xue, C.J. Sung, K.B. Brady, H.C. Mongia, P. Lee, The impact of venturi geometry on reacting flows in a swirl-venturi lean direct injection airblast injector. AIAA 2016-4560 (2016)

Effect of Alternative Fuels on Emissions and Engine Compatibility

Bhupendra Khandelwal, Charith J. Wijesinghe and Shabarish Sriraman

Abstract Given the increasing focus on climate change and emissions, alongside the motivation to combat these phenomena, it is prudent to consider alternative fuels for gas turbines, a significant source of emissions. Adopting some form of alternative fuels could reduce the carbon footprint as well as the emissions output from gas turbines to manageable levels, provided alternative fuels are coming from overall low life cycle emissions sources. In this chapter, the effects of alternative fuels on the gas turbines performance and their emissions are discussed. With respect to gaseous emissions, it has been found that alternative fuels provide no clear advantage in terms of emissions reduction compared to standard petroleum derived fuels. However, it has been found that the CO_2 emissions of a given fuel is contributed to by the H/C ratio of the fuel. An increase of the H/C ratio could lead to reduction in CO_2 emissions, though energy per unit mass of fuel goes down. The effect of alternative fuels on PM emissions however are more positive if alternative fuels are used, but PM emissions are dependent upon the aromatic content and its species in the fuel. The availability of alternative fuels from F-T processes, as well as bio-derived fuels with very low or no aromatic content, leads to very low PM emissions from alternative fuels. With respect to seal swell in fuel systems, it has been found that some alternative fuels may struggle to maintain good seal swell performance as seal swell has been historically related to aromatic content of the fuel. Therefore, it has been deemed that further research is required to find an alternative. When considering the noise and vibrations from a turbine, there appears to be insufficient data to draw clear correlations between fuel type and amount of noise and vibrations generated, however it has been noted that noise and vibration emitted is a function of the vapour pressure, surface tension and flame velocities used which in turn to a certain extent depend upon the fuel used. In terms of thermal stability, it has been noted that paraffinic fuels are better at absorbing heat and dissipating it without forming carbon deposits on the fuel system components.

B. Khandelwal (✉) · C. J. Wijesinghe · S. Sriraman
Low Carbon Combustion Centre, University of Sheffield, Sheffield, UK
e-mail: bhupendra.khandelwal@gmail.com

A. K. Runchal et al. (eds.), *Energy for Propulsion*, Green Energy and Technology,
https://doi.org/10.1007/978-981-10-7473-8_2

1 Introduction

With growth in the gas turbine industry increasing, demands for a cleaner and more sustainable fuel with lower emissions in needed. One that would also be a suitable replacement for Jet fuel and can satisfy the fuel consumption rate of this industry has increased. Emissions control has taken precedence and has driven the aviation industry to find better alternatives to the existing Jet fuel [1]. This also provides an opportunity to conduct extensive research on new fuel compositions which may provide a solution to the problem of higher emissions and unsustainability. Researchers have been going on a full-scale in the effort to introduce new types of fuels into the industry where there are also known as alternative fuels. To this end intensive research is being carried out to introduce drop-in alternative fuels to replace traditional petroleum-based fuels.

Alternative or advanced fuels can be formulated from a myriad of sources such as biomass, coal, and natural gas. One of the major processes for production of alternative jet fuel is the Fischer-Tropsch (F-T) process. In essence, fuel produced from any feedstock that conforms to strict fuels standard and has been through appropriate fuels approval process could be used as a jet fuel. SASOL of South Africa was the first company to get their 50% synthetic blend approved as a jet fuel. They were also the first company to supply alternative jet fuel commercially. Significant steps have been taken by the alternative fuels industry to make alternative jet fuel a reality (Fig. 1).

Emissions from gas turbines can be divided into two sections, gaseous and particulate emissions. Gaseous emissions are gases in the form of CO_2, NOx (Oxides of Nitrogen), SOx (Oxides of Sulphur) and particulate matter (PM). In western Europe it is estimated that diesel fuel combustion in transportation vehicles contributes 20% of all $PM_{2.5}$ emitted [2].

Whilst PM emissions contribute towards smog and human health risks, gaseous emissions such as CO_2 and NO_x which are detrimental to the environment as well. CO_2 contributes towards climate change as it is a GHG and NO_x is responsible for atmospheric phenomena such as acid rain, smog and ozone layer depletion. Moreover, it has been observed that inhalation of NO_x by human's due to air pollution

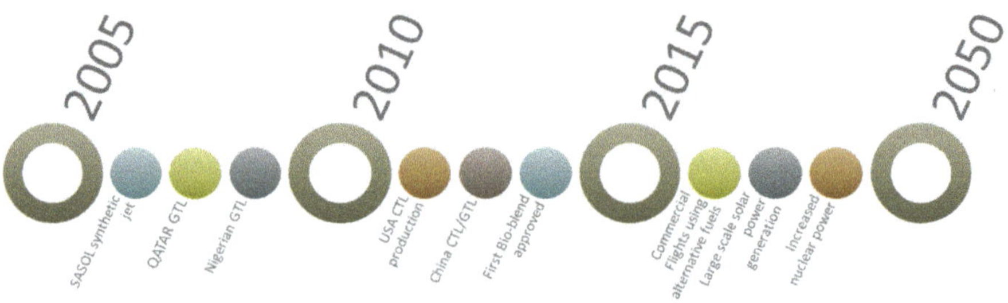

Fig. 1 Current and future development Road map of alternative fuels

lead to respiratory illnesses especially among vulnerable sections of the population such as the elderly and the young. Combustion dynamics have been under the scope for over a century, and many aspects of physical process and combustor geometrical arrangement have been researched earlier. With the emergence of new fuels, their varying chemical compositions may help alleviate or even eliminate combustion instabilities without implementing combustion control systems. This would reduce cost, weight and space required in a gas turbine engine. To the authors' knowledge no research other than author's own has been directed towards investigation of combustion instabilities, noise and vibrations through varied fuel compositions.

In this chapter, emissions production and engine performance while using alternative fuels in gas turbines will be discussed with a focus on their gaseous emissions, PM emissions, vibrations, noise and fuel system compatibility. This chapter has been divided into further sub-sections where all the above-mentioned topics have been discussed in detail.

2 Gaseous Emissions

The effects of alternative fuels on gaseous emissions is far less pronounced as opposed to particulate emissions. A study conducted by Cain et al. [3] using an Allison T63-A700 Turboshaft engine burning several alternative fuels and comparing them against JP-8. JP-8 (Jet Propellant-8) is a fuel similar to Jet-A1, but contains several fit-for-purpose (FPP) additives such as a lubricity enhancer, corrosion inhibiting additives and anti-icing additives. The alternative fuels tested were synthetic paraffinic kerosene (SPK) a F-T derived fuel and several other fuel blends whose main component C_{12} n-dodecane mixed with m-xylene (m-X), methylcyclohexane(MCH), iso-octane (i-C_8) or n-heptane (C_7). When these fuels were tested on the turboshaft engine at different power settings, CO_2 and CO was measured at the exhaust plane and the results are shown in Fig. 2.

As expected, for all fuels the EI (Emissions Index, grams of emission per kilogram of fuel burnt) for CO_2 increases with engine power. However, it should be noted that with respect to JP-8 the alternative fuels SPK, MCH and iC_8 consistently emit less CO_2. The study goes on to compare the CO output with respect to engine power as shown in Fig. 2 this shows a trend of CO decreasing with the increase of engine power, this is to be expected as CO is the product of incomplete combustion and as the engine starts to run at full power incomplete combustion is reduced if not entirely eliminated. Cain et al. [3] further goes on to stipulate that as the H/C (Hydrogen atoms to carbon atoms ratio) of the fuels increases the total amount of CO_2 emitted decreases. This is similar to CO_2 emissions trends observed on GTCP85 APU in CLEEN program [4].

Figure 3 shows the results from a study performed by Salvi et al. [5] showing the emissions variations for several bio-derived SPK blends with JP-8. It can be observed from the Fig. 3 that the bio blends lead to a reduction in $NO_{x,}$ especially for blend S8.

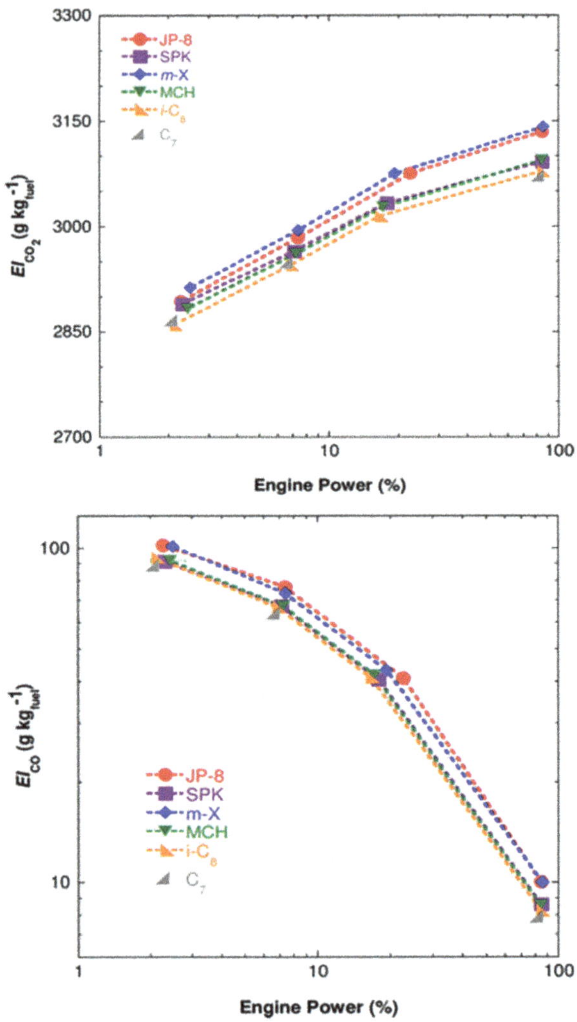

Fig. 2 Engine power setting versus EI of CO_2 for an Allison T63 Turboshaft burning alternative fuels [3]

The study conducted by Lobo et al. [6] measured the gaseous emissions from an Artouste Mk113 APU burning a CTL (F-T) and GTL fuel compared against Jet-A1. It was found that the NO_x emissions for Jet-A1 and the GTL were statistically indistinguishable. However, the CTL fuel showed a 5% decrease in NO_x with respect to Jet-A1 at full engine power. Moreover, a negative correlation between CO emissions and the fuels energy content is observed. As the fuel energy content increases the CO emissions reduced. Furthermore, the CTL fuel did not compare well against Jet A-1 in terms of unburnt hydrocarbon emissions (UHC) with CTL registering a 7%

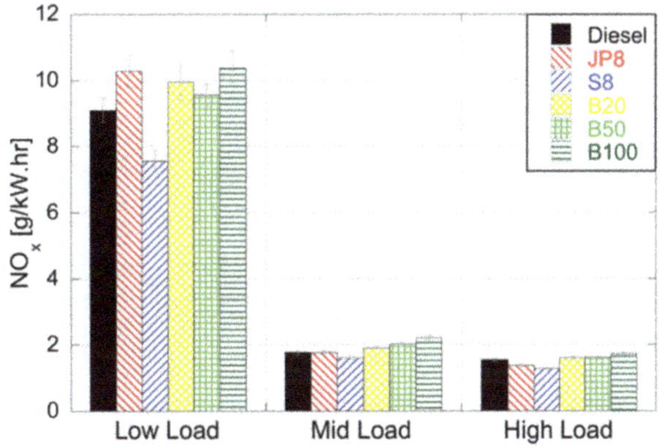

Fig. 3 Comparison bio-synthetic paraffinic kerosene (Bio-SPK) blends compared to reference JP-8 and Diesel [5]

increase in UHC at idle compared to Jet-A1. In testing done at the Low Carbon Combustion Centre (LCCC) under CLEEN program it was also observed that total NOx contents remain similar for all the alternative fuels and blends tested, but NO to NO_2 ratio changes. This observed phenomenon could have implications if only NO_2 or NO is being measured or regulated or used in further combustion research.

A Study conducted by Lee et al. [7] during the Alternative aviation fuel experiment (AAFEX) using a stationary McDonnell Douglas DC-8 fitted with CFM-56 turbines and measuring the emissions form the exhaust plumes 145 m downstream of the engines have observed several phenomenon as illustrated in Fig. 4.

It can be observed from the Fig. 4 that NO_x increases for all the fuels with increased engine power but the individual values of NO_x are all within the error bars indicating no statistically significant reduction in NO_x for the alternative fuels. It was also observed that the minor species of hydrogen peroxide is high at idle engine power and decreases as power is increased while nitrous acid shows the opposite correlation increasing with engine power. There are several other studies in literature which compares gaseous emissions of wide range of alternative jet fuel [3, 4, 8, 9].

It can be observed from Table 1 that the CO_2 emissions for all the fuels tested are within the uncertainty limits yielding no statistically significant trend. However, it can be seen that fuel 4 has significantly higher nitrogen dioxide output with respect to the other fuels and that the commercial alternatives have a slightly less NO_2 output with respect to Jet-A1.

Overall it can be said that CO_2 emissions depends on C/H ratio of fuel, as all the Carbon in fuel should get converted into CO_2, in most advanced engines. Provided the engine is operating on same power condition and its exhaust gas temperature and turbine entry temperatures are same for conventional fuels and alternative fuels then total NOx emissions normally should be similar. This is owing to the fact that NOx production depends on temperature, and fuel bound NOx is very low due to negligible amount of nitrogen being present in aviation fuels. Though in some studies it has been found that NO to NO_2 ratio in gaseous emissions change depending upon the source of the alternative fuel (i.e. If from CTL GTL or bio-derived). UHC and CO

Table 1 Gaseous emissions data for several Alternative fuels and potential drop-in alternatives fuel 1 is Jet-A1 fuels 2-4 are blends of Jet-A1 and SPK and fuels A-D are commercial alternatives [4]

	Uncertainty	Fuel 1	Fuel 2	Fuel 3	Fuel 4	Fuel A	Fuel B	Fuel C	Fuel D
CO_2 (EI)	±10	3169	3171	3172	3172	3170	3171	3170	3171
CO (EI)	±2	14.1	13.4	13.0	12.8	14	13.0	14.0	13.9
NOx (EI)	±0.1	4.1	4.2	4.2	4.2	4.4	4.1	4.0	4.2
NO (EI)	±0.1	2.9	2.9	3.0	2.9	2.8	2.9	2.8	2.9
NO_2 (PPM)	±0.1	34.1	33.4	33.6	47.9	32.4	33.2	32.3	33.3
THC PPM	±3	0	0	0	0	1	1	0	0
Oxygen (%)	±0.2	14.9	15.1	15.1	15.2	15.1	15.2	15.0	15.1

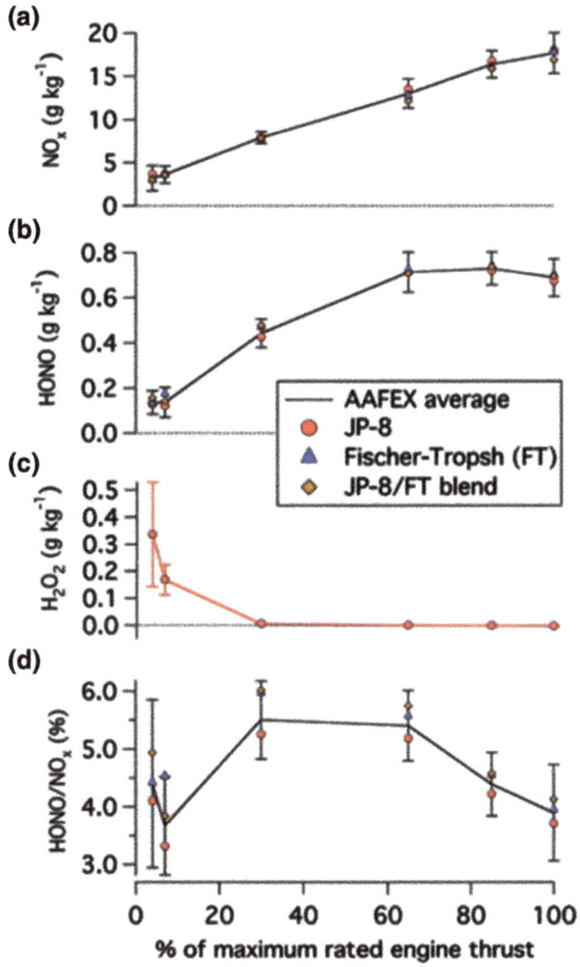

Fig. 4 Emission indices for NO_X HONO (Nitrous Acid), H_2O_2(Hydrogen Peroxide) and HONO/NO_X ratio obtained from the AAFEX experiment for several alternative fuels compared to JP-8 [7]

are products of incomplete combustion which could be due to a multitude of factors such as fuel properties like viscosity, surface tension, aromatic species and several others. Some studies have found that UHC and CO decreases when fuel has been changed to alternative fuels [3], whereas majority of the studies have found there is no significant change when fuel has been replaced with alternative fuels [7].

3 Particulate Emissions

Particulate emissions, when considered in the airborne context, conventionally can be into two subsections, $PM_{2.5}$ used for particles 2.5μm or less in diameter and PM_{10} for particles of diameter 10μm or less. Particle emissions, widely known as particulate matter (PM), refers to solids or liquids present in the exhaust gases after combustion. The particles can include carbonaceous particles, abraded metals, inorganic acids, as well as PM present in the ambient air generated from more mundane sources in the vein of soil and dust particles. Hence the shapes and sizes of the discrete particulates as well as their chemical composition can be irregular. To develop an accurate defection for particulate matter therefore would require clarification of their chemical composition, morphology and the abundance of each particle as a function of particle size. Therefore, some common describers of particulate matters include nvPM and vPM, non-volatile and volatile particulate matter. In general, nvPM are solid particles at the exit plane of the engine exhaust whereas vPM is liable to change state when it encounters the ambient conditions outside the engine, as the exhaust is at extremely elevated temperatures some gaseous emissions may condense into liquid and coat the solid particles when cooled down in the exhaust downstream of the turbine

The rate these gaseous emissions condense is somewhat dependent upon their vapour pressure and other ambient conditions such as temperature and humidity, a classic example of this phenomena are contrails from jet aircraft, which appear sometimes but not others. This is due to water being vapour being condensed due to the prevailing ambient conditions, as well as the exhaust temperature at the time.

Volatile particulate matter (vPM) are formed by the nucleation of gaseous precursors mainly consisting of sulphuric acid and other such organic compounds formed in the cooler exhaust gas downstream of the combustor [10–12]. Furthermore, it has been observed that these gaseous precursors condense to around the nvPM as illustrated in Fig. 6. The volatile PM definition also fits the condensable PM (CPM) terminology mostly used by the Environmental Protection Agency (EPA) (Figs. 5 and 6).

The main reason particulate emissions have garnered attention is due to the fact that it is an air pollutant, which among other things is mainly responsible for the smog that permeate industrial and heavy-traffic oriented cities such as Beijing and Delhi. Furthermore, particulates pose a significant health risk to humans, it is well-established in literature that humans exposed to particulate matter on a regular basis are subject to increased risk of mortality and loss of life expectancy due to respiratory and cardio pulmonary illnesses, such as lung-cancer and cardiac arrest [2, 14–17].

Therefore, it is imperative that particulate emissions be reduced to this end alternative fuels of diverse types have been scrutinized for their particulate emission levels. In one of the studies conducted by Lobo et al. [18] comparison of PM emissions from a commercial gas turbine (CFM-56) while using alternative fuel has been done. Different types of biomass and FT based fuels were used which were then compared with Jet A-1 as a standard. Several blends of FAME (Fatty-Acid-Methyl-Esters) and

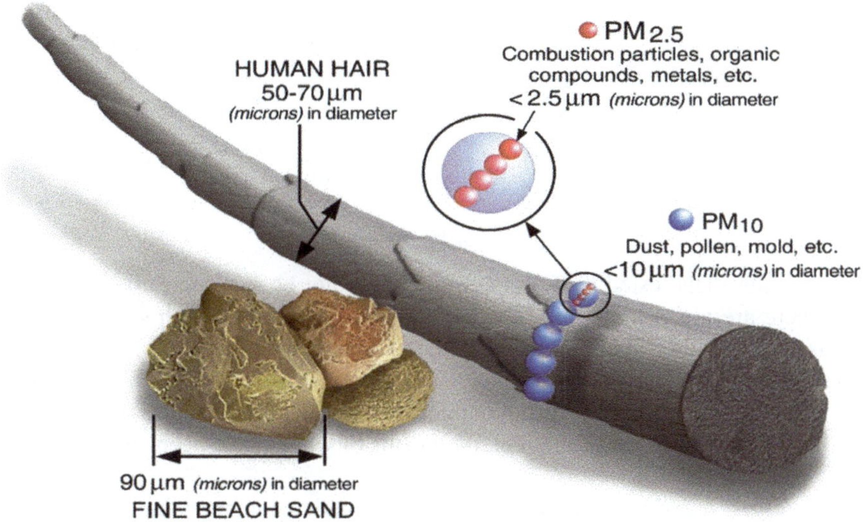

Fig. 5 Differentiation of particulate matter [13]

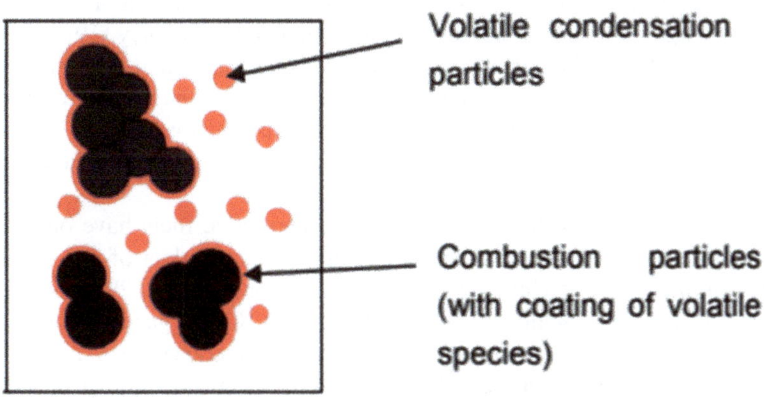

Fig. 6 Volatile particulate matter coating of soot [11]

Jet A-1 and 100% Fischer-Tropsch fuels were tested. The turbine was operated for full LTO (Landing and take-off) cycles for each blend of fuel.

The results of this these test show that PM emissions is reduced significantly when FAME blended fuels and FT fuels are used, as shown in the table below;

Table 2 presents the PM emissions reductions as a percentage when compared to standard Jet A-1, with 100% F-T fuel providing the greatest reduction in particulates matter, however all the alternative fuels tested had lower PM emissions number and size than Jet A-1, this can be attributed to the fact the fuels in this study has been chosen for their low aromatic content and high H/C ratios. Though it is to be noted

Table 2 PM mass and number reductions for alternative fuels with respect to Jet-A1 [18]

Alternative fuel	PM number reduction (%)	PM mass reduction (%)
20% FAME 80%jet-a1	22 ± 7	20 ± 8
40% FAME 60%JET-A1	35 ± 6	52 ± 5
50% F-T 50%JET-a1	34 ± 7	39 ± 7
100% F-T	52 ± 4	62 ± 4

that some the fuels tested by Lobo et al. [8] may not be suitable to be used as jet fuel. There are a substantial number of studies in literature which shows that the increase in aromatic content of a given fuel has a tendency to increase PM emissions in gas turbine exhausts. This effect has also been observed by Brem et al. [19] where an in-production high-bypass turbofan injected with fuel mixed with solvents which alter the aromatic content of the fuel, the results of which have been summarised in Fig. 2 [19].

From Fig. 7 it can be observed that as the aromatic content of the fuel increases the emission indices for nvPM also increases showing a clear causal relationship. Brem et al. [19] goes on to support the view that soot formation is the result of aromatic content in the fuel as opposed to incomplete combustion, as modern day turbines are highly efficient achieving 99.9% combustion efficiency [19]. Moreover a study conducted by DeWitt et al. [20] corroborates the fact that aromatic content of a fuel is proportional to PM emissions. The study measured the number of particles emitted and their size for JP-8 and F-T derived fuels and found that F-T derived fuels emitted particles that were a full order of magnitude smaller than those emitted whilst running JP-8.

Williams et al. [9] has described the effects alternative fuels have on vPM by measuring the organic matter concentrations in the exhaust duct of a Rolls-Royce Artouste Auxiliary Power Unit (APU) during tests conducted in 2009 [9]. During the course of this investigation several coal-to-liquid (CTL), gas-to-liquid (GTL), diesel and biodiesel fuels have been compared with the reference Jet A-1. It has been observed that the organic mass emitted by Jet A-1 is higher than that of the CTL and GTL blends used at the various power levels tested, this lends credence to the view that alternative fuels emit less vPM. Furthermore, the paper goes on to suggest that the vPM content in a given exhaust is sensitive to its measurement location as vPM is gaseous at first and condenses onto the nvPM particles downstream in the exhaust due to temperature drops. The resulting organic mass emissions results from the study are shown in Fig. 8.

In an another study conducted by Liati et al. [21] the size distributions of nvPM produced by a CFM-56 gas turbine with respect to engine power using electron microscopy was studied. It was found that at 100% engine static thrust the nvPM particles are larger and more numerous compared with 65% engine power. With lower engine settings, the amount of nvPM drops dramatically and also the mean size of the particles also drops, however these smaller particles are more oxidative and reactive with respect to larger particles. Reduction in PM emissions therefore

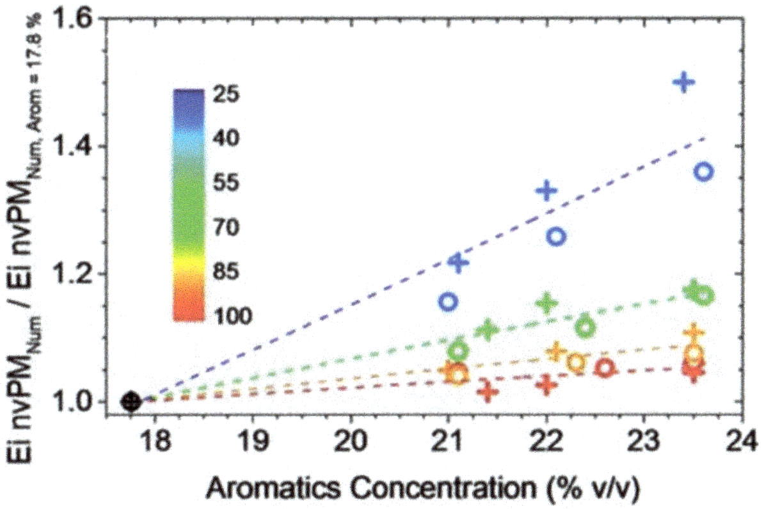

Fig. 7 Aromatic content of a fuel versus emission indices of the non-volatile particulate matter emitted upon the burning of the fuel in a gas turbine coloured lines indicate engine power setting as a percentage. [19]

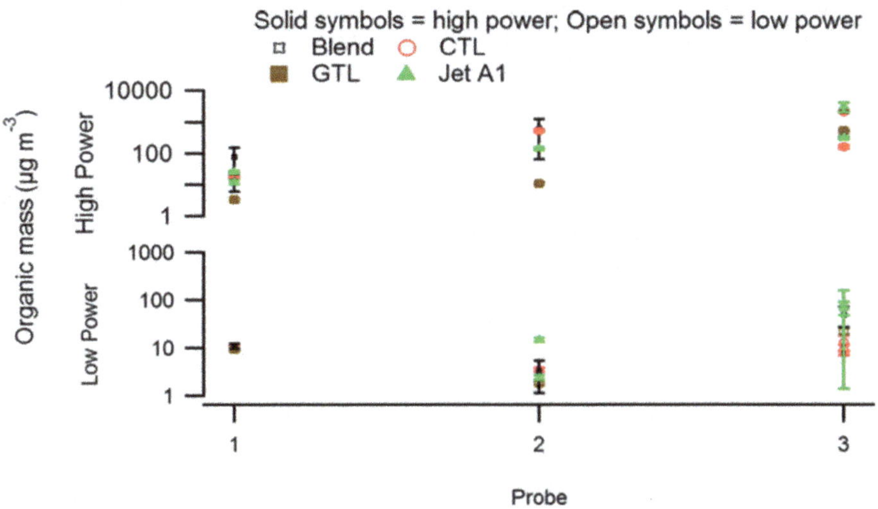

Fig. 8 Organic mass emitted by CTL, GTL and Jet-A1 against measurement locations and power settings [9]

can be achieved in several ways such as, combustor designs that limit the fuel rich areas in the combustion domain and reducing residence times of the fuel in very high temperature zones within the combustor [21].

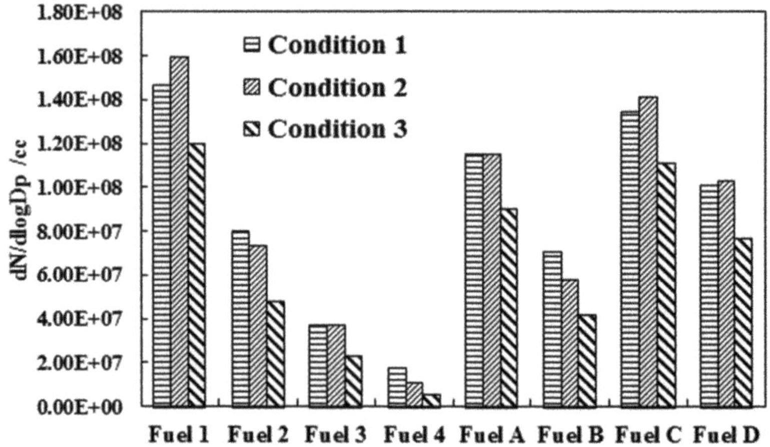

Fig. 9 Particulates measured for fuels at size of ~75 nm. Fuel 1 is Jet-A1 [4]

In another study under Continuous Lower Energy, Emissions and Noise (CLEEN) program at the University of Sheffield's Low Carbon Combustion Centre (LCCC), gaseous and PM emissions from gas turbines were measured for several alternative fuels and then compared to reference Jet-A1 where fuels 1-4 were blends of Jet-A1 and SPK (Synthetic Paraffinic Kerosene) and fuels A through D were potential alternative jet fuels. The turbine in used for the tests was a Honeywell GTCP85 APU [4]. Figure 9 shows the number of particles produced of size ~75 nm particulates for all the fuels tested in the study. Again, it can be observed that the fuels with lower aromatic content show reduced PM density has opposed to fuels with higher aromatic levels. Similar trends were attained for ~27 nm particulates, validating the pivotal role of aromatics in particulate emissions.

Dewitt et al. [20] studied various aromatic solvents, which are consistent with the molecular weight distribution shown by jet fuel used by military users (JP-8). These were then added to F-T fuels as blends and as individual components. The study observed an increased output of soot precursors which in turn indicated higher PM concentrations which was attributed to the increased PM emissions.

In conclusion it has been said from study of Dewitt et al. [20] that aromatic content of a fuel has a very strong impact upon the amount, and size distribution of particulate matter emitted from gas turbines. Furthermore, as the composition of alternative fuels such as those from the F-T process can be altered to reduce their aromatic content, they produce less particulate emissions.

4 Seals Compatibility

Although in the previous sections it has been determined that alternative fuels, on the whole, are beneficial to the aviation industry, as well as the environment, there remains the issue of whether these alternatives are compatible with existing fuel systems and infrastructure. Even though wholly alternative fuels have been approved for use in gas turbines, there exists a possibility of fuel leaks due to the varying composition of the alternative fuels. This happens because the seals in the engine are not compatible with the new fuels. One of the main reasons for the seals not to work is due the absence or reduction of aromatics in the new fuels. Low seal- swell or even seal-swell reduction has been attributed to the lack of Aromatic content in alternative fuels because of the increased particulate matter emissions [22]. The seals shrinking can cause seal failures thus damage in the fuel system and eventually leakages. On the other hand, aromatics are responsible for higher PM emissions.

In essence seal-swell is defined as the increase in volume experienced by a seal when in contact with a liquid and vice versa. This swelling normally means that the inner-diameter as well as the volume of the seal increases due to the absorption of fuel components such as aromatic content. Generally, naphthalene is considered a good hydrogen donor as opposed to alkanes or alkyl benzenes. DeWitt et al. [20] found that fuel component separation and assistance to seal-swell is in the following order:

alkanes < alkyl benzenes < naphthalene's

As observed by Thomas et al. [23] the swelling of the seal elastomers as a reaction against the fuel, moreover it has been determined by Qamar et al. [24] that seal swelling is caused by the seal absorbing hydrocarbons from the fuel. In the aviation field acceptable seal swell ranges from approximately 18–30% [25, 26] whereas in the automotive industry seal swell is at roughly 12%, this can be attributed to the fact that ground vehicles do not experience the same variation in ambient conditions as aircraft and hence require less seal-swell performance is required. When considered in greater detail it has been observed by Graham et al. [25] that several reaction takes place where intermolecular bonds of the fuel and polymer seal break and form new bonds with each other. Overall these reactions are in equilibrium and are energy balanced. On the contrary seal shrinking ensues in the event that particular molecules of the seals seep into the fuel causing the seal to reduce in volume, the lack of plasticizer in the seals can be a cause for seal shrinkage. This process is shown in the Fig. 10 Baltrus et al. [27] observed that the shrinking process involves the release of fuel components absorbed by the seals.

Figure 11 shows the effect on seal swell different aromatics has on nitrile seal using stress relaxation technique. It can be clearly observed from the figure that tetralin is giving significantly higher seal swell as compared to propyl benzene or P-xylene [22].

A study carried out by Liu and Wilson [28] a stress relaxation technique was utilised to observe the effects of several solvents including n-decane, iso-parrafins and cycloparaffins on seals composed of several materials.

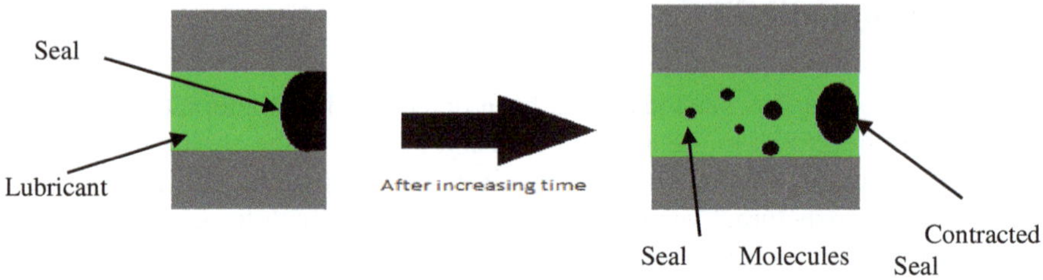

Fig. 10 Contracting process process of seal

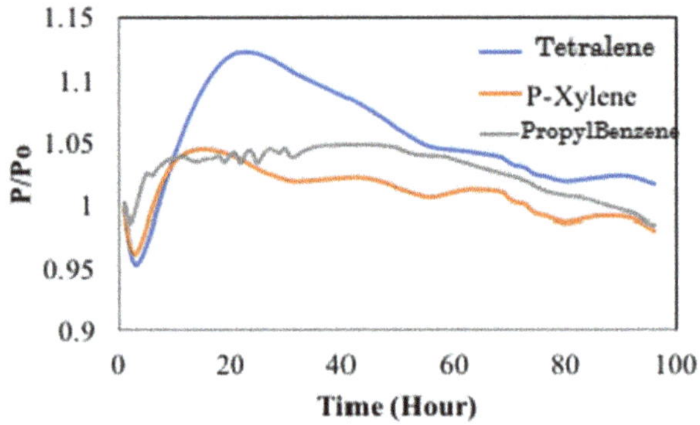

Fig. 11 Swelling effect of nitrile O-rings in mixture of 25% aromatic and SPK [22]

It was observed during this study that O-rings manufactured from fluorosilicone and fluorocarbons performed well in the presence of all the fuel blends tested. Furthermore, it has been found that nitrile O-rings are susceptible to substances other than aromatics and that n-decane causes seals to lose performance.

Figure 12 shows the amount of seal swell achieved while using different compositions of Decalin, Decane and Shellsol T. It can be clearly observed from the figure that there are compounds which leads to seal swell, while others may not take any part in swelling or lead of shrinkages. It is also found from the study that not just aromatics are responsible for swelling of seals. Similar patterns have also been observed by DeWitt et al. [20] and Graham et al. [29]. It was also found that several types of aromatics lead to different amount of seal swell. According to available literature, it can be comfortably said that further optimisation and research is required in alternative fuel industry so that appropriate seal swell can be achieved without compromising on other parameters.

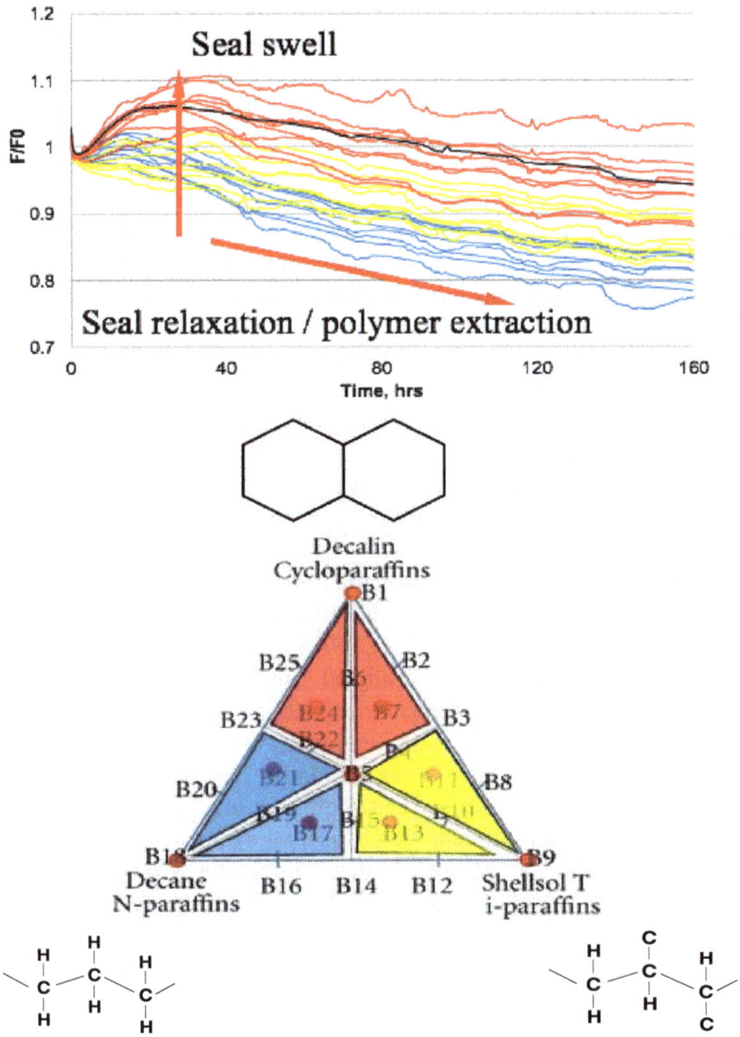

Fig. 12 Relaxation behaviour of nitrile O-rings in the triangle [28]

5 Thermal Stability

In this section, the ability of alternative fuels to withstand thermal stresses is discussed. Fuels must be able to withstand thermal stresses due to the customary practice in aircraft design of using fuel as heat sinks or coolants, moreover as the fuel approaches the engine and combustors through the fuel system its temperature begins to rise, and a range of chemical reactions begin to take place. These reactions may lead rise to particles in the fuel which may or may not be soluble [30, 31].

Particles such as these would have an obvious detrimental effect upon filters and valves in the fuel system which may lead to blockages and disruptions for fuel flow [32]. Increased pressure ratios in gas turbines of the 21st century pose several issues in terms of thermal stability, in that the higher-pressure ratios result in higher thermal loads on the lubrication system. This results in an increased heat sink demand [33]. The increased thermal efficiency of the turbine cycle caused by the increased pressure ratios result in decreased fuel flow rates which while ideal in when considering efficiency puts more strain upon the thermal characteristics of the fuel in that a lower volume of fuel must absorb more heat. This increased heat results in convection transfer into the mechanical components of the fuel system such as the swirlers and burner feed arms [33]. All these phenomenon results in the degradation of the fuel and may cause carbonaceous deposits to form in areas that encounter the fuel, much like atherosclerosis in human blood vessels. The causes for these deposits are multifactorial. Some of these causes are as follows; fuel composition, fuel temperature, duration of thermal exposure, flow characteristics, surface roughness of the fuel wetted areas [30].

In terms of alternative fuels, F-T derived fuels have been observed to perform well with respect to thermal stability whilst contributing far less PM emissions when compared to Jet-A1 [34–36]. When considering SPK fuels which have been derived from Syngas (H_2 CO) thermal stability is considered insignificant because of the reduced amount of impurities contained in syngas which has already been processed. Jet Fuel Thermal Oxidation Tester (JFTOT) is conventionally used to test the thermal stability of the fuels. Due to the JFTOT tubes being aluminium in construction the breakpoint temperature cannot be measured for fuels with high breakpoint temperature. In a recent study by Moses [37] different blends of semi-synthetic fuels were tested for their JFTOT breakpoint. It was observed that semi-synthetic jet fuel blends under study were having very high breakpoint, which indicates very high thermal stability. Moreover, when the depth at the conclusion of the test increased it was found that the tube temperature also increased. Which in turn enables the possibility of using SPK's to improve fuels which are on the verge of thermal stability.

In another study by Corporan et al. [38] thermal stability of 6 different paraffinic fuels was tested and compared with JP8. This shows that all the paraffinic fuels tested in the study have higher resistance to carbon formation and could be used in elevated temperature environment as a coolant. Figure 13 below shows headspace oxygen profiles and mass accumulation for all the fuels tested in the study. It can be observed from the figure that each fuel shows a different deposition and oxidation characteristic. It was also observed that oxidation profiles are very high in variance.

Alborzi et al. [33] investigated the effect of surface deposition on fuel injector feed arm which was simulated for sudden contraction and expansion. The study was conducted using an Aviation Fuels Thermal Stability Unit (AFTSTU), which can conduct a representative test at full scale 1000 flight hours for surface deposition, to determine how long it takes for surface deposition to start occurring and its associated performance impact upon the turbine. This is a different type of rig for testing ther-

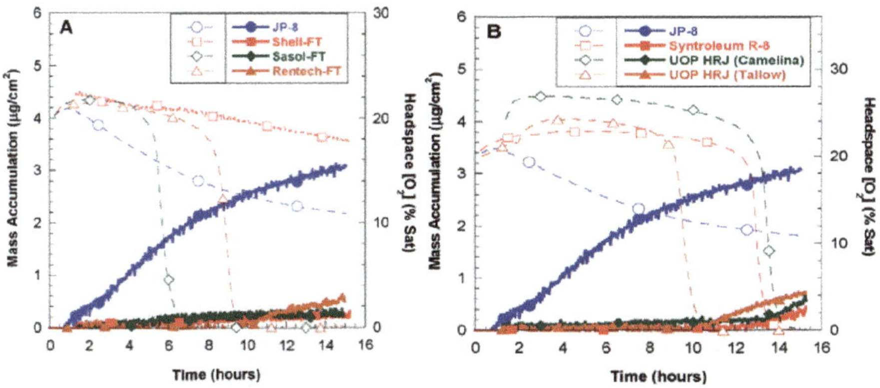

Fig. 13 Mass build-up (solid lines, closed markers) and head-space oxygen (dashed lines, open markers) profiles of alternative fuels. Corporan et al. [38]

mal stability of the fuels. It was observed that deposition in Contraction/expansion tube are significantly higher than straight tube. Balster et al. [39] presented thermal stability data for a novel coal derived fuel. It was observed that surface deposition for novel fuel was significantly lower than other fuels tested in the study which includes JP8. Overall it can be said that novel F-T process based fuels can give better thermal stability, though care needs to be taken while selecting and using these fuels due to other impacts they may have.

6 Combustion Vibration and Noise

In essence, combustion can be defined as a process where chemical energy is converted to heat energy in the company of oxygen [40]. During these reactions, the molecular bonds between the reactants are broken releasing energy, and bonds are formed to create the reaction products, the difference in energy between these reactions is dissipated into the surroundings, increasing the surrounding temperature. This rise in temperature is equivalent to the kinetic energy of the molecules in an object. Hence, according to the second law of thermodynamics these energies must reach equilibrium, generating noise and vibrations in the process. Noise is defined as the unwanted oscillation of air particles whilst vibration is the oscillation of solid material [41].

Combustion instabilities are significant amplitude oscillations that arise in gas turbine combustion. These give rise to thrust oscillations, thermal stresses, and more notably, resonant vibrations in mechanical components. Rayleigh's criterion is useful when understanding this phenomenon, as the namesake described the circumstances in which unsteady heat release oscillations result in acoustic oscillations, which in turn leads to fluctuations in the thermodynamic variables of the system [42, 43].

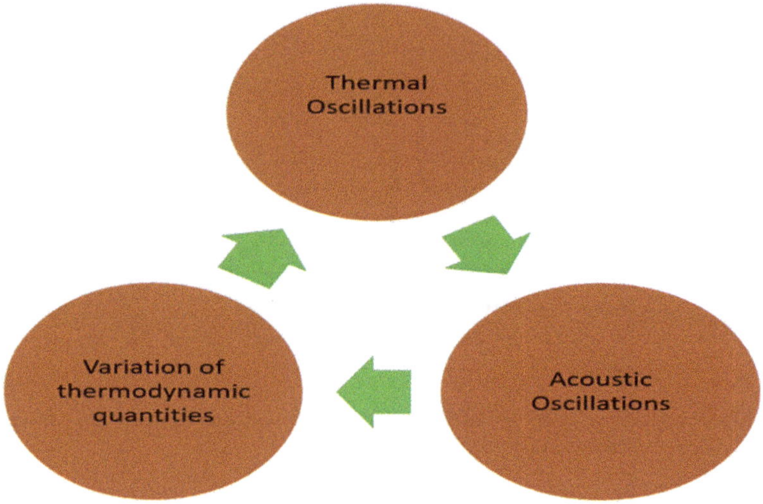

Fig. 14 Combustion instability feedback loop

Figure 14 is useful for visualizing how these factors compound together, thereby helping to rationalize the significant harms foreseeable, if ignored.

Due to the pressures and forces created by these oscillations there exist a possibility for the thrust produced by the turbine to oscillate, furthermore there can be possible interference of the engine control systems leading to malfunctions and premature wear of components due to cyclic fatigue [42]. In addition, if the vibrations caused matches the natural frequencies of the components being vibrated catastrophic failure could occur, however most aviation related components are tested for their natural frequencies and safe ranges of frequencies are specified for each component [44, 45]. Furthermore, if vibrations and instability is not handled correctly then damage to the combustors can happen as shown in Figs. 15 and 16.

The vibrations and noise frequencies generated by combustion are usually divided into 3 categories namely low frequency dynamics (LFD) or Helmholtz modes at less than 50 Hz, Intermediate frequency dynamics (IFD) at between 50 and 1000 Hz and finally high frequency dynamics (HFD) for vibrations above 1000 Hz (Fig. 17).

The extent to which vibration and noise manifest is partially dependent on the different properties of the fuel. Khandelwal et al. [46] investigated the role of different fuel composition and its impact on combustion vibrations [43]. Testing was done on a Honeywell GTCP85 APU using four different fuels. Fuel 1 and Fuel 2 are Jet A-1 sourced from two different sources, whereas fuel 3 and 4 are FT process produced fuel from different sources. It was observed that the FT process fuel which has lowest density from the fuels tested in this study produced higher frequency spectra of vibrations. Though highest amplitude of the vibration was produced by Jet A-1 from source 1. It is to be noted that Jet A-1 sourced from two different sources have similar frequency but significantly different vibrations (Fig. 18).

Fig. 15 Damaged Injectors [48]

Fig. 16 Damaged rocket motor injector shear [48]

Furthermore, the relationship between fuel density and vibration is more explicitly considered in the work done by How et al. [46], where a higher density, coconut oil blend, biodiesel was compared with conventional diesel [46]. In this study, the higher density fuel blend displayed a significant reduction in vibrational acceleration.

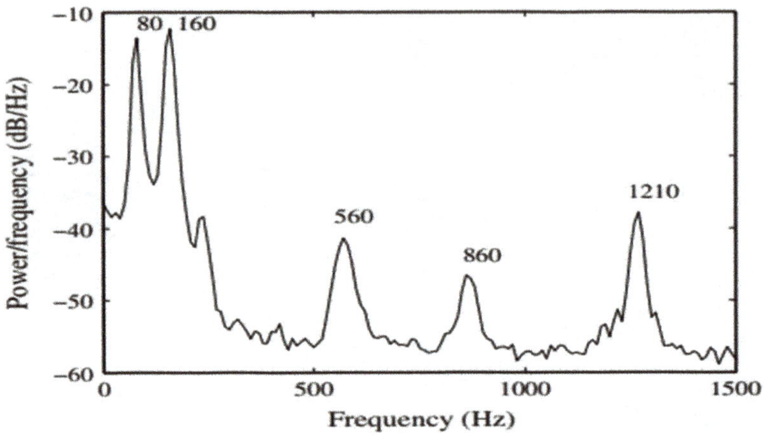

Fig. 17 Instabilities for combustors according to their frequency range [49]

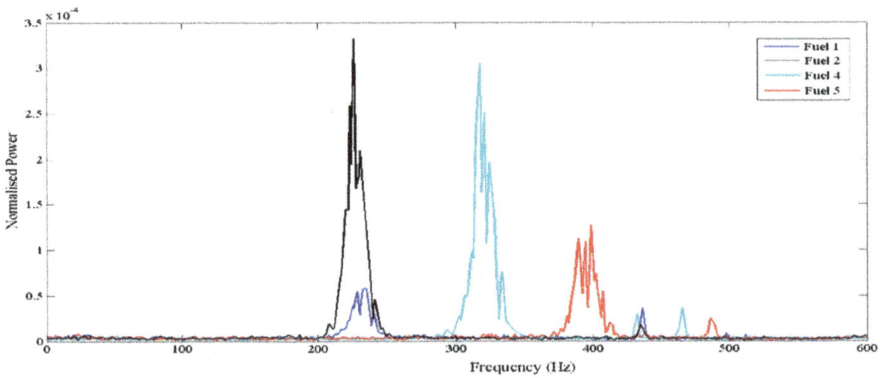

Fig. 18 Vibration amplitude and frequency with 4 different fuels [46]

Overall it is observed that there has been little research on the combustion instability, noise and variations caused by combustion induced vibration with alternative fuels. This pattern is reflected across all of the major public domain journals under current situations. Stricter emissions legislations demand the use of lean premixed combustion, but combustion instability is more likely in these types of lean combustors compared to current rich burn systems. It was observed that the instabilities may be due to oscillations in pressure, velocity, temperature or equivalence ratio of fuel. Among these variables, Rayleigh's criterion was made to be one of the primary conditions to be met for a self-excited combustion oscillation to occur. Equivalence ratio oscillations are a possible cause of combustion instabilities.

The underlying reason for change in combustion instability could be due to change in fuel's boiling point, viscosity, vapour pressure, flame speed, stoichiometric equivalence ratio, cetane rating, density, energy density and/or composition. So far, these

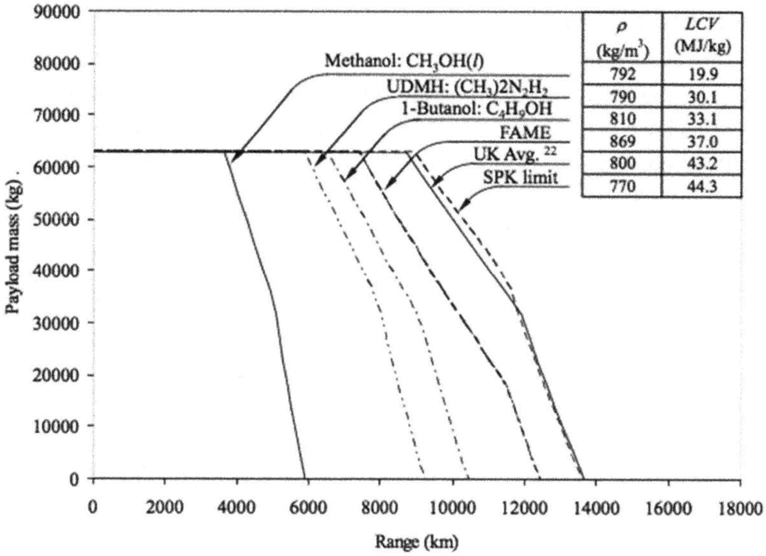

Fig. 19 Range versus Payload for B747-200B using alternative fuels [47]

instabilities are suppressed in existing engines by using Helmholtz resonators, Active combustion Control (ACC), noise suppressers, controlling the droplet size, changing the flame speed and length and varying the flame injector geometry and location. But with the arrival of new alternatives, if the instabilities can be understood better and designed out by fuel selection and combustion design optimisation.

7 Aircraft Range and Payload

Due to different properties of alternative fuels, not just emissions and engine performance change, but aircraft range can also change significantly. Blakey et al. [47] studied the effects on payload and range while using wide range of candidate fuels for aviation. This also includes variations in the blends of conventional hydrocarbon fuels. Short haul commercial flights tend to use low density fuels as range per volume of fuel is not relevant as for long haul or military aviation where the maximum range is required for a given volume of fuel. Figure 15 below shows the changes to range of a Boeing 747-200B can have while using a range of different fuels investigated under study (Fig. 19).

It was also observed that hydrocarbon blends could be suggested for each aircraft type, which could be designed for a maximum range while allowing maximum payload. It was also said that specific flight plans lower than the maximum range of the aircraft may be supplied with a fuel of lower specific energy.

8 Concluding Remarks

In this chapter, the effects of alternative fuels on the gas turbines performance and their emissions have been discussed. It has been found that alternative fuels provide no clear advantage in terms of emissions reduction compared to standard petroleum derived fuels though benefits in PM have been observed. Effect on seal performance, vibrations, noise and engine life still needs further work. Impact on range of an aircraft can be easily calculated by change in energy density. Overall further work is required to say which option is good and how to move ahead in this area.

References

1. S. Roy, *Combustion Instabilities and Emissions Analysis of Alternative Fuels and Their Compositions* (University of Sheffield, 2014)
2. J. Kuenen, B. Gschwind, Estimating particulate matter health impact related to the combustion of different fossil fuels, in *Proc. ...*, 2013
3. J. Cain et al., Characterisation of gaseous and particulate emissions from a turbo-shaft engine burning conventional, alternative, and surrogate fuels. Energy & Fuels **27**, 2290–2302 (2013)
4. Rolls-Royce, Rolls-Royce Alternative Fuels Program-Final Report (CLEEN), 2015
5. A.A. Salvi, D. Assanis, Z. Filipi, Impact of physical and chemical properties of alternative fuels on combustion, gaseous emissions, and particulate matter during steady and transient engine operation. Energy Fuels **26**(7), 4231–4241 (2012)
6. P. Lobo et al., Impact of alternative fuels on emissions characteristics of a gas turbine engine – part 1: gaseous and particulate matter emissions. Environ. Sci. Technol. **46**, 10805–10811 (2012)
7. B.H. Lee et al., Measurements of nitrous acid in commercial aircraft exhaust at the alternative aviation fuel experiment. Environ. Sci. Technol. **45**, 7648–7654 (2011)
8. A.A. Ivanov, A.N. Ermakov, R.A. Shlyakhov, On high suppression of NOx and CO emissions in gas-turbine plants with combined gas-and-steam cycles. Therm. Eng. **57**(13), 1132–1138 (2010)
9. P.I. Williams et al., Impact of alternative fuels on emissions characteristics of a gas turbine engine – part 2: volatile and semivolatile particulate matter emissions. Environ. Sci. Technol. **46**, 10812–10819 (2012)
10. M. Fiebig, S. Nyeki, C. Stein, A. Petzold, Emission of volatile and non-volatile ultrafine particles from a combustion source during part emis, in *European Conference on Aviation, Atmosphere and Climate (AAC)*, 2003
11. A. Kugele, F. Jelinek, R. Gaffal, in *Aircraft Particulate Matter Emission Estimation through all Phases of Flight EEC/SEE/2005/0014* (Brussels, 2005)
12. A. Petzold, C.W. Wilson, *Physical and Chemical Properties of Aircraft Engine Exhaust Particles, Oberpfaffenhofen*
13. US EPA, Particulate Matter (PM) Basics (2017). https://www.epa.gov/pm-pollution/particulate-matter-pm-basics#PM. Accessed 09 Nov 2017
14. D.W. Dockery, A.C. Pope, X. Xu, J.D. Spengler, An association between air pollution and mortality in six US cities. N. Engl. J. Med. **329**(24), 1753–1759 (1993)
15. J.D. Sacks et al., Particulate matter-induced health effects: Who is susceptible?, in *Environmental Health Perspectives*, vol. 119, no. 4 (National Institute of Environmental Health Science, Apr 2011), pp. 446–454
16. T. Reichhardt, EPA particulate matter health effects document criticized. Environ. Sci. Technol. **29**(10), 449A–449A (1995)

17. R. Bastiaans, A.W. Vreman, Numerical simulation of instabilities in lean premixed hydrogen combustion & quot; Unsteady RANS and scale adaptive simulations of a turbulent spray flame in a swirled- stabilized gas turbine model combustor using tabulated chemistry. Numer. Simul. Int. J. Numer. Methods Heat Fluid Flow **22**(5), 112–128 (2012)
18. P. Lobo, D.E. Hagen, P.D. Whitefield, Comparison of PM emissions from a commercial jet engine burning conventional, biomass, and fischer-tropsch fuels. Environ. Sci. Technol. **45**(24), 10744–10749 (2011)
19. B.T. Brem et al., Effects of fuel aromatic content on nonvolatile particulate emissions of an in-production aircraft gas turbine. Environ. Sci. Technol. **49**, 13149–13157 (2015)
20. M.J. Dewitt, E. Corporan, J. Graham, D. Minus, Effects of aromatic type and concentration in fischer-tropsch fuel on emissions production and material compatibility. Energy Fuels **22**, 2411–2418 (2008)
21. A. Liati et al., Electron microscopic study of soot particulate matter emissions from aircraft turbine engines. Environ. Sci. Technol. **48**, 10975–10983 (2014)
22. A. Anuar, *Effect of Fuels, Aromatics and Preparation Methods On Seal-Swell* (University of Sheffield, 2014)
23. E. Thomas, R. Fuller, K. Terauchi, Fluoroelastomer compatibility with biodiesel fuels. SAE 2007 Trans. J. Fuels ... (2007)
24. S.Z. Qamar, M. Akhtar, T. Pervez, M.S.M. Al-Kharusi, Mechanical and structural behavior of a swelling elastomer under compressive loading. Mater. Des. **45**, 487–496 (2013)
25. Boeing Company, "CLEEN report: impact of alternative jet fuel and fuel blends on non-metallic materials used in commercial aircraft fuel systems, in *Fuel*, 2011
26. G. Hemighaus, Synthetic fuels for aviation. Stand. News **35**(4) (2007)
27. J. Baltrus, D. Link, Screening of potential o-ring swelling additives for ultra-clean transportation fuels, ... Transp. Fuels (2007)
28. Y. Liu, C.W. Wilson, Investigation into the Impact of n-Decane, decalin, and isoparaffinic solvent on elastomeric sealing materials. Adv. Mech. Eng. **4**, 127430 (2012)
29. J.L. Graham, R.C. Striebich, K.J. Myers, D.K. Minus, W.E. Harrison, Iii, swelling of nitrile rubber by selected aromatics blended in a synthetic jet fuel. Energy Fuels **20**, 759–765 (2006)
30. R. N. Hazlett, *Thermal Oxidation Stability of Aviation Turbine Fuels*, 1991
31. E.G. Jones, W.J. Balster, Phenomenological study of the formation of insolubles in a Jet-A fuel. Energy Fuels **7**, 968–977 (1993)
32. W.F. Taylor, Development of high stability fuel. ESSO research and engineering report, 1972
33. E. Alborzi, S. Blakey, H. Ghadbeigi, C. Pinna, C. Lewis, Investigation of surface deposition in a simulated fuel injector feed arm with sudden expansion/contraction. Fuel **186**, 534–543 (2016)
34. T. Edwards et al., Fischer-tropsch jet fuels—characterization for advanced aerospace applications, in *40th AIAA/ASME/SAE/ASEE Joint Propulsion Conference and Exhibit*, 2004
35. W. Harrison, and S. Zabarnick, The OSD assured fuels initiative military fuels, in *31st International Technical Conference on Coal Utilization & Fuel Systems*, 2006
36. E. Corporan et al., Emissions characteristics of a turbine engine and research combustor burning a Fischer-Tropsch jet fuel. Energy Fuels **21**(5), 2615–2626 (2007)
37. C.A. *Moses, Semi-Synthetic Jet Fuels CRC Report*, Alpharetta, GA, 2008
38. E. Corporan et al., Chemical, thermal stability, seal swell, and emissions studies of alternative jet fuels. Energy Fuels **25**(3), 955–966 (2011)
39. L.M. Balster et al., *Development of an Advanced, Thermally Stable, Coal-Based Jet Fuel*, 2008
40. N. Hall, Combustion (2015). https://www.grc.nasa.gov/www/k-12/airplane/combst1.html. Accessed 21 Nov 2017
41. P. Le Grand, The energy transport by the propagation of sound waves in wave guides with a moving medium. J. Eng. Math. **11**(2) (1977)
42. T.C. Lieuwen and B.T. Zinn, *Combustion Instabilities* (AIAA, 2005)
43. B. Khandelwal, S. Roy, C. Lord, Effect of novel alternative fuels and compositions on vibrations of a gas turbine engine, in *50th AIAA/ASME/SAE/ASEE Joint Propulsion Conference*, 2014

44. Y. Huang, V. Yang, Dynamics and stability of lean-premixed swirl-stabilized combustion. Prog. Energy Combust. Sci. **35**(4), 293–364 (2009)
45. H.C. Mongia, G.C. Hsiao, Incorporation of combustion instability issues into design process: ge aeroderivative and aero engines experience, in *Combustion Instabilities in Gas Turbine Engines* (American Institute of Aeronautics and Astronautics, Reston, VA, 2006), pp. 43–63
46. B. Khandelwal, S. Roy, C. Lord, S. Blakey, Comparison of vibrations and emissions of conventional jet fuel with stressed 100% SPK and fully formulated synthetic jet fuel. Aerospace **1**(2), 52–66 (2014)
47. S. Blakey, C.W. Wilson, M. Farmery, R. Midgley, Fuel effects on range versus payload for modern jet aircraft. Aeronaut. J. **115**, 627–634 (2011)
48. J. Hulka, Chug and Buzz the neglected and disrespected combustion instabilities. *NASA* (2018), https://ntrs.nasa.gov/archive/nasa/casi.ntrs.nasa.gov/20170008957.pdf. Accessed 23 Feb 2018
49. J.E. Temme, P.M. Allison, J.F. Driscoll, Combustion instability of a lean premixed prevaporized gas turbine combustor studied using phase-averaged PIV. Combust. Flame **161**(4), 958–970 (2014)

Effect of Fuel Unsaturation on Emissions in Flames and Diesel Engines

Suresh K. Aggarwal

Abstract Due to the emergence of a new generation of renewable fuels and the need to accurately model the combustion chemistry of multi-component fuels, there is growing interest in examining the effect of fuel molecular structure on fuel reactivity. This book chapter provides an overview of research dealing with the effects of fuel unsaturation on the ignition, combustion, and emission characteristics. Results from both laboratory-scale configurations, such as shock tube, rapid compression machine, and laminar flames, as well as from high-pressure sprays in compression ignition engines are discussed. Experimental and kinetic modeling studies of homogeneous mixtures provide clear evidence that depending upon the number and position of C = C double bonds, and the reactivity of long-chain hydrocarbons is significantly affected by fuel unsaturation, especially at low to intermediate temperatures. Ignition data for saturated and unsaturated components indicate that the presence of double bond inhibits low-temperature reactivity, modifies the NTC behavior, and leads to reduction in CN number in diesel engines. This has important consequences regarding the effect of unsaturation on combustion and emission in practical systems. High-pressure spray simulations under diesel engine conditions indicate longer ignition delays for 1-heptene compared to those for n-heptane. In addition, the n-heptane spray flame contains two reaction zones, namely a rich premixed zone (RPZ) and a nonpremixed reaction zone (NPZ). In contrast, 1-heptene flame is characterized by three reaction zones, i.e., a lean premixed zone (LPZ) in addition to NPZ and RPZ. Simulations of laminar partially premixed flames (PPF) indicate higher amounts of NOx and soot precursor species (C_2H_2, C_6H_6, and $C_{16}H_{10}$) formed in 1-heptene flames than those in n-heptane flames. Consequently, the soot emission is higher in 1-heptene flames than that in n-heptane flames. Simulations of turbulent n-heptane and 1-heptene spray flames in diesel engines lead to similar conclusions, i.e., higher NOx and soot emissions in 1-heptene flames. The increased formation of PAH species can be attributed to the significantly higher amounts of 1,3-butadiene and allene formed due to β scission reactions resulting from the presence of double bond in 1-heptene.

S. K. Aggarwal (✉)
University of Illinois at Chicago, Chicago, IL, USA
e-mail: ska@uic.edu

A. K. Runchal et al. (eds.), *Energy for Propulsion*, Green Energy and Technology,
https://doi.org/10.1007/978-981-10-7473-8_3

Keywords Fuel unsaturation · Ignitability · Two-stage ignition · Soot precursors NOx and soot emissions

1 Introduction

There is growing interest in investigating the effect of fuel molecular structure on fuel reactivity and thus on the ignition and combustion characteristics. This has been driven by several considerations, such as the emergence of a new generation of renewable fuels, and the recognition that the combustion chemistry of real fuels cannot be modeled using single-component fuels, and that many conventional and emerging fuels contain a number of saturated and unsaturated hydrocarbon compounds. For instance, biodiesel fuels produced using the trans-esterification process contain a relatively large amount of unsaturated ester components, and the molecular structure of these components can vary significantly with respect to the number and location of C=C bonds, depending upon the feedstock. As a result, their reactivity and cetane number (CN) can vary significantly [1], with each C=C double bond in the long carbon chain of component molecule reducing the CN value by a significant amount [2]. For example, palm oil methyl ester with high fraction of saturated components, such as methyl stearate ($C_{19}H_{38}O_2$), has a CN of 62, whereas linseed oil methyl ester containing high fraction of methyl linolenate ($C_{19}H_{32}O_2$) with three C=C double bonds has a CN of 39, implying significant reduction in ignitability. Another important issue pertains the effect of fuel molecular structure on the fuel sensitivity (S), which is defined as the difference between research octane number (RON) and motor octane number (MON). Several recent studies have shown that ignitability is determined by both fuel octane rating and sensitivity [3, 4]. Sensitivity provides a measure of aromatic and other non-paraffinic content of the fuel, and high sensitivity fuels exhibit different temperature dependence at low, medium, and high temperatures. Experimental studies using homogeneous mixtures in shock tube (ST) and rapid compression machine (RCM) have reported that these fuels have low reactivity at low temperatures, but react rapidly at high temperatures, and exhibit varying degrees of negative temperature coefficient (NTC) behavior [4, 5]. Previous research has also shown that primary reference fuels (PRF), for which $S=0$, are not able to emulate this temperature dependence. Moreover, sensitivity is strongly related not only to fuel composition but also to its molecular structure or level of unsaturation. For example, as reported by Tanaka et al. [6], $S=0$ for n-heptane (alkane) and 13 for 1-heptene (alkene with a C=C bond). Finally, there is also fundamental interest in examining the pyrolysis/oxidation chemistry of alkenes, since such compounds are formed during the combustion of alkanes.

This chapter provides an overview of research dealing with the effects of fuel unsaturation on the ignition, combustion, and emission characteristics. A considerable research exists on this topic. Many experimental investigations have reported ignition data from ST and RCM experiments for different fuels and over a wide range of conditions. Complementary kinetic studies have also been performed to examine the reaction pathways during the pyrolysis/oxidation of saturated and unsaturated fuels and develop reaction mechanisms for predicting the ignition behavior of these fuels under homogeneous condition. This work is briefly reviewed. In addition, research dealing with laboratory-scale laminar flames and diesel sprays using saturated and unsaturated surrogates is briefly discussed. Experimental engine studies concerning the emission characteristics of various biodiesel fuels are also mentioned. As expected, the scope of this chapter is limited due to the broad range of topics covered and also due to the page limitation. It does not include details of the experimental methods and the physical-computational models used in simulations. Most of the discussion and results are taken from previously reported work, which is appropriately cited.

2 Effect of Unsaturation on Ignition in Homogeneous Mixtures

Fundamental investigations on the ignition of homogeneous fuel–air mixtures provide valuable information for the development and validation of reaction mechanisms. Such data is also useful for understanding ignition delays and burning rates in homogeneous charge compression ignition (HCCI) engines and examining knocking behavior in spark-ignition engines. Numerical algorithms and software, such as CHEMKIN-Pro [7] and Cantera [8], are being extensively used to examine ignition characteristics, validate reaction mechanisms, and analyze reaction pathways under constant volume or constant pressure conditions. Ignition characteristics include the ignition delay time as a function of system properties, such as temperature, pressure, equivalence ratio (ϕ), fuel composition, and two-stage ignition and negative temperature coefficient (NTC) phenomena. While straight chain alkanes have been extensively investigated, some experimental and modeling studies have also examined the oxidation of straight chain C_5, C_6, and C_7 alkene isomers. Various surrogates[1] of current and emerging fuels, such as diesel, gasoline, JP-8, and biofuels, have also been analyzed. Ignition delay and speciation data have been reported from shock tube [6, 9–11], RCM [12, 13], and flow reactor [14] experiments. Detailed kinetic models have also been developed to provide insight into the effects of the presence and position of double bonds on fuel reactivity and ignitibility [15, 16].

Tanaka et al. [6] and Vanhove et al. [13] performed RCM experiments and examined the effects of fuel molecular structure on ignition. Tanaka et al. [6] reported ignition data for several saturated and unsaturated hydrocarbons and analyzed the two-stage ignition and NTC behavior. Representative results from this work are provided in Figs. 1 and 2. Figure 1 presents data on ignition delays for several hydrocarbons. An important observation is the existence of two-stage ignition and NTC phenomenon for n-heptane, 1-heptene, and 2-heptene, but only a single-stage ignition for 3-heptene. In addition, results indicate longer ignition delays for unsaturated

[1] Such surrogates are not discussed in this chapter as the main topic is the effect of fuel unsaturation.

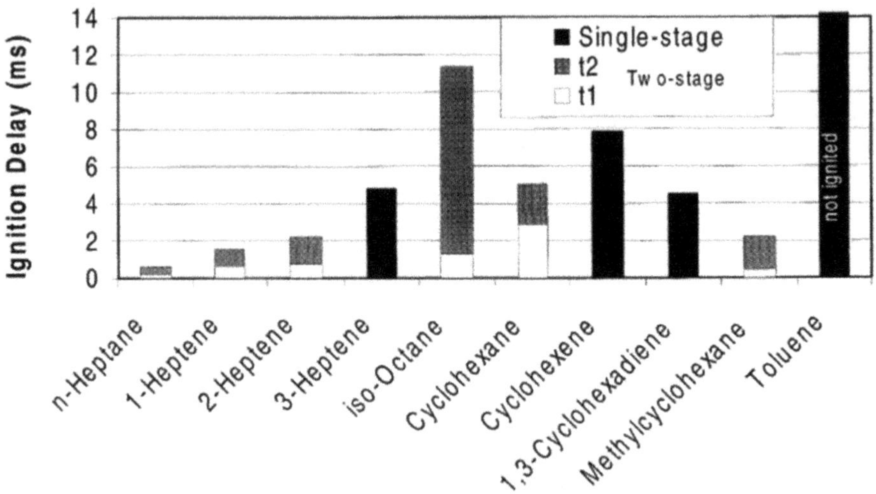

Fig. 1 Ignition delays for several fuels measured in RCM. Equivalence ratio: 0.4. Initial temperature and pressure are 318 K and 1 atm. Compression ratio: 16 From Tanaka et al. [6]

fuels (1-, 2-, and 3-heptene) compared to saturated fuel (n-heptane) and that the position of C=C bond strongly affects ignition delay. Transient ignition characteristics for these fuels in terms of pressure-time history in RCM are shown in Fig. 2. For the fuels that exhibit two-stage ignition, the 1st stage ignition is indicated by the first (smaller) peak in pressure, followed by the main or 2nd stage ignition indicated by the sharp rise in pressure. Also, the duration between the 1st stage and 2nd stage ignition becomes increasingly longer as the position of C=C bonds shifts inside for the unsaturated fuels.

Vanhove et al. [13] reported a systematic investigation on the effect of the position of the double bond on the ignition of 1-, 2-, and 3-hexene between 630 and 850 K. Figures 3 and 4 provide some results from this study, indicating generally similar ignition behavior as that observed by Tanaka et al. [6] for heptane isomers. Figure 3 plots the pressure and light emission traces after a rapid compression to 725 K and 9.4 bar for 4 different fuels. Results indicate a two-stage ignition with strong 1st stage ignition for 1-hexene, a two-stage ignition with a weak 1st stage ignition for 2-hexene, and only one-stage ignition for 3-hexene. Figure 4 shows the measured ignition delays versus gas temperature for these fuels. The data again indicates a strong effect of the position of the double bond on ignition. For 1-hexene, the behavior is generally similar to that of alkanes, i.e., two-stage ignition up to 800 K and a small NTC region between 750–830 K. For 2-hexene, these features become less clear showing a two-stage ignition with a weak 1st stage ignition that disappears above 730 K, and no NTC behavior but rather ignition delay decreasing slowly between 720 and 815 K. For 3-hexene, there is no two-stage ignition or NTC behavior, but only a faint inflection point near 730 K. Thus, the fuel ignitability is strongly influenced by the length of the saturated portion aside the double bond, with the longer alkyl

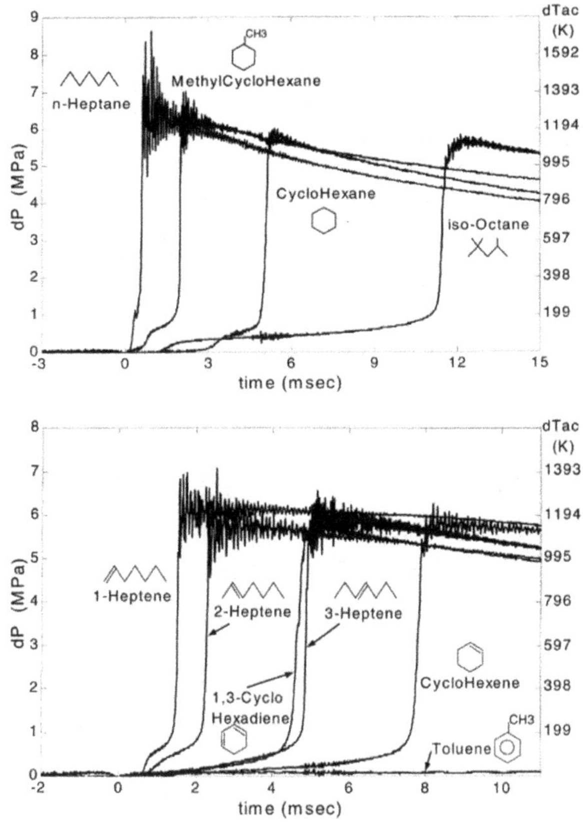

Fig. 2 RCM ignition and combustion characteristics of some saturated and unsaturated hydrocarbons. Initial temperature and pressure are 318 K and 1 atm. Compression ratio: 16 From Tanaka et al. [6]

chain yielding shorter ignition delay. Vanhove et al. [13] attributed this behavior of long-chain alkenes to a competition between the reactivity of double bond and that of alkenyl chain. For short alkenyl chains, such as in 3-hexene, the ignition is dominated by the reactivity of the double bond, while for long chains, such as in 1-hexene, it is dominated by the reactivity of alkenyl chain, resulting in two-stage ignition.

Mehl et al. [17] developed a detailed kinetic model for the oxidation of 1-, 2-, and 3-hexene and analyzed reaction pathways associated with their ignition chemistry in the temperature range of 660–1770 K. Consistent with experimental data, the reactivity of isomers was found to be strongly influenced by the position of the double bond, especially at low temperatures. Figure 5 compares their simulated ignition delays with the RCM data of Vanhove et al. [13], indicating a similar effect of the position of double bond on ignition as discussed above. Garner et al. [18] reported shock tube experiments and modeling study for the ignition of n-heptane, 1-heptene, 1, 6-heptadiene, and two C8 methyl esters, namely methyl octanoate and methyl trans-2-octenoate. Note that n-heptane and 1-heptane represent the hydrocarbon side chains of the C8 saturated and unsaturated methyl esters, respectively. Consistent with

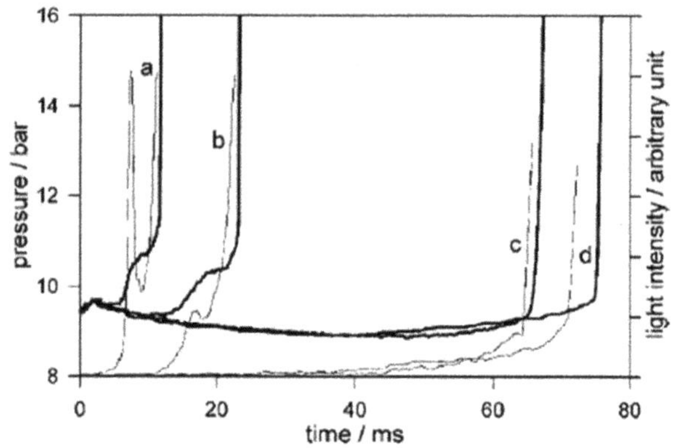

Fig. 3 Pressure (thick line) and light emission (thin line and arbitrary unit) traces after a rapid compression to 725 K and 9.4 bar for 1-hexene (**a**), 2-hexene (**b**), cyclohexene (**c**), and 3-hexene (**d**) From Vanhove et al. [13]

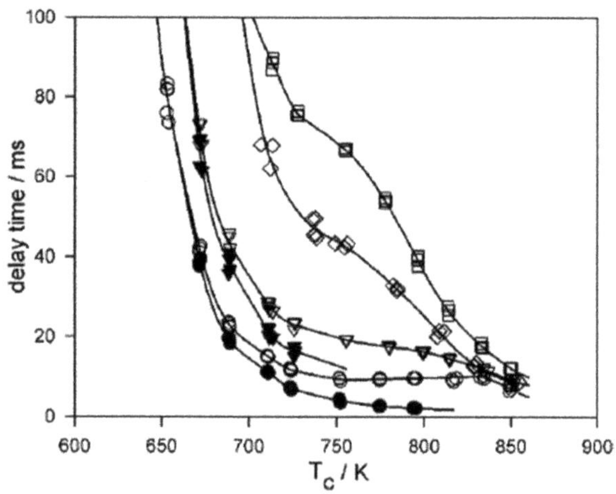

Fig. 4 1st stage (black symbols) and total (white symbols) delay times versus gas temperature for 1-hexene (circle), 2-hexene (triangle), cyclohexene (diamond), and 3-hexene (square) From Vanhove et al. [13]

previous studies, results indicated longer ignition delays as the degree of unsaturation was increased.

Westbrook et al. [19] employed a detailed kinetic mechanism to characterize the effect of C=C bond on the ignition behavior of five methyl ester compounds, namely methyl palmitate, methyl stearate, methyl oleate, methyl linoleate, and methyl linolenate, which are important components of many biodiesel fuels derived from different vegetable oils and animal fats. Methyl palmitate ($C_{17}H_{34}O_2$) and methyl stearate ($C_{19}H_{38}O_2$) contain a saturated straight chain of 15 and 17 C atoms respectively, whereas the other three components have straight chains of 17 C atoms with one C=C double bond in methyl oleate ($C_{19}H_{36}O_2$), two double bonds in methyl linoleate ($C_{19}H_{34}O_2$), and three double bonds in methyl linolenate ($C_{19}H_{32}O_2$).

Figure 6 compares the computed ignition delays for these components using stoichiometric fuel/air mixtures at 13.5 bar initial pressure. Ignition data is also included (shown as square and circle symbols for soy methyl ester (SME) and rapeseed methyl ester (RME) mixtures). Also for reference, the computed results are shown for

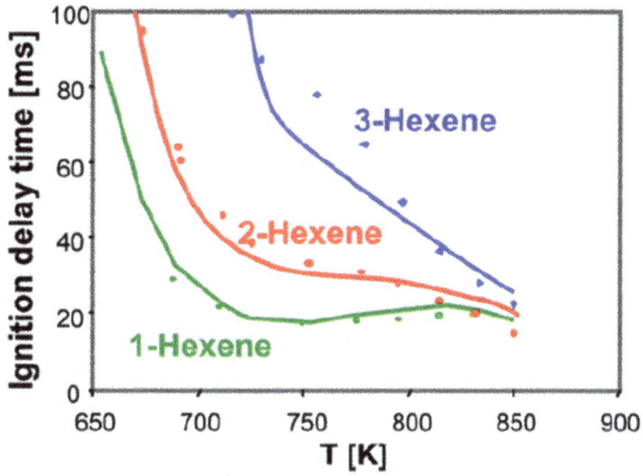

Fig. 5 RCM ignition delay times of hexene isomers ($p = 0.86$–1.09 MPa, $\Phi = 1$): experiments (symbols) and calculations (lines) From Mehl et al. [17]

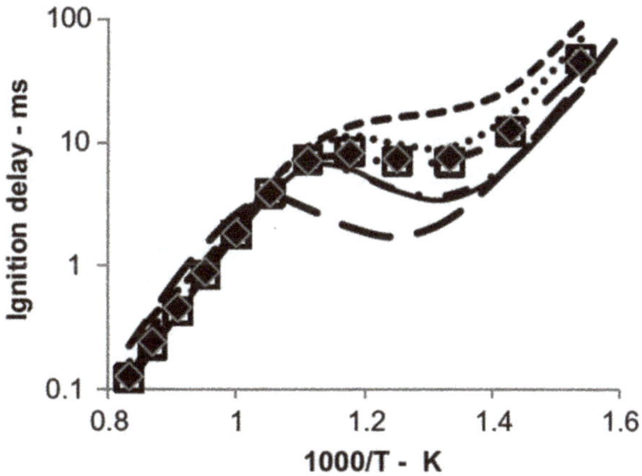

Fig. 6 Computed ignition delay times for biodiesel components, stearate (solid line), palmitate (dot-dash), oleate (dashes), linoleate (dotted), and linolenate (short dashes). Symbols show computed results for SME (squares) and RME (diamonds). Results for n-cetane are shown as long dashes. Mixtures are stoichiometric, at constant volume, and at 13.5 bar From Westbrook et al. [19]

n-cetane, which has much shorter ignite delay than any of the biodiesel components or composite fuels. Results for the five biodiesel components clearly indicate that their ignitability is adversely affected as the number of C = C bonds increases. Moreover, the effect is most pronounced in the NTC region. Thus, methyl linolenate with 3 C = C bonds is slowest to ignite, followed by methyl linoleate and methyl oleate. The ignition delays of both SME and RME are nearly the same as those of methyl oleate and methyl linoleate, which are the major components of these biodiesel fuels. As expected, both saturated components, methyl stearate and palmitate, are the fastest components to ignite, with ignition delays that are effectively identical.

Thus, ignition data for saturated and unsaturated components of petroleum-based and biodiesel fuels indicate that the number and position of the double bond has a strong effect on the reaction pathways responsible for the 1st stage ignition, total ignition delay, and the dependence of the ignition delay time on mixture temperature. In general, the presence of C = C double bond inhibits low-temperature reactivity, increases ignition delay times, and leads to reductions in CN number in diesel engines, compared with corresponding saturated fuel molecules.

3 Effect of Unsaturation on Ignition and Flame Structure in Sprays

Compared to homogeneous, premixed systems, the fuel unsaturation and chemistry effects on ignition are much less investigated and understood for nonpremixed systems, especially those involving droplets and sprays. For droplets, additional complexities are due to the coupled two-phase processes, gas-phase convection, droplet heating and vaporization. Much of the experimental and modeling research on droplet ignition has considered a spherically symmetric configuration [20]. As discussed in Refs. [21, 22], several of these studies have examined the temperature dependent chemistry effects, such as two-stage ignition and negative temperature coefficient (NTC) behavior. The two-stage ignition was defined through the temporal variation of temperature, with the first temperature rise marking the first-stage ignition and the second (sharper) rise indicating the second-stage ignition. However, the previous studies have not provided a clear evidence for the NTC region, and it has been surmised that the presence of non-homogeneous temperature and species field causes a transition from NTC to zero temperature coefficient (ZTC) behavior. While various researchers provide different explanations for this transition, relatively few studies provide an evidence of a ZTC region. It has been suggested that competition between the availability of fuel vapor and the reduction in mixture temperature due to evaporation plays a significant role in modifying the NTC behavior [23].

Similar to the research on droplet ignition, relatively few studies have been reported on the fuel unsaturation effects on spray ignition [24], although the general topic of spray ignition has been extensively investigated [25, 26]. This is somewhat surprising since ignition represents a critical process in CI engines, and strongly influences their combustion and emission characteristics [27, 28]. Moreover, autoignition in CI engines occurs at conditions where the two-stage ignition and NTC phenomenon are highly relevant. Fu and Aggarwal [29] investigated this phenomenon in sprays by performing 3-D, two-phase reacting flow simulations in a 1.9 L 4-cylinder engine. The CONVERGE software was used for simulating the processes of fuel injection, atomization, and spray ignition. As this engine is equipped with a 7-hole common rail injector in each cylinder, simulations considered a 1/7 (51.43°) sector of the cylinder using periodic boundary conditions at the front and back face of the sector. Additional details can be found in the cited work [29]. As the fuel is injected at a given crank angle (CA), processes of spray atomization, vaporization, and fuel–air mixing follow leading to ignition. Figure 7 from Ref. [29] depicts these processes prior to ignition by plotting the spray characteristics in the cylinder at different CA after the start of injection (SOI).

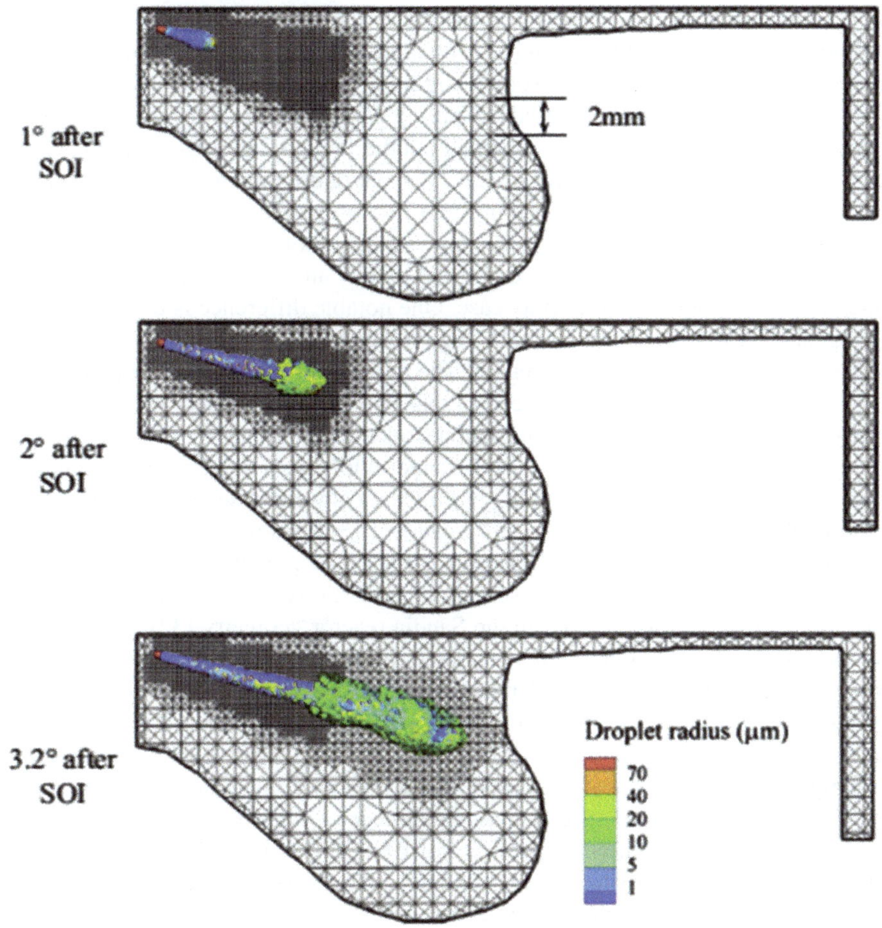

Fig. 7 A cross-section view of the cylinder through the spray, depicting the developing spray structure at 1°, 28°, and 3.28° CA after SOI, which is 8° from TDC. Each color dot represents a droplet radius in a parcel, and the size distribution is indicated by the droplet radius scale (1–70 μm) From Xu and Aggarwal [29]

The temporal and spatial behavior of the 1st and 2nd stage ignitions was analyzed through the evolution of QOOH (alkyl hydroperoxy) and OH fields, respectively, in both ϕ–T space and physical space. Note that the QOOH radicals play a key role in the chemistry of 1st stage ignition [29, 30], and its temporal evolution can be used to determine the corresponding ignition delay. Similarly, the 2nd stage ignition can be determined from the evolution of OH or HRR. A representative result from Ref. [29] is provided in Fig. 8a, which shows QOOH and OH mass profiles with respect of CA, depicting the 1st and 2nd stage ignitions for four different cases corresponding to SOI = 32°, 20°, 14°, 8° before TDC. The peak in QOOH profile determines the 1st stage ignition, while the sharp rise (after the first peak) in OH determines the 2nd stage (main) ignition. Figure 8b plots the 1st and 2nd stage ignition delay times versus SOI for both diesel sprays and homogeneous mixtures. Note that the effect of temperature is characterized by varying SOI. As SOI is delayed, the cylinder temperature increases, and, consequently, both the 1st and 2nd stage ignition delays decrease, but the effect is stronger on the 1st stage ignition, consistent with droplet ignition results. As further discussed in Ref. [29], results indicate global similarities between the ignition processes in diesel sprays and homogeneous mixtures, but also highlight differences between them due to temporally and spatially evolving temperature and species fields for the spray case. One notable difference is that ignition in homogeneous mixtures exhibits a NTC region (see Fig. 10 in Ref. [29]), while that in diesel spray shows a near ZTC region. The transition from NTC to ZTC behavior in diesel spray is mainly due to the evolving two-phase flow, and also due to the increase in pressure during the compression stroke. In addition, the 1st and 2nd stage ignitions in sprays occur over a wide ϕ range, implying a spatially wide ignition kernel. Also, in some cases, multiple ignition kernels were observed in sprays.

In a subsequent study, Sharma and Aggarwal [31] examined the effect of fuel unsaturation on the transient ignition and flame development in n-heptane and 1-heptene turbulent sprays under diesel engine conditions. 3-D, two-phase reacting flow simulations were performed in the Sandia reactor geometry [32] by using the CONVERGE software along with a validated reaction mechanism. Results demonstrated that the spray ignition and flame structure are strongly influenced by fuel unsaturation. Details are provided in Refs. [31, 33]. Figure 9 from Ref. [33] compares the transient ignition processes in n-heptane and 1-heptene sprays in terms of the temporal variations of heat release rate (HRR), volume-integrated fuel vapor mass (m_{fv}), and QOOH and OH species mass in the reactor. Results are shown for three cases corresponding to initial temperatures of 1000, 1100, and 1200 K. As discussed in Refs. [31, 33], similar to the ignition in homogeneous mixtures, the fuel ignitability is noticeably decreased due to fuel unsaturation. As indicated in Fig. 9, the ignition delay is much longer in 1-heptene sprays than that in n-heptane sprays, especially at lower temperatures. In addition, for the 1000 and 1100 K cases, the ignition in n-heptane sprays is characterized by two-stage ignition, while in 1-heptene sprays, only the 2nd stage or main ignition is observed for all three cases. For instance, for the 1000 K n-heptane case, the 1st stage ignition occurs at 0.185 ms, followed by the main ignition at 0.325 ms. The 1st stage ignition is indicated by the first decrease in fuel vapor mass, or by the first increase in HRR in Fig. 9a, or by a sharp decrease in QOOH mass in Fig. 9c. The 2nd stage or main ignition for all the cases is indicated by a sharp decrease in fuel vapor mass, or by a sharp rise in HRR or in OH mass profile. Another notable observation is that for the range of temperatures considered, spray ignition results do not indicate any NTC region, as the ignition delay time decreases monotonically as the initial temperature is increased. Another interesting feature of spray ignition, in contrast to gaseous mixtures, is the existence of multiple ignition locations, as noted earlier. For instance, OH scatter plot at 0.3 ms in Fig. 10b indicates two ignition kernels in T-ϕ space, one near $T \approx 1100$ K and $\Phi \approx 1.6$, and the other near $T \approx 1000$ K and $\Phi \approx 2.2$.

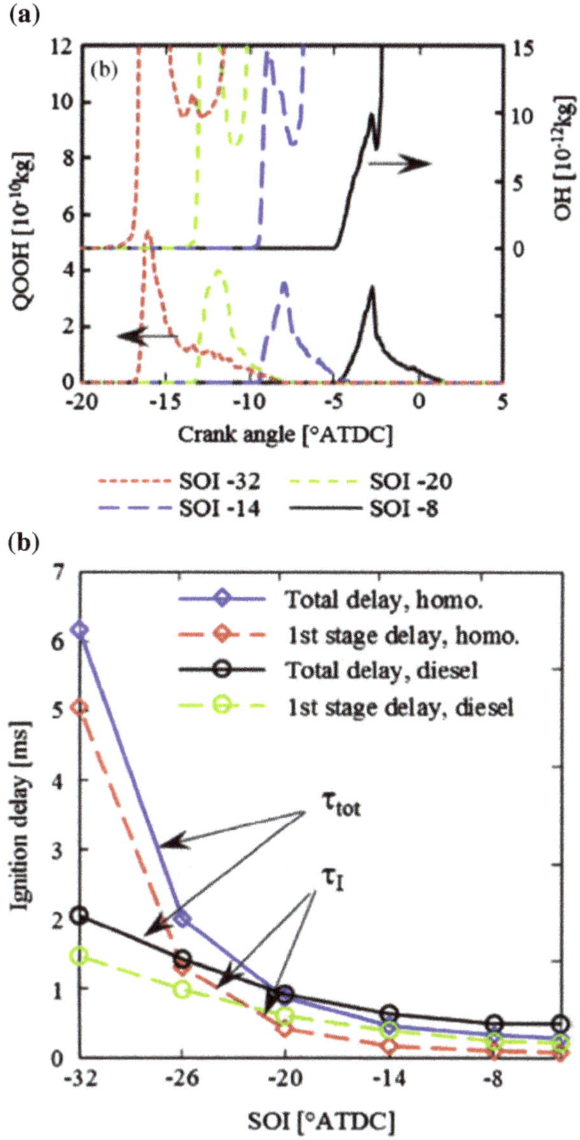

Fig. 8 **a** QOOH and OH mass profiles with respect to CA depicting the occurrence of 1st and 2nd stage ignitions for 4 cases corresponding to SOI = 32°, 20°, 14°, 8° before TDC. **b** 1st stage and 2nd stage (total) ignition delay times versus SOI for the diesel spray and homogeneous mixtures. From Xu and Aggarwal [29]

Further insight into the effect of fuel unsaturation on transient spray ignition and flame development is provided through OH scatter plots in Figs. 10 and 11 and through equivalence ratio (ϕ) and temperature (T) contours in Figs. 12 and 13. OH scatter plots in Fig. 10 again depict the two-stage ignition behavior in n-heptane spray, with the 1st and 2nd stage ignitions occurring at 0.185 ms and 0.325 ms, respectively. On the other hand, as indicated in Fig. 11, only the main ignition is observed in 1-heptene spray at 2.82 ms. Another important difference between the two fuels is

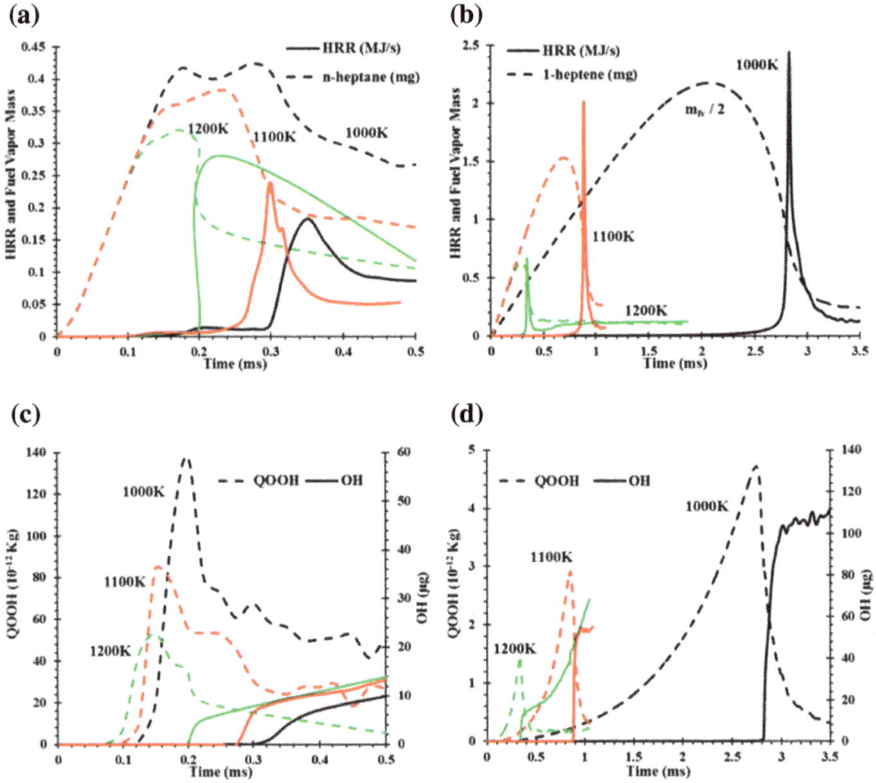

Fig. 9 Transient ignition processes in n-heptane (Fig. **a** and **c**) and 1-heptene (Fig. **b** and **d**) sprays, illustrated through the temporal variation of heat release rate (HRR), fuel vapor mass (m_{fv}), and QOOH and OH species mass for initial temperatures of 1000 K, 1100 and 1200 K. From Sharma [33]

that the main ignition in n-heptane spray occurs in rich mixtures ($1.4<\phi<1.6$), while that in 1-heptane involves lean mixtures ($\phi\approx0.7$). This can be attributed to the reduced ignitability of 1-heptene, resulting in longer ignition delays and thus enhancing fuel–air mixing in case of 1-heptene. This has important consequence for the subsequent spray flame structure for the two fuels. Thus, as indicated in Figs. 12 and 13, n-heptane flame structure is characterized by two reaction zones, namely a rich premixed zone (RPZ) and a nonpremixed reaction zone (NPZ), while the 1-heptane flame is characterized by three reaction zones, i.e., a lean premixed zone (LPZ), in addition to NPZ and RPZ. These reaction zones can be visualized from the ϕ and T contour plots shown in Figs. 12 and 13 and also from HRR scatter plots in Fig. 14. For instance, the NPZ is characterized by $\phi\approx1.0$ and high temperatures (≈2350 K), whereas RPZ is characterized by ϕ between 1.3–2.1 and T between 1500–2000 K and LPZ in 1-heptene flames by ϕ between 0.6–0.9 and T between 1500–1900 K. Note that the LPZ can be identified from the OH scatter plots in

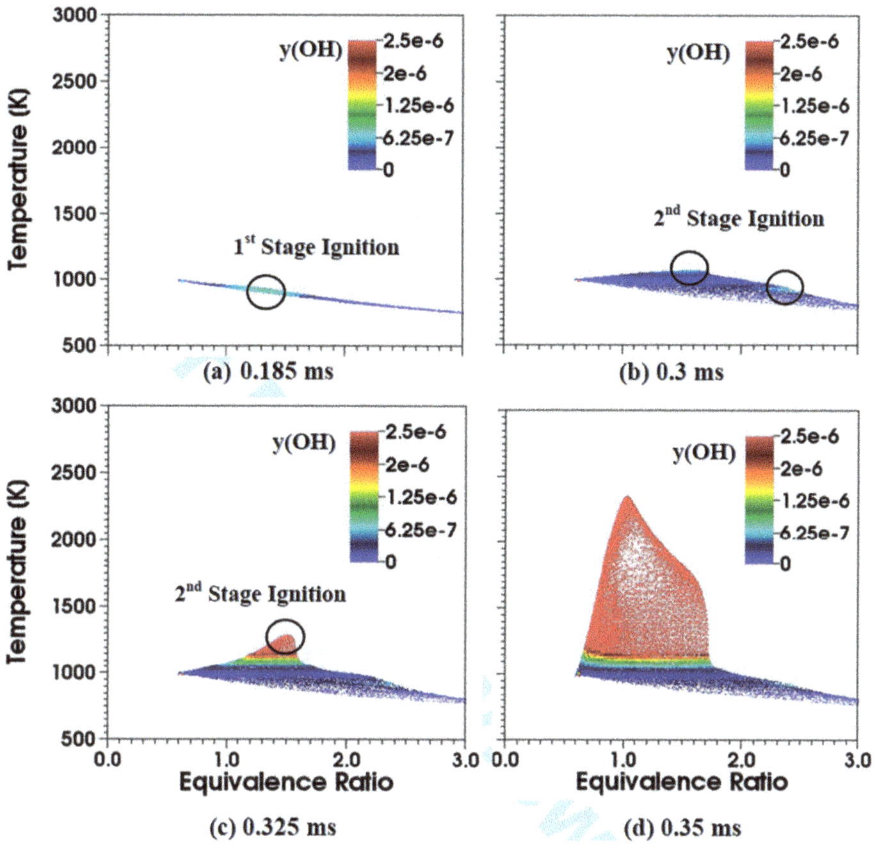

Fig. 10 Transient ignition and flame development in n-heptane spray at 1000 K; OH mass fraction scatter plots in ϕ-T space From Sharma [33]

Fig. 11, and from ϕ and T contour plots Fig. 13, and also from HRR scatter plots in Fig. 14. Thus, an important consequence of fuel unsaturation is the appearance of a lean premixed reaction zone in 1-heptene sprays, in addition to the nonpremixed and rich premixed zones.

Another important result pertains to the effect of fuel unsaturation on flame liftoff length. The flame liftoff length for various cases is indicated by a dotted vertical line in Figs. 12 and 13. The flame anchoring location downstream of the injector is determined by using a minimum OH mass fraction value of 0.02 of its peak value. For both fuels, as the ambient temperature is increased, the flame liftoff length decreases, i.e., the flame stabilization location moves closer to injector. Thus for n-heptane flames, the computed liftoff lengths are 24.5, 13, and 10.5 mm for 1000, 1100, and 1200 K, respectively. The corresponding values for 1-heptene flames are 27, 16, and 4.5 mm. It is also interesting to note that the flame liftoff position correlates with the ignition delay and ignition kernel location. Thus for 1000 and 1100 K, the liftoff

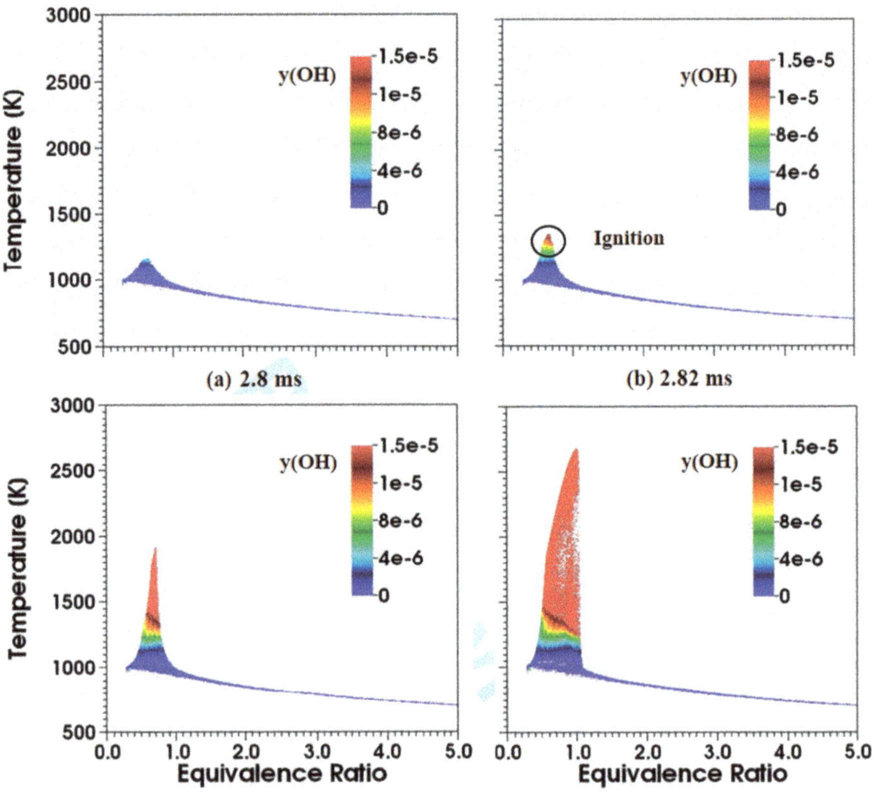

Fig. 11 Transient ignition and flame development in 1-heptene spray at 1000 K; OH mass fraction scatter plots in ϕ-T space. From Sharma [33]

lengths for 1-heptene are higher compared to those for n-heptane, since the ignition delays are longer for 1-heptene. On the other hand, for 1200 K, the liftoff length for 1-heptene is lower than that for n-heptane, since the ignition delay is shorter and the ignition kernel is located closer to injector for 1-heptene.

4 Effect of Fuel Unsaturation on NOx and Soot Emissions

This section provides a brief review of research concerning the effect of fuel unsaturation on NOx and soot emissions. Both laboratory-scale laminar flames and engine configurations are considered. Engine experiments using biodiesel fuels from different feedstock [34, 35] have observed an increase in NOx emission as the fatty acid chain length, and the number of double bonds in the fuel molecular structure

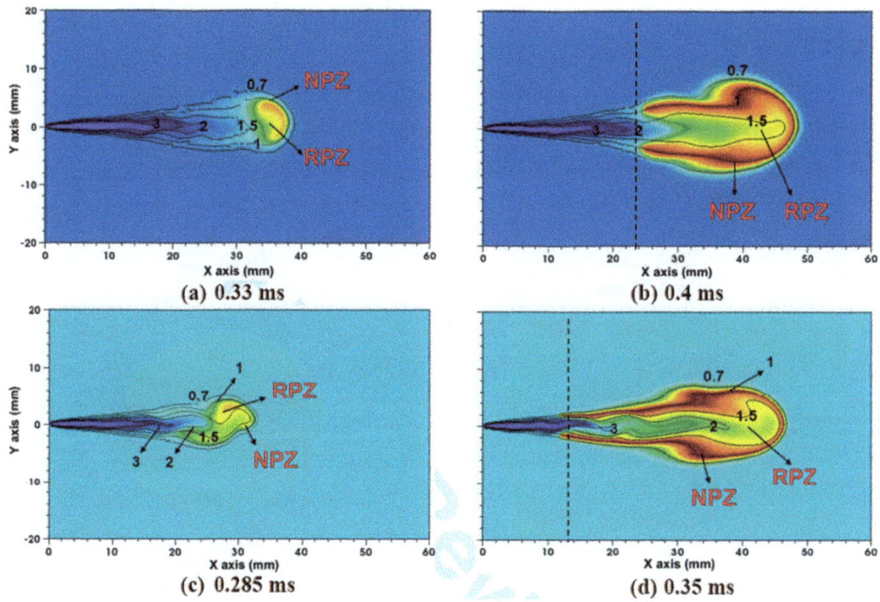

Fig. 12 Equivalence ratio (ϕ) and temperature (T) contours at different times showing the temporal evolution of n-heptane spray flames for initial temperatures of 1000 K (Fig. **a** and **b**), and 1100 K (Fig. **c** and **d**). Flooded contours indicate T between 600 to 2700 K. Contour lines represent ϕ between 0.7 and 3. Vertical line indicates the flame liftoff length. From Sharma [33].

is increased. In addition, experimental studies have reported that while PM (particulate matter) emissions are reduced using biodiesel fuels compared to conventional diesel, the amount of PM formed increases due to the presence of double bonds in the fuel molecular structure [35–37]. There have also been fundamental investigations on the emissions of NOx and soot (and their precursors) from the combustion of saturated and unsaturated biodiesel components. Garner et al. [11] reported shock tube experiments using n-heptane (n-C_7H_{16}) and 1-heptene (1-C_7H_{14}) and observed that 1-heptene produces more C_2H_2 acetylene than does n-heptane over intermediate temperatures. This has consequence for increased NOx formation through the prompt NO mechanism. Since n-heptane and 1-heptene represent the saturated and unsaturated hydrocarbon side chains of C_8 methyl esters, respectively, Garner and Brezinsky [38] and Garner et al. [18] extended the study to the oxidation of these esters and observed increased C_2H_2 formation in the case of unsaturated methyl ester. Note that C_2H_2 also provides a major route for soot particle surface growth through the "H-abstraction-C_2H_2-addition" (HACA) mechanism [39, 40].

Han et al. [41] simulated partially premixed laminar flames (PPFs) burning prevaporized n-heptane and 1-heptene fuels and observed higher amounts of acetylene and NOx in 1-heptene flames than that in n-heptane flames. As stated earlier, these fuels represent the hydrocarbon side chains of the saturated and unsaturated methyl esters, namely methyl octanoate and methyl *trans*-2-octenoate. Thus, the study was

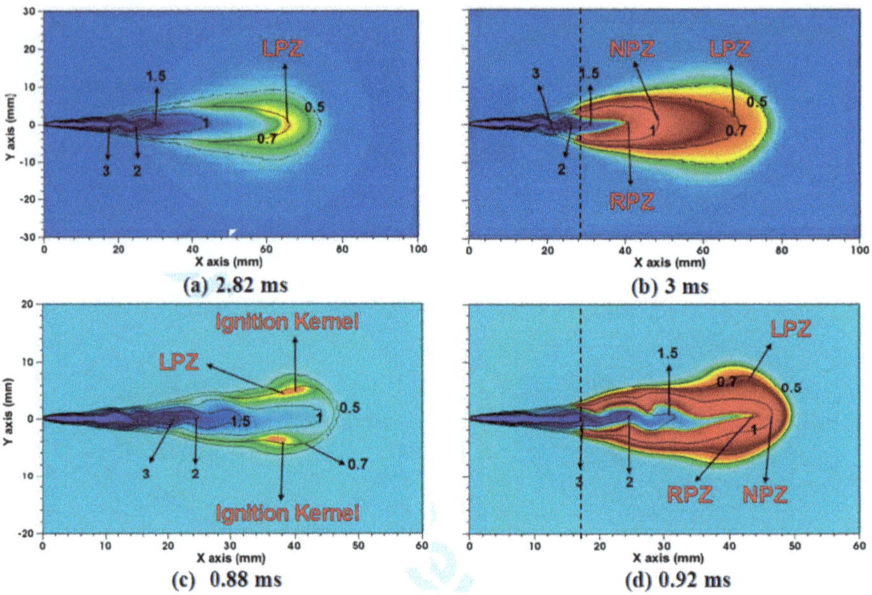

Fig. 13 Equivalence ratio (ϕ) and temperature (T) contours at different times showing the temporal evolution of 1-heptene spray flames for initial temperatures of 1000 K (Fig. **a** and **b**), and 1100 K (Fig. **c** and **d**). Flooded color contours indicate T between 600 to 2700 K. Contour lines represent ϕ between 0.7 and 3. Vertical line indicates the flame liftoff length. From Sharma [33]

also relevant to the understanding of NOx emissions from the combustion of biodiesel fuels. Triple flames were simulated in a counterflow configuration using the OPPDIF in CHEMKIN-Pro 15113 package [42]. The configuration involves two opposing, axisymmetric jets, one issuing a fuel-lean mixture and the other a fuel-rich mixture. A validated reaction mechanism [43] was used to model the n-heptane, 1-heptene oxidation chemistry. The mechanism was combined with a detailed NOx kinetics model that included sub-mechanisms for NO formation through the thermal, prompt, N_2O intermediate, NNH routes [44, 45].

A triple flame is characterized by the existence of three reaction zones, namely lean premixed zone (LPZ), rich premixed zone (RPZ), and nonpremixed reaction zone (NPZ), which are spatially separated but strongly coupled through the transport and chemical kinetic. Thus, the objective was to characterize the effect of the presence of the double bond on C_2H_2 and NOx emissions in partially premixed flames containing regions of lean premixed, rich premixed, and nonpremixed combustion. Moreover, using a counterflow geometry, the spatial separation between the three reaction zones can be controlled by varying the strain rate and the lean and rich equivalence ratios (ϕ_L and ϕ_R). Further, details are provided in Ref. [41]. One representative result from this study is shown in Fig. 15, which plots the emission indices of total NO and those formed via prompt, thermal, N_2O, and NNH routes in n-heptane and 1-heptene triple flames at different strain rates. An important observation is the higher NO emission

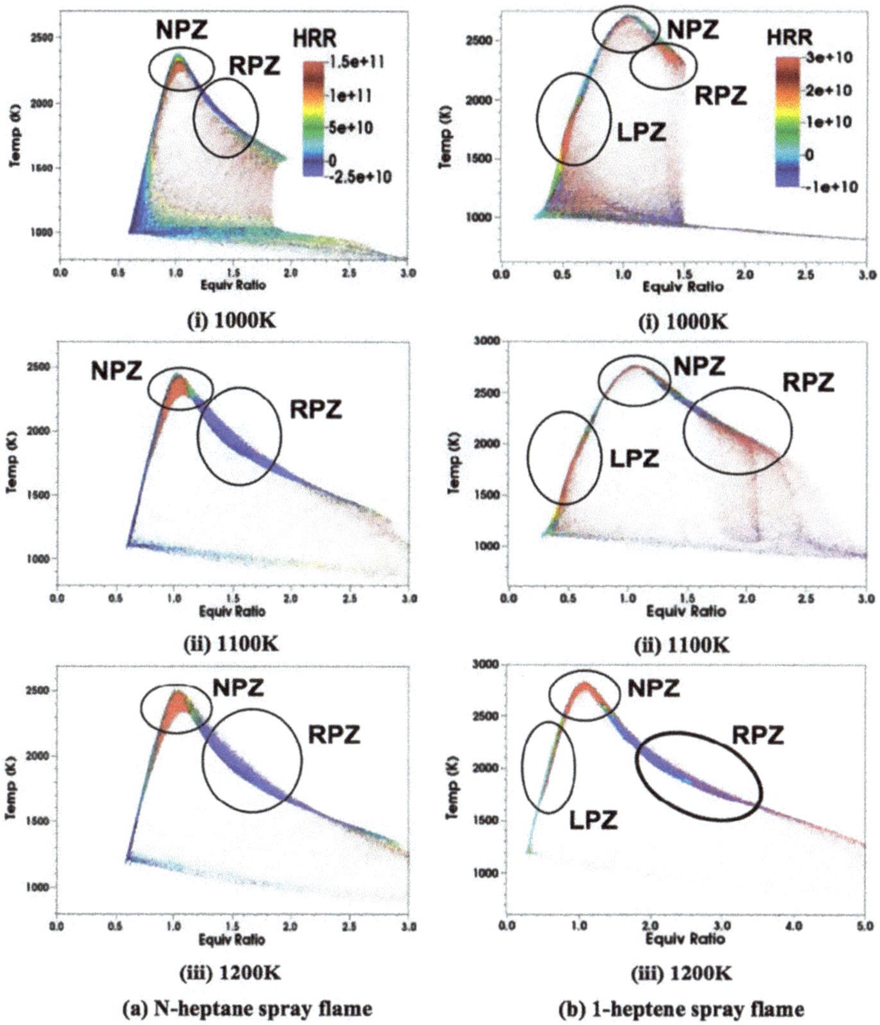

Fig. 14 HRR scatter plots in T-ϕ space for **a** n-heptane and **b** 1-heptene spray flames for initial temperatures of 1000 K (i), 1100 K (ii) and 1200 K (iii) From Sharma [33]

in 1-heptene flames compared to that in n-heptane flames irrespective of the strain rate. As discussed in Ref. [41], the reaction path analysis indicated that the β scission and oxidation reactions related to the double C=C bond lead to higher amount of C_2H_2 and thus increased NO through the prompt mechanism in 1-heptene flames compared to that in n-heptane flames. Moreover, as the strain rate is increased, the EINO first increases and then decreases, which was due to the variation of peak flame temperature with strain rate. The EI values for the prompt, thermal, N_2O, and NNH mechanisms follow a similar trend. However, the relative contribution of prompt NO

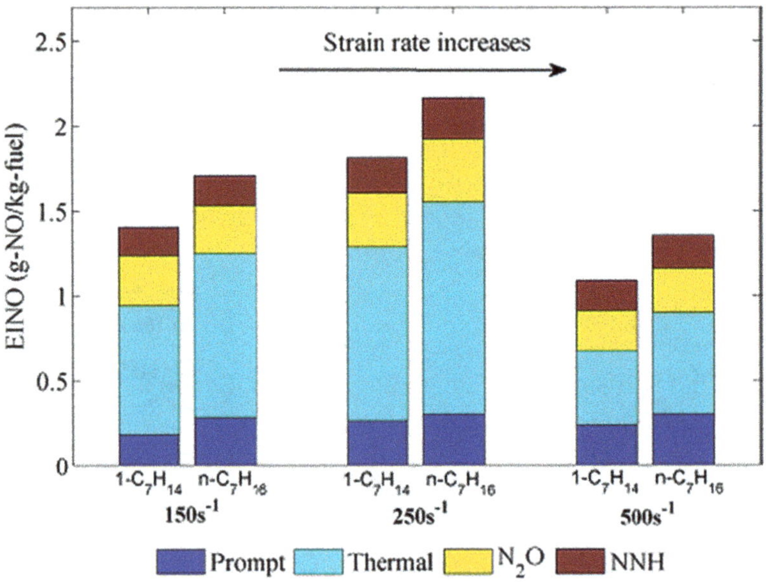

Fig. 15 Emission index for the total NO, and of NO produced through the prompt, thermal, N2O, and NNH mechanisms in n-heptane and 1-heptene triple flames established at strain rates of 150, 250, and 500 s^{-1}, and $\phi_L = 0.8$, and $\phi_R = 1.5$ From Han et al. [41]

becomes more pronounced at higher strain rates, while that of thermal NO is reduced to the lower residence time.

Fu et al. [46] extended the above investigation [41] and examined the effect of fuel unsaturation on NOx and PAH emissions in double PPFs, i.e., flames containing two reaction zones, RPZ and NPZ. The flames were again established in a counterflow geometry by issuing a fuel-rich mixture with a specified ϕ in one jet and air in the other jet. Depending upon the level of partial premixing (i.e., ϕ), a significant amount of soot formation is expected in the region between the rich premixed zone and the stagnation plane. Various PPFs were simulated by independently varying ϕ and global strain rate (a_G). One representative result from Ref. [46] is presented in Fig. 16, which shows the peak mole fractions of major soot precursor species, namely C_2H_2, benzene (C_6H_6), and pyrene ($C_{16}H_{10}$) in n-heptane and 1-heptene PPFs established at different strain rates and two levels of partial premixing with $\phi = 8$ and 2. As indicated, at any given ϕ and strain rate, the peak mole fractions of C_2H_2, C_6H_6, and $C_{16}H_{10}$ are noticeably higher in 1-heptene flames than those in n-heptane flames. Also, the effect of the double bond is more pronounced at lower ϕ (higher level of partial premixing) and higher strain rate.

Fu et al. [46] further extended this work to characterize the effect of fuel unsaturation on soot emissions in counterflow PPFs. The previously used fuel oxidation and NOx chemistry mechanism were combined with a detailed soot model [39, 40], and the combined model was validated against gaseous species measurements in n-

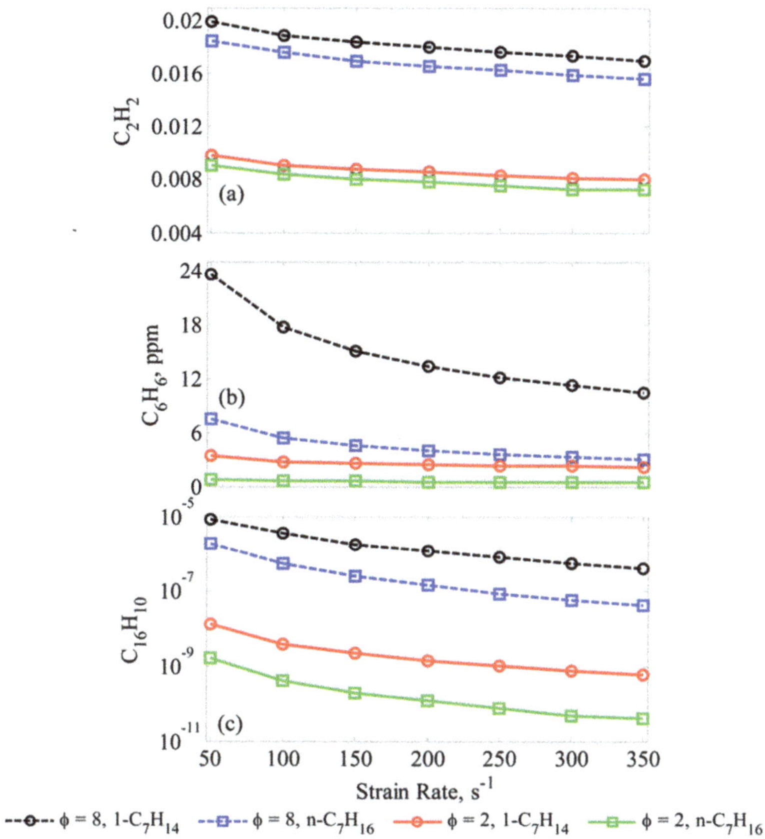

Fig. 16 Peak mole fractions of acetylene, benzene, and pyrene ($C_{16}H_{10}$) plotted versus strain rate for n-heptane and 1-heptene partially premixed flames PPFs at $\phi = 2$ and 8. Pyrene is plotted on a log scale From Fu et al. [46]

heptane PPFs and soot measurements in ethylene diffusion flames. Simulations were then performed to examine the effects of double bonds on PAHs and soot emissions at different strain rates and levels of premixing. A representative result from Ref. [46] is shown in Fig. 17, which plots the soot properties in n-heptane and 1-heptene PPFs established at different strain rates and two levels of partial premixing with $\phi = 8$ and 2. These results should be viewed along with those presented in Fig. 16, which indicated that at any given ϕ and strain rate, the amounts of soot precursor species (C_2H_2, C_6H_6, and $C_{16}H_{10}$) formed in 1-heptene flames are markedly higher than those in n-heptane flames. Consequently, as indicated in Fig. 17, the soot particle diameter, soot volume fraction, and number density are significantly higher in 1-heptene flames than those in n-heptane flames. Also, similar to soot precursor species, the effect of the double bond on soot becomes more significant as the strain rate and/or the level of premixing are increased (i.e., ϕ in the fuel jet is reduced).

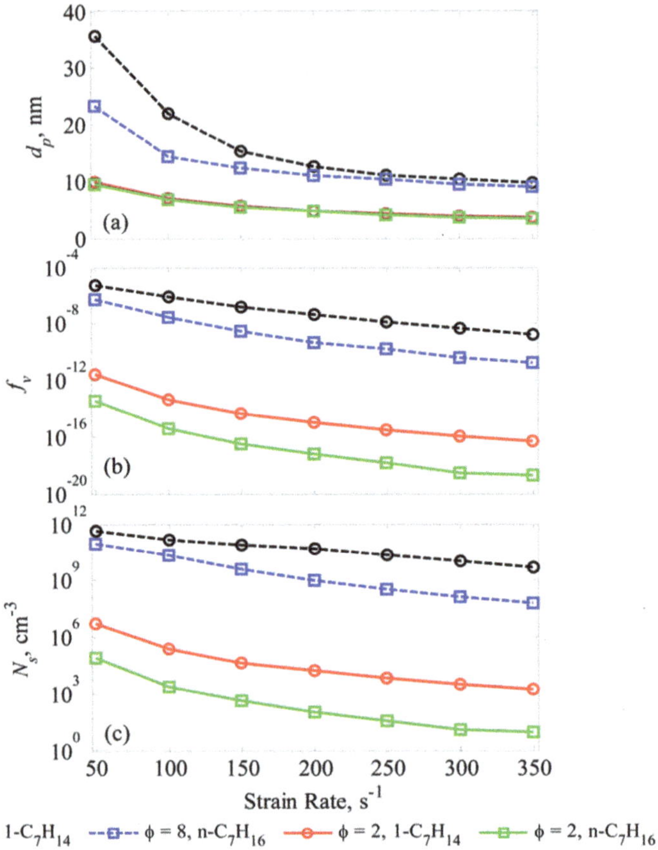

Fig. 17 Peak soot diameter (**a**), number density (**b**), and volume fraction (**c**) plotted versus strain rate for n-heptane and 1-heptene partially premixed flames at $\phi=2$ and 8. Number density and volume fraction are on a log scale, and soot diameter is on a linear scale. From Fu et al. [46]

Further, analysis indicated that the increased soot emission is related to the higher nucleation and surface growth rates, which are due to the increased production of $C_{16}H_{10}$ and C_2H_2 in 1-heptene flames compared to that in n-heptane flames. The increased production of $C_{16}H_{10}$ is due to the higher amount of C_6H_6 in 1-heptene flames.

A reaction path analysis was performed to identify the dominant routes for the formation of acetylene, benzene, and pyrene. Acetylene and benzene are known to be important precursors for larger PAH species, while acetylene also plays an important role in soot surface growth through the HACA mechanism. The analysis indicated that the major route for benzene formation in the RPZ is through the recombination reaction of propargyl radicals, which are mostly formed from allyl radicals. The other route is through the reaction of vinyl with butadiene. The presence of double bond leads to higher concentrations of propargyl and butadiene and thus increased benzene formation in 1-heptene flames than in n-heptane flames. The presence of double bond also increases the amount of C_2H_2 formed in 1-heptene flames due to the higher C_4H_5 concentration. Thus, the presence of double bond promotes scission β reactions leading to the increased production of C_2H_2, C_6H_6, and $C_{16}H_{10}$ and thus higher soot emissions in 1-heptene flames.

As mentioned earlier, there is relatively little work reported dealing with unsaturation effects in spray flames, except for engine experiments with biodiesel fuels. Fu and Aggarwal [48] recently reported a computational study on the effect of fuel unsaturation on NOx and PAH emission in turbulent spray flames under diesel engine conditions. N-heptane and 1-heptane spray flames in the Sandia reactor [32] were computed using the CONVERGE software. A reduced mechanism was developed starting from the detailed CRECK mechanism using the directed relation graph methodology. The mechanism was validated using the shock tube ignition data and reacting spray data from the Engine Combustion Network [32]. Further, details of the two-phase models and operating conditions are provided in Ref. [48]. Results indicated that the combustion in Sandia reactor is characterized by a double-flame structure with a rich premixed reaction zone (RPZ) near the flame stabilization region and a nonpremixed reaction zone (NPZ) further downstream. Most of NOx is formed in NPZ, while PAH species are mainly formed in RPZ. The presence of double bond leads to higher flame temperature and thus higher NO in 1-heptene flame than that in n-heptane flame. It also leads to the increased formation of PAH species, implying increased soot emission in 1-heptene flame than that in n-heptane flame. A representative result from Ref. [48] is shown in Fig. 18, which compares the temporal variation of volume-integrated NO, benzene and pyrene mass in n-heptane and 1-heptene flames. As indicated, 1-heptene flame produces significantly more NO and PAH species compared to n-heptane flame. For instance, at $t = 1.4$ ms, there is 20% more NO, 22.0% more benzene, and 21.9% more pyrene in 1-heptane flames. Figure 19 shows the spatial distribution of benzene in terms of its mass fraction contours for the two flames at $t = 1.4$ ms. The benzene formation region is located within the rich premixed zone for both flames, and the amount of benzene formed in 1-heptene is significantly higher. The peak mass fractions are 0.0118 and 0.0087 for 1-heptene and n-heptane, respectively, i.e., 36% higher benzene for 1-heptene. Similar trends were observed for the formation of heavier PAH, such as pyrene. Reaction path analysis indicates that the increased formation of PAH species can be attributed to the significantly higher amounts of 1,3-butadiene and allene formed due to β scission reactions resulting from the presence of double bond in 1-heptene.

5 Conclusions

A review of research dealing with the effects of fuel unsaturation on the ignition, combustion, and emission characteristics has been provided. Results from both laboratory-scale configurations, such as shock tube (ST), rapid compression machine (RCM), and laminar flames, as well as from high-pressure sprays in compression

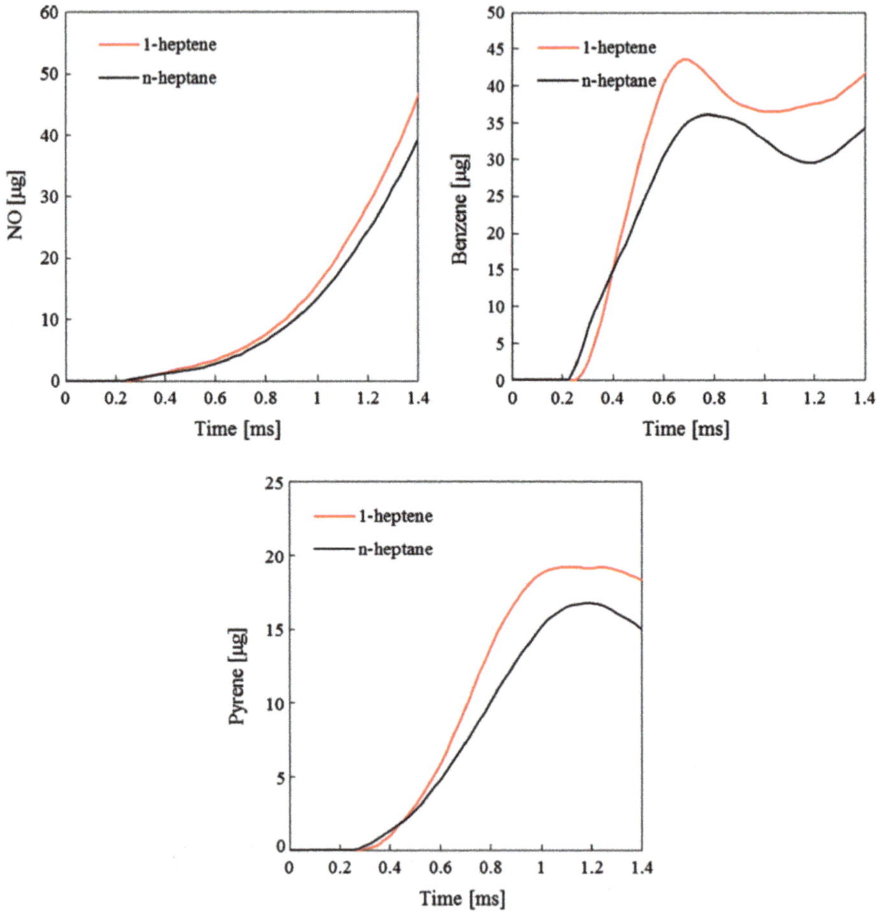

Fig. 18 Volume-integrated NO, benzene, and pyrene mass plotted versus time for n-heptane (black) and 1-heptane (red) spray flames in the Sandia reactor. Initial temperature is 1300 K From Fu and Aggarwal [48]

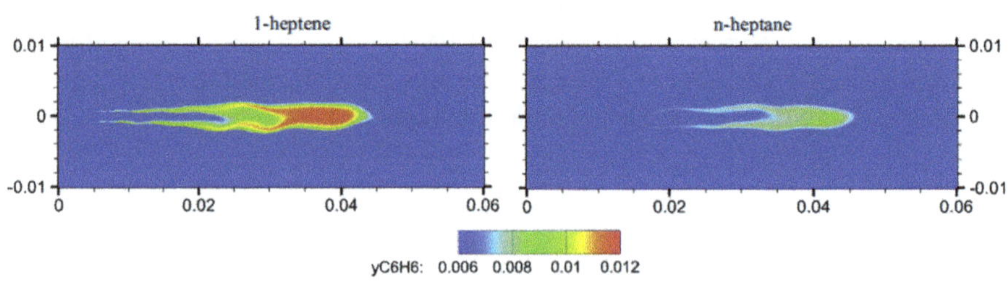

Fig. 19 Benzene mass fraction contours for 1-heptene and n-heptane flames at 1.4 ms. Mass fractions are between 0.006 and 0.012. Dimensions are in m Fu and Aggarwal [48]

ignition engines are discussed. Experimental and kinetic modeling studies provide clear evidence that depending upon the number and position of C = C double bonds, and the reactivity of long-chain hydrocarbons is significantly affected by fuel unsaturation, especially at low to intermediate temperatures that include the NTC region. Ignition data for homogeneous mixtures indicate that the presence of double bond has a strong effect on the 1st stage ignition, reduces low-temperature reactivity significantly, increases total ignition delay time, and leads to reduction in CN number in diesel engines. This has important consequences regarding the effect of unsaturation on the combustion and emission behavior in engines.

Results for n-heptane and 1-heptene diesel sprays indicate that at low to intermediate temperatures, ignition delays are longer for 1-heptene compared to those for n-heptane. Also, ignition in n-heptane sprays occurs in fuel-rich mixtures and is characterized by a two-stage ignition process. On the other hand, ignition in 1-heptene sprays occurs in lean mixtures, without any evidence of two-stage ignition. Consequently, the n-heptane flame contains two reaction zones, i.e., a rich premixed zone (RPZ) and a nonpremixed reaction zone (NPZ), while the 1-heptene flame contains three reaction zones, i.e., a lean premixed zone (LPZ) in addition to NPZ and RPZ. Also at lower temperatures, the flame liftoff length in 1-heptene spray is greater than that in n-heptane spray, while at higher temperatures, the liftoff length is smaller in 1-heptene spray.

Results for counterflow, partially premixed flames (PPF) indicate higher amounts of NOx and soot precursor species (C_2H_2, C_6H_6, and $C_{16}H_{10}$) formed in 1-heptene flames than those in n-heptane flames. Consequently, the soot emission (i.e., soot particle diameter, volume fraction, and number density) in 1-heptene flames is higher than that in n-heptane flames. Moreover, the effect of double bond on soot emission becomes more pronounced as the strain rate or the level of premixing is increased. Simulations of turbulent n-heptane and 1-heptene spray flames in diesel engines lead to similar conclusions regarding the effect of fuel unsaturation on emission, i.e., higher NOx and soot emissions in 1-heptene flames. Reaction path analysis indicates that the increased formation of PAH species can be attributed to the significantly higher amounts of 1,3-butadiene and allene formed due to β scission reactions resulting from the presence of double bond in 1-heptene.

Acknowledgements Most of the results presented in this chapter are from the work of my graduate students as part of their thesis research. In particular, I acknowledge the contributions of Dr. Sibendu Som, Dr. Xiao Fu, Mr. Xu Han, and Mr. Saurabh Sharma. The help provided by Dr. P. K. Senecal and his colleagues at Convergent Science in using the Converge code is also greatly appreciated. Many of the simulations were performed at the UIC High Performance Computing Cluster.

References

1. C.K. Westbrook, W.J. Pitz, S.M. Sarathy, M. Mehl, Detailed chemical kinetic modeling of the effects of CC double bonds on the ignition of biodiesel fuels. Proc. Combust. Inst. **34**, 3049–3056 (2013)
2. M.S. Graboski, R.L. McCormick, Combustion of fat and vegetable oil derived fuels in diesel engines. Prog. Energy Combust. Sci. **24**, 125–164 (1998)
3. G. Kalghatgi, H. Babiker, J. Badra, A simple method to predict knock using toluene, n-heptane and iso-octane blends (TPRF) as gasoline surrogates. SAE Int. J. Eng. **8**, 505–519 (2015)
4. T. Javed, C. Lee, M. Al Abbad, K. Djebbi, M. Beshir, J. Badra, H. Curran, A. Farooq, Ignition studies of n-heptane/iso-octane/toluene blends. Combust. Flame **171**, 223–233 (2016)
5. T. Javed, E.F. Nasir, A. Ahmed, J. Badra, K. Djebbi, M. Beshir, W. Ji, S.M. Sarathy, A. Farooq, Ignition delay measurements of light naphtha: a fully blended low octane fuel. Proc. Combust. Inst. **36**(1), 315–322 (2017)
6. S. Tanaka, F. Ayala, J.C. Keck, J.B. Heywood, Two-stage ignition in HCCI combustion and HCCI control by fuels and additives. Combust. Flame **132**, 219–239 (2003)
7. CHEMKIN-PRO 15141, Reaction Design, San Diego, 2015
8. D.G. Moffat, K. Harry, R.L. Speth, Cantera: an object- oriented software toolkit for chemical kinetics, thermodynamics, and transport processes (2017), http://www.cantera.org. Version 2.3.0. https://doi.org/10.5281/zenodo.170284
9. S. Touchard, F. Buda, G. Dayma, P.A. Glaude, R. Fournet, F. Battin-Leclerc, Int. J. Chem. Kinet. **37**(8), 451–463 (2005)
10. M. Mehl, W.J. Pitz, C.K. Westbrook, K. Yasunaga, C. Conroy, H.J. Curran, Autoignition behavior of unsaturated hydrocarbons in the low and high temperature regions. Proc. Combust. Inst. **33**, 201–208 (2011)
11. S. Garner, R. Sivaramakrishnan, K. Brezinsky, The high-pressure pyrolysis of saturated and unsaturated C7 hydrocarbons. Proc. Combust. Inst. **32**, 461–467 (2009)
12. R. Minetti, A. Roubaud, E. Therssen, M. Ribaucour, L.R. Sochet, The chemistry of pre-ignition of n-pentane and 1-pentene. Combust. Flame **118**, 213–220 (1999)
13. G. Vanhove, M. Ribaucour, R. Minetti, On the influence of the position of the double bond on the low-temperature chemistry of hexenes. Proc. Combust. Inst. **30**, 1065–1072 (2005)
14. S.K. Prabhu, R.K. Bhat, D.L. Miller, N.P. Cernansky, 1-Pentene oxidation and its interaction with nitric oxide in the low and negative temperature coefficient regions. Combust. Flame **104**, 377–391 (1996)
15. M. Pelucchi, M. Bissoli, C. Cavallotti, A. Cuoci, T. Faravelli, A. Frassoldati, E. Ranzi, A. Stagni, Improved kinetic model of the low-temperature oxidation of n-heptane. Energy Fuels **28**(11), 7178–7193 (2014)
16. E. Ranzi, A. Frassoldati, R. Grana, A. Cuoci, T. Faravelli, A.P. Kelley, C.K. Law, Hierarchical and comparative kinetic modeling of laminar flame speeds of hydrocarbon and oxygenated fuels. Prog. Energy Combust. Sci. **38**(4), 468–501 (2012)
17. M. Mehl, G. Vanhove, W.J. Pitz, E. Ranzi, Oxidation and combustion of the n-hexene isomers: a wide range kinetic modeling study. Combust. Flame **155**, 756–772 (2008)
18. S. Garner, T. Dubois, C. Togbe, N. Chaumeix, P. Dagaut, K. Brezinsky, Biologically derived diesel fuel and NO formation: Part 2: model development and extended validation. Combust. Flame **158**, 2302–2313 (2011)
19. C.K. Westbrook, W.J. Pitz, S.M. Sarathy, M. Mehl, Detailed chemical kinetic modeling of the effects of C = C double bonds on the ignition of biodiesel fuels. Proc. Combust. Inst. **34**(2), 3049–3056 (2013)
20. S.K. Aggarwal, Single droplet ignition: theoretical analyses and experimental findings. Prog. Energ Combust. Sci. (2014). https://doi.org/10.1016/j.pecs.2014.05.002
21. O. Moriue, C. Eigenbrod, H.J. Rath, J. Sato, K. Okai, M. Tsue, M. Kono, Effects of dilution by aromatic hydrocarbons on staged ignition behavior of n-decane droplets. Proc. Combust. Inst. **28**, 969–975 (2000)

22. A. Cuoci, M. Mehl, G. Buzzi-Ferraris, T. Faravelli, D. Manca, E. Ranzi, Autoignition and burning rates of fuel droplets under microgravity. Combust. Flame **143**, 211–226 (2005)
23. Z. Bouali, C. Pera, J. Reveillon, Numerical analysis of the influence of two-phase flow mass and heat transfer on n-heptane autoignition. Combust. Flame **159**, 2056–2068 (2012)
24. M.C. Wolff, J. Meisl, R. Koch, S. Wittig, The influence of evaporation on the autoignition-delay of n-heptane air mixtures under gas turbine conditions. Proc. Combust. Inst. **27**, 2025–2031 (1998)
25. S.K. Aggarwal, A review of spray ignition phenomenon: present status and future research. Prog. Energy Combust. Sci. **24**, 565–600 (1998)
26. E. Mastorakos, Ignition of turbulent non-premixed flames. Prog. Energy Combust. Sci. **35**, 57–97 (2009)
27. L.M. Pickett, S. Kook, H. Persson, O. Andersson, Diesel fuel jet lift-off stabilization in the presence of laser-induced plasma ignition. Proc. Combust. Inst. **32**, 2793–2800 (2009)
28. S. Som, S.K. Aggarwal, Effects of primary breakup modeling on spray and combustion characteristics of compression ignition engines. Combust. Flame **157**, 1179–1193 (2010)
29. X. Fu, S.K. Aggarwal, Two-stage ignition and NTC phenomenon in diesel engines. Fuel **144**, 188–196 (2015)
30. H.J. Curran, W.J. Pitz, C.K. Westbrook, C.V. Callahan, F.L. Dryer, Oxidation of automotive primary reference fuels at elevated pressures. Proc. Combust. Inst. **27**, 379–387 (1998)
31. S. Sharma, S.K. Aggarwal, Effects of fuel unsaturation on transient ignition and flame development in sprays. Combust. Sci. Technol. (2017). https://doi.org/10.1080/00102202.2017.1378649
32. Engine Combustion Network (ECN), diesel sprays data search utility (2015), http://www.sandia.gov/ecn/cvdata/dsearch/frameset.php. Accessed Feb 2015
33. S. Sharma, A numerical study of transient ignition and flame structure in diesel sprays in a constant volume reactor, M.S. thesis, Department of Mechanical and Industrial Engineering, University of Illinois at Chicago, Chicago, Illinois, 2017
34. R.L. McCormick, M.S. Graboski, T.L. Alleman, A.M. Herring, K.S. Tyson, Impact of biodiesel source material and chemical structure on emissions of criteria pollutants from a heavy-duty engine. Environ. Sci. Technol. **35**(9), 1742–1747 (2001)
35. S. Puhan, N. Saravanan, G. Nagarajan, N. Vedaraman, Effect of biodiesel unsaturated fatty acid on combustion characteristics of a DI compression ignition engine. Biomass Bioenerg. **34**, 1079–1088 (2010)
36. M. Lapuerta, J.M. Herreros, L.L. Lyons, R. García-Contreras, Y. Briceño, Effect of the alcohol type used in the production of waste cooking oil biodiesel on diesel performance and emissions. Fuel **87**, 3161–3169 (2008)
37. A. Schönborn, N. Ladommatos, J. Williams, R. Allan, J. Rogerson, The influence of molecular structure of fatty acid monoalkyl esters on diesel combustion. Combust. Flame **156**, 1396–1412 (2009)
38. S. Garner, K. Brezinsky, Biologically derived diesel fuel and NO formation: an experimental and chemical kinetic study, Part 1. Combust. Flame **158**, 2289–2301 (2011)
39. M. Frenklach, H. Wang, Proc. Combust. Inst. **23**, 1559 (1991)
40. M. Frenklach, Reaction mechanism of soot formation in flames. Phys. Chem. Chem. Phys. **4**, 2028–2037 (2002)
41. X. Han, S.K. Aggarwal, K. Brezinsky, Effect of unsaturated bond on NOx and PAH formation in n-Heptane and 1-Heptene triple flames. Energy Fuels **27**, 537–548 (2013)
42. A.E. Lutz, R.J. Kee, J.F. Grcar, FM. Rupley, OPPDIF: a FORTRAN program for computing opposed flow diffusion flames. Sandia Natl. Lab. [Tech. Rep.] SAND 96−8243, UC-1404 (1997)
43. E. Ranzi, M. Dente, A. Goldaniga, G. Bozzano, T. Faravelli, Lumping procedures in detailed kinetic modeling of gasification, pyrolysis, partial oxidation and combustion of hydrocarbon mixtures. Prog. Energy Combust. Sci. **27**, 99–139 (2001)
44. J.A. Miller, C.T. Bowman, Mechanism and modeling of nitrogen chemistry in combustion. Prog. Energy Combust. Sci. **15**, 287–338 (1989)

45. J.W. Bozzelli, A.M. Dean, O+NNH: a possible new route for NOx formation in flames. Int. J. Chem. Kinet. **27**, 1097–1109 (1995)
46. X. Fu, S. Garner, S.K. Aggarwal, K. Brezinsky, Numerical study of NO_x emissions from n-heptane and 1-heptene counterflow flames. Energy Fuels **26**, 879–888 (2012)
47. X. Fu, X. Han, K. Brezinsky, S.K. Aggarwal, Effect of fuel molecular structure and premixing on soot emissions from n-heptane and 1-heptene flames. Energy Fuels **27**(10), 6262–6272 (2013)
48. X. Fu, S.K. Aggarwal, Fuel unsaturation effects on NOx and PAH formation in spray flames. Fuel **160**, 1–15 (2015)

Investigation and Development of Modified Ross Yoke Mechanism on Alpha-Parallel-Cylinder Stirling Engine

Sutapat Kwankaomeng

Abstract This paper presents the redesign and modification of a Ross yoke mechanism driving an alpha Stirling engine with parallel cylinder arrangement. Engine sealing is a crucial factor affecting engine operation, power, and maintenance. Friction and lateral force acting on piston seals induce major wear and finally lead to leakage and losses in both power and operating cost. To reduce these problems, linear reciprocating and balancing motion of both piston and connecting rod are preferred. Mechanical inversion is introduced to invert general motion of connecting rod to rectilinear translation. The original Ross yoke part is altered from pin joints to slot pin joints allowing piston rod to be driven straight. Length of the slot groove was adjusted and optimized. Motion of the modified Ross yoke was investigated theoretically and experimentally. Both analytical model and prototype have the same operating conditions and swept volume of 25 cm^3. The ambient air was used as the working gas which was heated by LPG at flow rate of 0.6 kg/h and cooled by water. The maximum speed of 977 rpm was attained. Recorded maximum power and torque were 0.549 W at 486 rpm and 0.014 Nm at 260 rpm, respectively. The modified Ross yoke Stirling engine has operated smoothly when its piston rods were redesigned to slide linearly.

Keywords Stirling engine · Alpha-type · Ross yoke

1 Introduction

Global warming and climate change are prompting seriously concerns and drawing globally attention and collaboration for carbon footprint reduction, renewable energy utilization and air purification. Stirling engine, invented in 1816 and patented in 1817

S. Kwankaomeng (✉)
Faculty of Engineering, Department of Mechanical Engineering,
King Mongkut's Institute of Technology Ladkrabang, 1 Chalong Krung Rd. Ladkrabang,
Bangkok 10520, Thailand
e-mail: sutapat.kw@kmitl.ac.th

A. K. Runchal et al. (eds.), *Energy for Propulsion*, Green Energy and Technology,
https://doi.org/10.1007/978-981-10-7473-8_4

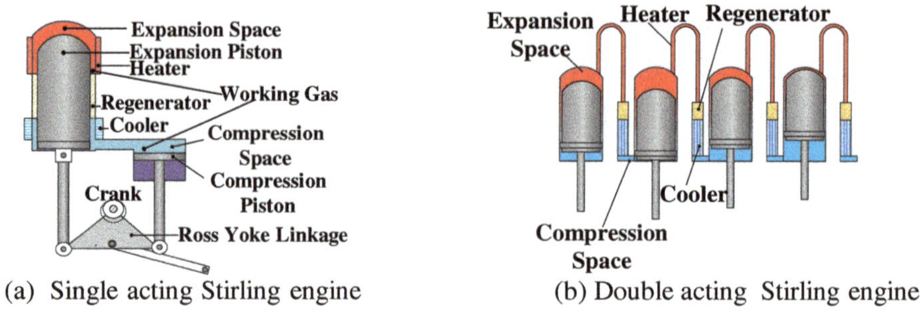

Fig. 1 Single acting and double acting Stirling engine configurations

by Sir Robert Stirling, is a potential option that meets these criteria as a solution. The engine is a sealed engine based on gas expansion and contraction shuttling between hot and cold spaces without internal combustion but external heating. Hence, it is eco-friendly and can be easily applied with various choices of heat sources including low grade fuel, recovery heat and waste. Heat and power are produced from the engine, making it an excellent component in combined heat and power (CHP) production system. It can be used either the prime mover or cooling engine.

1.1 Classification of Stirling Engines

There are many existing Stirling cycle machines with different configurations, mechanisms and functionalities. To categorize the Stirling engine is based on various considerations such as cylinder or piston arrangement, driving configuration, and characteristic of power production. Stirling engines can be assorted into two distinct categories consisting of (a) single acting or (b) double acting as depicted in Fig. 1. The single-acting engine shown in Fig. 1a is the assembly of hot or expansion space, cold or compression space through heater, regenerator and cooler with one or two cylinders containing two sliders which can be both pistons or displacer and piston. Double acting engine is the assembly of multiple cylinders that are arranged to have one piston each which functions as displacer and power piston. A series of heat exchanger is coupled between two cylinders at the expansion space of one cylinder and the compression space of the other cylinder as illustrated in Fig. 1b. The working gas, therefore, is circulated and shuttled from the hot section of one cylinder through the exchanger set to the cold section of the adjacent cylinder.

Single acting Stirling engines can be broadly classified into two groups: two piston machines and piston—displacer machines Walker [1]. Such crank-driven Stirling engines are usually categorized by their configuration. Three basic types are known as alpha, beta, and gamma represented in Fig. 2. The alpha type in Fig. 2a is a two-piston machine having hot and cold cylinder separately, enabling convenient ways to heat and cool the working gas from outside heat sources and heat sinks. The hot and

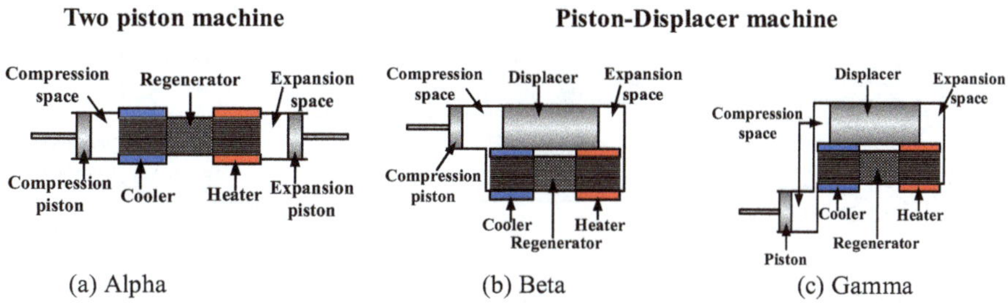

Fig. 2 Stirling engine configurations

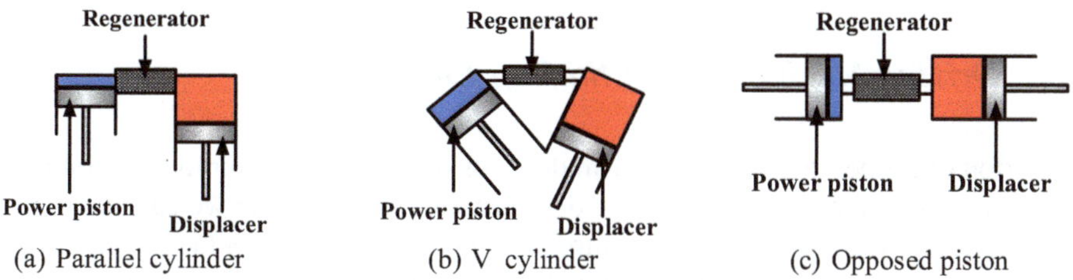

Fig. 3 Alternative arrangements of the single-acting alpha Stirling engine

cold pistons are assembled in the expansion and compression space, respectively. Nevertheless, to prevent gas leakage, both hot and cold pistons and their rods must be fit tightly with engine cylinder wall by seals. Beta and gamma patterns are the displacer-piston machines. The beta arrangement consists of displacer and piston reciprocating concentrically within the same cylinder or in-line cylinder as illustrated in Fig. 2b. For the last configuration, the gamma engine has two eccentric cylinders attaching and sharing cold space between them as depicted in Fig. 2c.

Stirling engine typically consists of two reciprocating pistons called displacer and piston or expansion piston and compression piston. The displacer or the hot piston does no work but to displace the working gas between hot and cold spaces. The working piston or cold piston produces power. The working gas is shuttled between the hot and cold sections called expansion and compression spaces, respectively.

For stationary cylinder, single acting alpha Stirling engine is typically arranged in three forms such as parallel cylinder pattern (Rider), V cylinder, and opposed piston in Fig. 3a–c, consecutively. This work focuses on the parallel configuration because of compact, simple and possible design for double acting multi-cylinder as presented in Fig. 1b.

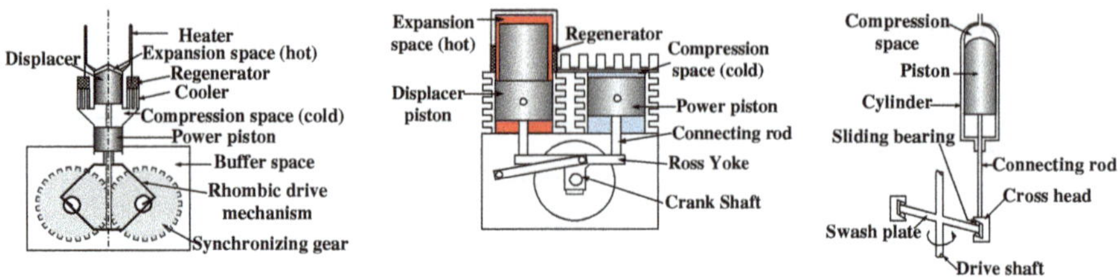

(a) Rhombic drive (b) Ross yoke drive (c) Swash plate drive

Fig. 4 Sample driving mechanisms in Stirling engines

1.2 Driving Mechanisms of Stirling Engines

Selection of proper driving mechanism depends on engine configuration, durability and reliability. Some mechanisms are demonstrated in Fig. 4. As we can see in the figure, slider-crank mechanism is the most common linkage used for the reciprocating engines. Based on the linkage principal, a rhombic drive mechanism in Fig. 4a was invented intentionally for beta type engine by Meijer 1950s and since then has been widely applied. Because two pistons are moved concentrically in the same cylinder, the rhombic drive couples two slider crank components consisting of two synchronizing gears, yokes, and connecting rods. Two slider crank linkages were also combined successfully by Ross [2] known as Ross yoke mechanism in driving alpha Stirling engine. The Ross yoke part is pinned to two con rods of the pistons as shown in Fig. 4b.

For multiple and parallel cylinder engines, the swash plate is one of the driving mechanisms as schematized in Fig. 4c used in double acting engine. However, the swash plate driven engine is complex and difficult to manufacture. The Ross linkage has been developed and used as driving mechanism in both single acting and double acting of an alpha Stirling engine. This linkage is simpler and lighter if compared to that of swash plate mechanism for a double-acting four-piston Stirling engine.

Many Stirling engines have been built with various design and developed worldwide by inspired researchers. Homutescu and Bălănescu [3] proposed the constructive optimization of an alpha-type Stirling engine by parameterizing diameters of the two pistons. Their results indicated that the diameter of the displace piston must be greater than the diameter of the power piston. The effect of geometrical and physical parameters on Ross yoke Stirling engine performance was investigated by Tlili and Musmar [4]. Their analysis indicated that the performance of Ross yoke Stirling cycle engine with air as working gas depends critically on the geometrical parameters and heat transfer. Martaj and Rochelle [5] showed the 1D modelling of an alpha double-acting Stirling engine and carried out the optimal values of the engine geometry. Moreover, dual cylinder and double acting arrangement were designed to enhance the engine power. Scollo et al. [6] studied the development of a twin cylinder alpha type Stirling engine model and also redesigned their prototype. The engine performance was very satisfactory. In Thailand, Silpsakoolsook et al. [7] presented their alpha-type Stirling engine prototype with crank-slider mechanism of the parallel cylinder pattern. The engine can be started when it is heated up to 190 °C at the heater head. The maximum output power was 83 mW at a temperature of 560 °C.

Although Stirling engines have been invented two centuries ago, they are still not a desirable choice for practical uses or as commercial products. The greatest challenge preventing Stirling engines from being the optimal choice of power generation applications is its low power density or high cost of power production. To overcome such obstacles, higher operating pressure by pressurization of the working gas must be accomplished. This certainly requires a perfectly sealing system of gas inside the engine in both working and bounce spaces. There are at least two seals on pistons in a single acting machine including rod seal in some cases. However, leakage and loss in operating cost and power are unavoidable since seals always wear off under moments of forces from the touching parts, especially when unbalanced force exits at the seal between dynamic pistons and cylinder wall during operation. To get seal that is more effective is not possible due to material limitation. Instead of seal replacement, more seal protection is important. Finding the major cause of seal problem is really needed. Rolling seal was introduced in rhombic drive Stirling engine. Such driving mechanism provides balanced force on both rods and pistons in beta type configuration. However, crank-slider mechanism is mostly used in driving Stirling engine. Oscillation of connecting rod causes imbalance. Peripheral force acting on seals can be eliminated and balanced by linear sliding motion of both piston and rod. Therefore, to lengthen service life and improve engine efficiency, a strategy to mitigate the sealing problem is proposed in this research work by redesigning the mechanical parts and linkage of alpha type Stirling engine using kinematic inversion concept.

2 Ross Yoke Redesign and Analysis

Effects of general motion, a combination of translation and rotation, of the connecting rods on seals are depicted in Fig. 5. The traditional Ross yoke has four pin joints. One joint is pinned the yoke to the crank shaft and rotated circularly around the crank shaft as shown in Fig. 5a. During the yoke oscillation, the other two joints, which are attached to the piston rods, move in general motion as demonstrated in Fig. 5b. The translation and rotation of the con rods result to periodic side force acting on piston or rod seal as depicted in Fig. 5c. Such fatigue loading causes wear and damage on seal. Finally, leakage is introduced decreasing engine performance and stopping engine operation. This effect requires more maintenance and seal replacement frequently because of shorter lifetime of sealing.

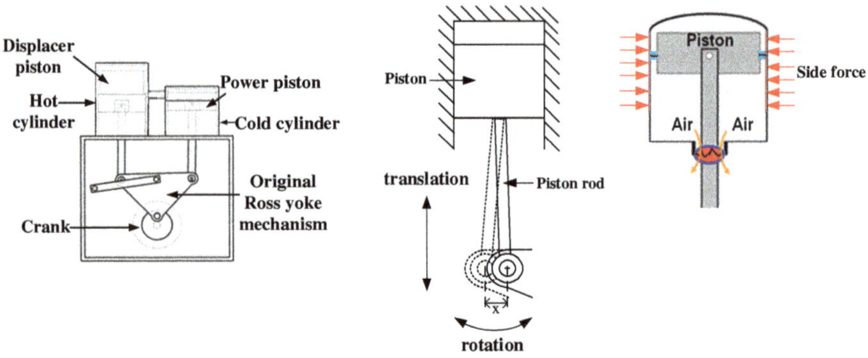

(a) Ross yoke engine (b) General motion of piston rod (c) Side force on piston seal and fatigue on rod seal

Fig. 5 Oscillation of piston rod on Ross yoke Stirling engine

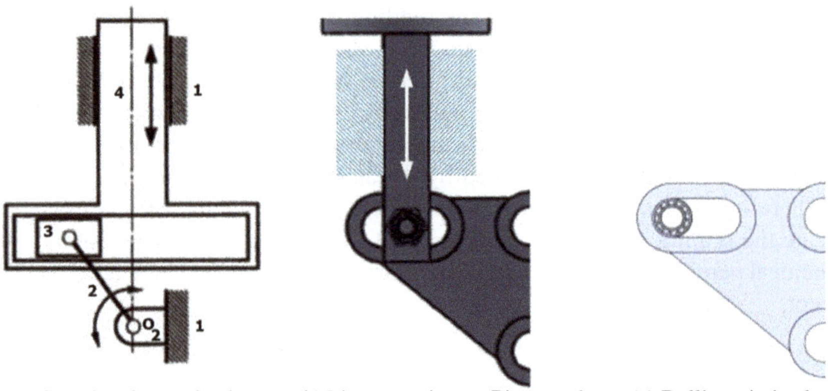

(a) Scotch yoke mechanism (b) Linear motion on Piston rod (c) Rolling pin in slot

Fig. 6 Scotch yoke mechanism and mechanical inversion on yoke

2.1 Ross Yoke Modification

To increase power output from the Stirling engine, one way that can be immediately done is to pressurize the engine. However, pressurized engine may encounter the seal problem. At least two piston seals are required. Rod seal is required when the displacer rod passes through the piston. However, oscillation of the piston rods creates side force on piston and seal. Therefore, decreasing peripheral force can be achieved by maintaining equilibrium lateral force by linear motion of the piston rod. Like Scotch yoke mechanism, known as slotted link mechanism, it gives advantage on absence of lateral force on the slider because of straight reciprocating motion as shown in Fig. 6a.

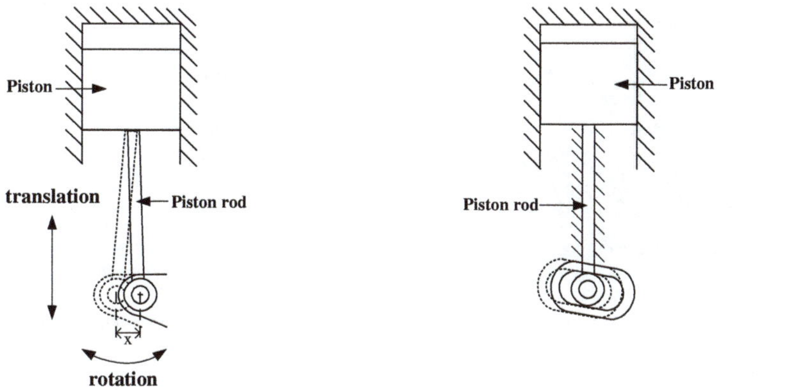

(a) General motion of piston rod (b) Linear motion of piston rod

Fig. 7 Motion of piston rod on the Ross yoke engine and the redesigned Ross yoke

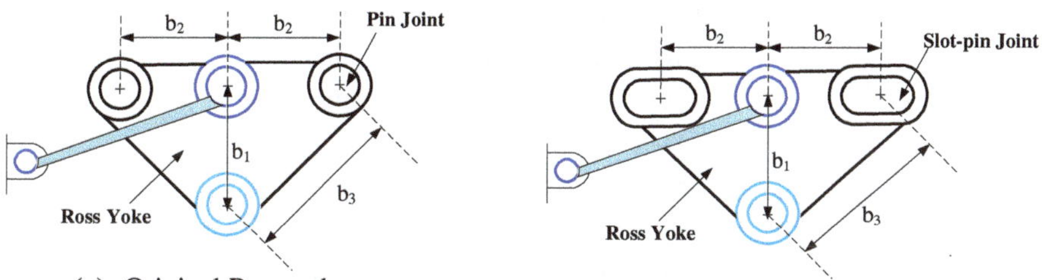

(a) Original Ross yoke (b) Modified Ross yoke

Fig. 8 Redesign of Ross yoke

Using the concept of Scotch yoke mechanism can obtain linear motion of the piston rod (4) by coupling with a slot that engages a pin on the rotating part (2). Therefore, the original Ross yoke part was redesigned and modified. Kinematic inversion is introduced. In Fig. 6b, pin coupling with rolling bearing is attached on piston rod end to eliminate rod rotation. Slot groove for rolling pin dwelling are conformed on the yoke instead of piston rod as depicted in Fig. 6c. Hole of the pin joint on the common Ross yoke is replaced by slot groove for pin sliding.

Constraint pin joints on the typical Ross yoke as in Fig. 7a causing piston rods oscillation are inverted to confine the piston rods in straight reciprocation by letting pin joints slide in slot instead as shown in Fig.7b. The original Ross yoke and the redesigned Ross yoke are represented in Fig. 8. The designed engine with modified Ross yoke is depicted in Fig. 9.

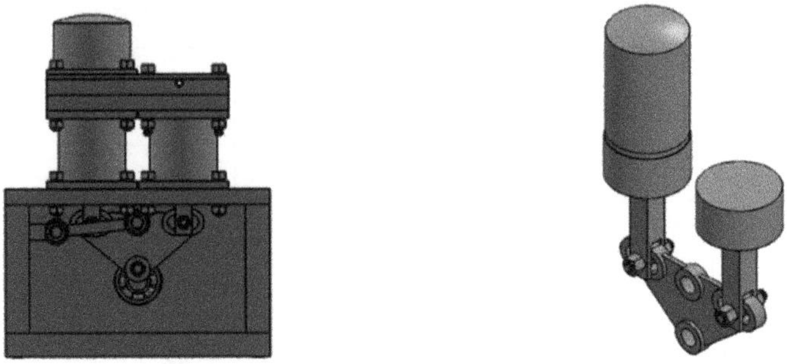

(a) Designed engine (b) Modified Ross yoke components

Fig. 9 Designed engine with yoke modification

Table 1 Symbols used for calculations

Symbols	Parameter name	Unit
X	Displacement in x direction	mm
Y	Displacement in y direction	mm
R	Crank radius	mm
θ	Crank angle	radian
b1, b2, b3	Geometry on Ross yoke	mm
V	Volume	m^3
T	Temperature	K
W	Work	J
n	Engine speed	rps
S, B, a, c	Parameter	
Subscripts		
R	Regenerator	
E	Expansion	
D	Dead (Volume)	
S	Swept (Volume)	
p	piston	
d	displacer	

2.2 *Geometry and Power Analysis*

Mechanical inversion of the Ross mechanism is used to restrain the piston rod moving in a straight line in order to achieve balanced force on piston seal. Therefore, the displacement of pin joints attached in piston rods on yoke is evaluated numerically in order to find the slot joint dimension. Tables 1 and 2 show symbols and engine specifications.

Table 2 Engine dimensions

Geometric parameters	Values
Crank radius (r)	7.6 mm
b_1	29 mm
b_2	29 mm
b_3	41.01 mm
Diameter of piston (d_p)	38.75 mm
Diameter of cylinder (d_c)	39.05 mm
Regenerator volume (V_r)	6 cm^3

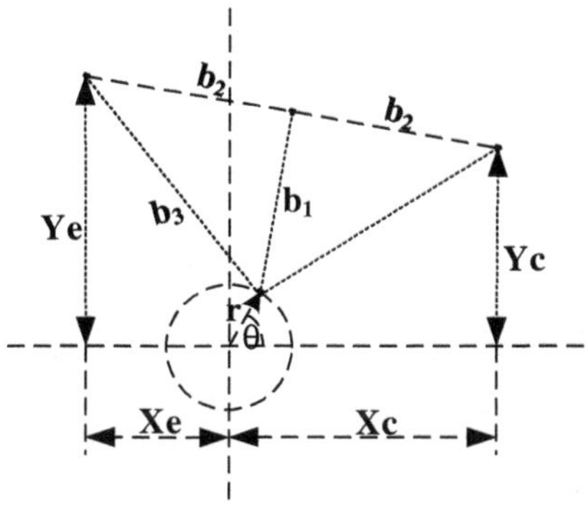

Fig. 10 Yoke geometry and parameter

Numerical analysis of an original Ross yoke mechanism in Fig. 10 can be formulated by using these components and parameters, OHIO University [8]. According to the original Ross yoke mechanism, displacements of displacer and power piston in y-axis are defined in expansion space (y_E) and compression space (y_C) in Eqs. (1) and (2), respectively. Operating conditions are described in Table 3.

$$Y_E = \left[r\left(\sin\theta + b_2 / b_1 \cos\theta\right)\right] + b\theta \tag{1}$$

$$Y_C = \left[r\left(\sin\theta - b_2 / b_1 \cos\theta\right)\right] + b\theta \tag{2}$$

where $b\theta = \sqrt{b_1^2 - (r\cos\theta)^2}$

2.2.1 Motion Verification

In order to test the redesign concept, motions of the piston rods on both typical Ross yoke mechanism and the modified one were observed and visualized by engine animation as in Fig. 11. Using the reference line to represent cylinder center line as the guide line in comparison to the axial line of piston rod, it can be seen that the

Table 3 Operating conditions of the modified Ross yoke Stirling engine

Type and operating conditions	Values/type
Engine type	Alpha
Fuel	LPG
Working fluid	Air
Coolant	Water
Mean engine pressure	101.325 kPa
Expansion Temperature (T_H)	923 K
Compression Temperature (T_C)	350 K

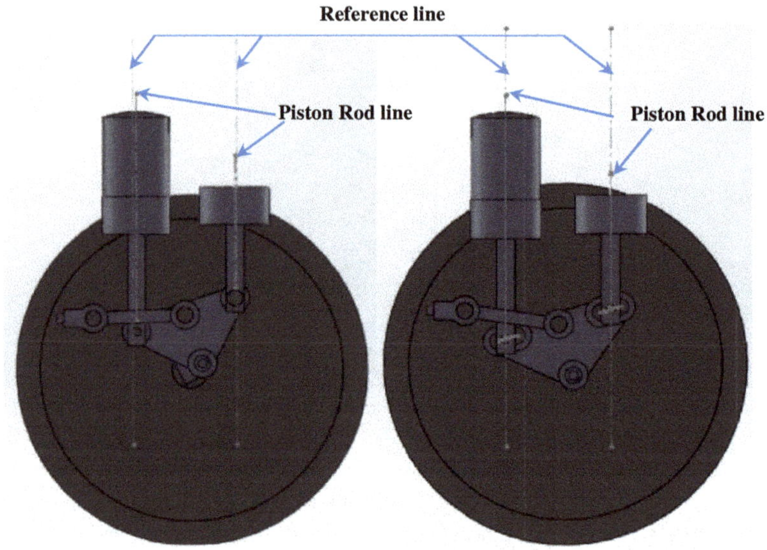

Fig. 11 Visualization of piston rod line with respect to the reference line

engine with the original Ross yoke has a gap between centerline of the piston rod while the modification Ross yoke maintains the coincide line between cylinder center line and rod line. The axial offset on the original Ross yoke reveals that the piston rod vibrates which causes side force on piston and seal while that of the modified Ross yoke moves in concentric and balanced direction. Hence, the adapted Ross yoke provides no lateral force on piston and seal.

2.2.2 Position Analysis and Groove Length of the Slot-Joints

Before conversion of a typical Ross yoke, lateral vibration of pin joints connecting piston rods was numerically investigated. Graphic position analysis was conducted by the vector loop-closure for the original Ross yoke presented in Fig. 12 and evaluated

Fig. 12 Vector diagram on Ross yoke mechanism

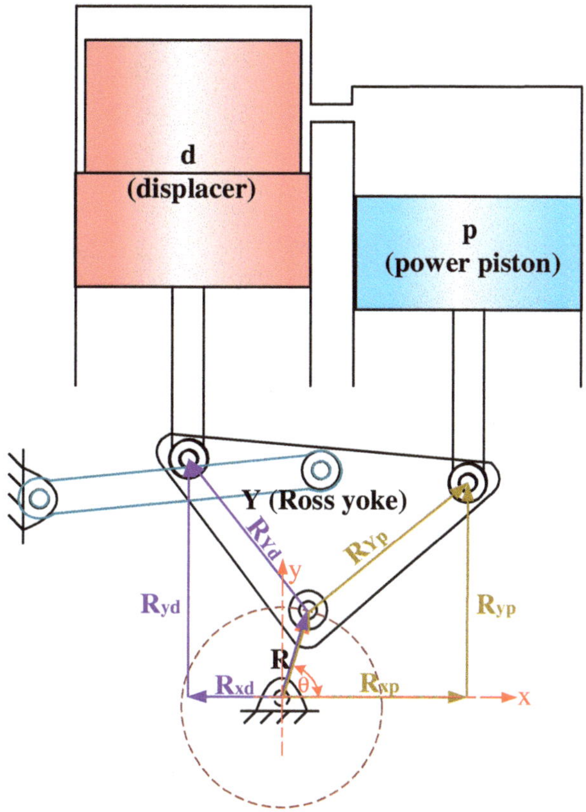

by Eqs. (3) and (4). Figure 13 presents the variation of lateral displacement of pin joints on the original Ross yoke along x axis in range of 5 mm each.

Length of two slot grooves must be sufficient for sliding pins in both right and left hand sides during the yoke oscillation. The identical grooves of 10 mm length were designed on the modified Ross yoke as shown in Fig. 14.

$$\vec{R}_{xd} + \vec{R}_{yd} = \vec{R} + \vec{\mathrm{R}}_{Yd} \tag{3}$$

$$\vec{R}_{xp} + \vec{R}_{yp} = \vec{R} + \vec{\mathrm{R}}_{Yp} \tag{4}$$

2.2.3 Displacement of the Expansion and Compression Spaces

Expansion and compression pistons are reciprocated in engine cylinders. The piston displacements in y direction along hot and cold cylinders with the variation of crank angle were calculated and plotted as shown in Fig. 15. Piston stroke can be derived

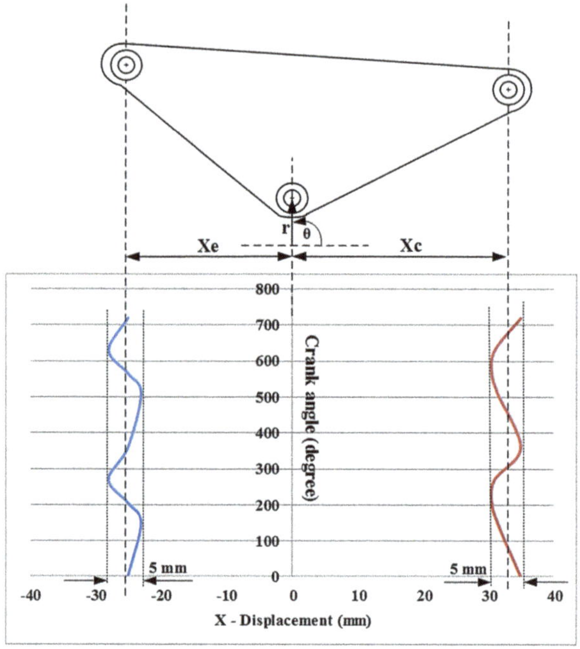

Fig. 13 Horizontal displacement of pin joints

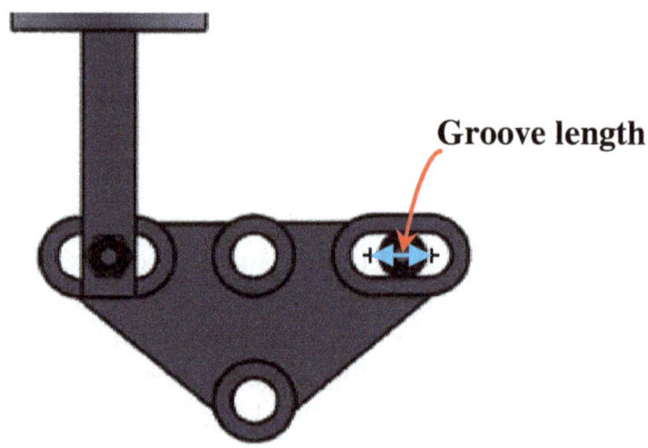

Fig. 14 Modified Ross yoke and groove length

from the displacement relationship. The piston stroke, therefore, is 21.6 mm and the hot piston displacement (Y_E) leads the cold piston displacement (Y_C) with 80° phase angle.

2.2.4 Engine Work and Power

Work produced by the engine can be calculated from P-V diagram and average power also derived by quotient of work by period. The Schmidt theory for Stirling engines

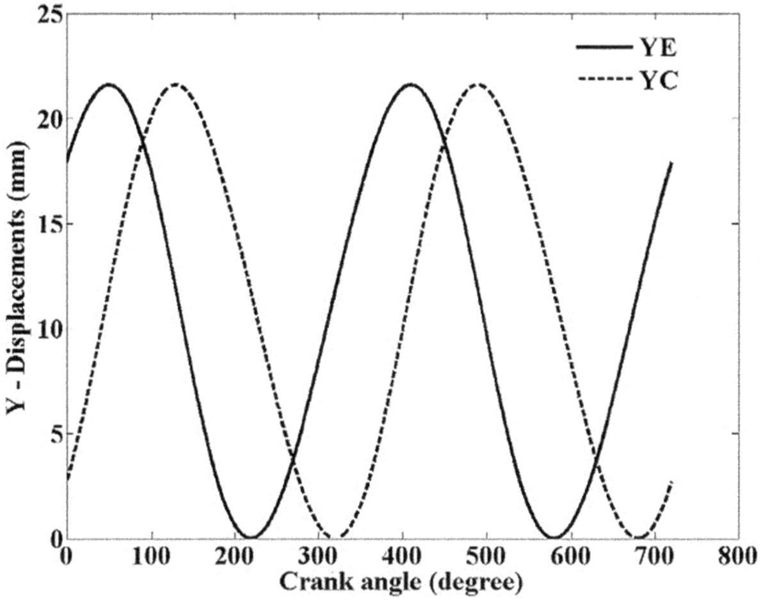

Fig. 15 Piston displacements in expansion and compression spaces with crank rotation

was presented by Hirata [9]. The indicated energy per one cycle of the engine, W_i, is derived by Eq. (5). Parameters a, c, B and S can be found in Eqs. (6)–(9). Rations of swept volumes, v, and temperatures, t, are expressed in Eqs. (10)–(11).

$$W_i = \frac{P_{mean} V_E \pi c(1-t)\sin a}{1+\sqrt{1-c^2}} \tag{5}$$

$$a = \tan^{-1}\frac{v \sin dx}{t + \cos dx - 1} \tag{6}$$

$$c = B\big/S \tag{7}$$

$$B = \sqrt{t^2 + 2tv\cos\phi + v^2} \tag{8}$$

$$S = t + 2tX_{DE} + (4tX_R/1+t) + v + 2X_{DC} \tag{9}$$

$$t = \frac{T_C}{T_E} \tag{10}$$

$$v = \frac{V_c}{V_E} \tag{11}$$

The indicated power of the Stirling engine is finally defined in Eq. (4) where n is the engine speed (rps).

$$\mathrm{P_i} = W_i \times n \tag{12}$$

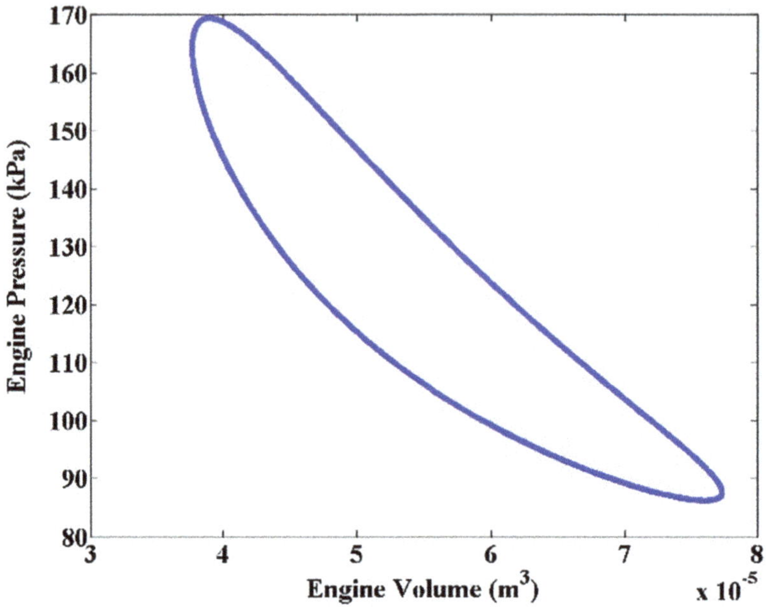

Fig. 16 PV diagram with Schmidt's analysis

The Stirling engine prototype was theoretically investigated using Schmidt's theory. Work can be evaluated from PV diagram as in Fig. 16. The maximum and minimum pressures were 169 kPa and 86.4 kPa, respectively. The indicated energy per one revolution of this engine is 0.743 J. Based on the prototype of Chalermpong et al. [10], the engine speed of 8 rps and using Eq. (12), the indicated power of engine is 6 W.

3 Prototype and Testing

For the proof of modification of concept device, the prototype engine with the adapted Ross yoke was built as shown in Fig. 17a. Assembly of engine and cooling system are arranged in Fig. 17b. The slot grooves were constructed on the yoke represented in Fig. 18a and assembled with rolling pins coupling piston rods in Fig. 18b.

3.1 Experimental Set up

Figure 19 shows the engine and experimental set up. The preliminary tests were conducted in order to verify engine operation. Unpressurized engine was primary

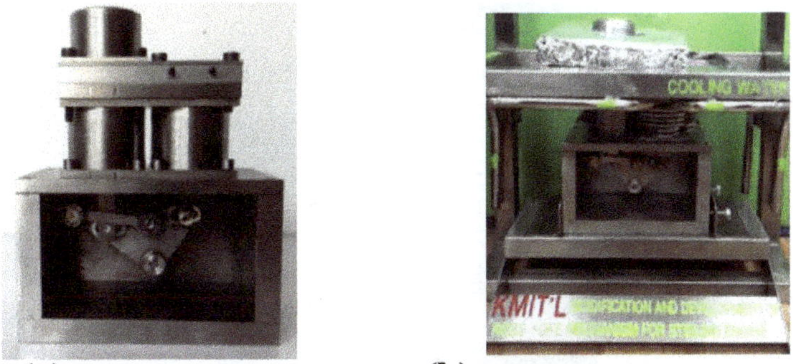

(a) Prototype engine **(b)** Prototype with cooling system

Fig. 17 Structure of Ross yoke prototype

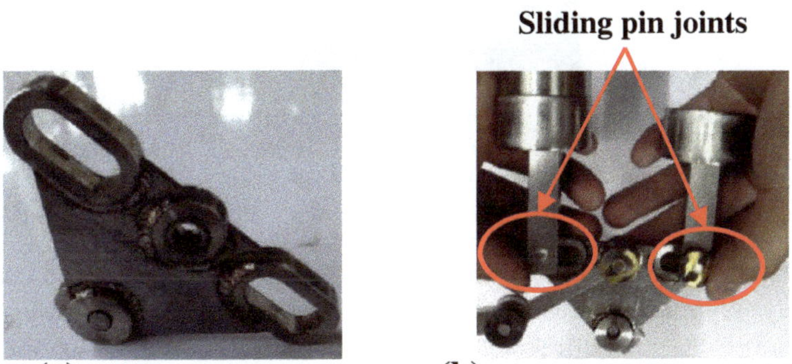

(a) Modified Ross yoke **(b)** Assembly of sliding pin joints

Fig. 18 Slot-pin joints on Ross yoke

investigated. A gas burner was equipped with LPG tank, pressure regulator, and adjusted valve. Gas flow rate was measured by the gas flowmeter model ST75 V. LPG fueled the engine with 0.6 kg/h. The inlet and outlet temperatures of cooling water were probed by K-type thermocouples. Engine speed is detected by digital tachometer for rotational revolution observation. Brake power and torque were measured by rope brake dynamometer. It consists of a rope wound around the rotating drum attached to the output shaft. Weight is applied on the opposite direction of flywheel rotation. One side of the rope is connected to a digital scale and the other side to a loading device or weight. The power is absorbed in friction between the rope and the drum.

3.2 Experimental Results

The prototype was built and tested for the concept proof of the modified Ross yoke mechanism. The engine has swept volume of 25 cm^3. LPG was fueled at flow rate

Fig. 19 Rope brake dynamometer test set up on the prototype

of 0.6 kg/h. The ambient air and water are used as the working gas and coolant, respectively. With the sufficient slot length on a new Ross yoke, the piston rods inside the engine can be operated in linear motion. Figure 18 shows the variation of engine power with speed. The variation of power on engine speed is shown in Fig. 20. Power was increased with a speed of 250 rpm until the maximum power of 0.55 W was achieved at 480 rpm and then reduced with the higher speed.

Figure 21 shows the variation of engine torque with speed. The maximum engine torques of 0.014 N m at 260 rpm and decrease with increasing of engine speed.

4 Summary and Conclusions

The objective of this modification is to prove that the converted joints on Ross yoke are working properly. The modified Ross yoke mechanism of an alpha-type Stirling engine was numerically analyzed for an appropriate length of slot joints yielding linear translation of piston rods. The numerical results showed that the proper slot length on the modified Ross yoke part is at least 5 mm each for this engine scale. The stroke of piston is 21.6 mm. The indicated energy and power using Schmidt's theory are 0.743 J and 6 W, respectively. The Stirling engine prototype with modified Ross yoke was tested and engine characteristic and performance were evaluated. The prototype was heated by LPG burner and cooled by water. The highest power of the engine was 0.55 W at engine speed of 486 rpm while the maximum torque was 0.014

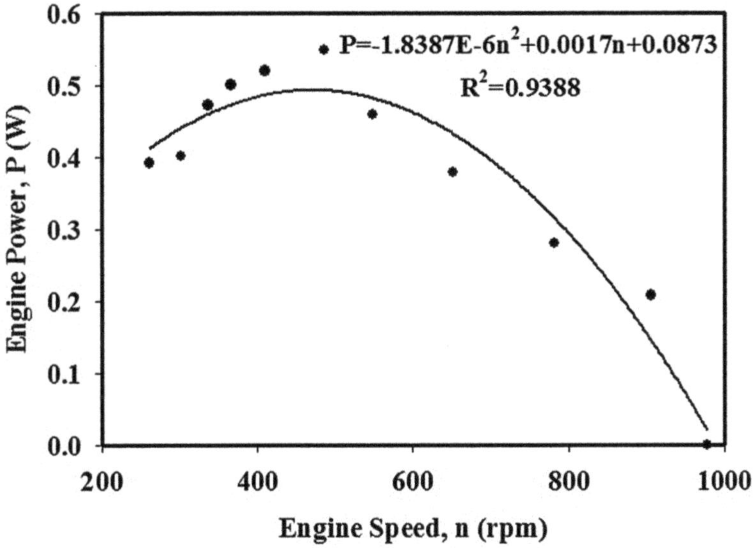

Fig. 20 Power of the engine with speed

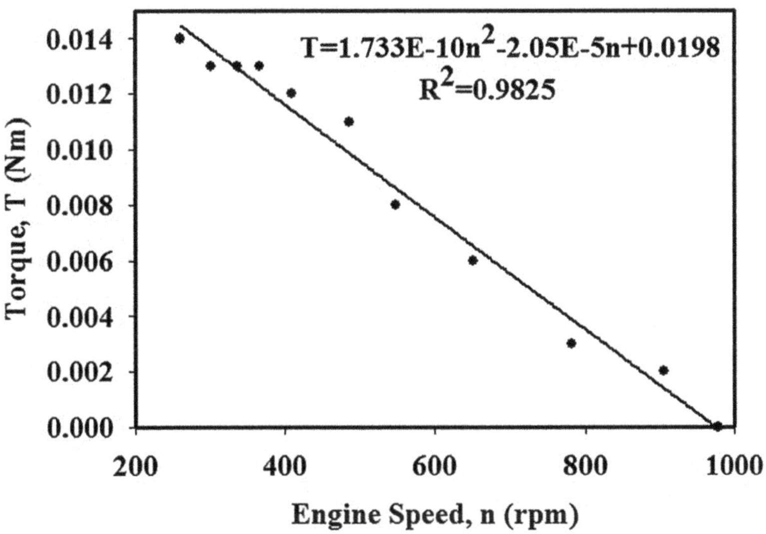

Fig. 21 Torque versus engine speed

Nm at engine speed of 260 rpm. The highest engine speed was achieved at 977 rpm. Both hot and cold piston rods were reciprocated smoothly and in linear motion. Further experiments on pressurized engine will be conducted in order to investigate and improve engine performance.

Acknowledgements The author gratefully acknowledges the financial support provided by the Office of the Higher Education Commission (OHEC) in Thailand and Research team in Green Energy Engine Research (GEER) Laboratory. Thanks are also due to the Department of Mechanical engineering and the Department of Agricultural Engineering, Faculty of Engineering, King Mongkut's Institute of Technology Ladkrabang, for their kind cooperation.

References

1. G. Walker, *Stirling Engine* (Oxford University Press, 1980)
2. A. Ross, *Making Stirling Engine*, 1997
3. V.M. Homutescu, D.-T. Bălănescu, Optimization of diameter Ratio for Alpha-Type Stirling engines (2010). http://www.agir.ro/buletine/808.pdf. Accessed 5 June 2016
4. I. Tlili, S.A. Musmar, Thermodynamic evaluation of a second order simulation for Yoke Ross Stirling engine. Energy Convers. Manag. **68**, 149–160 (2013)
5. N. Martaj, P. Rochelle, 1D modelling of an alpha type Stirling engine. Int. J. Simul. Multi. Des. Optim. **5**, A07 (2014)
6. L.S. Scollo et al., Twin cylinder alpha Stirling engine combined model and prototype redesign. Int. J. Hydrog. Energy **38**, 1988–1996 (2013)
7. B. Silpsakoolsook, et al., Manufacturing of an Alpha type Stirling engine, in *Proceedings of the 22th Conference of Mechanical Engineering Network of Thailand*, 2008
8. OHIO University, Volume variations—Ross Yoke-drive engine (2011). https://www.ohio.edu/mechanical/stirling/engines/yoke_vol.html. Accessed 7 Sept 2016
9. K. Hirata, Schmidt Theory for Stirling Engines (2002). https://www.nmri.go.jp/eng/khirata/stirling/schmidt/schmidt_e.pdf. Accessed 22 Sept 2016
10. K. Chalermpong et al., Modification and development of Ross Yoke Mechanism for Stirling engine. thesis, King Mongkut's Institute of Technology Ladkrabang, 2013

Part II
Flame and Combustor Dynamics

Dynamic Systems Approach for Laminar Ducted Flames

Uddalok Sen, Achintya Mukhopadhyay and Swarnendu Sen

Abstract Combustion in physical systems is always affected by dynamic instabilities, most of them being of thermoacoustic origin. The interaction of the acoustics of the flame enclosure and the unsteady heat release from the flame are responsible for such instabilities. During such unstable operation, the flame often changes its dynamic state, with transition across several dynamic states being also quite common. The present chapter presents a brief review on the recent developments of dynamic systems approach applied to laminar ducted flames. The different tools of nonlinear time series analysis that are commonly used for this purpose have been described. Representative case studies of the technique applied to ducted non-premixed, premixed, and inverse diffusion flames have been presented. Finally, the promising nature of the complex networks-based approach for dynamic characterization of combustion systems has been highlighted.

1 Introduction

Combustion is the most widely used source of energy in the industrial world, it being the primary heat release mechanism in land-based and aviation gas turbines, tanks, maritime vessels, boilers, and furnaces. Most combustion applications require the flame to be enclosed by a combustion chamber, which often leads to the coupling of the heat release from the flame and the chamber acoustics. Such flame–acoustic interactions lead to self-excited oscillations within the combustor—oscillations that are detrimental to the performance as well as the structural integrity of the device.

U. Sen
Department of Mechanical and Industrial Engineering, University of Illinois at Chicago, Chicago, IL 60607, USA
e-mail: uddalok.sen.us@gmail.com

A. Mukhopadhyay (✉) · S. Sen
Mechanical Engineering Department, Jadavpur University, Kolkata 700032, India
e-mail: achintya.mukho@gmail.com

S. Sen
e-mail: sen.swarnendu@gmail.com

A. K. Runchal et al. (eds.), *Energy for Propulsion*, Green Energy and Technology,
https://doi.org/10.1007/978-981-10-7473-8_5

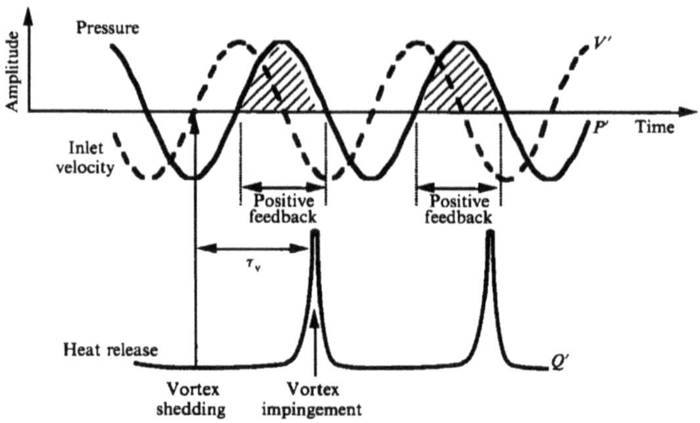

Fig. 1 Positive feedback loop established between duct acoustics and unsteady heat release results in the increase in amplitude of the vibrations, as predicted by the Rayleigh criterion. The figure has been reproduced, with permission from [7]

Researchers have studied combustion instabilities in land-based gas turbines [1], aeroengines [2], premixed gas turbine combustors [3], aeroengine afterburners [4, 5], rocket motors [6], ramjets [7], boilers, and furnaces. Lord Rayleigh [8] postulated that an unsteady heat release in phase with the chamber acoustics increases the amplitude of the acoustic vibrations. In a practical situation, an unsteady heat release is commonly caused due to local changes in the fuel–air ratio [3] and/or hydrodynamic instabilities [9, 10]. Both these factors, however, can be affected by acoustic pressure fluctuations. The most general setup used in the laboratory scale for studying thermoacoustic oscillations is the ducted flame, where a flame is enclosed by a coaxial cylindrical duct—the duct mimicking the chamber. A positive feedback loop (Fig. 1) is established when a flow or combustion process couples with one or more modes of the duct acoustics [11], which leads to self-sustained pressure oscillations of large amplitudes [12].

Nonlinear interactions between acoustics and the flame cause thermoacoustic instability [13], which justifies the use of tools of nonlinear dynamics for the study of combustion instabilities. However, the combustion process itself is highly nonlinear, with the difficulties of computation being compounded several folds for turbulent flames. The propagation velocity for a laminar flame is governed by the thermal and chemical properties of the reactants, whereas for a turbulent flame, in addition to those two factors, the flame propagation velocity depends on the character of the flow as well [14]. Hence, as a starting platform, researchers [15, 16] have focused on the dynamic characteristics of the flame–acoustic interaction of laminar flames in the presence of an enclosing duct. One advantage of laminar flames is the absence of inherent fluctuations of turbulent flows, which simplifies the characterization of the system dynamics. However, studies [17, 18] also exist where the same tools have been used for analyzing turbulent combustion and spray combustion [19] in the laboratory scale. Moreover, dynamical systems analysis has also been employed for investigating the dynamic characteristics of thermal pulse combustors [20, 21] and detonations [22], which highlights the versatility of such an approach. The tools being referred to here are extremely robust, and besides theoretical studies [23, 24], they have been shown to perform exceptionally well for experimental combustion systems [25–27] also. Several review articles [6, 12, 28–30] already exist in the literature that highlight the salient developments in this field.

In the present chapter, a brief review on the developments of dynamical systems analysis applied to laminar ducted flames is presented. The next section is an introduction into the various commonly used tools of nonlinear time series analysis, followed by a section on some representative case studies of non-premixed, premixed, and inverse diffusion flames. The chapter finally concludes by summarizing the key findings and also highlighting an emerging direction in the application of dynamics systems analysis to combustion systems—the complex networks-based approach.

2 Nonlinear Time Series Analysis

Nonlinear time series analysis, an essential toolbox of dynamical systems analysis, has been used for several years in studying combustion systems. Linear methods such as autocorrelation and power spectra often fail to reveal any significant information about the underlying physics of the system. Nonlinear time series analysis, on the other hand, projects the entire dynamics of the complex system onto the phase space or the state space. Certain invariants are calculated based on the projection that allows one to form an idea of the dynamical system and also sheds light on the governing equations [31]. The only limitation of such methods is the requirement of determinism, which can be easily tested using established methods [32]. In the present section, some of the predominant tools of nonlinear dynamics used for studying laminar ducted flames have been introduced.

2.1 Phase Space Reconstruction

The primary requisite of using nonlinear time series analysis is to represent the time series data into an appropriate phase space. However, two major roadblocks for such representation, particularly for experimental data, are the lack of the knowledge of the number of governing variables or degrees of freedom of the system, and the extreme cost of sensors and data acquisition systems to experimentally measure each of those variables [33], if at all possible. Takens' embedding theorem [34] provides an effective work-around by allowing one to use the time series of a single experimentally observed variable to reconstruct the dynamics of a multidimensional, multivariable system. The phase space is reconstructed first by forming a vector $x(t), x(t+\tau), x(t+2\tau), \ldots, x(t+(d_E-1)\tau)$ from the time (t) series of the experimentally observed variable, x. The optimum delay time, τ, and the embedding dimension, d_E, are the two unknowns in the above reconstruction. Several methods exist for the determination of the optimum time delay—the two most popular being the autocorrelation function [35] and the average mutual information [36]. The autocorrelation function is defined as

$$C_L(\tau) = \frac{\frac{1}{N}\sum_{m=1}^{N}(x(m+\tau)-\bar{x})(x(m)-\bar{x})}{\frac{1}{N}\sum_{m=1}^{N}(x(m)-\bar{x})^2} \tag{1}$$

where N is the total number of data points in the time series and $\bar{x} = \frac{1}{N}\sum_{m=1}^{N} x(m)$. The optimum time delay, τ, is determined by the first value of τ at which the autocorrelation function vanishes. The average mutual information, on the other hand, is calculated as

$$I(\tau) = \sum_{i=1}^{N} P(x_i(t), x_i(t+\tau)) log_2 \frac{P(x_i(t), x_i(t+\tau))}{P(x_i(t))P(x_i(t+\tau))} \tag{2}$$

where each time delayed vector is considered as an experiment with the individual probabilities $P(x_i(t))$, $P(x_i(t+\tau))$ and the joint probability $P(x_i(t), x_i(t+\tau))$ being calculated. The optimum delay time, τ, corresponds to the location of the first minimum of $I(\tau)$.

Phase space reconstruction is essentially the mapping of the original multidimensional phase space onto a subspace of reduced dimensions such that the system invariants remain constant. However, such reconstruction is always governed by the optimum time delay and the embedding dimension. Two popular methods for determination of the embedding dimension are the Grassberger–Procaccia algorithm [37] and the method of false nearest neighbors [38]. The Grassberger–Procaccia algorithm proceeds by first calculating the correlation integral. The neighboring vectors, $x(j)$, of every vector, $x(i)$, within the m-dimensional hypersphere of radius ε are identified, and the correlation integral is calculated as

$$C^m(\varepsilon) = \frac{1}{N^2}\sum_{\substack{i,j=1 \\ i\neq j}}^{N} H(\varepsilon - ||\mathbf{x}_i - \mathbf{x}_j||) \tag{3}$$

where $H(\cdot)$ represents the Heaviside function and $||\mathbf{x}_i - \mathbf{x}_j||$ denotes the Euclidean distance between the two vectors $x(i)$ and $x(j)$. The correlation integral, $C^m(\varepsilon)$, scales with the hypersphere radius, ε, as $C^m(\varepsilon) \sim \varepsilon^{D_C(m)}$, where $D_C(m)$ corresponds to the correlation dimension for the chosen dimension m. Hence, the correlation dimension can be obtained as

$$D_C(m) = \lim_{\varepsilon \to 0} \frac{log C^m(\varepsilon)}{log \varepsilon} \tag{4}$$

$D_C(m)$ is plotted against m, and the value of m where $D_C(m)$ saturates is identified as the embedding dimension d_E. The method of false nearest neighbors, on the other hand, is based on the fact that the unfolding of an attractor onto a subspace having dimension less than d_E results in false crossing of trajectories, leading to a large number of false nearest neighbors. A gradual increase in the subspace dimension results in a decrease of the false nearest neighbors until at a particular dimension, the percentage of false nearest neighbors becomes zero. That dimension is chosen to be the embedding dimension of the time series.

2.2 Determination of Chaos

Thermoacoustic oscillations often exhibit chaotic behavior [39], where the dynamics of the system are extremely sensitive to initial conditions. It is often useful to determine whether a time series relates to chaotic behavior or not, and if so, what is the route to chaos that the system follows. A fractional value of the correlation dimension obtained from the Grassberger–Procaccia algorithm confirms the presence of a strange attractor, with only the possibility of the system actually exhibiting chaotic dynamics. Confirmation of chaotic behavior of a time series can be achieved by calculating its Lyapunov exponents [40]. The Lyapunov exponents measure the exponential divergence, in time, of two neighboring phase space trajectories, and a positive value implies that the two trajectories, in fact, diverge from each other as time progresses, thus indicating chaotic behavior. The average separation between neighboring trajectories in the reconstructed phase space is calculated, and an exponential trend is searched for in the average separation. The average separation, calculated as a function of the temporal separation Δn, is given by [41]

$$S(\Delta n) = \frac{1}{\tau} \sum_{t=1}^{\tau} log \left(\frac{1}{|\mathcal{U}_t|} \sum_{i \in \mathcal{U}_t} |\mathbf{x}_{t+\Delta n} - \mathbf{x}_{i+\Delta n}| \right) \tag{5}$$

where $\mathcal{U}_t$ denotes the neighborhood of any point $\mathbf{x}_t$ in the phase space. The maximal Lyapunov exponent is obtained by noting the slope of the variation of $S(\Delta n)$ with Δn. Another confirmatory test for the detection of chaos is the so-called $0 - 1$ test [42], which is not based on phase space reconstruction. The interested reader is referred to articles [43, 44] for the implementation of $0 - 1$ test to experimental data.

2.3 Recurrence Quantification Analysis (RQA)

Recurrence is inherently observed in any deterministic system and can be further utilized for its dynamical characterization. Recurrence plots [45] can be useful visual aids for estimation of the time instants at which a trajectory revisits approximately the same area in the vicinity of reconstructed phase space. The recurrence matrix, R, is defined as a binary matrix whose elements, R_{ij}, are equal to 1 if $\mathbf{x}_i$ is a neighbor of $\mathbf{x}_j$, and 0 otherwise. This can be expressed mathematically as [46]

$$R_{ij} = H(\delta - ||\mathbf{x}_i - \mathbf{x}_j||) \quad i, j = 1, 2, \ldots, M \tag{6}$$

where $M = N - d_E \tau f_s$ corresponds to the total number of data points in the reconstructed phase space, δ corresponds to the maximum distance between any two recurrent data points, and f_s is the sampling rate of data acquisition. R, therefore, is a square symmetric matrix with all its elements either 0 or 1. For visual representation of the recurrence matrix in the form of the recurrence plot, the zeros are represented as white dots and the ones are represented as black dots. It is to be noted here that $R_{ii} = 1$ due to self-recurrence, thus making the principal diagonal of any recurrence plot to be composed solely of black dots. Thus, the principal diagonal, in the context of recurrence plots, is also known as the Line of Identity (LOI). For a purely periodic signal, the recurrence plot consists of equispaced diagonal lines parallel to the LOI, with the horizontal or vertical spacings between the lines being representative of the time period. On the other hand, lines parallel to the LOI having unequal spacings represent quasiperiodic systems [47]. Disconnected short segments represent aperiodic behavior stemming from noise or chaos, while the vertical lines denote intermittent steady behavior or 'laminar' states [48]. RQA has been shown to be effective for a diverse range of applications such as flow identification in multiphase systems [49–51], analysis of financial markets [52, 53], in biomedical systems [54, 55], and modeling game sports [56]. Recurrence quantification has been used in experimental combustion studies by various researchers as well [57–64].

Recurrence quantification measures are metrics based on the line structures in the recurrence plots, which often reveal significant features about the dynamical system. The commonly used RQA measures are [46]:

Recurrence rate (RR):

$$RR = \frac{1}{M^2} \sum_{i,j=1}^{M} R_{ij} \tag{7}$$

The density of the black dots in the recurrence plot is denoted by the recurrence rate, which is indicative of the recurrence in the phase space.

Determinism (DET):

$$DET = \frac{\sum_{l=l_{min}}^{M} lP(l)}{\sum_{i,j=1}^{M} R_{ij}} \tag{8}$$

The fraction of recurrence points that form diagonal structures in the recurrence plot is given by the measure determinism, and it directly correlates to the periodic behavior of the system. $P(l)$ denotes the number of diagonal lines of length l, and the minimum number of consecutive points that form a diagonal line is denoted by l_{min}.

Average diagonal line length (L):

$$L = \frac{\sum_{l=l_{min}}^{M} lP(l)}{\sum_{l=l_{min}}^{M} P(l)} \tag{9}$$

The average time that two segments of a trajectory remain in proximity of each other can be estimated by the average diagonal line length. The divergence of trajectories is inversely related to the length of the diagonal line, which implies that the Lyapunov exponent is proportional to the inverse of the diagonal line lengths [16]. Hence, a decreasing L would signify gradually increasing chaotic behavior.

Laminarity (LAM):

$$LAM = \frac{\sum_{v=v_{min}}^{M} vP(v)}{\sum_{v=1}^{M} vP(v)} \tag{10}$$

The relative proportion of recurrence plots forming vertical structures is given by laminarity. Vertical structures are indicative of close proximity of states at consecutive time instants, which points to the presence of laminar states or intermittency. The notations indicate the same corresponding quantities for the vertical lines as the formulation for determinism.

Trapping time (TT):

$$TT = \frac{\sum_{v=v_{min}}^{M} vP(v)}{\sum_{v=v_{min}}^{M} P(v)} \tag{11}$$

Trapping time quantitatively estimates the average duration that the systems spend at a particular state, and is also representative of the average length of the vertical structures in the recurrence plot.

Entropy (ENT):

$$ENT = -\sum_{l=l_{min}}^{M} P(l)log(P(l)) \tag{12}$$

Entropy is based on the Shannon information entropy and reflects upon the complexity of the recurrence plot.

3 Case Studies

Some representative cases of dynamical systems analysis applied to the three major types of flame used in the literature are presented in this section. We discuss the non-premixed flames first, which is still the predominant mode of combustion in the industry, followed by the premixed or lean premixed flame, and finally the inverse diffusion flame—a flame whose hybrid nature is the reason behind its recent growth in popularity.

3.1 Non-premixed Flames

Non-premixed or diffusion flames are the predominant types of flame used in practice. Their high stability and relatively large operating limits make them attractive for applications where reliability is paramount. The existing literature [24, 65, 66] on thermoacoustic instabilities of ducted non-premixed flames suggests that such flames indicate several interesting dynamical features such as limit cycles and Hopf bifurcations [67]. In this context, Balasubramanian and Sujith [23], through a modeling approach, identified the presence of non-normality in the thermoacoustic oscillations of ducted laminar non-premixed flames and highlighted its consequences. A two-dimensional non-premixed flame in the Burke–Schumann configuration [14] was considered, with the combustion being modeled using the infinite rate chemistry thermodiffusive model [68], where the flame is considered as a thin sheet. The detailed problem geometry can be found in [23]. Duct acoustics was modeled as a one-dimensional field, and the heat release was obtained by solving for the Schvab–Zeldovich variable [14]. The Schvab-Zeldovich variable is governed by an advection–diffusion equation, the operator of which is known to be non-normal [69, 70]. A non-normal operator and its adjoint do not commute [71] and have non-orthogonal eigenvectors, which leads to transient growth of oscillations in the short term before they eventually decay in the long term. Linear stability analysis fails to predict stable operation limits in case of non-normal systems [72], where non-linearity closes the system feedback loop by converting a part of the system output into input [73]. Non-normality in physical systems is not uncommon and has been also observed in case of instability of magnetic plasmas [74], formation of cyclones [75], and in transition to turbulence [73]. The Galerkin technique [76] was used to convert the partial differential equations in their system to ordinary differential equations, and the fourth-order Runge–Kutta scheme was used to numerically solve them. The linearized evolution equations were found to be non-normal, and they yield non-orthogonal eigenvectors responsible for transient growth.

The acoustic modes were found to be orthogonal in an isothermal situation. However, in the presence of combustion, velocity perturbations result in unsteady combustion, which in turn instigate acoustic oscillations. The in-phase mode of acoustic oscillations is driven by combustion oscillations, as defined by the Rayleigh criterion [8].

However, the phase lag between these two oscillations is not constant and evolves in time depending upon the phase lag between the two said fields at some earlier time. Hence, the mode which is excited at a particular instant of time depends upon some other mode that was excited at an earlier time, leading to the non-orthogonal behavior of the eigenmodes. The authors attributed the non-normality of the combustion process to entropy disturbances moving from regions of high entropy to low entropy. Moreover, the system was found to be non-normal in addition to it being nonlinear. This indicates that even if individual eigenvalues indicate linear stability, the system as a whole may not be so.

The time evolutions of the non-dimensional acoustic velocity ($u\prime$) and the non-dimensional heat release rate ($Q\prime$) for a case where the initial conditions lead to decay of oscillations have been presented in Fig. 2a and 2b, respectively. The waveforms clearly indicate that although there are instances of transient growth, those are not strong enough to trigger nonlinearities. The phase portrait (Fig. 2c) represents an inward spiral, which points to an almost linear response of the heat release rate to the velocity fluctuations. The elliptical shape also indicates the presence of a phase difference. For a different set of initial conditions, the time evolutions of the non-dimensional acoustic velocity and the non-dimensional heat release rate are shown in Fig. 2d and Fig. 2e, respectively. The situation is entirely different in this case compared to Fig. 2a, b. The transient growth occurring in this case is strong enough to trigger the nonlinear instabilities, which leads to growth of oscillations. The phase portrait (Fig. 2f) exhibits fractal behavior with folds, similar to the Rössler bands [67]. Such folds were also observed for premixed flames by Fichera et al. [77]. One must mention in this situation that classical linear stability analysis based on the eigenvalues would have shown all the eigenmodes of the coupled system to be stable.

Orthogonal eigenmodes do not exchange energy among themselves, but non-normal systems do redistribute energy among the various modes since the modes interact with each other. The time evolution of the dimensionless acoustic velocity is shown for a particular case in Fig. 3a. The projections of the acoustic energy onto the first three Galerkin modes are shown in Fig. 3b–d. The decay of energy of the first mode coincides with the injection of energy onto the second and third modes. After sufficient exchange of energy in this direction, the second and third modes inject energy back into the first mode, resulting in growth of energy of the first mode. This phenomenon is known as 'bootstrapping' and has been observed in the context of turbulence [73, 78]. Balasubramanian and Sujith [23] also investigated the evolution of thermoacoustic oscillations in the presence of acoustic damping. The damping model proposed by Matveev [79] was used, and it was shown that triggered oscillations also grow initially in the presence of damping but ultimately saturate, whereas un-triggered oscillations decay when damped.

Triggering is, in fact, quite common in experimental combustion systems, which occurs when a combustor operating in an otherwise stable regime exhibits self-sustained oscillations in the presence of small finite amplitude perturbations. In the context of dynamic systems, this phenomenon is termed as triggering instability or subcritical transition to instability. For any system that exhibits such triggering

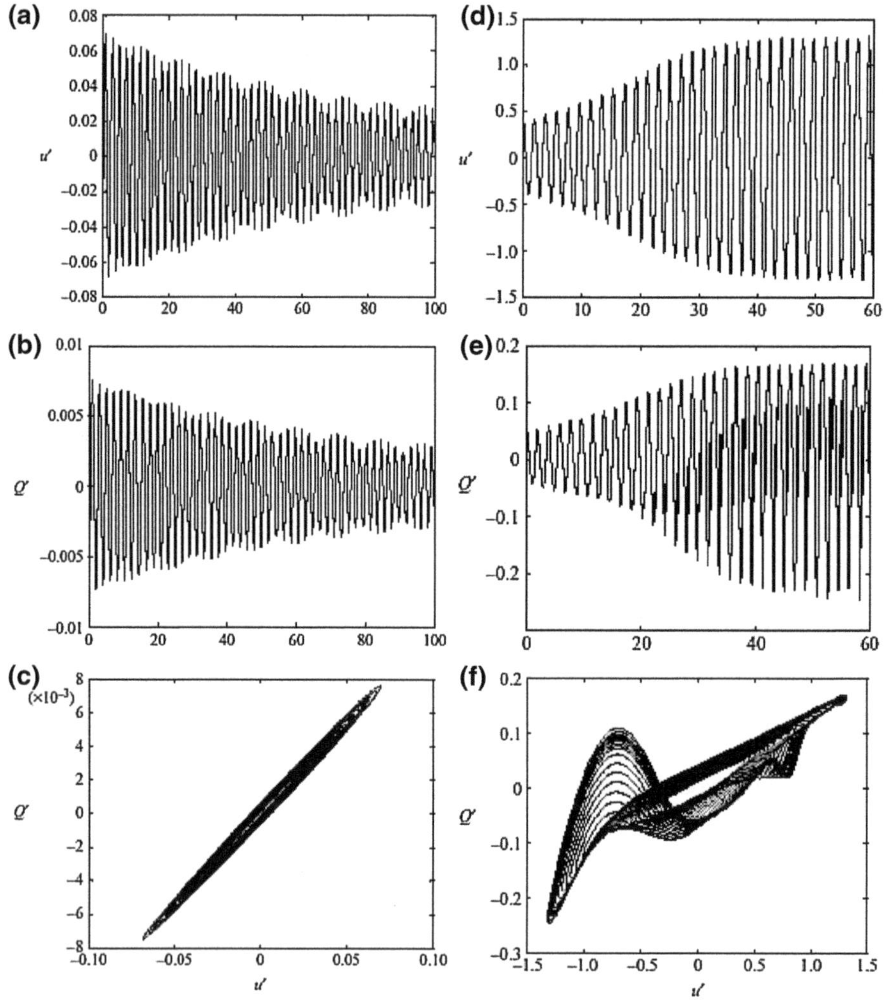

Fig. 2 Evolution of **a** acoustic velocity, **b** heat release rate, and **c** phase portrait when the initial conditions correspond to decay of oscillations; the evolution of **d** acoustic velocity, **e** heat release rate, and **f** phase portrait when the initial conditions correspond to growth of oscillations. The figures have been reproduced, with permission from [23]

instability, there exists a threshold amplitude of perturbation ('triggering amplitude'), below which the system asymptotically approaches a stable state and above which the system exhibits self-sustained oscillations. Kim and Hochgreb [81] and Noiray et al. [82] have also observed triggering instabilities in experimental combustion systems, while Juniper [83] identified that the initial energy required for triggering is significantly less than the energy of the oscillations itself. Triggering may occur in combustion systems due to the noise within the system itself, termed as 'noise-induced

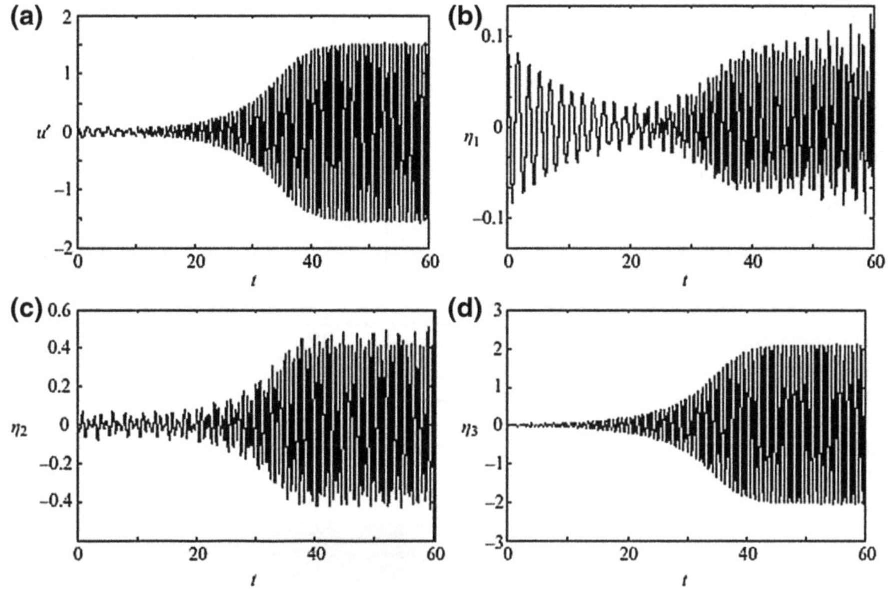

Fig. 3 **a** Evolution of the acoustic velocity at the flame location with time; **b–d** the projection of the acoustic velocity onto the first three Galerkin modes. The figures have been reproduced, without permission from [23]

triggering' (NIT). As mentioned before, thermoacoustic systems are inherently non-normal, which also happens to be an important factor for noise-induced triggering [84]. Noise-induced triggering in thermoacoustic systems has already been shown by other researchers [85, 86], where the window of stable operation for a combustor was found to decrease with increasing noise level [87]. Jegadeesan and Sujith [80] performed experiments to study the dynamic states associated with noise-induced triggering in experimental combustion systems. A subwoofer was used to perturb the fuel flow rate of a ducted laminar non-premixed Burke–Schumann type flame, the experimental setup being described in detail in [80]. A sinusoidal wave at the resonant frequency of the system was used as the perturbation field. The bifurcation diagram of the acoustic pressure (Fig. 4a) clearly indicates the hysteretic behavior of the system, with the transitions being a subcritical Hopf bifurcation followed by a saddle-node bifurcation. The oscillating state is a limit cycle, with triggering instability being observed in the bistable region within the hysteresis loop. The pressure time series in Fig. 4b clearly highlights the role played by the threshold amplitude or the triggering amplitude during instability. When the amplitude of disturbance is below the triggering amplitude, the system asymptotically approaches a stable state, as evident from the phase portrait (Fig. 4d). The system in this case tries to attain the threshold loop but only manages to hover in its vicinity before spiralling back to the fixed point. The oscillations, on the other hand, grow if the system is perturbed at the triggering amplitude (Fig. 4b). Upon forcing, the system migrates from the stable

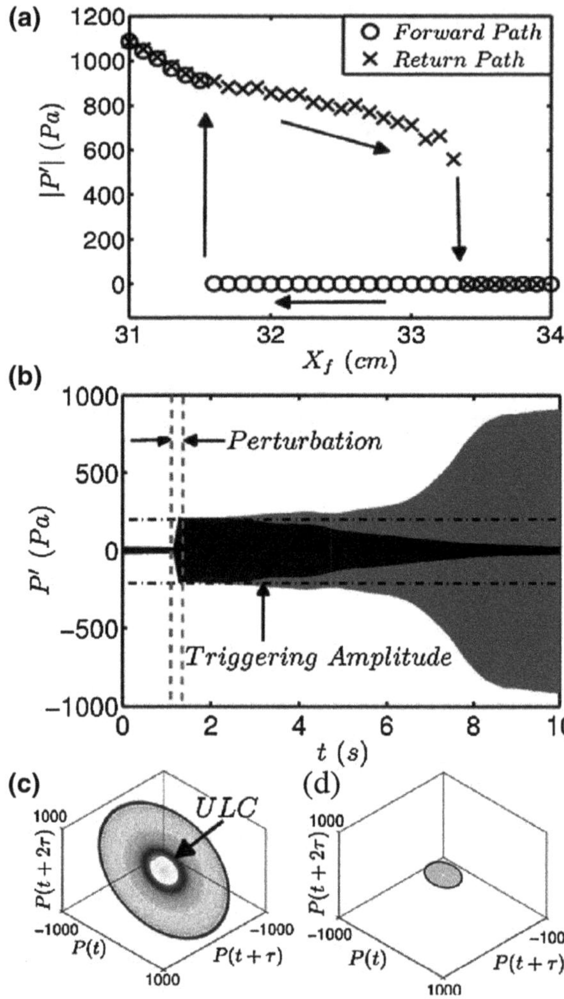

Fig. 4 **a** Bifurcation diagram for acoustic pressure; **b** gray pressure trace represents system evolving to a limit cycle, while black pressure trace represents system evolving to a fixed point; **c** phase portrait of a system that just exceeds the triggering threshold, evolving to a limit cycle, with 'ULC' signifying unstable limit cycle; **d** phase portrait of a system asymptotically reaching a stable state. The figures have been reproduced, with permission from [80]

fixed point and spirals toward the inner loop in the phase portrait (Fig. 4c) before spiralling out to a stable limit cycle again. Juniper [83] proposed that upon triggering, a transient growth toward an unstable limit cycle (ULC) is observed before eventual growth to a stable limit cycle, which also happens to be the case in the experiments of Jegadeesan and Sujith [80]. The unstable limit cycle lies on the surface of the basin boundary separating the steady state and the stable limit cycle [88]. An initial perturbation outside the basin of attraction forces the system to limit cycle oscillations, while one within the basin of attraction results in decay of oscillations.

3.2 Premixed Flames

Although non-premixed diffusion flames are quite robust toward dynamic instabilities, they generate significant amounts of pollutants such as NOx and soot. Increasingly, stringent environmental norms, however, impose strict restrictions on the operating limits of such flames. The premixed or lean premixed flame provides a viable alternative since they have significantly less NOx emissions. However, lean premixed flames are more prone to thermoacoustic instability and lean blowout [89, 90].

Kabiraj and Sujith [16] experimentally investigated the nonlinear thermoacoustic oscillations of a ducted conical laminar premixed flame. Details of the experimental setup and procedure are provided in [16]. The flame location with respect to the duct was varied, and the pressure oscillations at each location were measured by a pressure sensor to generate a bifurcation plot (as shown in [16]).

Starting from a steady-state condition, the flame location was gradually varied and the amplitude of the local maxima of the acoustic pressure time series at each location is recorded. The reconstructed phase plot (Fig. 5a) having a single loop, and the frequency spectrum (Fig. 5d) showing a single frequency and its higher harmonics, confirm limit cycle oscillations. This zone of operation (in terms of flame position) is denoted as regime 'A' in the bifurcation plot. The bifurcation plot shows that on increasing the flame location to 56.5 cm, a subcritical Hopf bifurcation [67] is observed. At 62 cm, another bifurcation occurs, followed by another bifurcation at 63.5 cm. At each of those bifurcations, a spread in the pressure amplitude was observed, with the width of the spread increasing with each subsequent bifurcation. The reconstructed phase plot (Fig. 5b) shows a toroidal structure with the trajectory never returning to the initial state, while the frequency spectrum (Fig. 5e) shows the

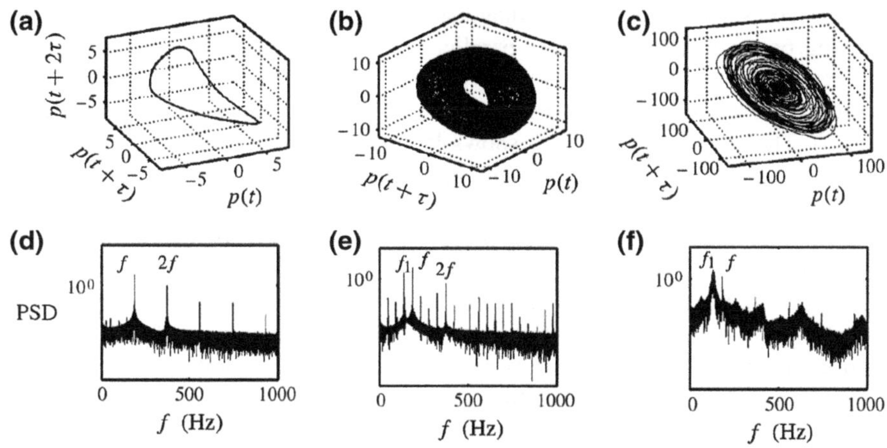

Fig. 5 **a–c** Reconstructed phase space attractor and **b–d** corresponding frequency spectrum for regions A, B, and C of the bifurcation plot, as shown in [16]. The figures have been reproduced, with permission from [16]

presence of multiple incommensurate frequencies, which indicate that the system is, in fact, in a quasiperiodic state. A quasiperiodic state emerges from a limit cycle due to a secondary Hopf bifurcation, also known as the Neimark-Sacker bifurcation [91]. This regime has been referred to as regime 'B' in the bifurcation plot [16]. Beyond 63.5 cm, the distribution of the pressure amplitudes on the bifurcation plot was observed to be highly irregular, and no further bifurcations were observed. This type of behavior of the flame continued until blowout occurred due to violent oscillations in the acoustic pressure and the flame. This regime is marked as 'C' in the bifurcation plot. Regime 'C' is, in fact, characterized by irregular and 'intermittent' bursts. The appearance of alternating intervals of chaotic/burst state and steady/periodic behavior is known in the literature as 'intermittency' and has been observed in other physical systems such as binary fluid convection [92], biochemical systems [93], coupled neural oscillators [94], and pH oscillators [95]. Moreover, intermittent behavior has been observed in other combustion studies [96–98] as well. Pomeau and Manneville [99] classified intermittent behavior into three types based on the type of transition. A type I intermittency is related to a saddle-node bifurcation, a type II intermittency is the result of a Hopf bifurcation and is observed in tandem with the appearance of a quasiperiodic state, and a type III intermittency occurs during a reverse period-doubling bifurcation [100]. The frequency spectrum (Fig. 5f) corresponding to regime 'C' of the bifurcation plot shows a broadband spectrum, while the reconstructed phase space (Fig. 5c) exhibits two dense, concentric disks. The authors postulated that the trajectories are pushed to the outer disk during the bursts, followed by a re-injection back into the inner disk. Moreover, since the intermittency was observed as a quasiperiodic attractor disintegrates, the intermittency was identified to be of type II.

The recurrence plots corresponding to regimes A, B, and C, along with the time series of the pressure oscillations, have been shown in Fig. 6. The limit cycle oscillations are represented by equispaced diagonal lines parallel to the LOI in the recurrence plot (Fig. 6c), while diagonal lines parallel to the LOI but having unequal spacing represent the quasiperiodic oscillations (Fig. 6d). Dense black patches signifying the laminar states are observed in the recurrence plot for the intermittent oscillations (Fig. 6f). The time series (Fig. 6e) shows laminar states interrupted by oscillatory bursts, the system dynamics deviating significantly from recurrence during these bursts. A zoomed-in view of the recurrence plot (Fig. 6g) shows the black patches in the recurrence plot being composed of kite-like structures, which are indicative of type II intermittency [101].

Apart from investigating the dynamics of thermoacoustic instability for a single ducted laminar premixed flame, Kabiraj et al. [39, 102] experimented with multiple premixed laminar flames enclosed by a duct, the burner configuration being one that was used by other researchers [82, 103] as well. As the flame position within the duct was changed, the system was observed to transition from a steady state to a periodic state via a subcritical Hopf bifurcation. The periodic state underwent a Neimark–Sacker bifurcation to a quasiperiodic state, which ultimately transitioned to chaotic oscillations through the Ruelle–Takens route [33, 91, 104], where a quasiperiodic torus breaks and chaotic oscillations ensue.

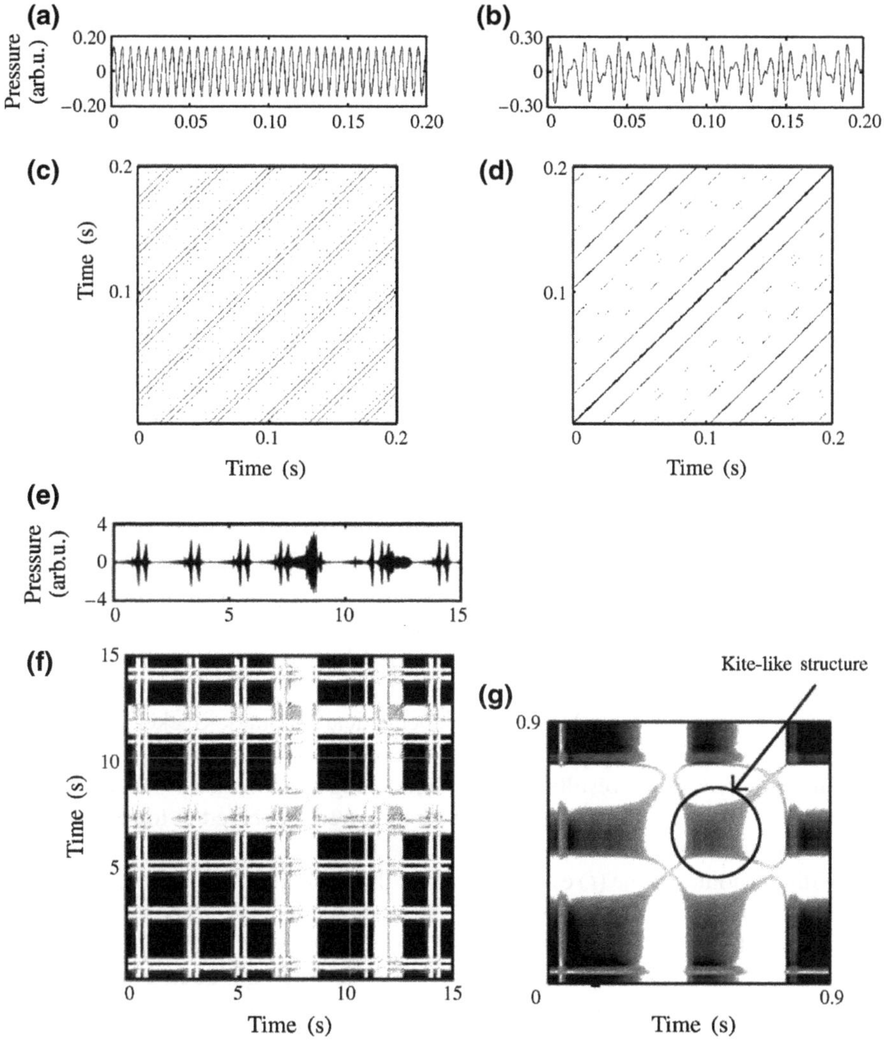

Fig. 6 **a** Pressure time series and **c** recurrence plot for limit cycle oscillations; **b** pressure time series and **d** recurrence plot for quasiperiodic oscillations; **e** pressure time series and **f** recurrence plot for intermittent oscillations; **g** kite-like structures observed in the recurrence plot for intermittent oscillations. The figures have been reproduced, with permission from [16]

Kashinath et al. [105] used a modeling approach to investigate the dynamic characteristics of ducted laminar premixed flames. The details of the problem formulation and solution procedure are provided in [105]. The acoustic field was modeled as one-dimensional, with the governing equations discretized using the Galerkin technique. The Matveev model [79] was used for the acoustic damping. The G-equation approach [14] was used to model the flame. The evolution equations were solved using a weighted essential non-oscillatory (WENO) fifth-order scheme [106] in space and a third-order total variation diminishing (TVD) Runge-Kutta scheme [107] in time. The bifurcation diagram of the dimensionless acoustic velocity (u_f) as the dimensionless flame location (x_f) is varied as shown in Fig. 7a. The bifurcation diagram was constructed by time marching simulations with the final state of the previous simulation being the initial condition for the next one. With decreasing x_f, the stable fixed point of mode 1 undergoes a subcritical Hopf bifurcation at $x_f = 0.48$ to an unstable period-1 oscillation, which is followed by a fold bifurcation of the unstable period-1 oscillation to a stable period-1 oscillation. The stable period-1 oscillation then transitions to quasiperiodic oscillations via a supercritical Neimark–Sacker bifurcation at $x_f = 0.438$, beyond which the mode 1 loses stability (at $x_f = 0.42$). A stable branch of mode 2 commences from a fold bifurcation to period-1 oscillations at $x_f = 0.456$, which further undergoes a supercritical Neimark–Sacker bifurcation to quasiperiodic oscillations at $x_f = 0.434$. For x_f values in the range of 0.415–0.422, the mode 2 branch possesses period-k states, while for $0.347 < x_f < 0.415$, quasiperiodic oscillations are observed in the system, which transition to period-1 oscillation at $x_f = 0.347$ via a supercritical Neimark–Sacker bifurcation. A subcritical period-doubling bifurcation followed by a fold bifurcation to period-2 oscillations at $x_f = 0.311$ results in the system losing stability at $x_f = 0.289$ with an abrupt end of the period-2 branch. Similar trends were observed at lower values of x_f, with chaotic bands for $0.03 < x_f < 0.1$ and $0.27 < x_f < 0.295$.

The dimensionless time (t) evolutions of the dimensionless acoustic velocity (u_f) and the dimensionless heat release rate (q) are shown in Fig. 7. The system is initially attracted toward an unstable period-1 oscillatory state in Fig. 7b ($30 < t < 60$), followed by a repulsion from it, before ultimately being attracted to a period-1 attractor ($180 < t$). In Fig. 7c, an unstable quasiperiodic attractor initially attracts the system ($20 < t < 40$) before repelling it, with the system finally being attracted to a period-2 attractor ($120 < t$). The system is attracted toward a quasiperiodic attractor in Fig. 7d, while it is first attracted toward an unstable quasiperiodic attractor ($15 < t < 35$) in Fig. 7e before being repelled from it, and finally being attracted toward a period-5 attractor ($50 < t$). Finally, the system tends toward a chaotic attractor in Fig. 7f.

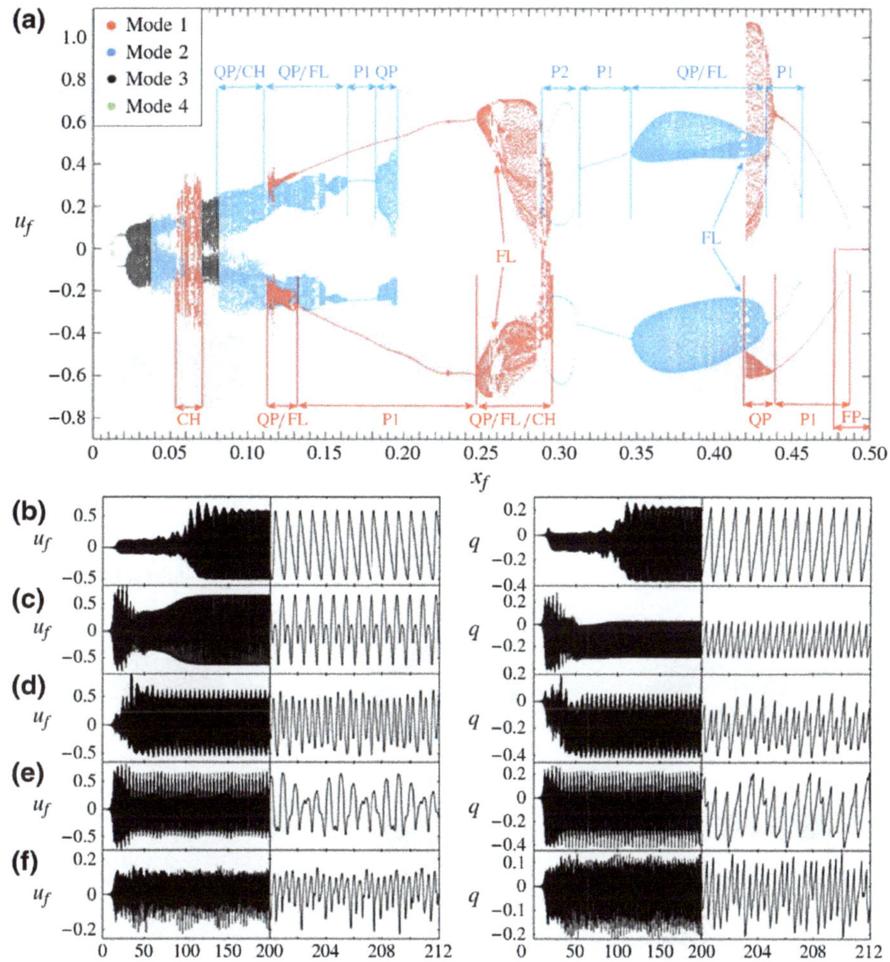

Fig. 7 **a** Bifurcation diagram with flame position as the control parameter, with salient features such as stable fixed point (FP), stable period-1 oscillations (P1), period-2 oscillations (P2), frequency-locked or period-k oscillations (FL), quasiperiodic oscillations (QP), and chaotic oscillations (CH) labeled; time series of velocity and heat release rate for **b** period-1 oscillations, **c** period-2 oscillations, **d** quasiperiodic oscillations, **e** period-k oscillations, and **f** chaotic oscillations. The figures have been reproduced, with permission from [105]

3.3 *Inverse Diffusion Flames*

The high emissions associated with diffusion flames and the relatively unstable operation of premixed flames have led to an impetus on the research on inverse diffusion flames, where a central air jet is surrounded by a coaxial fuel jet. The literature [108, 109] bears testimony to the fact that inverse diffusion flames have much less NOx and soot emissions, like a premixed flame, in addition to it having a wide stability range and adjustable flame lengths, akin to normal diffusion flames. Higher temperatures [110] and lower soot emissions [111], as compared to normal diffusion flames, have also been observed for inverse diffusion flames. These advantages have led to the inverse diffusion flame being used nowadays in rocket engines and staged combustion systems.

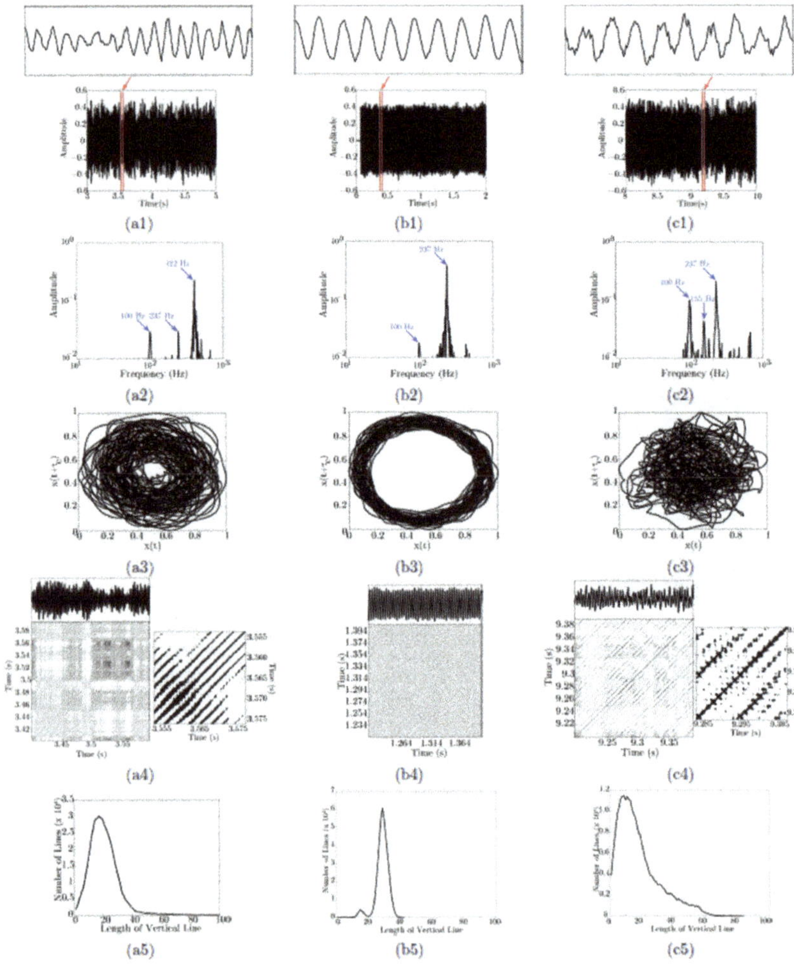

Fig. 8 (**a1**) Audio signal, (inset) high magnification, (**a2**) frequency spectrum, (**a3**) reconstructed phase portrait, (**a4**) recurrence plot, (inset) high magnification, (**a5**) frequency distribution of vertical lines in the recurrence plot at $x = 24$ cm and $\phi = 0.86$; (**b1**) audio signal, (inset) high magnification, (**b2**) frequency spectrum, (**b3**) reconstructed phase portrait, (**b4**) recurrence plot, (**b5**) frequency distribution of vertical lines in the recurrence plot at $x = 36$ cm and $\phi = 0.86$; (**c1**) audio signal, (inset) high magnification, (**c2**) frequency spectrum, (**c3**) reconstructed phase portrait, (**c4**) recurrence plot, (inset) high magnification, (**c5**) frequency distribution of vertical lines in the recurrence plot at $x = 50$ cm and $\phi = 0.86$. The figures have been reproduced, with permission from [98]

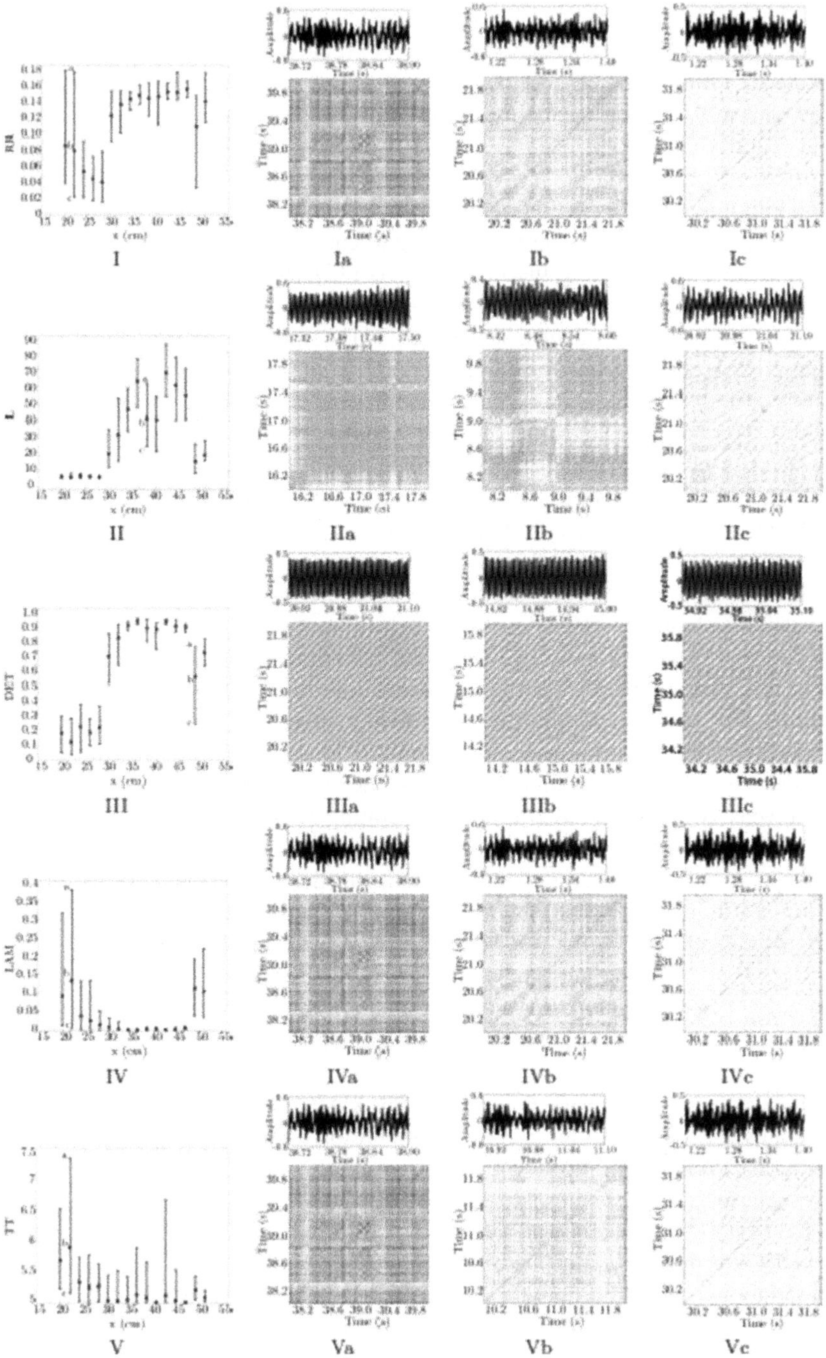

Fig. 9 Variation of **I**: recurrence rate, **II**: average diagonal line length, **III**: determinism, **IV**: laminarity, and **V**: trapping time with x; **a**, **b**, and **c** indicate the recurrence plots at the instants shown in the error bars. The figures have been reproduced, with permission from [98]

A recent study [98] investigated the dynamic characteristics of a laminar ducted inverse diffusion flame. The flame location (x) relative to the duct at a fixed global equivalence ratio (ϕ) and the flame equivalence ratio at a fixed flame position were varied, and tools of nonlinear time series analysis were employed on the acoustic signal captured from the system. For $x < 30$ cm, the sound emitted from the flame comprised of broadband noise, with the waveform (Fig. 8a1) comprising of both low and high amplitude oscillations characteristic of intermittent systems. The frequency spectrum (Fig. 8a2) shows three significant peaks at 100, 237, and 422 Hz. A theoretical analysis along the lines of [76, 112] was performed by the authors, and it was found that the 100 Hz frequency corresponds to the first isothermal eigenfrequency of the duct acoustics, while the 237 Hz signifies the first non-isothermal (in the presence of combustion) eigenfrequency, and the 422 Hz being the second non-isothermal eigenfrequency. The non-isothermal eigenfrequency appearing in the frequency spectrum is clear signs of flame–acoustic coupling being established. The reconstructed phase portrait (Fig. 8a3) points to the presence of a homoclinic orbit [33], which is formed when the stable and unstable manifolds of a system intersect. Homoclinic orbits are commonly observed in intermittent systems [113, 114] and are characterized by switching from one periodic orbit to another as the system evolves. Nair and Sujith [115] had also observed homoclinic orbits in experimental combustion systems having intermittent behavior, which is not surprising as these orbits appear in situations where there are phenomena that evolve over multiple time scales, and a flame exhibiting thermoacoustic oscillations clearly satisfies that condition. For such a flame, the acoustic field evolves in a time scale governed by the speed of sound, while the hydrodynamic field operates over multiple different time scales governed by the underlying flow phenomena. These two fields, however, exhibit weak coupling in the $x < 30$ cm operating range, leading to the homoclinic orbit. Confirmatory tests for the presence of a homoclinic orbit exist in the literature, one such being based on the passage time of trajectories near the saddle point. In case of a homoclinic orbit, given an initial distribution of points near the saddle point, the passage time will always adhere to a skewed distribution with an exponential tail [116, 117]—the aforementioned behavior being independent of the initial distribution of points [118]. For an experimental time series, the distribution of the lengths of the vertical lines in the recurrence plot directly correlates with the passage time of trajectories [115]. The recurrence plot (Fig. 8a4) for $x = 24$ cm at a global equivalence ratio (ϕ) of 0.86 shows dense black patches corresponding to the 'laminar' states [48], and the dense black structure on a significantly lighter structure also signifies either type II or type III intermittency. The frequency distribution of the vertical lines in the recurrence plot (Fig. 8a5) is skewed with an exponential tail, which confirms the existence of homoclinic orbits and also the presence of a type II or type III intermittency [101], as reported by other researchers as well [119, 120]. It is important to note in this context that the vertical lines in the recurrence plot would follow a bimodal distribution in case of a type I intermittency [121, 122]. Moreover, the black patches in the recurrence plot are observed to have perforated upper and right edges, thus confirming the intermittent behavior to be of type II [115]. For x in the range of 30–46 cm, a drastic change in the flame behavior was observed, with the sound emitted being a sharp tone of distinct frequency as opposed to a broadband noise. The strong flame–acoustic coupling results in a periodic acoustic signal in this regime (Fig. 8b1), with the instability-generated 237 Hz frequency dominating over the cold natural frequency of 100 Hz in the frequency spectrum (Fig. 8b2). A limit cycle with large width of periodic orbits, often observed in experimental periodic signals [59], is observed in the reconstructed phase portrait (Fig. 8b3), and the equispaced parallel diagonal lines in the recurrence plot (Fig. 8b4) also attest to that fact. The strong flame–acoustic coupling was observed to disappear for $x > 46$ cm, with the waveform (Fig. 8c1) suggesting intermittent behavior. In addition to the cold natural frequency (100 Hz) and the instability-generated frequency (237 Hz), another significant peak at 155 Hz is observed in the power spectral density plot (Fig. 8c2), which suggests quasiperiodic behavior as a result of a Neimark–Sacker bifurcation (since quasiperiodic behavior emerges from a limit cycle oscillation). The phase plot (Fig. 8c3) depicts an experimental quasiperiodic system, and the frequency distribution of the vertical lines in the recurrence plot (Fig. 8c5) confirms the intermittent behavior to be of either type II or type III. A similar behavior of the flame was observed by the same group when the global equivalence ratio of the flame, at a fixed flame position, was changed by varying either the air flow rate [98] or the fuel flow rate [64].

Sen et al. [98] used the different RQA measures to quantitatively differentiate the dynamic states. The variation of the different RQA measures with x is shown in Fig. 9. As expected, measures based on the diagonal lines (DET and L) assumed lower values for the intermittent regimes and higher values for the limit cycle oscillation. On the other hand, LAM and TT are based on the length of the vertical lines and assumed higher values in the intermittent regimes as compared to the periodic regime. A similar trend was observed for the RQA measures when ϕ was varied instead of x [98]. The drastic difference in the RQA values between the intermittent regimes and the periodic regimes indicates that the different RQA measures can be quite effective precursors for the onset of thermoacoustic instability.

4 Summary and Future Directions

In this chapter, an overview of dynamical systems analysis applied to laminar ducted flames was presented. Enclosed flames are ubiquitous in the modern industrial world, and the ducted flame is a reasonably good laboratory scale approximation for studying the dynamics. A brief review of the different tools of nonlinear time series analysis was provided, followed by specific case studies of laminar ducted non-premixed, premixed, and inverse diffusion flames. Several interesting and diverse dynamical features were observed such as limit cycles, bifurcations, quasiperiodicity, intermittency, homoclinic orbits, and chaos. The usefulness of recurrence plots and recurrence quantification analysis for the detection of thermoacoustic instability was also highlighted.

In the context of discussing dynamical systems analysis, it is worthwhile to mention complex networks [123] and their effectiveness for the identification of different dynamical states. Physical systems are inherently complex, but diverse systems are capable of showing the same dynamical features. These complex systems are composed of several individual components, and the complex networks approach focuses on how these different components interact with each other. Researchers have shown that complex networks can be used for detection of dynamical features in biomedical applications [124, 125], World Wide Web networks [126, 127], power grid networks [128, 129], to distinguish various regimes of turbulent jet flow [130], and to characterize energy dissipation in fully developed turbulence [131]. A recent study [132] has also shown that ε-recurrence networks, which are based on the adjacency matrix (adjacency matrix = recurrence matrix − identity matrix of same order), are extremely robust in terms of identifying dynamical features from a signal in the presence of noise. The method was able to identify the dynamic states of the well-known Lorenz system [33] even when different colored noise levels of up to 50% were synthetically introduced into the system. Recently, several researchers [133–138] have shown the effectiveness of complex networks-based approaches for detection of combustion instability in turbulent combustion system. Hence, it can be concluded with extreme confidence that such approaches would reveal interesting insights when applied on laminar ducted flames. Complex networks-based techniques show particular promise for the ducted inverse diffusion flame mentioned previously in this chapter, primarily because of the dearth of research on its dynamics.

References

1. H.C. Mongia, T.J. Held, G.C. Hsiao, R.P. Pandalai, Challenges and progress in controlling dynamics in gas turbine combustors. J. Propul. Power **19**(5), 822–829 (2003)
2. F. Giuliani, P. Gajan, O. Diers, M. Ledoux, Influence of pulsed entries on a spray generated by an air-blast injection device: an experimental analysis on combustion instability processes in aeroengines. Proc. Combust. Inst. **29**, 91–98 (2002)
3. G.A. Richards, M.C. Janus, Characterization of oscillations during premix gas turbine combustion. J. Eng. Gas Turb. Power **120**, 294–302 (1998)
4. P.J. Langhorne, Reheat buzz: an acoustically coupled combustion instability. Part 1. Experiment. J. Fluid Mech. **193**, 417–443 (1988)
5. G.J. Bloxsidge, A.P. Dowling, P.J. Langhorne, Reheat buzz: an acousitcally coupled combustion instability. Part 2. Theory. J. Fluid Mech. **193**, 445–473 (1988)
6. A.P. Dowling, A.S. Morgans, Feedback control of combustion oscillations. Annu. Rev. Fluid Mech. **37**, 151–182 (2005)
7. K.H. Yu, A. Trouvé, J.W. Daily, Low-frequency pressure oscillations in a model ramjet combustor. J. Fluid Mech. **232**, 47–72 (1991)
8. J.W.S. Rayleigh, The explanation of certain acoustical phenomena. Nature **18**, 319–321 (1878)
9. T.J. Poinsot, A.C. Trouvé, D.P. Veynante, S.M. Candel, E.J. Esposito, Vortex-driven acoustically coupled combustion instabilities. J. Fluid Mech. **177**, 265–292 (1987)
10. P.-H. Renard, D. Thévenin, J.C. Rolon, S. Candel, Dynamics of flame/vortex interactions. Prog. Energy Combust. Sci. **26**, 225–282 (2000)
11. F.E.C. Culick, Unsteady motions in combustion chambers for propulsion systems. AGARDograph RTO-AG-AVT-039, 2006

12. K.R. McManus, T. Poinsot, S.M. Candel, A review of active control of combustion instabilities. Prog. Energy Combust. Sci. **19**, 1–29 (1993)
13. X. Wu, M. Wang, P. Moin, N. Peters, Combustion instability due to the nonlinear interaction between sound and flame. J. Fluid Mech. **497**, 23–53 (2003)
14. S.R. Turns, *An Introduction to Combustion: Concepts and Applications* (McGraw-Hill Book Co., Singapore, 2000)
15. N. Karimi, M.J. Brear, S.H. Jin, J.P. Monty, Linear and non-linear forced response of a conical, ducted, laminar premixed flame. Combust. Flame **156**, 2201–2212 (2009)
16. L. Kabiraj, R.I. Sujith, Nonlinear self-excited thermoacoustic oscillations: intermittency and flame blowout. J. Fluid Mech. **713**, 376–397 (2012)
17. H. Gotoda, H. Nikimoto, T. Miyano, S. Tachibana, Dynamic properties of combustion instability in a lean-premixed gas turbine combustor. Chaos **21**, 013124 (2011)
18. S. Domen, H. Gotoda, T. Kuriyama, Y. Okuno, S. Tachibana, Detection and prevention of blowout in a lean premixed gas-turbine model combustor using the concept of dynamical system theory. Proc. Combust. Inst. **35**, 3245–3253 (2015)
19. S.A. Pawar, R. Vishnu, M. Vadivukkarasan, M.V. Panchagnula, R.I. Sujith, Intermittency route to combustion instability in a laboratory spray combustor. J. Eng. Gas Turb. Power **138**, 041505 (2016)
20. S. Datta, S. Mondal, A. Mukhopadhyay, D. Sanyal, S. Sen, An investigation of nonlinear dynamics of a thermal pulse combustor. Combust. Theor. Model. **13**(1), 17–38 (2009)
21. C.S. Daw, J.F. Thomas, G.A. Richards, L.L. Narayanaswami, Chaos in thermal pulse combustion. Chaos **5**(4), 662–670 (1995)
22. H.D. Ng, A.J. Higgins, C.B. Kiyanda, M.I. Radulescu, J.H.S. Lee, K.R. Bates, N. Nikiforakis, Nonlinear dynamics and chaos analysis of one-dimensional pulsating detonations. Combust. Theor. Model. **9**, 159–170 (2005)
23. K. Balasubramanian, R.I. Sujith, Non-normality and nonlinearity in combustion-acoustic interaction in diffusion flames. J. Fluid Mech. **594**, 29–57 (2008)
24. M. Tyagi, S.R. Chakravarthy, R.I. Sujith, Unsteady combustion response of a ducted non-premixed flame and acoustic coupling. Combust. Theor. Model. **11**(2), 205–226 (2007)
25. T.C. Lieuwen, Experimental investigation of limit-cycle oscillations in an unstable gas turbine combustor. J. Propul. Power **18**(1), 61–67 (2002)
26. S. Mondal, A. Mukhopadhyay, S. Sen, Dynamic characterization of a laboratory-scale pulse combustor. Combust. Sci. Technol. **186**, 139–152 (2014)
27. S. Balusamy, L.K.B. Li, Z. Han, M.P. Juniper, S. Hochgreb, Nonlinear dynamics of self-excited thermoacoustic system subjected to acoustic forcing. Proc. Combust. Inst. **35**, 3229–3236 (2015)
28. S. Candel, Combustion dynamics and control: progress and challenges. Proc. Combust. Inst. **29**, 1–28 (2002)
29. S. Ducruix, T. Schuller, D. Durox, S. Candel, Combustion dynamics and instabilities: elementary coupling and driving mechanisms. J. Propul. Power **19**(5), 722–734 (2003)
30. W.A. Sirignano, Driving mechanisms for combustion instability. Combust. Sci. Technol. **162–205**, 2015 (1887)
31. E. Bradley, H. Kantz, Nonlinear time-series analysis revisited. Chaos **25**, 097610 (2015)
32. J. Theiler, S. Eubank, A. Longtin, B. Galdrikian, J.D. Farmer, Testing for nonlinearity in time series: the method of surrogate data. Physica D **58**, 77–94 (1992)
33. R.C. Hilborn, *Chaos and Nonlinear Dynamics*, 2nd edn. (Oxford University Press, New York, 2000)
34. F. Takens, Detecting strange attractors in turbulence, in *Dynamical Systems and Turbulence (Lecture Notes in Mathematics)*, ed. by D.A. Rand, L.S. Young (Springer, Berlin, 1981)
35. H.D.I. Abarbanel, R. Brown, J.J. Sidorowich, L.S. Tsimring, The analysis of observed chaotic data in physical systems. Rev. Mod. Phys. **65**(4), 1331–1392 (1993)
36. A.M. Fraser, H.L. Swinney, Independent coordinates for strange attractors from mutual information. Phys. Rev. A **33**(2), 1134–1140 (1986)

37. P. Grassberger, I. Procaccia, Measuring the strangeness of strange attractors. Physica D **9**, 189–208 (1983)
38. M.B. Kennel, R. Brown, H.D.I. Abarbanel, Determining embedding dimension for phase-space reconstruction using a geometrical construction. Phys. Rev. A **45**(6), 3403–3411 (1992)
39. L. Kabiraj, A. Saurabh, P. Wahi, R.I. Sujith, Route to chaos for combustion instability in ducted laminar premixed flames. Chaos **22**, 023129 (2012)
40. A. Wolf, J.B. Swift, H.L. Swinney, J.A. Vastano, Determining Lyapunov exponents from a time series. Physica D **16**, 285–317 (1985)
41. H. Kantz, A robust method to estimate the maximal Lyapunov exponent of a time series. Phys. Lett. A **185**, 77–87 (1994)
42. G.A. Gottwald, I. Melbourne, A new test for chaos in deterministic systems. Proc. R. Soc. Lond. A **460**, 603–611 (2004)
43. I. Falconer, G.A. Gottwald, I. Melbourne, K. Wormnes, Application of the 0–1 test for chaos to experimental data. SIAM J. Appl. Dyn. Syst. **6**(2), 395–402 (2007)
44. G.A. Gottwald, I. Melbourne, On the implementation of the 0–1 test for chaos. SIAM J. Appl. Dyn. Syst. **8**(1), 129–145 (2009)
45. J.P. Eckmann, S.O. Kamphorst, D. Ruelle, Recurrence plots of dynamical systems. Europhys. Lett. **4**(9), 973–977 (1987)
46. N. Marwan, M.C. Romano, M. Thiel, J. Kurths, Recurrence plots for the analysis of complex systems. Phys. Rep. **438**, 237–329 (2007)
47. Y. Zou, Exploring recurrences in quasiperiodic dynamical systems. Ph.D. thesis, University of Potsdam, 2007
48. E. Ott, *Chaos in Dynamical Systems* (Cambridge University Press, 1993)
49. G. Górski, G. Litak, R. Mosdorf, A. Rysak, Two phase flow bifurcation due to turbulence: transition from slugs to bubbles. Eur. Phys. J. B. **88**, 239 (2015)
50. R. Mosdorf, G. Górski, Detection of two-phase flow patterns using the recurrence network analysis of pressure fluctuations. Int. Commun. Heat Mass Transf. **64**, 14–20 (2015)
51. M.F. Llop, N. Gascons, F.X. Llauró, Recurrence plots to characterize gas-solid fluidization regimes. Int. J. Mult. Flow **73**, 43–56 (2015)
52. J.A. Bastos, J. Caiado, Recurrence quantification analysis of global stock markets. Physica A **390**, 1315–1325 (2011)
53. S. Nakano, Y. Hirata, K. Iwayama, K. Aihara, Intra-day response of foreign exchange markets after the Tohoku-Oki earthquake. Physica A **419**, 203–214 (2015)
54. J. Yan, Y. Wang, G. Ouyang, T. Yu, X. Li, Using max entropy ratio of recurrence plot to measure electrocorticogram changes in epilepsy patients. Physica A **443**, 109–116 (2016)
55. J. Schlenker, V. Socha, L. Riedlbauchová, T. Nedělka, A. Schlenker, V. Potočková, Š. Malá, P. Kutílek, Recurrence plot of heart rate variability signal in patients with vasovagal syncopes. Biomed. Signal Process. Control **25**, 1–11 (2016)
56. M. Stöckl, D. Plück, M. Lames, Modeling game sports as complex systems—application of recurrence analysis to golf and soccer. Math. Comput. Model. Dyn. Syst. **23**(4), 399–415 (2017)
57. L.-P. Yang, S.-L. Ding, G. Litak, E.-Z. Song, X.-Z. Ma, Identification and quantification analysis of nonlinear dynamics properties of combustion instability in a diesel engine. Chaos **25**, 013105 (2015)
58. L. Kabiraj, A. Saurabh, H. Nawroth, C.O. Paschereit, Recurrence analysis of combustion noise. AIAA J. **53**(5), 1199–1210 (2015)
59. H. Gotoda, Y. Okuno, K. Hayashi, S. Tachibana, Characterization of degeneration process in combustion instability based on dynamical systems theory. Phys. Rev. E **92**, 025906 (2015)
60. V. Nair, R.I. Sujith, A reduced-order model for the onset of combustion instability: physical mechanisms for intermittency and precursors. Proc. Combust. Inst. **35**, 3193–3200 (2015)
61. H. Kinugawa, K. Ueda, H. Gotoda, Chaos of radiative heat-loss-induced flame front instability. Chaos **26**, 033104 (2016)
62. L. Christodoulou, L. Kabiraj, A. Saurabh, N. Karimi, Characterizing the signature of flame flashback precursor through recurrence analysis. Chaos **26**, 013110 (2016)

63. L.-P. Yang, E.-Z. Song, S.-L. Ding, R.J. Brown, N. Marwan, X.-Z. Ma, Analysis of the dynamic characteristics of combustion instabilities in a pre-mixed lean-burn natural gas engine. Appl. Energy **183**, 746–759 (2016)
64. U. Sen, T. Gangopadhyay, C. Bhattacharya, A. Misra, S. Karmakar, P. Sengupta, A. Mukhopadhyay, S. Sen, Investigation of ducted inverse nonpremixed flames using dynamic systems approach, in *Proceedings of ASME Turbo Expo*, p. V04BT04A059, 2016
65. M. Tyagi, N. Jamadar, S.R. Chakravarthy, Oscillatory response of an idealized two-dimensional diffusion flame: analytical and numerical study. Combust. Flame **149**, 271–285 (2007)
66. S.J. Illingworth, I.C. Waugh, M.P. Juniper, Finding thermoacoustic limit cycles for a ducted Burke-Schumann flame. Proc. Combust. Inst. **34**, 911–920 (2013)
67. S.H. Strogatz, *Nonlinear Dynamics and Chaos* (Perseus Books, Reading, MA, 1994)
68. V.N. Kurdyumov, M. Matalon, Dynamics of an edge flame in a mixing layer. Combust. Flame **139**, 329–339 (2004)
69. S.C. Reddy, L.N. Trefethen, Pseudospectra of the convection-diffusion operator. SIAM J. Appl. Math. **54**(6), 1634–1649 (1994)
70. L.N. Trefethen, Pseudospectra of linear operators. SIAM Rev. **39**(3), 383–406 (1997)
71. P.J. Schmid, D.S. Henningson, *Stability and Transition in Shear Flows* (Springer, New York, 2001)
72. L.N. Trefethen, A.E. Trefethen, S.C. Reddy, T.A. Driscoll, Hydrodynamic stability without eigenvalues. Science **261**, 578–584 (1993)
73. J.S. Baggett, T.A. Driscoll, L.N. Trefethen, A mostly linear model of transition to turbulence. Phys. Fluids **7**(4), 833–838 (1995)
74. W. Kerner, Large-scale complex eigenvalue problems. J. Comput. Phys. **85**, 1–85 (1989)
75. B.F. Farrell, Optimal excitation of baroclinic waves. J. Atmos. Sci. **46**(9), 1193–1206 (1989)
76. A.P. Dowling, The calculation of thermoacoustic oscillations. J. Sound Vib. **180**(4), 557–581 (1995)
77. A. Fichera, C. Losenno, A. Pagano, Clustering of chaotic dynamics of a lean gas-turbine combustor. Appl. Energy **69**, 101–117 (2001)
78. T. Gebhardt, S. Grossmann, Chaos transition despite linear stability. Phys. Rev. E **50**(5), 3705–3711 (1994)
79. K. Matveev, Thermoacoustic instabilities in the Rijke tube: experiments and modeling, Ph.D. thesis, California Institute of Technology, 2003
80. V. Jegadeesan, R.I. Sujith, Experimental investigation of noise induced triggering in thermoacoustic systems. Proc. Combust. Inst. **34**, 3175–3183 (2013)
81. K.T. Kim, S. Hochgreb, Measurements of triggering and transient growth in a model lean-premixed gas turbine combustor. Combust. Flame **159**, 1215–1227 (2012)
82. N. Noiray, D. Durox, T. Schuller, S. Candel, A unified framework for nonlinear combustion instability analysis based on the flame describing function. J. Fluid Mech. **615**, 139–167 (2008)
83. M.P. Juniper, Triggering in the horizontal Rijke tube: non-normality, transient growth, and bypass transition. J. Fluid Mech. **667**, 272–308 (2011)
84. S. Fedotov, I. Bashkirtseva, L. Ryashko, Stochastic analysis of a non-normal dynamical system mimicking a laminar-to-turbulent subcritical transition. Phys. Rev. E **66**, 066310 (2002)
85. V.S. Burnley, F.E.C. Culick, Influence of random excitations on acoustic instabilities in combustion chambers. AIAA J. **38**(8), 1403–1410 (2000)
86. I.C. Waugh, M.P. Juniper, Triggering in a thermoacoustic system with stochastic noise. Int. J. Spray Combust. Dyn. **3**(3), 225–242 (2011)
87. T. Lieuwen, A. Banaszuk, Background noise effects on combustor stability. J. Propul. Power **21**(1), 25–31 (2005)
88. Y. Duguet, A.P. Willis, R.R. Kerswell, Transition in pipe flow: the saddle structure on the boundary of turbulence. J. Fluid Mech. **613**, 255–274 (2008)
89. S. Nair, T. Lieuwen, Acoustic detection of blowout in premixed flames. J. Propul. Power **21**(1), 32–39 (2005)

90. S.J. Shanbhogue, S. Husain, T. Lieuwen, Lean blowoff of bluff body stabilized flames: scaling and dynamics. Prog. Energy Combust. Sci. **35**, 98–120 (2009)
91. A.H. Nayfeh, B. Balachandran, *Applied Nonlinear Dynamics: Analytical, Computational, and Experimental Methods* (Wiley-VCH Verlag GmbH & Co. KGaA, Weinheim, Germany, 2004)
92. O. Batiste, E. Knobloch, I. Mercader, M. Net, Simulations of oscillatory binary fluid convection in large aspect ratio containers. Phys. Rev. E **65**, 016303 (2001)
93. L. Holden, T. Erneux, Understanding bursting oscillations as periodic slow passages through bifurcation and limit points. J. Math. Biol. **31**, 351–365 (1993)
94. S.K. Han, D.E. Postnov, Chaotic bursting as chaotic itinerancy in coupled neural oscillators. Chaos **13**(3), 1105–1109 (2003)
95. R. Straube, D. Flockerzi, M.J.B. Hauser, Sub-Hopf/fold-cycle bursting and its relation to (quasi-)periodic oscillations. J. Phys.: Conf. Ser. **55**, 214–231 (2006)
96. V. Nair, G. Thampi, R.I. Sujith, Intermittency route to thermoacoustic instability in turbulent combustors. J. Fluid Mech. **756**, 470–487 (2014)
97. V. Nair, R.I. Sujith, Intermittency as a transition state in combustor dynamics: an explanation for flame dynamics near lean blowout. Combust. Sci. Technol. **187**, 1821–1835 (2015)
98. U. Sen, T. Gangopadhyay, C. Bhattacharya, A. Mukhopadhyay, S. Sen, Dynamic characterization of a ducted inverse diffusion flame using recurrence analysis. Combust. Sci. Technol. **190**(1), 32–56 (2018)
99. Y. Pomeau, P. Manneville, Intermittent transition to turbulence in dissipative dynamical systems. Commun. Math. Phys. **74**, 189–197 (1980)
100. H. Okamoto, N. Tanaka, M. Naito, Intermittencies and related phenomena in the oxidation of formaldehyde at a constant current. J. Phys. Chem. A **102**, 7353–7361 (1998)
101. K. Klimaszewska, J.J. Żebrowski, Detection of the type of intermittency using characteristic patterns in recurrence plots. Phys. Rev. E **80**, 026214 (2009)
102. L. Kabiraj, R.I. Sujith, P. Wahi, Bifurcations of self-excited ducted laminar premixed flames. J. Eng. Gas Turbines Power **134**, 031502 (2012)
103. Y. Matsui, An experimental study on pyro-acoustic amplification of premixed laminar flames. Combust. Flame **43**, 199–209 (1981)
104. D. Ruelle, F. Takens, On the nature of turbulence. Commun. Math. Phys. **20**, 167–192 (1971)
105. K. Kashinath, I.C. Waugh, M.P. Juniper, Nonlinear self-excited thermoacoustic oscillations of a ducted premixed flame: bifurcation and routes to chaos. J. Fluid Mech. **761**, 399–430 (2014)
106. G.-S. Jiang, D. Peng, Weighted ENO schemes for Hamilton-Jacobi equations. SIAM J. Sci. Comput. **21**(6), 2126–2143 (2000)
107. S. Gottlieb, C.-W. Shu, Total variation diminishing Runge-Kutta schemes. Math. Comput. **67**(221), 73–85 (1998)
108. A. Sobiesiak, S. Rahbar, H.A. Becker, Performance characteristics of the novel low-NOx CGRI burner for use with high air preheat. Combust. Flame **115**, 93–125 (1988)
109. J. Miao, C.W. Leung, C.S. Cheung, Z. Huang, W. Jin, Effect of H_2 addition on OH distribution of LPG/air circumferential inverse diffusion flame. Int. J. Hydrog. Energy **41**, 9653–9663 (2016)
110. L.K. Sze, C.S. Cheung, C.W. Leung, Appearance, temperature, and NOx emission of two inverse diffusion flames with different port design. Combust. Flame **144**, 237–248 (2006)
111. C.R. Kaplan, K. Kailasanath, Flow-field effects on soot formation in normal and inverse methane-air diffusion flames. Combust. Flame **124**, 275–294 (2001)
112. S.R. Stow, A.P. Dowling, Thermoacoustic oscillations in an annular combustor, in *Proceedings of ASME Turbo Expo*, p. V002T02A004, 2001
113. H. Herzel, P. Plath, P. Svensson, Experimental evidence of homoclinic chaos and type-II intermittency during the oxidation of methanol. Physica D **48**, 340–352 (1991)
114. D. Parthimos, D.H. Edwards, T.M. Griffith, Shil'nikov chaos is intimately related to type-III intermittency in isolated rabbit arteries: Role of nitric oxide. Phys. Rev. E **67**, 051922 (2003)

115. V. Nair, R.I. Sujith, Identifying homoclinic orbits in the dynamics of intermittent signals through recurrence quantification. Chaos **23**, 033136 (2013)
116. P. Holmes, Can dynamical systems approach turbulence? In J. Lumley, editor, *Whither Turbulence? Turbulence at the Crossroads*, (Springer, 1990), pp. 195–249
117. E. Stone, M. Gorman, M. el Hamdi, K.A. Robbins, Identification of ordered patterns as heteroclinic connections. Phys. Rev. Lett. **76**(12), 2061–2064 (1996)
118. E. Stone, P. Holmes, Unstable fixed points, heteroclinic cycles and exponential tails in turbulence production. Phys. Lett. A **155**(1), 29–42 (1991)
119. P.W. Hammer, N. Platt, S.M. Hammel, J.F. Heagy, B.D. Lee, Experimental observation of on-off intermittency. Phys. Rev. Lett. **73**(8), 1095–1098 (1994)
120. M. Frank, M. Schmidt, Time series investigations on an experimental system driven by phase transitions. Phys. Rev. E **56**(3), 2423–2428 (1997)
121. H.G. Schuster, W. Just, *Deterministic Chaos: An Introduction*, 4th edn. (Wiley-VCH Verlag GmbH & Co. KGaA, Weinheim, Germany, 2005)
122. D.L. Feng, J. Zheng, W. Huang, C.X. Yu, W.X. Ding, Type-I-like intermittent chaos in multicomponent plasmas with negative ions. Phys. Rev. E **54**(3), 2839–2843 (1996)
123. A.-L. Barabási, The network takeover. Nat. Phys. **8**, 14–16 (2011)
124. E. Alm, A.P. Arkin, Biological networks. Curr. Opin. Struct. Biol. **13**, 193–202 (2003)
125. A.-L. Barabáasi, Z.N. Oltvai, Network biology: undersanding the cell's functional organization. Nat. Rev. Genet. **5**, 101–113 (2004)
126. R. Albert, H. Jeong, A.-L. Barabási, Diameter of the World-Wide Web. Nature **401**, 130–131 (1999)
127. R. Albert, H. Jeong, A.-L. Barabási, Error and attack tolerance of complex networks. Nature **406**, 378–382 (2000)
128. G.A. Pagani, M. Aiello, The power grid as a complex network: a survey. Physica A **392**, 2688–2700 (2013
129. S. Arianos, E. Bompard, A. Carbone, F. Xue, Power grid vulnerability: a complex network approach. Chaos **19**, 013119 (2009)
130. A.K. Charakopoulos, T.E. Karakasidis, P.N. Papanicolaou, A. Liakopoulos, The application of complex network time seris analysis in turbulent heated jets. Chaos **24**, 024408 (2014)
131. C. Liu, W.-X. Zhou, W.-K. Yuan, Statistical properties of visibility graph of energy dissipation rates in three-dimensional fully developed turbulence. Physica A **389**, 2675–2681 (2010)
132. R. Jacob, K.P. Harikrishnan, R. Misra, G. Ambika, Characterization of chaotic attractors under noise: a recurrence network perspective. Commun. Nonlinear Sci. Numer. Simulat. **41**, 32–47 (2016)
133. Y. Okuno, M. Small, H. Gotoda, Dynamics of self-excited thermoacoustic instability in a combustion system: pseudo-periodic and high-dimensional nature. Chaos **25**, 043107 (2015)
134. M. Murugesan, R.I. Sujith, Combustion noise is scale-free: transition from scale-free to order at the onset of thermoacoustic instability. J. Fluid Mech. **772**, 225–245 (2015)
135. M. Murugesan, R.I. Sujith, Detecting the onset of an impending thermoacoustic instability using complex networks. J. Propul. Power **32**(3), 707–712 (2016)
136. H. Gotoda, H. Kinugawa, R. Tsujimoto, S. Domen, Y. Okuno, Characterization of combustion dynamics, detection, and prevention of an unstable combustion state based on a complex-network theory. Phys. Rev. Appl. **7**, 044027 (2017)
137. V. Godavarthi, V.R. Unni, E.A. Gopalakrishnan, R.I. Sujith, Recurrence networks to study dynamical transitions in a turbulent combustor. Chaos **27**, 063113 (2017)
138. J. Singh, R.B. Vishwanath, S. Chaudhuri, R.I. Sujith, Network structure of turbulent premixed flames. Chaos **27**, 043107 (2017)

Synchronization Transition in a Thermoacoustic System: Temporal and Spatiotemporal Analyses

Sirshendu Mondal, Samadhan A. Pawar and R. I. Sujith

Abstract The occurrence of thermoacoustic instability has been a major concern in the combustors used in power plants and propulsive systems such as gas turbine engines, rocket motors. A positive feedback between the inherent processes such as the acoustic field and the unsteady heat release rate of the combustor can result in the occurrence of large-amplitude, self-sustained pressure oscillations. Prior to the state of thermoacoustic instability, intermittent oscillations are observed in turbulent combustors. Such intermittent oscillations are characterized by an apparently random appearance of bursts of large-amplitude periodic oscillations interspersed between epochs of low-amplitude aperiodic oscillations. In most of the earlier studies, the pressure oscillations alone have been analyzed to explore the dynamical transition to thermoacoustic instability. The present chapter focuses on the instantaneous interaction between the acoustic field and the unsteady heat release rate observed during such a transition in a bluff-body-stabilized turbulent combustor. The instantaneous interaction of these oscillations will be discussed using the concepts of synchronization theory. First, we give a brief introduction to the synchronization theory so as to summarize the concepts of locking of phase and frequency of the oscillations. Then, the temporal and spatiotemporal aspects of the interaction will be presented in detail. We find that, during stable operation, aperiodic oscillations of the pressure and the heat release rate remain desynchronized, whereas synchronized periodic oscillations are noticed during the occurrence of thermoacoustic instability. Such a transition happens through intermittent phase-synchronized oscillations, wherein synchronization and desynchronization of the oscillators are observed during the periodic and the aperiodic epochs of the intermittent oscillations, respectively. Further, the spatiotemporal analysis reveals a very interesting pattern in the reaction zone. Phase asynchrony among the local heat release rate oscillators is observed during the stable operation, while they become phase-synchronized during the onset of

S. Mondal (✉) · S. A. Pawar · R. I. Sujith
Indian Institute of Technology Madras, Chennai 600036, India
e-mail: sirshendumondal13@gmail.com

S. A. Pawar
e-mail: samadhanpawar@ymail.com

R. I. Sujith
e-mail: sujith@iitm.ac.in

A. K. Runchal et al. (eds.), *Energy for Propulsion*, Green Energy and Technology,
https://doi.org/10.1007/978-981-10-7473-8_6

thermoacoustic instability. Interestingly, the state of intermittent oscillations corresponds to a simultaneous existence of synchronized periodic and desynchronized aperiodic patterns in the reaction zone. Such a coexistence of synchrony and asynchrony in the reactive flow field mimics a chimera state.

Nomenclature

R_a	Rayleigh integral
Δf	Frequency detuning
ω_0	Natural angular frequency
E	Embedding dimension
T	Optimum time delay
V	Volume of the combustion zone
Θ	Heaviside step function
$\mathbb{R}_{i,j}$	Recurrence Matrix
ε	Cutoff threshold
$P(\tau)$	Probability of recurrence
τ	Time lag
$\bar{u}$	Mean flow velocity
p'	Pressure oscillations
$\dot{q}'$	Heat release rate oscillations
$R(t)$	Kuramoto order parameter
$\bar{R}$	Time-averaged Kuramoto order parameter
$\zeta(t)$	Analytic signal
RP	Recurrence Plot
PMT	Photomultiplier tube
PSD	Power spectral density
PS	Phase synchronization
GS	Generalized synchronization
IPS	Intermittent phase synchronization
FWHM	Full width at half maximum

1 Introduction

The complex interaction between hydrodynamics, combustion, and acoustic fields in practical combustors such as gas turbine engines and rocket motors often results in the occurrence of large-amplitude pressure oscillations, popularly known as combustion instability or thermoacoustic instability. Such large-amplitude oscillations emerge as a result of a three-way coupling between these processes, which are the fundamental components of a turbulent reactive flow [40]. A coupled interaction of these processes can lead to the unstable operation of a combustor [43, 75].

This is a major problem faced during the development of high-performance combustion systems [10]. The presence of undesirable pressure oscillations can have catastrophic consequences such as structural damage due to excessive heat transfer and vibrations, thrust oscillations, damage of the electronics in aircraft and satellites, and even the complete failure of engines. Over the years, researchers have been trying to explore such interactions so that new techniques can be devised to mitigate the amplitude of oscillations [43, 60, 67], to forewarn their onset [19, 36, 51, 52], or to control the oscillations in exceptional cases of pulse combustors [11, 44, 45, 68].

1.1 Thermoacoustic Instability

The transfer of energy to the acoustic field from the reacting flow field primarily depends on the coupled interaction between the acoustic modes of the combustor and the unsteady flame dynamics. Such addition or removal of acoustic energy has been usually quantified in terms a measure known as Rayleigh integral [40], R_a, as computed through Eq. 1.

$$R_a = \int_V \int_T p'(\mathbf{x}, t)\dot{q}'(\mathbf{x}, t)dtdv \quad (1)$$

The sign of R_a is determined by the average phase difference between the heat release rate oscillations ($\dot{q}'$) and the acoustic pressure oscillations (p') which are functions of space ($\mathbf{x}$) and time (t). Here, V is the total volume of the combustor and T is the time period of the oscillations. If the average phase difference lies between $-90°$ and $90°$, the integral becomes positive, leading to the addition of energy to the acoustic field from the combustion processes. This necessary condition for the onset of thermoacoustic instabilities is popularly known as the Rayleigh criterion [60, 61]. However, this is not a sufficient condition as there is always energy dissipation (acoustic damping) present in the combustion system. Once the energy added to the acoustic field exceeds the amount of energy dissipated (due to sound radiation at the boundaries, viscous dissipation, etc.), the growth in pressure oscillations occurs. However, due to the nonlinearities present in the system, the amplitude of pressure oscillations does not grow indefinitely but saturates at a certain value to constitute limit cycle oscillations, wherein acoustic driving and damping balance each other [40].

In turbulent combustors, the feedback loop responsible for the onset of thermoacoustic instabilities consists of hydrodynamics, acoustic oscillations, and heat release rate oscillations in the flame. Fluctuations in the flow velocity cause oscillations in heat release rate. The heat release rate oscillations thus generated excite acoustic oscillations. Finally, the acoustic oscillations, in turn, affect the velocity fluctuations, thereby closing the feedback loop. The different mechanisms which can lead to thermoacoustic instability in gas turbine engines are fuel line-acoustic coupling

[7, 31], equivalence ratio oscillations [38], spray-acoustic coupling [7, 31], flame area oscillations [4], and large-scale vortex shedding [21, 67]. The detailed discussions of these mechanisms can be found in Lieuwen and Yang [40].

1.2 Dynamical Systems Theory Approach

The framework of linear stability analysis has been extensively used to analyze thermoacoustic instability in the past. However, through linear stability analysis the amplitude of limit cycle oscillations cannot be predicted [40]. Further, large-amplitude disturbances destabilizing a linearly stable system are known as triggering which is a nonlinear phenomenon. In short, there is a significant role of nonlinearities in the occurrence of thermoacoustic instabilities in gas turbine combustors [8, 35]. Therefore, in order to a get deeper insight into such phenomena, one must look for the solutions of the nonlinear models of combustion processes [40].

As the amplitude of acoustic oscillations is small compared to the mean pressure in the combustor, the acoustic wave propagation is generally presumed to be linear [13]. Nonlinearities in the combustor dynamics therefore mainly arise through the interaction between the heat release rate fluctuations and the acoustic field [13]. Such nonlinearities that stem from these coupled interactions lead to the saturation of amplitude of the oscillations, and thus, the system dynamics attains a state of constant-amplitude periodic oscillations, known as limit cycle oscillations [9].

Dynamical systems theory (also known as nonlinear dynamics) finds its application in diverse fields such as physics, chemistry, and biology, even in economics and social sciences [17, 41]. The viewpoint of dynamical systems theory has also aided in understanding the different nonlinear phenomena exhibited by thermoacoustic systems [11, 18, 24, 26, 27, 47, 50, 71]. For instance, we can characterize the transition to thermoacoustic instability using bifurcation theory [3, 18, 27, 39, 74].

Jahnke and Culick [24] observed a torus-birth bifurcation in the model of a solid rocket motor that exhibits a transition from limit cycle oscillations to a quasiperiodic state. On the other hand, modeling studies by Sterling [71], Lei and Turan [34], and Datta et al. [11] have shown the presence of period doubling route to chaos, for different models exhibiting the transition to combustion instability. Kabiraj and co-workers have experimentally shown the existence of these different dynamical states [26, 27] and different routes to chaos [28] in a laminar thermoacoustic system. They observed various dynamical states such as period-2, quasiperiodic, mode-locked, period-k, and chaotic oscillations, in addition to limit cycle oscillations. Thereafter, Kashinath et al. [29] verified the evidence of such states in the numerical simulation of a premixed flame, for different settings of the control parameters. In short, the approach of dynamical systems theory has provided more insight in exploring the transitions and the dynamical behaviors exhibited by a thermoacoustic system [25, 75].

1.3 Transition to Thermoacoustic Instability

A traditional approach was to treat the onset of thermoacoustic instability as a Hopf bifurcation, wherein the system dynamics directly change its behavior from steady (stable operation) to periodic (unstable operation) oscillations at the bifurcation point [39]. Such a description of the transition to thermoacoustic instability is valid mostly for laminar combustion systems [27, 73]. However, in the case of complex turbulent combustors, such a simple description of the onset may not be very useful. A recent study in turbulent combustors [18] has reported that the transition to thermoacoustic instability from stochastic fluctuations happens through a state of lower-dimensional chaotic oscillations. Subsequently, a study by Nair, Thampi, and Sujith [53] suggests that the transition to thermoacoustic instability happens gradually through the presence of an intermediate state known as intermittency. Intermittency, here, refers to an alternate switching of the combustor dynamics from low-amplitude aperiodic to relatively higher-amplitude bursts of periodic oscillations. The phenomenon of intermittency, however, was first reported in the context of thermoacoustics by Kabiraj and Sujith [28] in a laminar combustor. They observed intermittency as the occurrence of high-amplitude bursts of aperiodic oscillations among the regions of periodic oscillations prior to blowout of the flame. Henceforth, various studies have reported the presence of such intermittent oscillations prior to the onset of thermoacoustic instability [12, 19, 55, 78, 79]. The existence of intermittency has led various studies to develop control strategies that can forewarn the onset of thermoacoustic instability in turbulent combustors [51–53].

Although some success has been achieved in characterizing the transition to thermoacoustic instability in the past, most of these studies were primarily based on analyzing the properties of a single variable, i.e., the acoustic pressure fluctuations. Since the onset of thermoacoustic instability is a coupled response of both the acoustic pressure and the heat release rate fluctuations, none of the previous studies characterized the coupled temporal behavior of these variables for the intermittency route to thermoacoustic instability. Furthermore, we know that spatiotemporal dynamics of the coupled turbulent reactive flow with the duct acoustic field play a major role in exhibiting different dynamical states of the combustor. Studies, hitherto, were based on characterizing the instantaneous or the average dynamics of the combustor flow field during stable and unstable operations. These studies conclude that the formation of large-scale vortices plays a dominant role in the onset of thermoacoustic instability in turbulent combustion systems [30, 59, 70]. However, the coupled dynamics of the reactive flow field and the acoustic pressure in the combustor is not yet explored during the intermittency route to thermoacoustic instability. We observe the perfect locking between the acoustic pressure and the heat release rate fluctuations during the state of thermoacoustic instability and their unlocking during the state of combustion noise. This observation further motivates us to adopt a framework of synchronization in studying such a transition of coupled dynamics to thermoacoustic instability [46, 48, 49, 56, 57]. A brief introduction of synchronization theory is provided in the next section.

2 Synchronization

Synchronization is an emergent phenomenon that occurs in a broad range of dynamical systems, including neural signals, beating of the heart, and flashing of fireflies. Its history dates back to seventeenth century when Christiaan Huygens first observed synchronization of two pendulum clocks suspended from a common wooden beam [58]. He described the phenomena as "*sympathy of two clocks*." Later, in nineteenth century, Lord Rayleigh described the phenomenon of synchronization between two similar but distinct organ-pipes which began to sound in unison while kept side by side. Further, synchronization in a large population of oscillating systems was also observed, e.g., synchronous clapping in a large audience. It is important to note that synchronization is essentially a nonlinear phenomenon and occurs only in the self-sustained oscillators.

2.1 *Definition and Overview*

In classical terms, synchronization is defined as "*an adjustment of rhythms of two or more oscillating objects due to their weak interaction*" [58]. Oscillating objects, here, refer to the self-sustained oscillators which are essentially nonlinear and their rhythms indicate the period or frequency of their oscillations. When two oscillating objects interact, they may adjust their frequency and oscillate with a common frequency. This phenomenon of their eventual synchronization is also known as frequency entrainment. Whether the oscillators synchronize or not depends on the strength of the coupling and the value of the frequency detuning ($\Delta f = f_1 - f_2$) between them.

The ability of self-sustained oscillators to be synchronized is based on the instantaneous phase (ϕ) of the oscillators. It grows uniformly with time as $d\phi/dt = \omega_0$ (constant for periodic oscillators), where ω_0 is the natural angular frequency of oscillations. The special feature of the variable, phase (ϕ), is that it is neutrally stable. In other words, any perturbation in the phase neither grows nor decays. Therefore, it can be easily adjusted by any external perturbation (periodic forcing or coupling to another system). As a consequence, the oscillators synchronize [63].

Modern research on synchronization covers both periodic and chaotic oscillators. Synchronization of periodic oscillators can be anticipated, but the fact that chaotic oscillators can also show synchronization behavior has fascinated various researchers. Synchronization of coupled chaotic oscillators has been mainly categorized into phase synchronization [64], lag synchronization [65], complete synchronization [16], and generalized synchronization [66]. During phase synchronization, instantaneous phases of both the oscillators are perfectly locked; however, their amplitudes remain uncorrelated. In the case of complete synchronization, both amplitudes and phases of the coupled oscillators' signals are perfectly locked, whereas for lag synchronization, these signals are locked with a constant phase shift.

During generalized synchronization, both the signals from coupled oscillators have a functional relationship. Sometimes, a state of intermittent synchronization, wherein signals are alternately phase-locked and unlocked, is observed prior or after the state of perfect synchronization [2].

2.2 Calculation of Instantaneous Properties of the Signal

2.2.1 Instantaneous Phase and Frequency

Synchronization behavior of coupled oscillators can be understood by calculating the instantaneous properties of the individual signals. Phase is one of such measure that is often used in synchronization theory to quantify synchronous characteristics of such oscillators. Various methods have been devised to quantify the instantaneous phase characteristics of any signal [58]. One such method is based on the Hilbert transform [64]. Calculation of phase from the Hilbert transform requires the construction of analytic signal, a complex signal in which the real part is the original signal, $s(t)$ and the imaginary part is Hilbert transform of it. The Hilbert transform of any signal can be calculated as $\left(s_H(t) = \frac{1}{\pi} P.V. \int_{-\infty}^{\infty} \frac{s(\tau)}{(t-\tau)} d\tau\right)$, where $P.V.$ is the Cauchy principle value of the integral. Thus, the analytic signal based on the Hilbert transform can be obtained as,

$$\zeta(t) = s(t) + i s_H(t) = A(t) e^{i\phi(t)} \tag{2}$$

where $\phi(t)$ signifies the instantaneous phase, $d\phi/dt = \omega(t)$ is the instantaneous frequency, and $A(t)$ is the instantaneous amplitude of the signal. Now, the synchronization behavior of coupled oscillators can further be understood by analyzing the temporal variation of the instantaneous phase difference between the signals as $\Delta\phi_{i,j}(t) = \phi_j(t) - \phi_i(t)$. The condition for phase locking (or synchronization) is $|\Delta\phi_{i,j}(t)| =$ constant (where constant $<2\pi$). This suggests that when the phase difference between the signals is bounded or exhibits fluctuations around a constant phase shift, a synchronization is seen to be achieved in the system. For an ideal condition, the relative phase between the signals should be constant; in contrast, for practical situations, the relative phase fluctuates around this constant value of the phase shift.

Although the analytic signal approach is more common to calculate the instantaneous properties of the signals, it carries a few limitations. First, the application of this approach is limited to narrowband signals, and second, the calculated phase is physically meaningful only if the analytic plane (plot between real and imaginary parts of the analytic signal) of such signals owns a unique center of rotation [80]. While different methods have been developed to encounter this issue of phase calculation from the Hilbert transform [22, 54], we, in this study, use a measure based on recurrence plots [62] to characterize the synchronization properties of acoustic pressure and heat release rate signals. The recurrence-based method will be discussed next.

2.2.2 Probability of Recurrence $P(\tau)$

Recurrence is a property of deterministic dynamical systems [14]. It quantifies the recurrence behavior of a trajectory at a given state point after some divergence in the phase space. Such recurrence properties of a dynamical system can be effectively represented in a two-dimensional binary matrix, known as the recurrence plot (RP) [42]. Further, the construction of RP depends on the reconstruction of the phase space using Takens' delay embedding theorem [76]. Therefore, the parameters that are required to construct RP are maximum embedding dimension (E), optimum time delay (T), and cutoff threshold (ε). The RP is then constructed using the following equation [42],

$$\mathbb{R}_{i,j} = \Theta(\varepsilon - ||\mathbf{x}_i - \mathbf{x}_j||); \qquad i, j = 1, 2, \ldots, N_1 \tag{3}$$

where Θ is the Heaviside step function, $N_1 = N - (E - 1)T$ is the total number of points in the delayed vector, N is the total number of points in the time series, and $|| \cdot ||$ is the Euclidean norm. Whenever the recurrence of phase space trajectory falls inside the threshold, it is marked as 1; otherwise, it is marked as 0 in the recurrence matrix.

A synchronization measure developed on the recurrence property of the phase space trajectory is known as the probability of recurrence, $P(\tau)$. It calculates the probability of recurrence of the phase space trajectory to its neighborhood position after a time lag τ [62]. The value of $P(\tau)$ can be calculated as,

$$P(\tau) = \frac{1}{(N_1 - \tau)} \sum_{i=1}^{N_1-\tau} \Theta(\varepsilon - ||\mathbf{x}_i - \mathbf{x}_{i+\tau}||) \tag{4}$$

The value of $P(\tau)$ peaks to 1 for a perfect recurrence of the phase space trajectory; otherwise, its value will be less than 1. The synchronization features of two signals can be analyzed from the plots of $P(\tau)$ against τ. The matching of the positions and heights of such peaks in variation of $P(\tau)$ with τ aids in detecting the type of synchronization. In the case of desynchronization of oscillators, the $P(\tau)$ plots of these signals display a complete mismatch. When the oscillators are phase-synchronized, since their phases are perfectly locked but not amplitudes, such behavior manifests in the locking of positions of the peaks alone and not their heights in $P(\tau)$ plot. Conversely, if the oscillators are in a state of generalized synchronization, their recurrence plots become nearly identical, and hence, their plots of $P(\tau)$ show a perfect locking in positions and also in the heights of the peaks [33, 62].

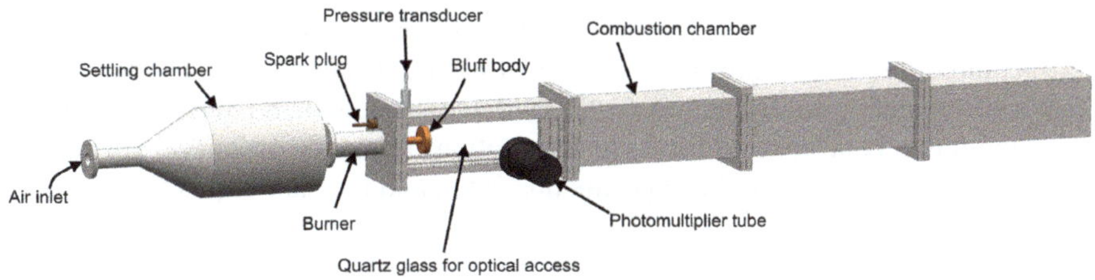

Fig. 1 Schematic of a backward facing step, bluff-body-stabilized, turbulent flame combustor [49, 56]. Reproduced with permission from Pawar et al. [55]

3 Experiments with a Laboratory-Scale Turbulent Combustor

A laboratory-based combustion system consisting of a partially premixed turbulent flame is used for the experiments (see Fig. 1). The main parts of the experimental setup are: (i) a settling chamber, (ii) a burner, and (iii) a combustor. Dimensions of the combustor duct are: 1400 mm (length) $\times$ 90 mm (width) $\times$ 90 mm (height). A backward facing step (or dump plane) is present at the inlet of the combustor. A circular bluff body (diameter 47 mm and thickness 10 mm) is located at a distance of 4.5 cm from the dump plane. The bluff body serves the purpose of anchoring the flame in a high-speed turbulent flow field of the combustor. Fuel (LPG, containing 40% propane and 60% butane, by volume) is partially mixed in the burner section with the airflow entering to the settling chamber. During experiments, the mean flow velocity is varied, by changing airflow rate and maintaining the fuel flow rate constant, in such a way that equivalence ratio (φ) of air–fuel mixture decreases from a near-stoichiometry value ($\varphi = 0.98$) to a value corresponding to fuel lean condition ($\varphi = 0.49$). Such mixture is initially ignited using a spark plug located at the dump plane of the combustor. The controlled air and fuel flow rates are supplied separately through the mass flow controllers (Alicat Scientific, MCR 2000SLPM—for air, and MCR 100SLPM—for fuel; uncertainty is $\pm 0.8\%$ of measured reading $+0.2\%$ of full-scale reading). The fuel flow rate is fixed at 25 SLPM, while the airflow rate is varied from 400 slpm to 940 slpm. This results in a change in equivalence ratio from 0.98 ± 0.02 to 0.46 ± 0.01.

Three types of measurements are performed during the experiments. A piezoelectric transducer (PCB piezotronics, PCB103B02), located at the dump plane, is used to measure acoustic pressure fluctuations (p') present in the combustor. The global unsteady heat release rate fluctuations ($\dot{q}'$) are captured by using a photomultiplier tube, PMT, (Hamamatsu H10722-01). The local heat release rate fluctuations from the reactive field of the flame are also acquired using a Phantom v12.1 high-speed camera (not shown in the schematic). A CH^* chemiluminescence imaging of the flame is performed using a CH^* band-pass filter ($\lambda = 435$ and 10 nm FWHM) placed in front of the PMT and the camera lens. The chemiluminescence intensity represents the heat release rate from the flame [20, 23]. The data of acoustic pressure fluctuations (using transducer) and global heat release rate oscillations (using PMT) were acquired for 3 s at a sampling frequency of 10 kHz. Data acquisition is performed using a 16-bit analog-to-digital (NI-6143) card. On the other hand, high-speed imaging of the flame is performed for 3 s at a sampling frequency of 2 kHz. For further details on the description of the experimental setup and the data acquisition, refer to [49, 56].

3.1 Temporal Analysis

In this section, we discuss the temporal synchronization behavior of the acoustic pressure (p') and the global heat release rate ($\dot{q}'$) fluctuations obtained from the combustor, while the system dynamics transition from the state of combustion noise to limit cycle oscillations. Since the effects of underlying hydrodynamic flow field and the combustion processes are difficult to separate out in the combustor, we consider both these processes as one oscillator, named "turbulent reactive flow oscillator." Furthermore, as the flow field in the combustor is turbulent, acoustic oscillations are always present in such systems. When mean flow velocity, $\bar{u} = 9.4$ m/s, we notice low-amplitude aperiodic oscillations in the signals of both p' and $\dot{q}'$ (Fig. 2a). Such oscillations have traditionally been referred to as combustion noise [72]. By using various tools from dynamical systems theory, Nair et al. [52] and Tony et al. [77] have recently shown that combustor dynamics observed during combustion noise state has properties of high-dimensional chaos contaminated with white and colored noise. As the flow velocity is increased further ($\bar{u} = 11.9$ m/s), we observe an alternate occurrence of bursts of high-amplitude periodic oscillations among regions of low-amplitude aperiodic oscillations in both p' and $\dot{q}'$ signals (Fig. 2b). Such a state of combustion dynamics is recently discovered and called as intermittency [53]. The inset of Fig. 2b reveals that these signals are seemingly locked during the epochs of periodic oscillations and unlocked during the epochs of aperiodic oscillations.

When the velocity of the flow is increased further ($\bar{u} = 13.2$ m/s), we observe weakly periodic oscillations throughout the length of the signal (Fig. 2c). Here, we notice a wide variation in cycle-to-cycle amplitudes of the both p' and $\dot{q}'$ oscillations. As the value of $\bar{u}$ is further increased ($\bar{u} = 17.2$ m/s), the dynamics transitions to a state of high-amplitude strongly periodic oscillations in both the signals (Fig. 2d). During this state, we observed minimal variation in the cycle-to-cycle amplitudes of both p' and $\dot{q}'$ signals. In Fig. 2e, we plot the variation of *rms* value of the acoustic pressure (p'_{rms}) for various values of $\bar{u}$. The plot reasserts a continuous increase in the amplitude of acoustic pressure fluctuations; however, the slope of the plot does not remain same in different regions of the combustor dynamics.

The coupled interaction of p' and $\dot{q}'$ signals can further be understood through a frequency domain analysis. In Fig. 3a, a plot of waterfall diagram for demonstrating the variation of power spectral density (PSD) calculated for the acoustic pressure and the heat release rate oscillations with the change in mean velocity of the flow is shown. The variation of dominant frequencies of p' and $\dot{q}'$ as a function of $\bar{u}$ is

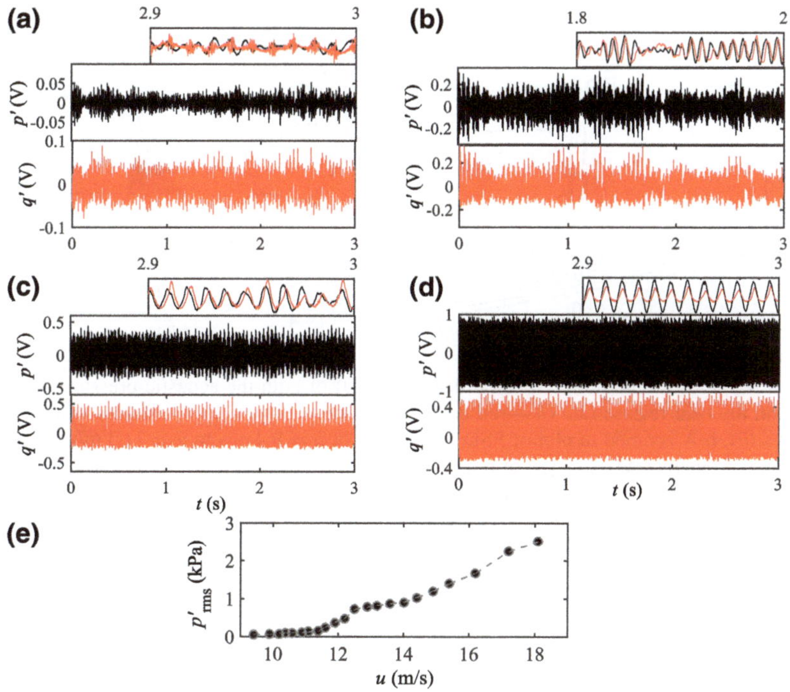

Fig. 2 **a–d** Acoustic pressure (p'—black color) and the heat release rate ($\dot{q}'$—red color) signals acquired for flow velocities of $\bar{u}$ = 9.4 m/s, 11.9 m/s, 13.2 m/s, and 17.2 m/s, respectively. Insets in each plot demonstrate the dynamics corresponding to different states of the combustor operation such as **a** a low-amplitude aperiodic oscillations, **b** an intermittency, **c** a weakly periodic limit cycle oscillations, and **d** a strongly periodic limit cycle oscillations. **e** A plot showing the variation of root-mean-square (rms) value of the acoustic pressure signal (p'_{rms}) with different values of $\bar{u}$ ranging from 9.4 to 18.1 m/s. Reproduced with permission from Pawar et al. [55]

plotted in Fig. 3c. We notice the presence of distinct shallow peaks of dominant frequency around 130.7 Hz and 25.3 Hz in p' and $\dot{q}'$ signals, respectively, during the state of combustion noise. Whereas, during the transition of system dynamics from combustion noise to intermittency, we observed an emergence of a secondary band of frequencies in between the previously observed individual frequency bands of p' and $\dot{q}'$ (Fig. 3b). However, the dominance of frequencies still remains at the acoustic and heat release rate frequency bands. When the periodicity in the intermittency signals is sufficiently strong, the dominance of these coupled signals shifts to a secondary band of frequency. This is in contrast to the study by Chakravarty et al. [5] which suggests that the locking of dominant frequencies happens during the onset of thermoacoustic instability. Conversely, our study shows that the dominant frequencies of p' and $\dot{q}'$ signals get locked for the first time during the state of intermittency and not during the onset of thermoacoustic instability for the turbulent combustors. With further increase in $\bar{u}$, we notice a continuous increase in the value

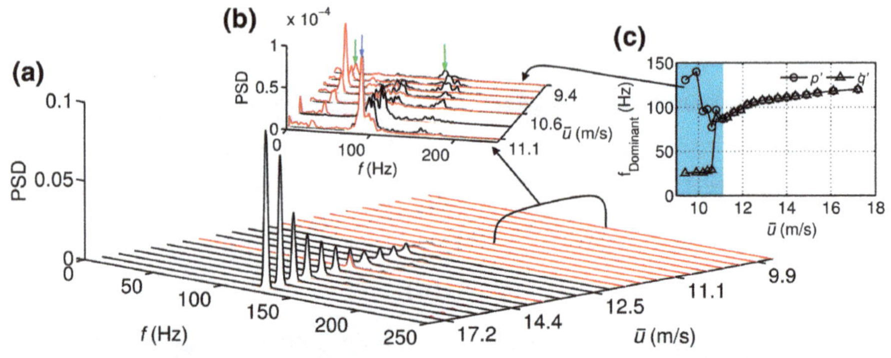

Fig. 3 **a** Variation of the power spectral density (PSD) of both the acoustic pressure (p'—black color) and the heat release rate ($\dot{q}'$—red color) fluctuations obtained for different values of flow velocities ($\bar{u}$). **b** A zoomed in plot of PSD for a range of $\bar{u} = 9.4$ m/s to $\bar{u} = 11.1$ m/s. **c** Variation of dominant frequencies ($f_{Dominant}$) of both p' and $\dot{q}'$ with $\bar{u}$. Green color arrows in **b** indicate the distinct unlocked frequencies of p' and $\dot{q}'$ oscillations observed at $\bar{u} = 9.4$ m/s, and a blue color arrow represents the frequency band observed at first time locking of dominant frequencies of both p' and $\dot{q}'$ oscillations. Reproduced with permission from Pawar et al. [55]

of locked dominant frequencies of these signals (see the non-highlighted portion of Fig. 3c). As this value of dominant frequency approaches the fundamental mode of acoustic field in the system, oscillations in both p' and $\dot{q}'$ signals grows to a state of high-amplitude periodic oscillations. Thus, the presence of separate frequency peaks in both acoustic pressure and heat release rate signals during the combustion noise state, and locking of these frequencies at the emergent secondary band of response frequency, observed in between these frequencies, during the intermittency state suggests the existence of mutual synchronization [58] between these signals in the thermoacoustic system.

The synchronization behavior of coupled p' and $\dot{q}'$ signals can be determined by simultaneously plotting the variation in the probability of recurrence, $P(\tau)$ (Sect. 2.2.2), for different values of delay time (τ) [56]. Figure 4 shows the plots of $P(\tau)$ against τ for the state of combustion noise, weakly correlated and strongly correlated limit cycles. Figure 5, further, shows a plot for the intermittency state of the combustion dynamics. In such plots, the locking of positions as well as heights of peaks of $P(\tau)$ depicts the different types of synchronization present in the system of coupled oscillators [62]. When the system dynamics is in a state of aperiodic oscillations (see Fig. 2a), we observe no correspondence between the locking of positions or heights of peaks of $P(\tau)$ of both p' and $\dot{q}'$ signals, suggesting their desynchronization behavior. Instead, when the system dynamics transition to the states of weakly and strongly correlated limit cycle oscillations (see Fig. 2c, d), we notice a perfect locking of the positions of the peaks of $P(\tau)$ for both p' and $\dot{q}'$ signals (see Fig. 4b, c). Such a locking of the positions of peaks further reasserts synchronization of the signals; however, the absence of perfect locking in the heights of $P(\tau)$ of acoustic pressure and heat release rate oscillations (Fig. 4b) indicates phase synchronization

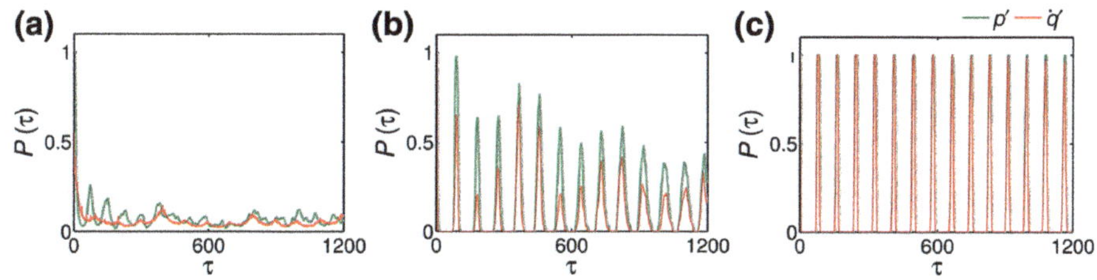

Fig. 4 Plots of probability of recurrence ($P(\tau)$) against different values of delay time (τ) for acoustic pressure (p'—green color) and heat release rate ($\dot{q}'$—red color) fluctuations. **a–c** Different states of synchronization such as desynchronization (aperiodic oscillations), phase synchronization (periodic oscillations), and generalized synchronization (periodic oscillations) obtained at $\bar{u} = 9.4$ m/s, $\bar{u} = 13.2$ m/s, and $\bar{u} = 17.2$ m/s, respectively. The parameters that are being fixed: $E = 12$, $T = 2$ ms, $\varepsilon = 25\%$ of the maximum size of the attractor, and $N = 3000$. Reproduced with permission from Pawar et al. [55]

(PS) of these oscillators in the system [62]. Locking of the heights of peaks in the plot of $P(\tau)$ for both p' and $\dot{q}'$ signals suggests the presence of generalized synchronization (GS) between the oscillators (Fig. 4c).

During the state of intermittency, we detect the occurrence of bursts of periodic oscillations amidst the regions of aperiodic oscillations in both the signals. The synchronization characteristics of such signals can further be understood using a plot of $P(\tau)$ as a function of τ obtained during the periodic and aperiodic epochs of the signals (see Fig. 5). The plot of $P(\tau)$ for a periodic region of the signal (refer Fig. 5b) demonstrates the perfect locking of positions of peaks of $P(\tau)$ but unlocking of their heights confirms phase synchronization during the bursts of the periodic oscillations, whereas, during aperiodic epochs of the oscillations, $P(\tau)$ plots for both the signals show unlocking in locations and heights confirming the desynchronization of these oscillations. Thus, system dynamics during intermittency suggests the intermittent phase synchronization (IPS) of both p' and $\dot{q}'$ signals during intermittency. Therefore, in turbulent combustion system examined in the present study, we observe the transition of coupled acoustic pressure and heat release rate fluctuations from a state of complete desynchronization (aperiodic oscillations) to that of generalized synchronization (periodic oscillations) through the occurrence of intermittent phase synchronization (intermittency) and phase synchronization (periodic oscillations).

Further, we characterize the distinct features of two limit cycle oscillations observed in the system. We differentiate them into two types of synchronization as phase synchronization (PS—weakly correlated in amplitude) and generalized synchronization (GS—strongly correlated in amplitude). In Fig. 6a, b, we plot the temporal variation of the instantaneous phase difference between p' and $\dot{q}'$ signals ($\Delta\phi_{p',\dot{q}'}$) observed during PS and GS, respectively. The Hilbert transform (explained in Sect. 2.2.1) is used to calculate the instantaneous phases of these signals. The presence of synchronization in between the signals can be inferred from the fluctuations of their relative phases around a constant phase difference. We observe difference in the properties the relative phase signals observed during the different states of

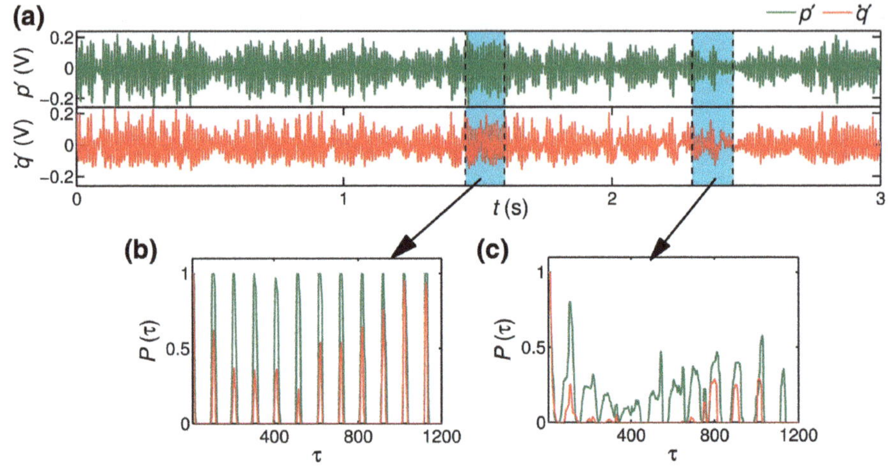

Fig. 5 **a** Time series plots of acoustic pressure (p'—green color) and heat release rate fluctuations ($\dot{q}'$—red color) acquired during the state of intermittency for $\bar{u} = 11.6$ m/s. **b**, **c** The plots of $P(\tau)$ versus τ obtained for periodic and aperiodic epochs of the p' and $\dot{q}'$ signals, respectively. The parameter values those are constant are: $E = 12$, $T = 2$ ms, $\varepsilon = 25\%$ of the maximum size of the attractor, and $N = 3000$. Reproduced with permission from Pawar et al. [55]

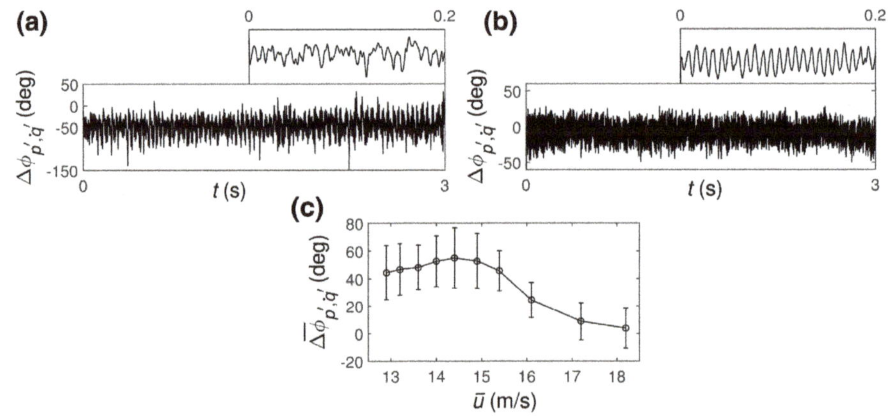

Fig. 6 **a**, **b** The plots of instantaneous phase difference ($\Delta\phi_{p',\dot{q}'}$) between p' and $\dot{q}'$ signals calculated using Hilbert transform. **a** The state of phase synchronization ($\bar{u} = 13.2$ m/s) and **b** the state of generalized synchronization ($\bar{u} = 17.2$ m/s). **c** A plot showing the variation of mean relative phase ($\overline{\Delta\phi_{p',\dot{q}'}}$) between p' and $\dot{q}'$ signals for different values of flow velocities observed during transition from PS to GS state. Reproduced with permission from Pawar et al. [55]

synchronization. For the PS state, signal of relative phase shows noisy behavior (Fig. 6a), whereas that of the GS state shows the existence of seemingly periodic oscillations (Fig. 6b). Furthermore, we plot the variation in the mean value of the relative phase ($\overline{\Delta\phi_{p',\dot{q}'}}$), wrapped in the interval $-180°$ to $180°$, obtained between p' and $\dot{q}'$ during the transition from PS to GS state. We observe that such a value of the mean relative phase is close to 50° for the PS state and it gradually decreases to a value close to 0° for the GS state. Thus, we note that the mean phase shift between the signals of acoustic pressure and heat release rate fluctuations reduces as the system dynamics transition from PS to GS state. A reduced order model exhibiting such a synchronization transition in acoustic pressure and circulation of the underlying hydrodynamic field can be found in [56]. Having discussed the instantaneous interaction between p' and $\dot{q}'$ in the temporal domain, let us turn our attention to investigate their spatiotemporal interactions.

3.2 *Spatiotemporal Analysis*

The spatiotemporal dynamics during the transition to thermoacoustic instability is studied by analyzing the chemiluminescence (CH^*) images of reaction zone of the combustor, captured with a high-speed camera. Although the reactive flow field is three-dimensional, we analyzed a two-dimensional flow field by imaging the central plane in the combustor. The plane of imaging had a depth of focus of 9 mm out of 90 mm width of the combustor. Such images are then analyzed to extract the information of local heat release rate oscillations at their individual pixel locations. The coupled analysis of the local heat release rate oscillations with the unsteady pressure data is performed to explore the transitions in spatiotemporal dynamics of the reaction zone due to change in the control parameter.

The unsteady acoustic pressure, $p'(t)$, is measured at a single location near the dump plane of the combustor. The single point measurement suffices as the acoustic pressure along the reaction zone is nearly uniform. As the length of the combustion zone being analyzed is much less than the acoustic wavelength, the pressure variation observed across the flame zone is very small [37]. Further, the local heat release rate field, $\dot{q}'(x, y, t)$, is obtained from the chemiluminescence images. The instantaneous phase difference between the local heat release rate oscillations, $\dot{q}'(x, y, t)$, and the pressure oscillations, $p'(t)$, is calculated using a Hilbert transform. The phase field thus obtained is analyzed to study the change in the spatial pattern at the onset of thermoacoustic instability. Comparison of phase fields from other approaches can be found in Appendix-A of Ref. [49].

Figure 7 shows typical instantaneous phase fields obtained for different states of combustor dynamics such as combustion noise, intermittency, and thermoacoustic instability. The spatial distribution of instantaneous phases observed during the stable operation is shown in Fig. 7a. The asynchrony in the local heat release rate field is apparent from the grainy nature of the instantaneous phase field. This asynchronous behavior of the local heat release rate causes the amplitude of the pressure

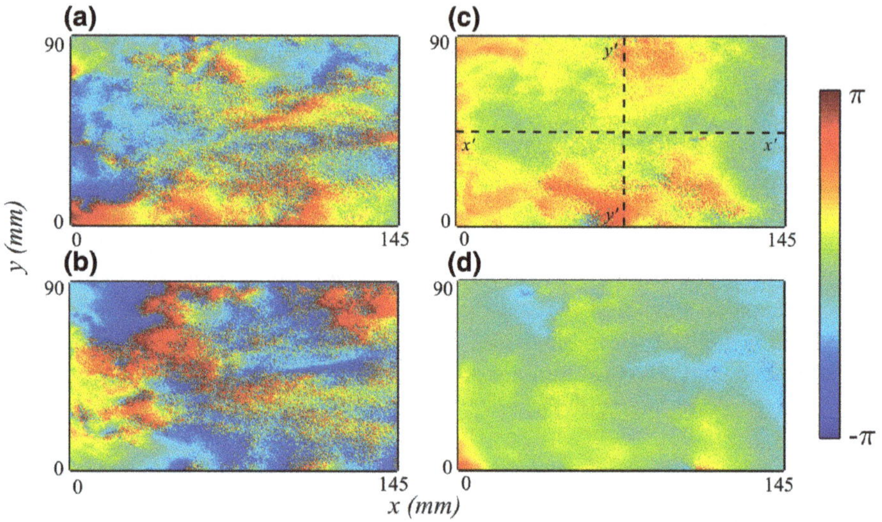

Fig. 7 Typical instantaneous phase fields are shown at different dynamical states: **a** during combustion noise ($\varphi = 0.98$), **b** during low-amplitude aperiodic pressure oscillation in intermittency ($\varphi = 0.72$), **c** during large-amplitude periodic burst in intermittency ($\varphi = 0.72$), and **d** during thermoacoustic instability ($\varphi = 0.58$). The asynchrony in the local heat release rate oscillations can be inferred from the grainy nature in (**a**), whereas thermoacoustic instability corresponds to a coherence in the field of local heat release rate (**d**). During intermittency (**b** and **c**), we observe the regions of synchrony and asynchrony to coexist. Further, synchronous regions during large amplitude periodic oscillations **c** are mostly "in-phase" (between $-\pi/2$ and $\pi/2$, indicated by light colors). Reproduced with permission from Mondal et al. [49]

oscillations not to grow, which, in turn, results in low-amplitude aperiodic (broadband) oscillations in the temporal dynamics of both p' and $\dot{q}'$ signals observed during the state of combustion noise (Fig. 2a). The underlying turbulence in the flow field may result in such incoherent pattern observed in the local heat release rate fluctuations. Chaté and Manneville [6] have shown that the turbulence can lead to loss of spatial correlation in the flow field.

Intermittency, a state observed prior to thermoacoustic instability, is referred to as an alternate (apparently random) occurrence of sequence of low-amplitude aperiodic oscillations and high-amplitude bursts of periodic oscillations (Fig. 2b). To examine the phase field during intermittency, instantaneous phase fields at two time instants during periodicity and aperiodicity are shown in Fig. 7b, c, respectively. In both the figures, there is a spatial coexistence of phase synchrony and phase asynchrony of oscillating heat release rate (Fig. 7b, c). However, the difference lies in the phase values of the synchronous regions. During high-amplitude burst, the phase values lie between $-\pi/2$ and $\pi/2$ in synchronous regions (Fig. 7c), whereas, during low-amplitude aperiodic oscillations, they do not. During thermoacoustic instability, the spatial distribution of phases shows a few coherent patches in the phase field (Fig. 7d), indicating phase synchrony.

The Rayleigh criterion says that when the heat release rate and the acoustic pressure oscillate in phase or their phase difference lies between $-\pi/2$ and $\pi/2$ radians, the net acoustic driving becomes positive. Therefore, acoustic driving originates from the "in-phase" regions (shown by light colors in Fig. 7) and acoustic damping comes from the "out-of-phase" regions (shown by dark colors in Fig. 7) of the instantaneous phase field. During the stable operation (combustion noise), the phases are randomly distributed over the reactive flow field (Fig. 7a). The spatial asynchrony of phases causes very low acoustic driving, which means that the total energy transfer into the acoustic field is low, and hence, the large-amplitude acoustic oscillations are not exhibited during the state of combustion noise. Combustion noise, therefore, happens to be aperiodic and with low-amplitude oscillations (Fig. 2a). On the other hand, we observed the large amplitude periodic oscillations during the state of thermoacoustic instability (Fig. 2d), wherein the coherent regions exhibit nearly "in-phase" oscillations (Fig. 7d).

During intermittency, the regions of phase synchrony and phase asynchrony exist simultaneously in the reaction zone. The reaction zone is mostly dominated by phase asynchronous patches during low-amplitude aperiodic pressure oscillations (Fig. 7b). On the contrary, during the bursts of periodic oscillations, the reaction zone is dominated by regions of coherent oscillations which are nearly "in-phase" (Fig. 7c). The positive driving due to "in-phase" coherent regions in the combustor causes intermittent periodic bursts. At other instances, when incoherent regions prevail in the reaction zone, the system exhibits low-amplitude aperiodic oscillations.

Figure 8 shows a transition from asynchrony to synchrony as the system approaches thermoacoustic instability. The instantaneous phases between p' and $\dot{q}'$ are depicted with arrows in the phasor plots. Spatial asynchrony can be inferred from randomly distributed phasors during the stable operation (Fig. 8a). Such spatial asynchrony, also known as phase turbulence [69], can be attributed to the inhomogeneity due to the underlying hydrodynamic turbulence in the system. Spatial phase synchrony

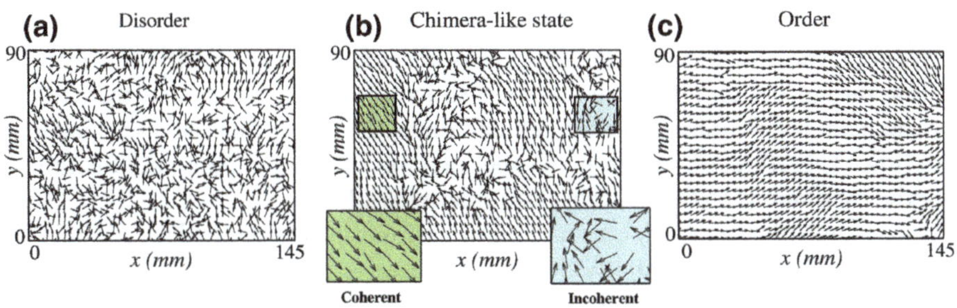

Fig. 8 Typical phasor fields are shown for different dynamical states at the onset of thermoacoustic instability. An order is observed to be emerged from phase turbulence. The randomly oriented phasors at $\varphi = 0.98$ **a** indicates the phase asynchrony, whereas at $\varphi = 0.58$ (**c**), the phasors are mostly aligned to their neighbors, showing phase synchrony. An interesting pattern is observed during intermittency (**b**, $\varphi = 0.72$). Groups of ordered and disordered phasors coexist simultaneously in the reaction zone. Reproduced with permission from Mondal et al. [49]

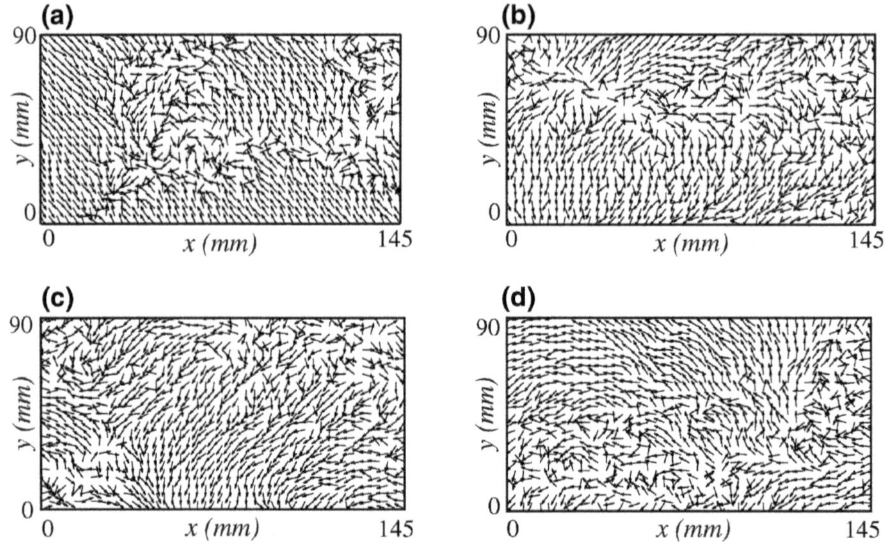

Fig. 9 Instantaneous phasor fields at different instants of time are shown during intermittency. Time instances are **a** $t = 0.05$ s, **b** $t = 0.1$ s, **c** $t = 0.15$ s, **d** $t = 0.2$ s. The regions of ordered and disordered phasors change their location in the flow domain with time. This indicates the transition from aperiodic to periodic state as they synchronize intermittently. Reproduced with permission from Mondal et al. [49]

emerges during the occurrence of thermoacoustic instability (Fig. 8c) due to the enhanced acoustic driving in the system.

At intermediate equivalence ratios, the regions of phase asynchrony and the regions of phase synchrony, coexist, resembling a chimera state (Fig. 8b) which was observed in an ensemble of coupled nonidentical oscillators. Such a chimera-like state is found to be a characteristic of intermittency. The chimera state with varying global synchrony in time is known as breathing chimera [1].

The snapshots of the instantaneous phasor fields obtained at different instants of time during the chimera-like state are shown in Fig. 9. We notice that, in the flow domain, the regions of disordered and ordered phasors change their locations with time. Therefore, desynchronized aperiodic oscillations undergo a transition to periodic oscillations as they synchronize intermittently.

Space–time plots and instantaneous phase distribution for 1D field are further presented in Fig. 10. The 1D spatially extended fields ($x' - x'$ and $y' - y'$) chosen for further analysis are marked in Fig. 7c. Phase asynchrony is observed in the reaction zone (Fig. 10a-I) and the instantaneous phase distribution becomes scattered (Fig. 10a-II) during the occurrence of combustion noise. On the contrary, phase synchrony is observed (Fig. 10c-I) as a dense phase distribution over the field (Fig. 10c-II).

The coexistence of regions of phase synchrony and asynchrony in the space and time (Fig. 10b-I) observed during intermittency was interesting. Such concurrence

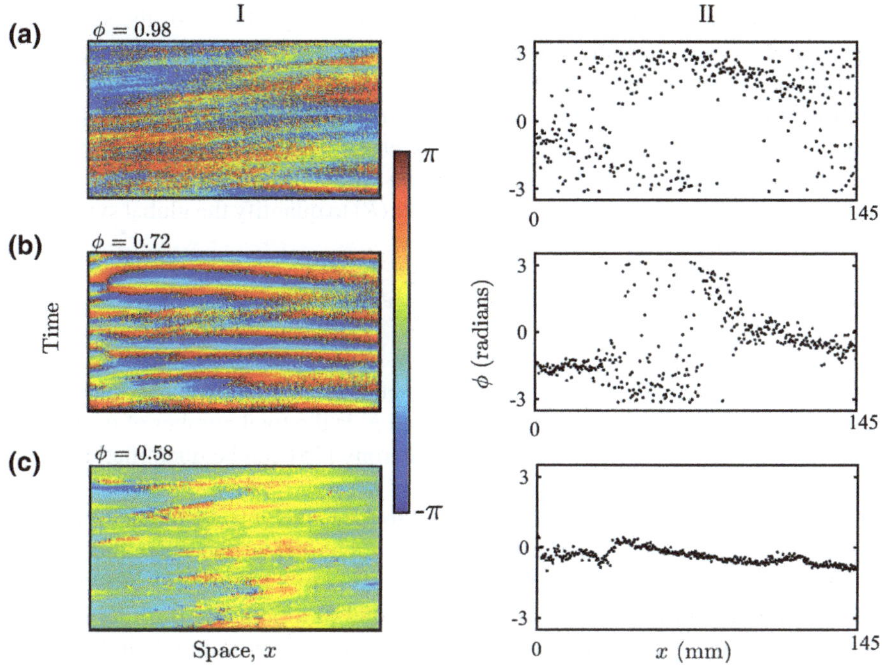

Fig. 10 (**I**) Space–time plots and (**II**) spatial distribution of instantaneous phase for 1D spatially extended field along a $(x' - x')$ line shown in Fig. 7c. The grainy structure in the space–time plot (**a-I**) and the scattered instantaneous phases in (**a-II**) indicate the existence of phase turbulence during the occurrence of combustion noise. The phase synchrony is apparent in (**c-I**) and also from the congregated phases around a line in (**c-II**) during thermoacoustic instability. During intermittency, the simultaneous existence of phase synchrony and phase asynchrony in the reaction zone is observed (**b-I, b-II**). Reproduced with permission from Mondal et al. [49]

of synchrony and asynchrony in the reaction zone resembles a chimera state. The distribution of the instantaneous phases shown in Fig. 10b-II clearly reveals the simultaneous existence of phase synchrony and phase asynchrony in 1D spatially extended field. The regions of phase synchrony give rise to the intermittent bursts of periodic oscillations, whereas the low-amplitude aperiodic pressure oscillations is a consequence of phase asynchrony in the reactive flow field.

During the state of intermittency, the system exhibits a novel spatiotemporal dynamics. This refers to concurrence of regions of phase synchronous periodic oscillations and phase asynchronous aperiodic oscillations of local heat release rate. Further, an increase in acoustic driving is caused by "in-phase" synchronous patches that emerge out of the completely asynchronous local heat release rate oscillations having "out-of-phase" phase relation. Consequently, this intermittent occurrence of energy transfer from the reacting flow field to the acoustic field results in the bursts of large-amplitude periodic oscillations in the temporal dynamics of acoustic pressure (Fig. 2b). Similar space–time plot with the section $y' - y'$ can be found in [49].

Further, we characterize the transition in the global synchrony of the system as it approaches thermoacoustic instability. Such synchronization transition in thermoacoustics is similar to the phase transition reported in thermodynamics and is quantified through a measure called order parameter developed by Kuramoto [32]. An order parameter is defined as the degree of order in a system undergoing a phase transition in thermodynamics. The values of the order parameter range from zero to one. We compute the Kuramoto order parameter (R) to quantify the global synchrony of the phase field as,

$$R(t) = N^{-1}\left|\sum_{j=1}^{N} e^{i\theta_j(t)}\right| \tag{5}$$

where $R(t)$ is the time-dependent order parameter, $\theta_j(t)$ is the instantaneous phase of jth spatially extended oscillator at time t, and N is the total number of oscillators in the network. A transition in the global synchrony [15] can be quantified through the Kuramoto order parameter.

Each $e^{i\theta_j(t)}$ term in Eq. 5 can be associated to a vector lying on the unit circle in the complex plane. The order parameter (R) is a measure of the magnitude of the resultant of these vectors, normalized with the total number of vectors. R is close to zero for randomly oriented vectors that correspond to a perfectly disordered state, whereas, when they are all aligned in one direction, R approaches 1, indicating a perfectly ordered state.

For all the dynamical states we considered, the fluctuations in R are evident in the time series of the order parameter. These fluctuations might be due to the unsteadiness caused by the underlying turbulence. However, the gradual increase in the mean value of the order parameter, $\bar{R}$, indicates a transition from a disordered to ordered state (Fig. 11) which is more evident in Fig. 12a, wherein the variation of $\bar{R}$ is plotted against the equivalence ratio. The growth in global synchrony shown in Fig. 12a is apparent at the onset of thermoacoustic instability.

During thermoacoustic instability, the value of $\bar{R}$ attains 0.64 which is not close to 1. This happens due to a couple of reasons: (1) the turbulent fluctuations exist even in the ordered state, (2) all phasors in the flow field of the ordered state do not align in the same direction, but, they exhibit local synchrony (see Fig. 8c). In contrast, the fluctuations in R become lower when the system is either in the ordered or disordered state (Fig. 12b). Prior to the ordered state, large fluctuations in the value of R are observed, which indicate a state mimicking a breathing chimera [1]. Such breathing nature of the chimera-like state is further confirmed by plotting the time series of $\cos(\theta_j)$ for a few arbitrarily chosen spatial locations (bottom of Fig. 11b). The phase synchrony in this plot is shown through overlapping of the time series, and the asynchrony is indicated through the non-overlapping of the time series. This features of $\cos(\theta_j)$ reaffirms the breathing nature of the chimera-like state.

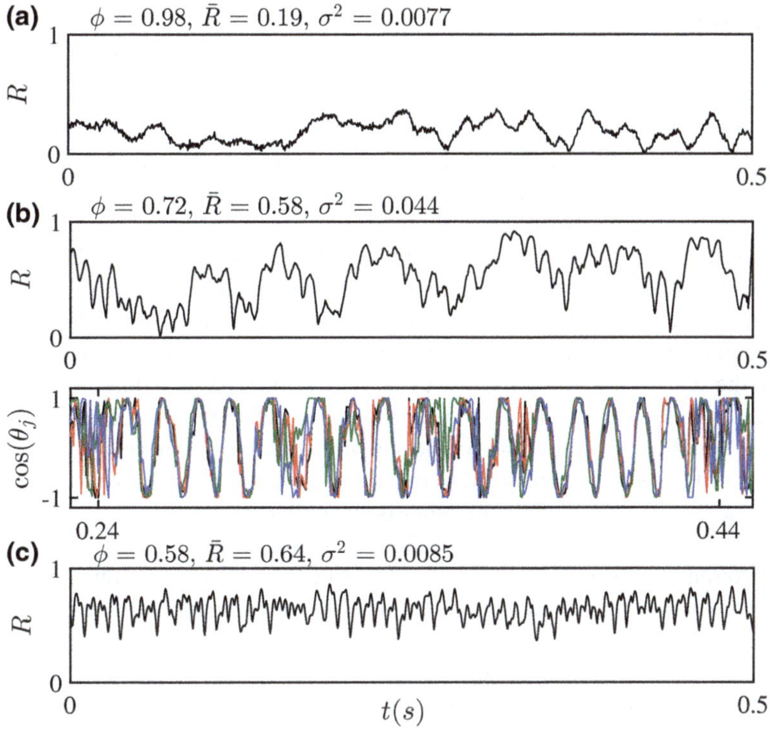

Fig. 11 Variation of Kuramoto order parameter (R) in time indicates the change in the global phase synchrony observed during the states of **a** combustion noise, **b** intermittency, and **c** thermoacoustic instability. Time-averaged value ($\bar{R}$) and variance (σ^2) of R are shown. The breathing nature chimera-like state is further confirmed through the temporal variation of $\cos(\theta_j)$ obtained for randomly chosen spatial locations in the combustion zone. Different spatial locations are indicated through different colors. Reproduced with permission from Mondal et al. [49]

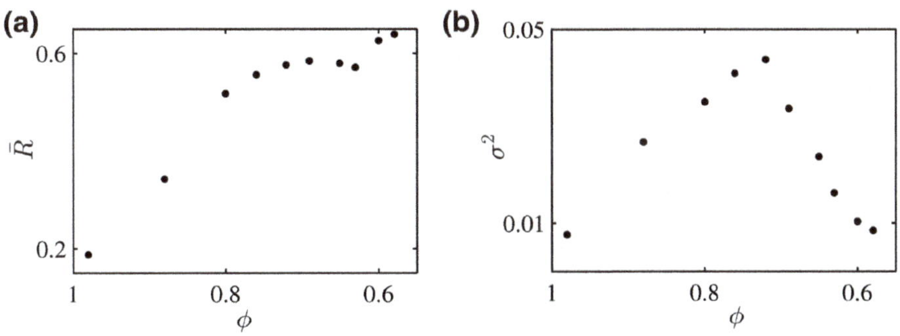

Fig. 12 Variation of **a** mean ($\bar{R}$) and **b** variance (σ^2) of the Kuramoto order parameter (R) are plotted with the range of equivalence ratios. Higher value of σ^2 obtained during intermittency indicates a state which mimics breathing chimera. Reproduced with permission from Mondal et al. [49]

4 Summary

In this chapter, we discussed the characterization of temporal and spatiotemporal transition of a thermoacoustic system of a turbulent combustor using the concepts of synchronization theory. The temporal analysis reveals that the large-amplitude periodic pressure oscillation during thermoacoustic instability is an outcome of the synchronization between the acoustic field and the turbulent reactive flow inside the combustor. There exists a bidirectional coupling between such oscillators. Thus, synchronization theory applied to a turbulent thermoacoustic system provides a new description to the intermittency route to thermoacoustic instability. Here, the system dynamics finally reach to a state of generalized synchronization, having passed through the state of intermittent phase synchronization (IPS) and phase synchronization. During IPS, the oscillators are synchronized in the periodic epochs, while, in the aperiodic epochs, the oscillations are desynchronized. Furthermore, we observe that the states of PS and GS correspond to two different types of limit cycle oscillations: one is weakly correlated, while the other is strongly correlated.

The spatiotemporal dynamics is studied by obtaining a field of the instantaneous phase difference between the acoustic field and the local heat release rate fluctuations. The instantaneous phase field shows that the state of combustion noise corresponds to phase asynchronous aperiodic oscillations, while thermoacoustic instability corresponds to phase synchronous periodic oscillations in the reaction zone. The emergence of phase synchrony in the spatial field results in an increase in the acoustic driving, leading to the onset of thermoacoustic instability in the system. Such transition is preceded by an intermediate state, wherein regions of phase asynchronous aperiodic oscillations and regions of phase synchronous periodic oscillations coexist simultaneously, resembling a chimera-like state similar to that was reported for a network of coupled identical oscillators. Further, we observed that, during the occurrence of the chimera-like state, the spatial regions of synchronous and asynchronous phasors change their locations in the reaction zone with time, being qualitatively similar to characteristics of "breathing chimera."

To summarize, our approach to viewing the onset of thermoacoustic instability as a synchronization phenomenon is the first step toward understanding the complex nonlinear interactions. Such interactions in combustors with turbulent flow can be observed temporally as well as spatiotemporally between the coupled subsystems of such combustors. Furthermore, the use of synchronization framework can be extended to other practical fluid mechanical systems involving turbulent non-reacting flows passed over a bluff body or flows involving density stratified reacting wakes. In such systems, the coupled response of interacting oscillators leads to the unstable phenomenon of self-sustained oscillations, where the quantitative analysis of synchronization is still not a much explored field of research.

Acknowledgements The authors would like to thank Office of Naval Research Global (ONRG) for the funding (grant no. N62909-14-1-N 299); Dr. R. Kolar is the contract monitor from ONRG. SM gratefully acknowledges the institute postdoctoral fellowship from IIT Madras. The authors gratefully acknowledge the valuable discussions with Mr. A. Seshadri, Dr. V. R. Unni, Dr. D. V. Senthilkumar, Dr. V. K. Chandrasekar, and Prof. M. Lakshmanan. The authors would also like to acknowledge the help provided by Mr. N. Babu, Mr. Thilagraj, Mr. Manikandan, and Mr. Syam for conducting the experiments. We are grateful to Mr. Thilagraj for providing the schematic of the experimental setup and to Dr. T. Komarek and Prof. W. Polifke for providing the design, which was adapted to fabricate this setup.

References

1. D.M. Abrams, R. Mirollo, S.H. Strogatz, D.A. Wiley, Solvable model for chimera states of coupled oscillators. Phys. Rev. Lett. **101**(8), 084–103 (2008)
2. S. Ahn, C. Park, L.L. Rubchinsky, Detecting the temporal structure of intermittent phase locking. Phys. Rev. E **84**(1), 016–201 (2011)
3. N. Ananthkrishnan, S. Deo, F.E. Culick, Reduced-order modeling and dynamics of nonlinear acoustic waves in a combustion chamber. Combust. Sci. Technol. **177**(2), 221–248 (2005)
4. S. Candel, Combustion dynamics and control: progress and challenges. Proc. Combust. Inst. **29**(1), 1–28 (2002)
5. S.R. Chakravarthy, O.J. Shreenivasan, B. Boehm, A. Dreizler, J. Janicka, Experimental characterization of onset of acoustic instability in a nonpremixed half-dump combustor. J. Acoust. Soc. Am **122**(1) 120–127 (2007)
6. H. Chaté, P. Manneville, Transition to turbulence via spatio-temporal intermittency. Phys. Rev. Lett. **58**(2), 112 (1987)
7. L. Crocco, S.I. Cheng, Theory of combustion instability in liquid propellant rocket motors, Technical report. Princeton Univ, NJ, 1956
8. F. Culick, Nonlinear behavior of acoustic waves in combustion chambersi. Acta Astronaut. **3**(9–10), 715–734 (1976)
9. F. Culick, Some recent results for nonlinear acoustics in combustion chambers. AIAA J. **32**(1), 146–169 (1994)
10. F. Culick, P. Kuentzmann, Unsteady motions in combustion chambers for propulsion systems, Technical report. Nato Research and Technology Organization Neuilly-Sur-Seine (France), 2006
11. S. Datta, S. Mondal, A. Mukhopadhyay, D. Sanyal, S. Sen, An investigation of nonlinear dynamics of a thermal pulse combustor. Combust. Theory Model. **13**(1), 17–38 (2009)
12. S. Domen, H. Gotoda, T. Kuriyama, Y. Okuno, S. Tachibana, Detection and prevention of blowout in a lean premixed gas-turbine model combustor using the concept of dynamical system theory. Proc. Combust. Inst. **35**(3), 3245–3253 (2015)
13. A.P. Dowling, Nonlinear self-excited oscillations of a ducted flame. J. Fluid Mech. **346**, 271–290 (1997)
14. J.P. Eckmann, S.O. Kamphorst, D. Ruelle, Recurrence plots of dynamical systems. EPL (Europhysics Letters) **4**(9), 973 (1987)
15. E.M. Essaki Arumugam, M.L. Spano, A chimeric path to neuronal synchronization. Chaos: Interdiscip. J. Nonlinear Sci. **25**(1), 013–107 (2015)
16. H. Fujisaka, T. Yamada, Stability theory of synchronized motion in coupled-oscillator systems. Prog. Theor. Phys. **69**(1), 32–47 (1983)
17. J. Gleick, *Chaos: Making a New Science*, Enhanced edn. (Open Road Media, 2011)
18. H. Gotoda, H. Nikimoto, T. Miyano, S. Tachibana, Dynamic properties of combustion instability in a lean premixed gas-turbine combustor. Chaos: Interdiscip. J. Nonlinear Sci. **21**(1), 013–124 (2011)

19. H. Gotoda, Y. Shinoda, M. Kobayashi, Y. Okuno, S. Tachibana, Detection and control of combustion instability based on the concept of dynamical system theory. Phys. Rev. E **89**(2), 022–910 (2014)
20. F. Guethe, D. Guyot, G. Singla, N. Noiray, B. Schuermans, Chemiluminescence as diagnostic tool in the development of gas turbines. Appl. Phys. B **107**(3), 619–636 (2012)
21. U. Hegde, D. Reuter, B. Daniel, B. Zinn, Flame driving of longitudinal instabilities in dump type ramjet combustors. Combust. Sci. Technol. **55**(4–6), 125–138 (1987)
22. N.E. Huang, Z. Shen, S.R. Long, M.C. Wu, H.H. Shih, Q. Zheng, N.C. Yen, C.C. Tung, H.H. Liu, The empirical mode decomposition and the hilbert spectrum for nonlinear and non-stationary time series analysis, in *Proceedings of the Royal Society of London A: mathematical, physical and engineering sciences*, vol. 454 (The Royal Society, 1998), pp. 903–995
23. Y. Ikeda, J. Kojima, H. Hashimoto, Local chemiluminescence spectra measurements in a high-pressure laminar methane/air premixed flame. Proc. Combust. Inst. **29**(2), 1495–1501 (2002)
24. C.C. Jahnke, F.E. Culick, Application of dynamical systems theory to nonlinear combustion instabilities. J. Propuls. Power **10**(4), 508–517 (1994)
25. M.P. Juniper, R. Sujith, Sensitivity and nonlinearity of thermoacoustic oscillations. Ann. Rev. Fluid Mech. **50**(1) (2017)
26. L. Kabiraj, R. Sujith, Nonlinear self-excited thermoacoustic oscillations: intermittency and flame blowout. J. Fluid Mech. **713**, 376–397 (2012)
27. L. Kabiraj, A. Saurabh, P. Wahi, R. Sujith, Route to chaos for combustion instability in ducted laminar premixed flames. Chaos: Interdiscip. J. Nonlinear Sci. **22**(2), 023–129 (2012)
28. L, Kabiraj, R. Sujith, P. Wahi, Investigating the dynamics of combustion-driven oscillations leading to lean blowout. Fluid Dyn. Res. **44**(3), 031–408 (2012)
29. K. Kashinath, I.C. Waugh, M.P. Juniper, Nonlinear self-excited thermoacoustic oscillations of a ducted premixed flame: bifurcations and routes to chaos. J. Fluid Mech. **761**, 399–430 (2014)
30. J.O. Keller, L. Vaneveld, D. Korschelt, G. Hubbard, A. Ghoniem, J. Daily, A. Oppenheim, Mechanism of instabilities in turbulent combustion leading to flashback. AIAA J. **20**(2), 254–262 (1982)
31. D.W. Kendrick, T.J. Anderson, W.A. Sowa, T.S. Snyder, Acoustic sensitivities of lean-premixed fuel injectors in a single nozzle rig, in *ASME 1998 International Gas Turbine and Aeroengine Congress and Exhibition* (American Society of Mechanical Engineers, 1998)
32. Y. Kuramoto, *Chemical Oscillations, Waves, and Turbulence*, vol. 19 (Springer Science & Business Media, 2012)
33. M. Lakshmanan, D.V. Senthilkumar, *Dynamics of Nonlinear Time-Delay Systems* (Springer Science & Business Media, 2011)
34. S. Lei, A. Turan, Nonlinear/chaotic behaviour in thermo-acoustic instability. Combust. Theory Model. **13**(3), 541–557 (2009)
35. T. Lieuwen, Modeling premixed combustion-acoustic wave interactions: a review. J. Propuls. power **19**(5), 765–781 (2003)
36. T. Lieuwen, Online combustor stability margin assessment using dynamic pressure data. Trans. ASME-A-Eng. Gas Turbines Power **127**(3), 478–482 (2005)
37. T. Lieuwen, B.T. Zinn, Application of multipole expansions to sound generation from ducted unsteady combustion processes. J. Sound Vib. **235**(3), 405–414 (2000)
38. T. Lieuwen, H. Torres, C. Johnson, B.T. Zinn, A mechanism of combustion instability in lean premixed gas turbine combustors. Trans.-Am. Soc. Mech. Eng.-J. Eng. Gas Turbines Power **123**(1), 182–189 (2001)
39. T.C. Lieuwen, Experimental investigation of limit-cycle oscillations in an unstable gas turbine combustor. J. Propuls. Power **18**(1), 61–67 (2002)
40. T.C. Lieuwen, V. Yang, in *Combustion Instabilities in Gas Turbine Engines(operational experience, fundamental mechanisms and modeling)*. Progress in Astronautics and Aeronautics, 2005
41. E.N. Lorenz, Deterministic nonperiodic flow. J. Atmos. Sci. **20**(2), 130–141 (1963)
42. N. Marwan, M.C. Romano, M. Thiel, J. Kurths, Recurrence plots for the analysis of complex systems. Phys. Rep. **438**(5), 237–329 (2007)

43. K. McManus, T. Poinsot, S. Candel, A review of active control of combustion instabilities. Prog. Energy Combust. Sci. **19**(1), 1–29 (1993)
44. S. Mondal, A. Mukhopadhyay, S. Sen, Effects of inlet conditions on dynamics of a thermal pulse combustor. Combust. Theory Model. **16**(1), 59–74 (2012)
45. S. Mondal, A. Mukhopadhyay, S. Sen, Dynamic characterization of a laboratory-scale pulse combustor. Combust. Sci. Technol. **186**(2), 139–152 (2014)
46. S. Mondal, V.R. Unni, R. Sujith, Chimera-like states observed during the transition to thermoacoustic instability in turbulent combustor, in *Conference on Nonlinear Systems & Dynamics IISER Kolkata*, vol. 16, p. 18, 2016
47. S. Mondal, A. Mukhopadhyay, S. Sen, Bifurcation analysis of steady states and limit cycles in a thermal pulse combustor model. Combust. Theory Model. **21**(3), 487–502 (2017)
48. S. Mondal, S. Pawar, R. Sujith, Synchronous behaviour of two interacting oscillatory systems undergoing quasiperiodic route to chaos. Chaos: Interdiscip. J. Nonlinear Sci. **27**(10), 103–119 (2017)
49. S. Mondal, V.R. Unni, R. Sujith, Onset of thermoacoustic instability in turbulent combustors: an emergence of synchronized periodicity through formation of chimera-like states. J. Fluid Mech. **811**, 659–681 (2017)
50. S. Mondal, A. Mukhopadhyay, S. Sen, Characterization of turbulent combustion systems using dynamical systems theory, in *Modeling and Simulation of Turbulent Combustion* (Springer, 2018), pp. 543–567
51. V. Nair, R. Sujith, Multifractality in combustion noise: predicting an impending combustion instability. J. Fluid Mech. **747**, 635–655 (2014)
52. V. Nair, G. Thampi, S. Karuppusamy, S. Gopalan, R. Sujith, Loss of chaos in combustion noise as a precursor of impending combustion instability. Int. J. Spray Combust. Dyn. **5**(4), 273–290 (2013)
53. V. Nair, G. Thampi, R. Sujith, Intermittency route to thermoacoustic instability in turbulent combustors. J. Fluid Mech. **756**, 470–487 (2014)
54. G.V. Osipov, B. Hu, C. Zhou, M.V. Ivanchenko, J. Kurths, Three types of transitions to phase synchronization in coupled chaotic oscillators. Phys. Rev. Lett. **91**(2), 024–101 (2003)
55. S.A. Pawar, R. Vishnu, M. Vadivukkarasan, M. Panchagnula, R. Sujith, Intermittency route to combustion instability in a laboratory spray combustor. J. Eng. Gas Turbines Power **138**(4), 041–505 (2016)
56. S.A. Pawar, A. Seshadri, V.R. Unni, R.I. Sujith, Thermoacoustic instability as mutual synchronization between the acoustic field of the confinement and turbulent reactive flow. J. Fluid Mech. **827**, 664–693 (2017). https://doi.org/10.1017/jfm.2017.438
57. S.A. Pawar, S. Mondal, N.B. George, R. Sujith, Synchronization behaviour during the dynamical transition in swirl-stabilized combustor: temporal and spatiotemporal analysis, in *2018 AIAA Aerospace Sciences Meeting*, p. 0394, 2018
58. A. Pikovsky, M. Rosenblum, J. Kurths, *Synchronization: A Universal Concept in Nonlinear Sciences*, vol. 12 (Cambridge University Press, 2003)
59. T.J. Poinsot, A.C. Trouve, D.P. Veynante, S.M. Candel, E.J. Esposito, Vortex-driven acoustically coupled combustion instabilities. J. Fluid Mech. **177**, 265–292 (1987)
60. A.A. Putnam, *Combustion Driven Oscillations in Industry* (Elsevier Publishing Company, 1971)
61. J.W.S. Rayleigh, The explanation of certain acoustical phenomena. Nature **18**(455), 319–321 (1878)
62. M.C. Romano, M. Thiel, J. Kurths, I.Z. Kiss, J. Hudson, Detection of synchronization for non-phase-coherent and non-stationary data. EPL (Europhysics Letters) **71**(3), 466 (2005)
63. M. Rosenblum, A. Pikovsky, Synchronization: from pendulum clocks to chaotic lasers and chemical oscillators. Contemp. Phys. **44**(5), 401–416 (2003)
64. M.G. Rosenblum, A.S. Pikovsky, J. Kurths, Phase synchronization of chaotic oscillators. Phys. Rev. Lett. **76**(11), 1804 (1996)
65. M.G. Rosenblum, A.S. Pikovsky, J. Kurths, From phase to lag synchronization in coupled chaotic oscillators. Phys. Rev. Lett. **78**(22), 4193 (1997)

66. N.F. Rulkov, M.M. Sushchik, L.S. Tsimring, H.D. Abarbanel, Generalized synchronization of chaos in directionally coupled chaotic systems. Phys. Rev. E **51**(2), 980 (1995)
67. K. Schadow, E. Gutmark, Combustion instability related to vortex shedding in dump combustors and their passive control. Prog. Energy Combust. Sci. **18**(2), 117–132 (1992)
68. S. Sen, S. Mondal, A. Mukhopadhyay, Dynamics of thermal pulse combustor, in *Energy Combustion and Propulsion* (New Perspectives, Athena Academic, 2015), pp. 269–312
69. B.I. Shraiman, Order, disorder, and phase turbulence. Phys. Rev. Lett. **57**(3), 325 (1986)
70. D.A. Smith, E.E. Zukoski, Combustion instability sustained by unsteady vortex combustion, in *AIAA Joint Propulsion Conference* (1985)
71. J.D. Sterling, Nonlinear analysis and modelling of combustion instabilities in a laboratory combustor. Combust. Sci. Technol. **89**(1–4), 167–179 (1993)
72. W.C. Strahle, Combustion noise. Prog. Energy Combust. Sci. **4**(3), 157–176 (1978)
73. P. Subramanian, *Dynamical Systems Approach to the Investigation of Thermoacoustic Instabilities*, 2011
74. P. Subramanian, S. Mariappan, R. Sujith, P. Wahi, Bifurcation analysis of thermoacoustic instability in a horizontal rijke tube. Int. J. Spray Combust. Dyn. **2**(4), 325–355 (2010)
75. R. Sujith, M. Juniper, P. Schmid, Non-normality and nonlinearity in thermoacoustic instabilities. Int. J. Spray Combust. Dyn. **8**(2), 119–146 (2016)
76. F. Takens et al., Detecting strange attractors in turbulence. Lect. Notes Math. **898**(1), 366–381 (1981)
77. J. Tony, E. Gopalakrishnan, E. Sreelekha, R. Sujith, Detecting deterministic nature of pressure measurements from a turbulent combustor. Phys. Rev. E **92**(6), 062–902 (2015)
78. V.R. Unni, R. Sujith, Multifractal characteristics of combustor dynamics close to lean blowout. J. Fluid Mech. **784**, 30–50 (2015)
79. J.M. Wilhite, B.J. Dolan, L. Kabiraj, R.V. Gomez, E.J. Gutmark, C.O. Paschereit, Analysis of combustion oscillations in a staged mldi burner using decomposition methods and recurrence analysis, in *54th AIAA Aerospace Sciences Meeting, AIAA SciTech Forum*, (AIAA 2016-1156) (American Institute of Aeronautics and Astronautics, 2016), pp. 1–17
80. T. Yalçınkaya, Y.C. Lai, Phase characterization of chaos. Phys. Rev. Lett. **79**(20), 3885 (1997)

Interaction of Water Spray with Flame

Sourav Sarkar, Joydeep Munshi, Santanu Pramanik, Achintya Mukhopadhyay and Swarnendu Sen

Abstract Increasing concerns about the role of halons on the depletion of ozone in the stratosphere have led to a search for alternate agents for suppression. Water, sprayed in the form of tiny droplets, has emerged as a potential fire suppressant. The present chapter presents a brief review of the recent studies on flame water spray interaction. The effects of water spray on both premixed and non-premixed flames are discussed. The significance of droplet size in flame suppression is explained in details. This understanding will lead to efficient atomizer design for fire suppression systems.

1 Introduction

Fire hazard is one of the major catastrophes that can cause a great damage to property and loss of life. It can occur in all three spaces—above land in a skyscraper, on the ground in the forest; below the ground in mines. Fire protection and safety research

S. Sarkar
Department of Mechanical Engineering, Jadavpur University, Kolkata 700032, India
e-mail: souravsarkar.iitm@gmail.com

J. Munshi
Department of Mechanical Engineering and Mechanics, Lehigh University, 19 Memorial Drive, West Bethlehem, USA
e-mail: jom317@lehigh.edu

S. Pramanik
Mechanical Engineering Department, Indian Institute of Science, Bangalore, Bengaluru 560012, India
e-mail: santanupramanik07@gmail.com

A. Mukhopadhyay · S. Sen (✉)
Mechanical Engineering Department, Jadavpur University, Kolkata 700032, India
e-mail: sen.swarnendu@gmail.com

A. Mukhopadhyay
e-mail: achintya.mukho@gmail.com

A. K. Runchal et al. (eds.), *Energy for Propulsion*, Green Energy and Technology,
https://doi.org/10.1007/978-981-10-7473-8_7

becomes a major concern due to rapid urban growth in present time. Fire research and development groups have been founded to address industry challenges in multiple areas, including detection and signaling, hazardous materials, electrical safety, fire suppression, storage of commodities, and fire-fighter protective clothing and equipment, among other issues. After Second World War, research in fire safety science increased greatly. However, fire suppression using water spray does not receive much attention due to extensive application of Halon gases. Increasing concerns about the role of halons on depletion of ozone in the stratosphere have led to a search for alternate agents for suppression. A dramatic change happened in the research trend when Halon 1301, Halon 1211 and Halon 2402 were announced illegal in developed countries under the Montreal Protocol (1987).

Waster mist system is widely used as fire extinguishment agent in building and compartment fire. Water sprinkler systems are commonly in engine room, machinery space, electronic equipment rooms and computer rooms. In recent time water mist technology has become one of the major interest of the research community in ships and aircrafts application due to availability, low cost and less storage space. In recent times fire suppression using water spray has been adopted as a technique to mitigate fires related to hydrogen in nuclear power plants. Fine sprays can be used along with the igniters in order to quench the flame when hydrogen concentration is below a threshold limit.

Water spray has emerged as a potential fire suppressant due to its enhanced thermal and physical properties. Water possesses several advantages like non-toxicity and environment friendliness. The physical properties like high specific and latent heats ensure that a small quantity of water can be used for extinguishing the fire. It has high heat capacity and latent heat of vaporization. 418 kJ of heat is required to raise the temperature of 1 L of water from 0° to 100 °C and further 2257 kJ is required to convert it from water to vapor. Water spray helps in suppressing fire in two ways:

1. Water droplets are efficient fire suppressing agent because of its rate of evaporation and high latent heat of evaporation, hence while interacting with the flame, it can absorb heat from the flame reducing the flame temperature. If the flame temperature is below the activation temperature, the flame cannot sustain itself.
2. When droplets vaporize water vapor gets added to the continuous phase displacing oxygen inside the reaction zone of the flame; hence it reduces the local equivalence ratio of the reactant mixture. Eventually, if the equivalence ratio becomes less than the lower flammability limit, flame extinction can occur.

2 Quantitative Characterization of Water Spray

The size of the water droplet is an important parameter that determines the effectiveness of the spray by directly affecting the heat transfer and evaporation. The dynamics of the droplets, that are influenced by the kinetic energy (~d^3) and the aerodynamic drag (~d), are also dependant on the diameter. The size of the droplets vary with the

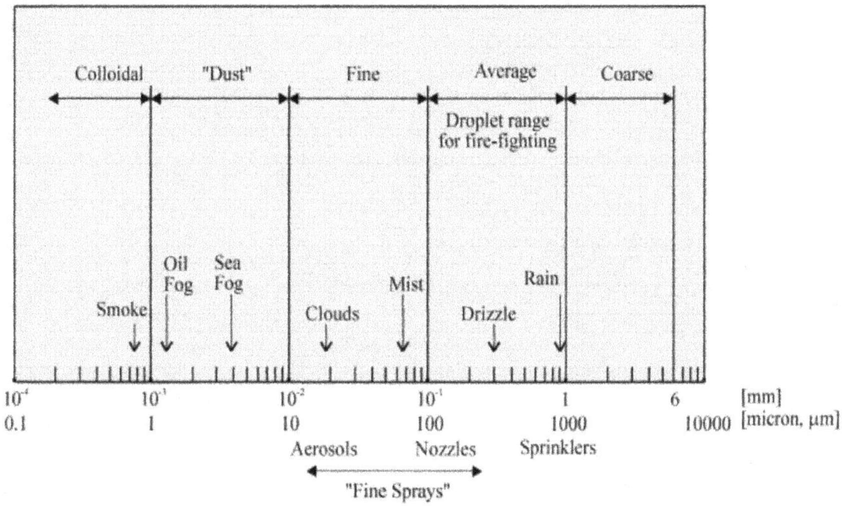

Fig. 1 Classification of droplets based on the diameter. The text below the x-axis shows the range defined as 'fine sprays', together with the approximate locations of 'aerosols', 'nozzles' and 'sprinklers' in the droplet spectrum [1]

mode of generation suited for a particular application and can be divided into several classes as shown in Fig. 1. While distinct boundaries exist for different classes of droplets, the boundary between 'sprays' and 'mists' is somewhat arbitrary. A droplet distribution with a mean diameter of 80–200 μm and a D_{V99} less than or equal to 500 μm has been reported as mist in the literature [1]. The size categories where the 'average' droplet diameter ranges from 100 to 1000 μm is the zone of interest for fire extinction.

Monodisperse droplets are very rare in practical applications as it requires an expensive droplet generator. Hence, practical sprays are mostly polydisperse in nature containing a wide distribution of droplet diameters. Several measures of droplet diameter have been introduced in the literature that represents some physical attribute of the spray as a whole. A standard notation for defining droplet diameters has been suggested by Mugele and Evans [2].

$$D_{mn} = \left(\frac{\sum D^m}{\sum D^n}\right)^{\frac{1}{m-n}}$$

The most commonly used measures of droplet diameters are the mean diameter (D_{10}) and the Sauter mean diameter (D_{32}). For the normal mean diameter, $m = 1$ and $n = 0$; for the Sauter Mean Diameter, $m = 3$ and $n = 2$. Hence, the magnitude of the Sauter Mean Diameter is expected to be higher than the Normal Mean Diameter. The Sauter Mean Diameter (SMD) is the ratio between the sum of the droplet volumes and the sum of the droplet surface areas in a spray. It signifies a particular droplet

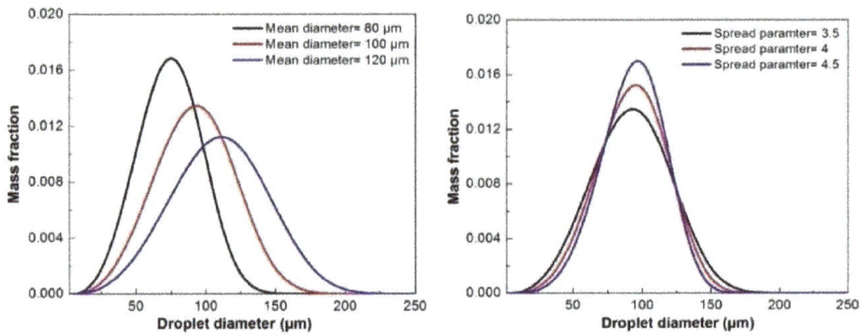

Fig. 2 Droplet distributions different mean diameter and spread parameter for same droplet range

which has the mean volume and surface area for the whole spray. The volume median diameter is another commonly used representative diameter. The number of droplets having diameter lesser than the Volume median diameter is half of the total number of droplets.

Along with the mean diameter, droplet diameter distribution is also an important characteristic of any spray. The empirically obtained diameter distribution in a spray can fit with different mathematical functions. The most popular one is Rosin-Rammler distribution and is given by:

$$Y_d = e^{-\left(\frac{d}{\bar{d}}\right)^n}$$

Here, Y_d is the mass fraction of droplets of diameter greater than d, $\boldsymbol{d}$ is the mean diameter and n is an exponent known as the spread parameter. Figure 2 illustrates the effect of the mean diameter and the spread parameter on the droplet distribution.

3 Premixed and Non-premixed Flames

Flames can be broadly classified into two types depending on the mixing between the fuel and the oxidizer: premixed flames and non-premixed (or diffusion) flames. Sometimes another type of flame known as partially premixed flame can also be observed which is intermediate to the above two types mentioned. Non-premixed flames occur where fuel and oxidizer enter through different inlets into the combustion chamber and combustion takes place at the stoichiometric mixing plane. Once ignited, a non-premixed flame will situate itself somewhere between the fuel and oxidizer streams in order to satisfy the stoichiometry requirement. Hence, the flame in case of non-premixed combustion is always stoichiometric in nature. This type of flame is also named as diffusion flame as the reaction rate is generally determined by the mixing caused by diffusion [3]. Several examples of diffusion flames can be observed such as the burning of match sticks, candles, lighters, diesel internal combustion engines etc. Premixed flames, on the other hand, are more like a wave-phenomenon. In this case, the oxidizer and the fuel are homogeneously mixed prior to ignition. Once ignited, the flame starts propagating with a finite speed towards the unburnt reactant mixture. This speed of propagation is known as the flame speed which is a function of the transport properties and the reaction kinetics. In its way of propagation, the flame consumes the unburnt reactant to sustain itself.

3.1 Non-premixed Flame

Several canonical configurations for non-premixed flames exist in the literature that greatly simplifies experimentation and validation of numerical models. One such configuration that has been studied extensively to investigate the physics of flame and water spray interaction is the counterflow diffusion flame. In this configuration a pure diffusion flame is established near the stagnation region of the two opposing jets: the fuel and the oxidizer. The counterflow diffusion flames can be divided into two groups: (1) the counterflow diffusion flame between two opposed gaseous jets of fuel and oxidizer, and (2) the counterflow diffusion flame established in the forward stagnation region of a porous burner [4]. The configurations pertaining to the different counterflow burners are presented in Fig. 3. Counterflow diffusion flames are essentially two-dimensional planar flames that can be reduced to one-dimension by a similarity transform [4]. This greatly reduces the computational costs associated with the validation of complex kinetic mechanisms with the experimental results. As a result, these flames have been studied extensively for understanding the extinction mechanism and complex chemical kinetics and transport processes. An extensive pool of literature consisting of both theoretical and experimental works exists on the structure and extinction of flat laminar counterflow diffusion flames. Smooke and coworkers [5–12] have computationally and experimentally investigated the chemical kinetics and transport processes in counterflow diffusion flames.

Studies on laminar counterflow flames are also used to model complex turbulent diffusion flames. A turbulent diffusion flame can be imagined as an ensemble of several stretched and curved laminar flamelets [13, 14]. The extinction studies on laminar flames are used to generate a flamelet library as a function of the scalar dissipation which is useful for turbulent combustion simulation.

3.2 Premixed Flame

Now if we consider the flame-stationary reference frame for the premixed flame propagation, then upstream mixture approaches the flame with a speed, which is same as the laminar flame speed (S_u) and temperature $\mathrm{T_u}$, whereas the product mixture leaves the surface with a speed U_b and temperature T_b. From continuity equation, we can realize that, as, across the flame, the temperature is going to increase, the

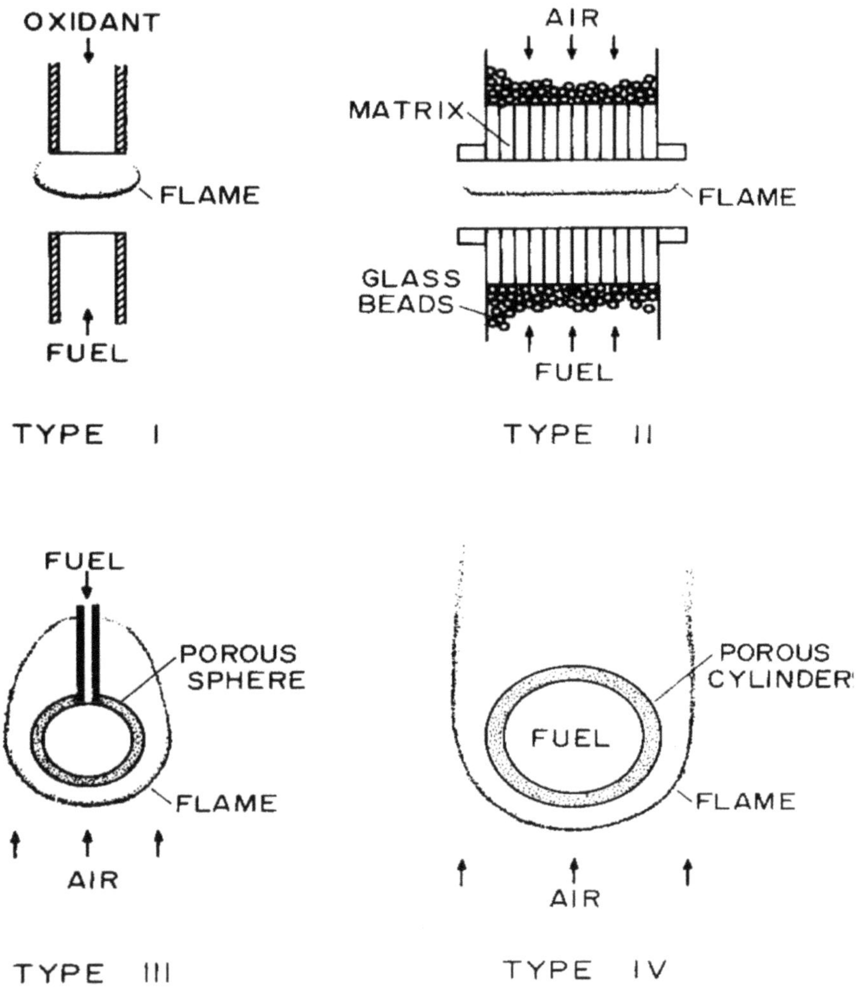

Fig. 3 Different types of counter flow burner [4]

density of the burnt product mixture will be much less than compared to reactant mixture; hence flow acceleration across the flame surface will be observed ($\rho_b \gg \rho_u$; hence $U_u = S_u \ll U_b$). Premixed flame structures can be considered at different levels of analysis. At the hydrodynamic level, premixed flames can be considered as an interface (wave front) across which discontinuities in temperature, species mass fraction, and density can be observed (Fig. 4a). However, these two different fluid dynamic states across the interface are related by the conservation of mass, momentum, species concentration and overall energy. At the flame interface the temperature changes from the reactant temperature (T_u) to adiabatic flame temperature (T_b) and species concentration changes from initial species fraction to zero in the product

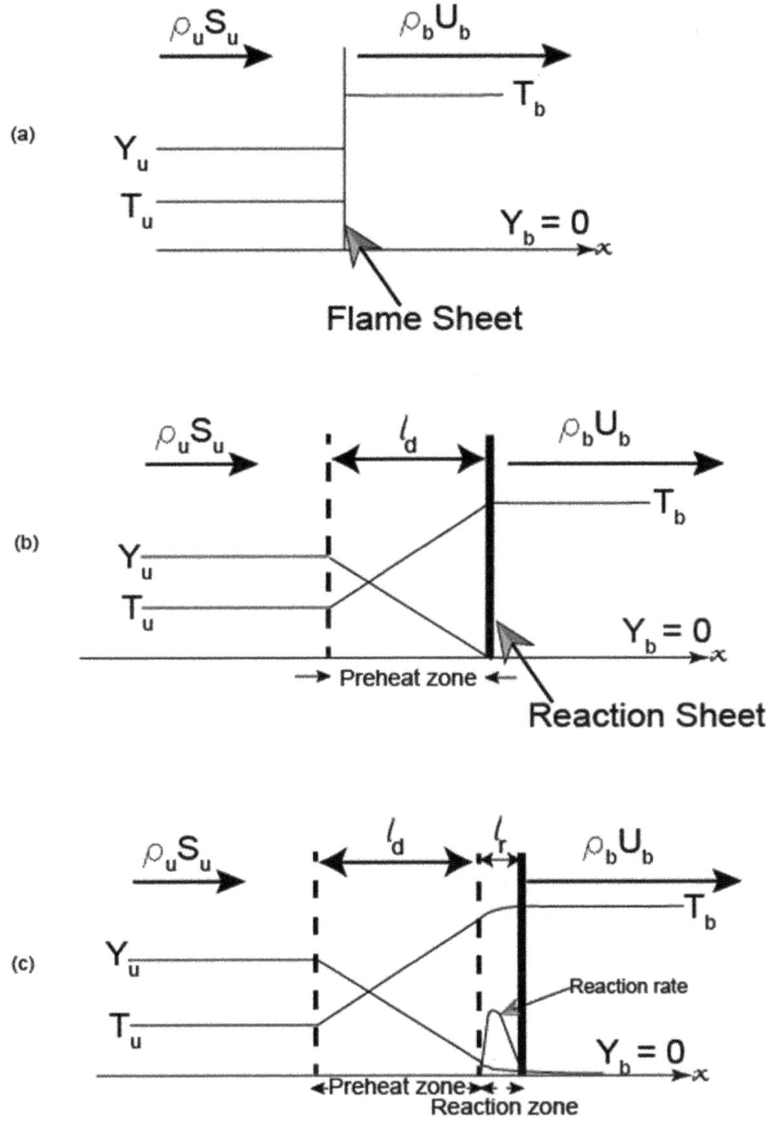

Fig. 4 Premixed flame structure

mixture. In the next level of transport-dominated analysis, the flame interface can be enlarged to get a zone called preheat zone of characteristic thickness l_d (Fig. 4b). The physics here is governed by a balance of heat and mass diffusion. Inside the preheat zone, the temperature gradually increases from reactant temperature T_u to T_b and similarly species concentration decreases from the reactant concentration to zero as shown in Fig. 4b. However, all of the reaction is still taking place in a very thin sheet like region named as reaction sheet. So, essentially in the preheat zone, we can observe a balance between convective heat and mass transfer and diffusion process. In the third level of analysis, we get a most detailed flame description by enlarging the reaction sheet itself (Fig. 4c). In this region, there is a balance between reaction and diffusion process. So the heat released due to the chemical reaction gets conducted upstream to the preheat zone where reactant heats up and finally reach the ignition point (activation temperature) and chemical reactions take place inside the reaction zone (length scale, $l_r \ll l_d$). It is to be noted that the reaction zone is very small as compared to the preheat zone and even the preheat zone is very small as compared to the whole domain (for example combustion chamber). Usually, preheat zone itself has a few millimeters of length scale. Hence, finally, to conclude, the flame structure can, therefore, be considered to have two distinct zones- preheat and reaction zone. In the preheat zone diffusion process balances with convection of the overall flow field and in the reaction zone diffusion process balances with chemical reaction because of the very high gradient of temperature and species concentration across this region. Overall, across the flame (combining both regions), mass, momentum, species, and energy must be conserved; due to high temperature jump, the density of the mixture reduces and the overall flow field gets accelerated quite a few orders of magnitude (6–7 times for 1800–2100 K if $T_u = 300$ K, less than 1 order of magnitude).

4 Non-premixed Flame and Water Spray Interaction

Flame extinction is considered as one of classical phenomena in combustion studies. Inert gas are often used as diluents for flame suppression. Carbon dioxide and nitrogen are commonly used as inert gas for extinction studies [15, 16]. Lock et al. [17, 18] is studied the effects of CO_2 dilution in fuel and air stream on extinction. Lock et al. [19] investigated the effects of nitrogen dilution and various levels of partial premixing in axisymmetric coflowing jets. Still the use of diluents for fire suppression application are limited to the specific applications due to high storage space and high cost. Water spray is widely applied for fire suppression for its efficient thermal and physical properties. Droplet size of water spray is also important parameter that affects performance of water spray.

4.1 Physical, Thermal, and Chemical Effects of Fine-Water Droplets

Water spray helps in fire suppression through its thermal and chemical effects on the flame. Though the understanding the exact contribution of different effects is not straight forward in large scale fire experiments and modelling. Often laboratory scale flames with a simple configuration such as the counter flow diffusion flame are chosen for theoretical and experimental study. In those cases, the conventional counterflow flame is modified by the provision for introducing gaseous or condensed-phase agents with the oxidizer stream (Fig. 5). Theoretical or numerical studies often preferred over experimental studies because it allows us to decouple different effects and to perform parametric studies. Lentati and Chelliah [20] numerically have investigated the effects of water spray on methane-air counterflow non-premixed flame based on a hybrid Eulerian–Lagrangian formulation for gas and droplet phase. They have considered mono disperse water droplet of 5–50 μm diameter which is introduced into the air stream of a steady laminar counterflow flame. To observe the physical effect of water vapor air stream with saturated vapor (3.51% by moles or 2.24% by mass) is supplied. 25% of reduction in extinction strain rate is observed due to dilution or displacement of oxygen from air stream due to water vapor addition. Further, same proportion of nitrogen is added instead of water vapor. The effect of this dilution on maximum flame temperature is presented in Fig. 6.

The difference in maximum flame temperature between water vapor and nitrogen dilution is due to different specific heat effects. Addition of water vapor can affect the flame temperature by oxygen displacement and take part in the chemical reactions. It is expected that water vapor can participate in chemical reaction as it has a high three-body collisional efficiency. At sufficient water vapor concentration also it can affect the water gas shift reaction. To isolate the chemical effect water vapor, H_2O is identified as a different compound which did not take part chemical reaction and simulation is performed. It is found from Fig. 6 that the chemical effect of water vapor is negligible.

The contribution of chemical effect is studied for evaporating droplets. It is found from Fig. 7 that the third body recombination effects have a minor contribution in flame suppression. In numerical model, discrete droplet phase interacts with continuous gas phase through source term of mass, momentum, and energy conservation equation. Source term energy conservation equation represents the latent heat of vaporization and increase in the sensible enthalpy of water droplets. To isolate the thermal and chemical effects, four different cases are simulated. In the Fig. 8 symbol (□) denotes when thermal and chemical effects are included. Symbol (Δ) is used to indicate dilution effect when both thermal (inclusion of energy source term) and chemical effects are excluded. In the same figure, symbol (+) and symbol (*) denotes the chemical effects (including dilution and excluding thermal effect) and thermal effects (including dilution and excluding chemical effect) respectively. It was found that extinction strain rate reduces 255–176 s^{-1} due to combined thermal, dilution and chemical effect. If the chemical effect is excluded it drops down to 190 s^{-1}. It is clearly seen that chemical effect has a very small contribution. A reduction of thermal effects is observed for higher droplet size because of incomplete evaporation of larger droplets due to higher evaporation time scale.

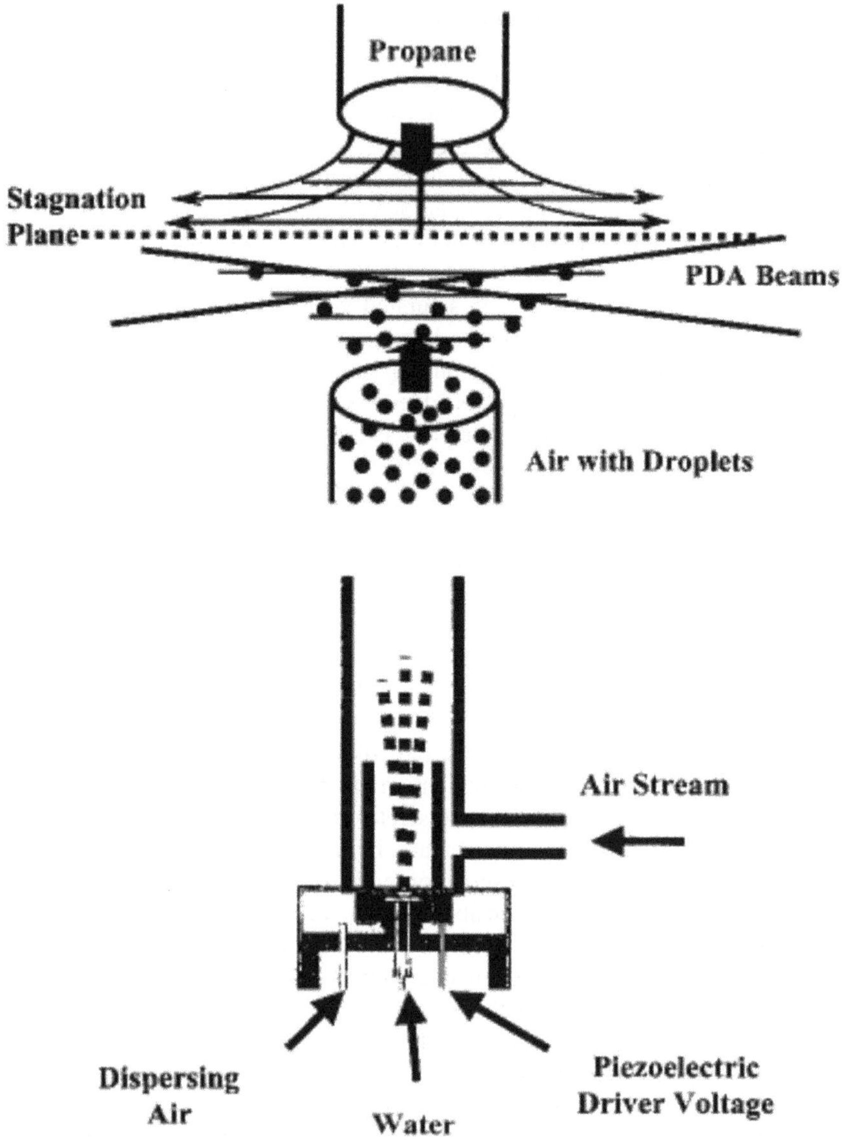

Fig. 5 Experimental setup [21]

4.2 Dynamics of Water Droplets in a Counterflow Field

Droplet size plays a crucial role in flame extinction. Fine water spray is preferred over larger droplets as it can suspend longer period in air than larger droplets. The smaller droplets also have higher surface area to volume ratio than larger droplets

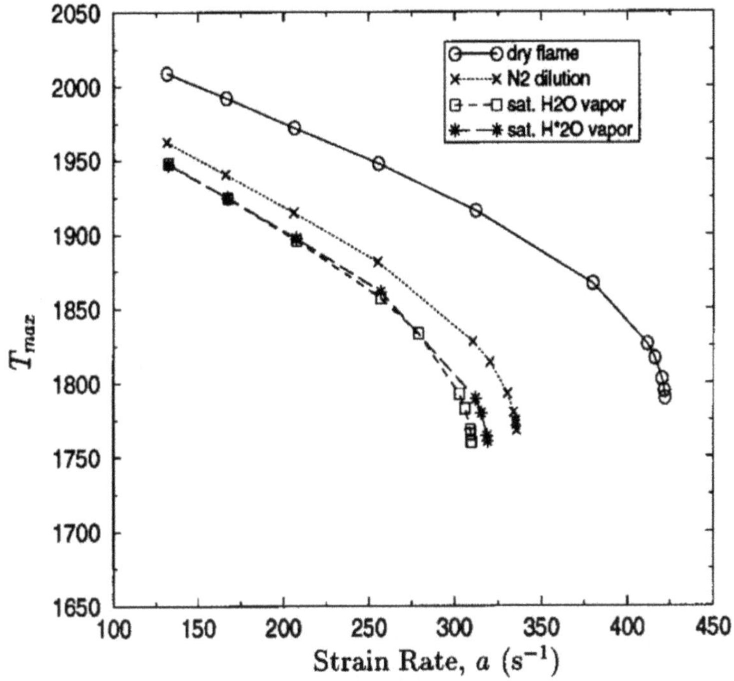

Fig. 6 Comparison of T_{max} versus a, for dry cases (o), diluted cases with saturated water vapor (□), water vapor excluded from chemistry (*), and nitrogen (×) [20]

which enhances the evaporation process. However, it is found that effectiveness of water spray is not solely governed by the surface area. Droplet dynamics near flame region also plays an important role during fire suppression. Larger droplets may not completely evaporate beyond the flame zone reducing the effectiveness of fire suppression. On the other hand, low droplet/gas velocities and/or very small sized droplets may lead to evaporation far ahead of the flame. In such cases, the full quenching potential of the spray is not utilized. Depending on the competition between the residence and evaporation times of the droplets, various flame regimes like oscillatory flames may arise.

Lentati and Chelliah [22] numerically studied the dynamics of water droplets in a counterflow field and their effect on flame extinction. They used a hybrid Eulerian Lagrangian formulation to model gas and droplet phase. Several monodisperse sizes of water droplets ranging 5–50 μm is chosen for simulation. The strain rate is kept 130 s^{-1} in the simulation. Velocity at different axial location presented in the following Fig. 9. It is observed 5 mm droplet completely follow the gas phase velocity due to low inertia where as large droplets significantly deviate from the gas phase velocity. It is also observed that 5 μm droplets completely evaporate before reaching the flame front whereas 50 μm penetrate inside the flame and enter in the fuel side then further reverse its direction. A non-dimensional number referred as Stokes number

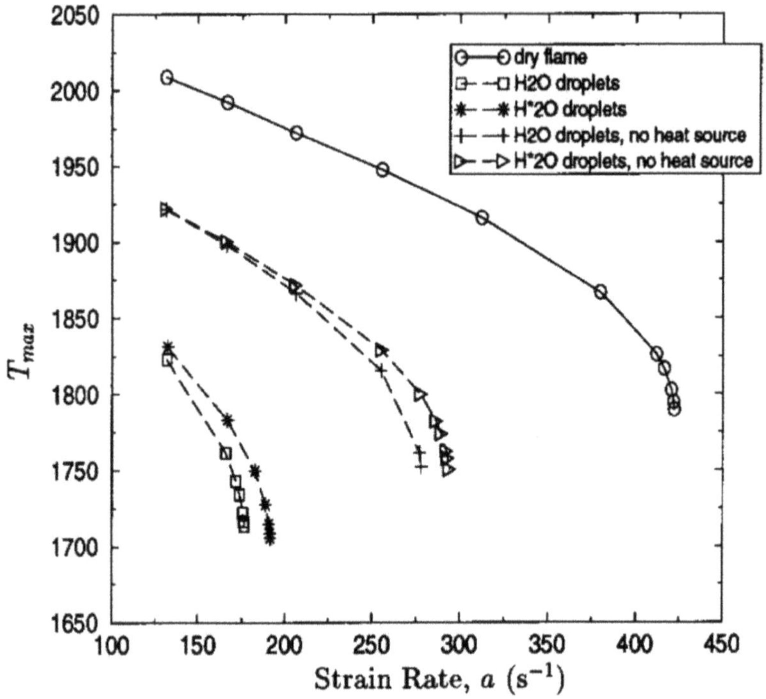

Fig. 7 Comparison of $T_{\max}$ versus a, for dry cases (o) and 2% by mass of 20-μm water droplets, with chemical effects modified (*, Δ) and thermal effects modified (+, Δ) [2]

in the literature is often used to characterize the droplet motion is spatiotemporally varying gas flow field. Stokes number is defined as the ratio of characteristic droplet response time to the characteristic flow time. It is expressed as,

$$St = \frac{\rho_o D_0^2 a}{18\mu_g}$$

where is the ρ_0 droplet density and μ_g is the gas dynamic viscosity. Droplet inertia can be a measure of characteristic droplet response time. Higher the inertia lesser it will deviate from initial trajectory; hence higher will be the droplet response time. Characteristics droplet response time increases with the increase of droplet diameter. So for a constant strain rate, Stokes number increases with the increase of droplet diameter. When Stokes number approaches the value 0.1 the deviation of the droplets are almost negligible. Figure 10 shows the droplet temperature, T_d, of different droplet sizes. It is observed that 5 μm droplet is almost in thermal equilibrium with the gas phase. A small thermal lag is observed for larger droplets. The source term in the gas phase continuous phase continuity equation due evaporation of different

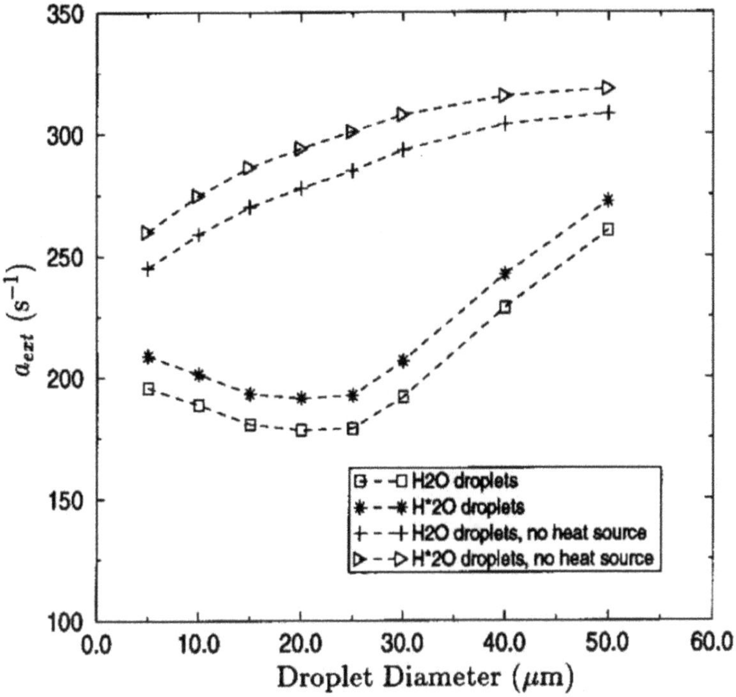

Fig. 8 Variation of extinction strain rate as a function of droplet size, with chemical and thermal effects modified [21]

droplet size is presented in Fig. 11. It is clearly observed that 5 μm droplet completely evaporated before reaching stagnation plane (Figs. 12 and 13).

Zegers et al. [21] have experimentally studied droplet velocity and evolution of droplet number density in the vicinity of the flame zone of a non-premixed counter-flow propane/air flames. Piezoelectrically generated a monodispersed water mists of initial droplet size ranging from 14 and 42 μm is seeded into the air stream to investigate the flame suppression effectiveness. The number density and average velocity of droplet size 30 and 18 μm are plotted versus axial position. It is observed that number density increases initially then starts to drop in front of the flame both 30 and 18 μm droplets. The variation of number density is explained using drag force and evaporation. Gas velocity starts to drop in the axial direction as it approaches the stagnation plane. In this region, droplets start to decelerate which causes a spatial accumulation of droplets and number density increases. The droplets start to move in the radial direction due to diverging flow field at the exit of the lower air duct. Evaporation process also starts to dominate as it approaches flame front. These two reasons cause a rapid drop in number density. It is found that 18 μm droplets completely evaporated after reaching the flame front. Whereas 30 μm droplets have penetrated the flame front and traveled in the radial direction.

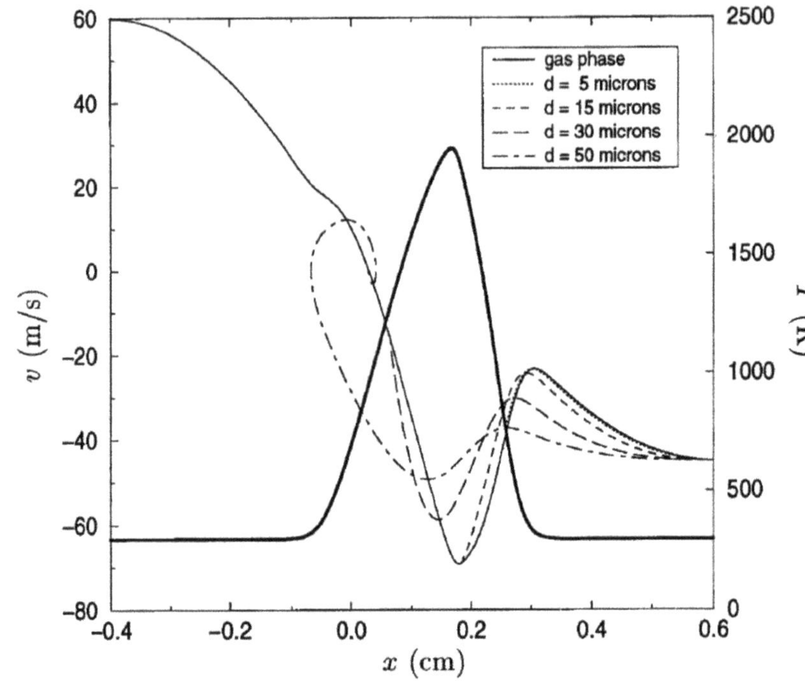

Fig. 9 Comparison of the gas velocity and droplet velocity of different sizes, with droplet source terms turned "off" in gas-phase calculations. Also shown is the gas temperature (thick line) [22]

4.3 *Optimum Droplet Diameter*

It is clearly understood that droplet size plays an important role in flame suppression. It is observed from the studies that there exists an optimum diameter for a monodisperse spray for which the effectiveness of spray is maximum. In observation can be explained on the basis of comparative effects between vaporization Damköhler number and Stokes number. Vaporization Damköhler number is defined as the ratio of characteristic flow time to vaporization time. It is expressed as

$$Da = K/D_0^2 a$$

where K is the vaporization rate, D_0 is the droplet diameter and a is the strain rate Characteristic flow time can be estimated by the inverse of strain rate. For a constant evaporation rate, vaporization time increases with the increase of droplet diameter. So for a constant strain rate, Damköhler number decreases with the increase of droplet diameter. Stokes number is defined as the ratio of characteristic droplet response time to the characteristic flow time. It is expressed as

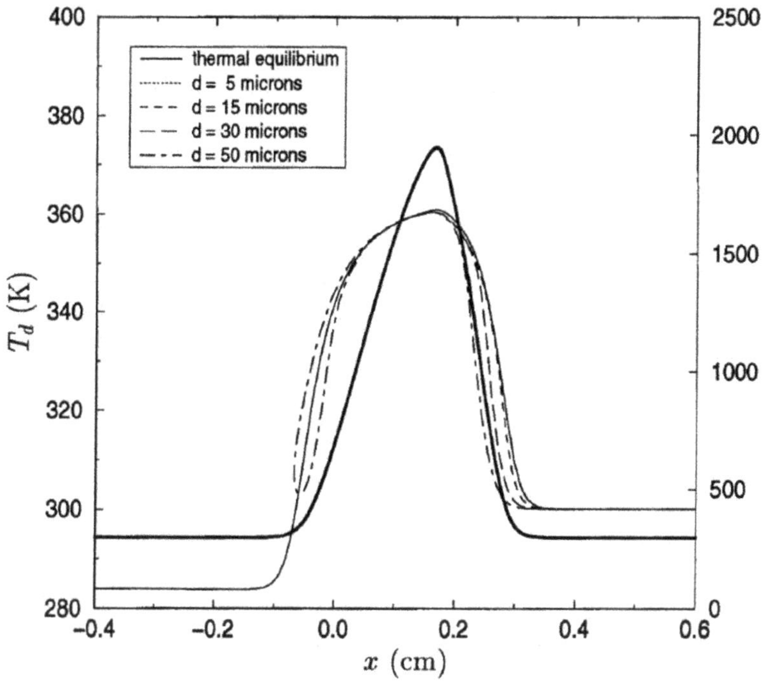

Fig. 10 Comparison of droplet temperature, *Td*, of different droplet sizes, with droplet source terms turned "off" in gas-phase calculations. Also shown is the gas temperature (thick line) [22]

$$St = \frac{\rho_0 D_0^2 a}{\mu_g}$$

where is the ρ_0 droplet density and μ_g is the gas dynamic viscosity. Droplet inertia can be a measure of characteristic droplet response time. Higher the inertia lesser it will deviate from initial trajectory; hence higher will be the droplet response time. Characteristics droplet response time increases with the increase of droplet diameter. So for a constant strain rate, Stokes number increases with the increase of droplet diameter. So higher stokes number will be more preferred due to penetration into the flame and higher Damköhler number will be preferred due to better evaporation. The simultaneous requirement of these to lead to an existence of an optimum droplet diameter. Chelliah [23] investigated the variation of optimal droplet diameter for different water mass loading for counter flow methane air non-premixed flame. 15–20-mm droplets are found to be the most effective. Zegers et al. [21] investigated the effect of 3 different size droplets on a propane air non-premixed flame. They also found an optimum droplet size for more effective flame suppression.

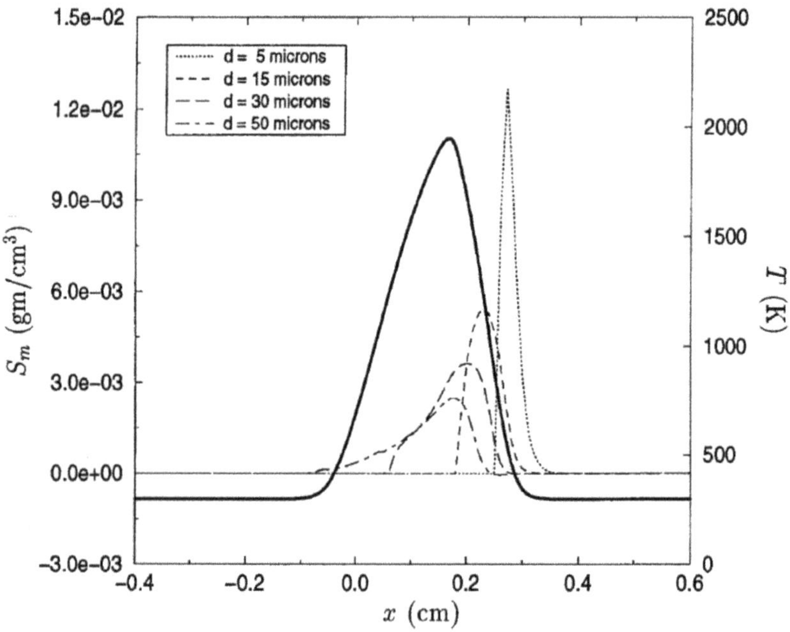

Fig. 11 Typical S_m profiles of different droplet sizes, with Y_O =0.02 and droplet source terms turned "off" in gas-phase calculations. Also shown is the gas temperature (thick line) [22]

4.4 Effect of Size Distribution of Polydisperse Water Spray

Practical atomizers produce a large spectrum of droplets. The common practice of representing the polydisperse spray by some suitable mean diameter (e.g., Sauter mean diameter) may not be effective in the present situation where the dynamics of the individual droplets are determined by their respective sizes. Dvorjetski and Greenberg [24] investigated the effect droplet size distribution on a counterflow diffusion flame. They considered six initial droplet size distributions for analysis. Three of them have the same Sauter mean diameter and rest of three have the same D_{20}. Three different distributions: (1) a quasi-monodisperse spray, (2) a bimodal spray and (3) a "normal"-spray are considered for both cases. In addition to these, 2 mono-sectional descriptions of the spray are considered for mathematical reference. The first mono-sectional spray has droplet size ranging from 20 to 84.7 and SMD 44.8 μm. The second mono-sectional has a droplet size of 20–80.1 mm and D_{20}= 34.4 μm. The effect size distribution with same SMD on flame temperature is presented in Fig. 14. The maximum flame temperature obtained without water spray is 1570 K. Maximum flame temperature obtained with bimodal water spray and full polydisperse water spray are 1480 K and 1470 K respectively. The quasi-monodisperse spray reduces the flame temperature a few more degree further than full polydisperse water spray. Mono-sectional model of the spray gives the lowest flame temperature.

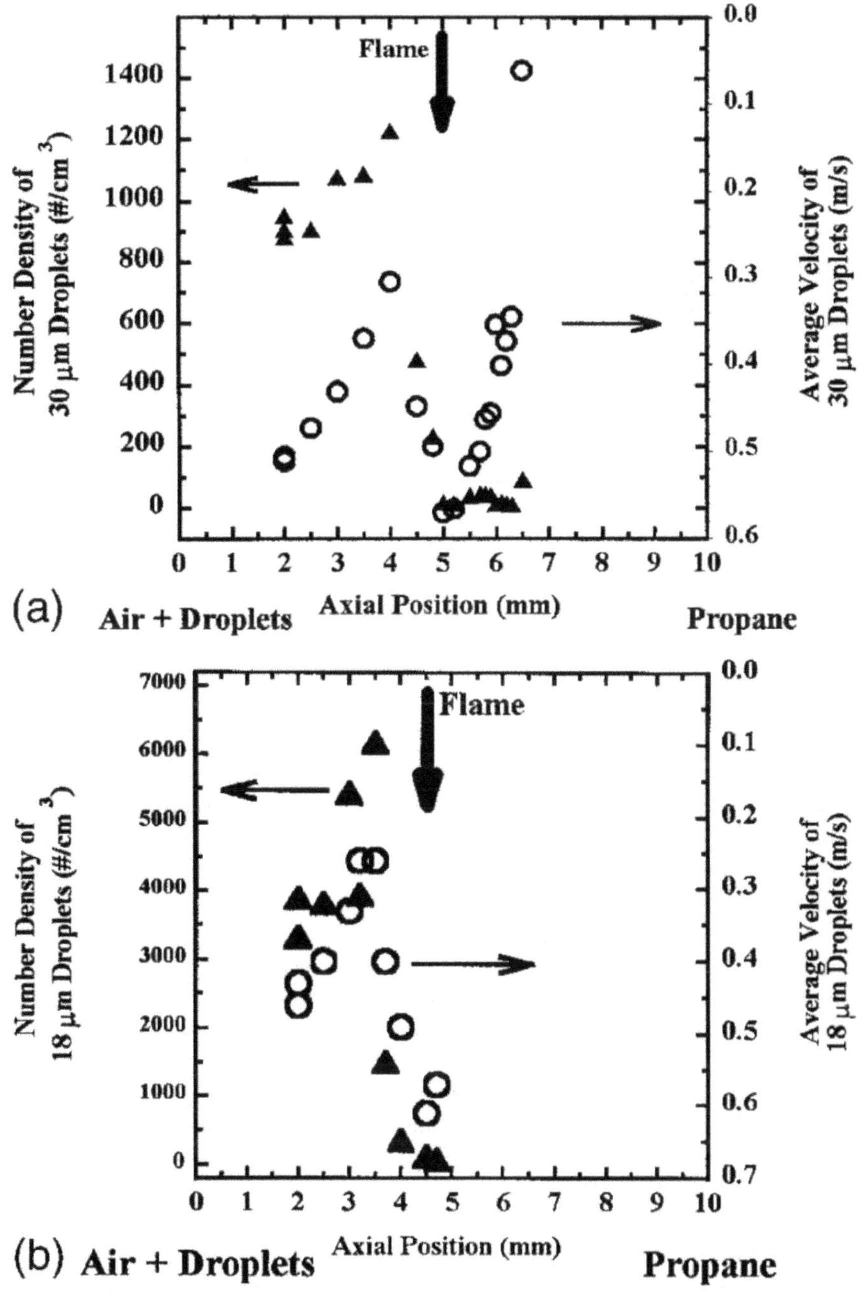

Fig. 12 **a** Profiles of number density (triangles) and velocity (circles) for 30 μm droplets versus location in a 170 s^{-1} strain rate propane/air 30 μm mist counterflow flame. **b** Profiles of number density (triangles) and velocity (circles) of 18 μm droplets versus location in a propane/ air 18 μm mist counterflow flame [21]

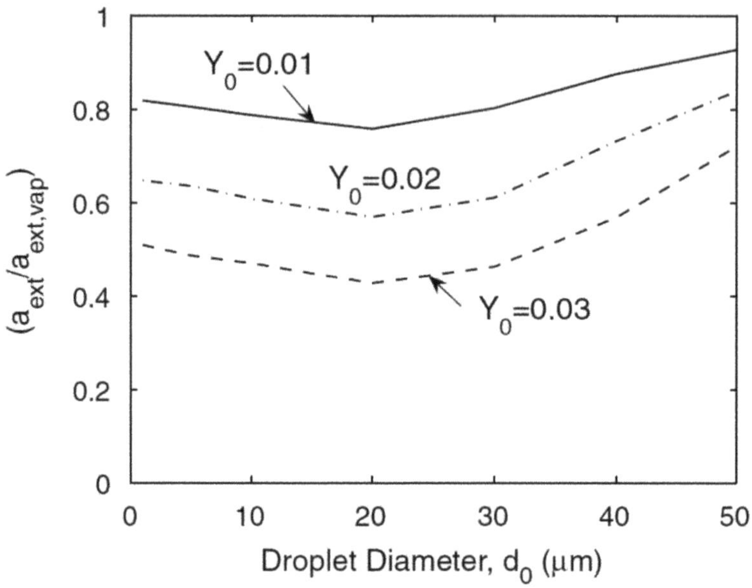

Fig. 13 Variation of the flame strength of a nonpremixed flame versus droplet diameter, for selected water mass fractions Y_0 [23]

Figures 15 and 16 represent the variation of the water mass fraction versus critical strain rate curves for the distributions with same SMD and same D_{20} respectively. It is clear that quasi mono-disperse spray is the most effective among three realistic distribution with same SMD. However full polydisperse water spray is found most effective among three realistic distribution with the same $D_{20.}$ So it can be concluded that representing a polydisperse by a single representative diameter (number mean, Sauter mean, D_{20}) is inaccurate (Figs. 17 and 18).

Pramanik and Mukhopadhyay [25] numerically investigated the effects of polydisperse and monodisperse spray on counterflow diffusion flame. They found that polydisperse spray is more effective than monodisperse spray after a certain droplet diameter. Sasongko et al. [26] experimentally investigated the extinction condition of a counterflow diffusion flame in presence of polydisperse water spray. Extinction condition is obtained by slowly reducing the oxygen concentration in the oxidizer stream. The effect of mean diameter of the polydisperse spray on the oxygen concentration at the extinction is studied experimentally. In the experiment a double cylinder concentric counterflow burner was used. The inner diameter of the burner was 23 mm which was surrounded by a 40 mm diameter outer cylinder. Methane and oxygen diluted by N_2 were supplied from top and bottom inner cylinder respectively. N_2 was supplied from the outer cylinder to prevent the interaction of atmospheric air with the flame. Strain rate was varied by changing the separation distance between the burners. In the present case strain rate of 160 s^{-1} was chosen for study. Water spray

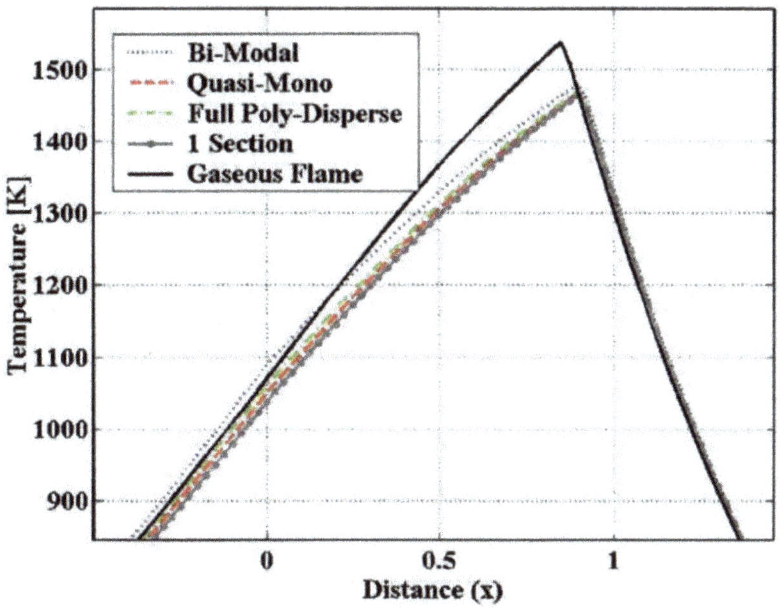

Fig. 14 Comparison of temperature profiles in opposed flow diffusion flames with and without various water sprays present: all sprays have same SMD [24]

produced by a twin fluid atomizer was introduced from the bottom. The atomizing gas flow rate was varied to obtain different droplet size distributions.

The burning behavior of counterflow diffusion flame with water spray for strain rate a = 160 s^{-1} is captured and presented in the figure. Oxygen concentration is varied from 0 and 0.30 to before extinction. Initially, the flame consists of two regions, a yellow luminous region on the fuel side and a blue region on the oxidizer side. The yellow luminous region starts to decrease with the decrement of oxygen concentration as flame temperature reduces. Flame completely turns into blue before extinction. Figure 19 presents the comparison of oxygen concentration at extinction for two different strain rate a = 160 s^{-1} and a = 320 s^{-1}. It is observed that optimum droplet diameter shift in the smaller droplet size with the increase of strain rate. The optimum droplet diameter exists around SMD 65–75 μm at 160 s^{-1} strain rate. For 320 s^{-1} strain rate, optimum droplet SMD lies between 40 and 50 μm (Fig. 20).

Sarkar et al. [27] numerically investigated extinction condition of flat and curved counterflow laminar diffusion flame with polydisperse water sprays. Two dimensional Navier-Stokes equation and energy equations were solved for the gas phase. A reduced reaction mechanism with 16 species and 46 reactions was used to model chemistry of the combustion phenomena. Rosin Rammler distribution was fitted with the experimentally obtained droplet distribution and 1.86 was found as the spread parameter. Discrete phase model with the obtained Rosin Rammler distribution parameters was used for present simulation. The O_2 mass fraction of the

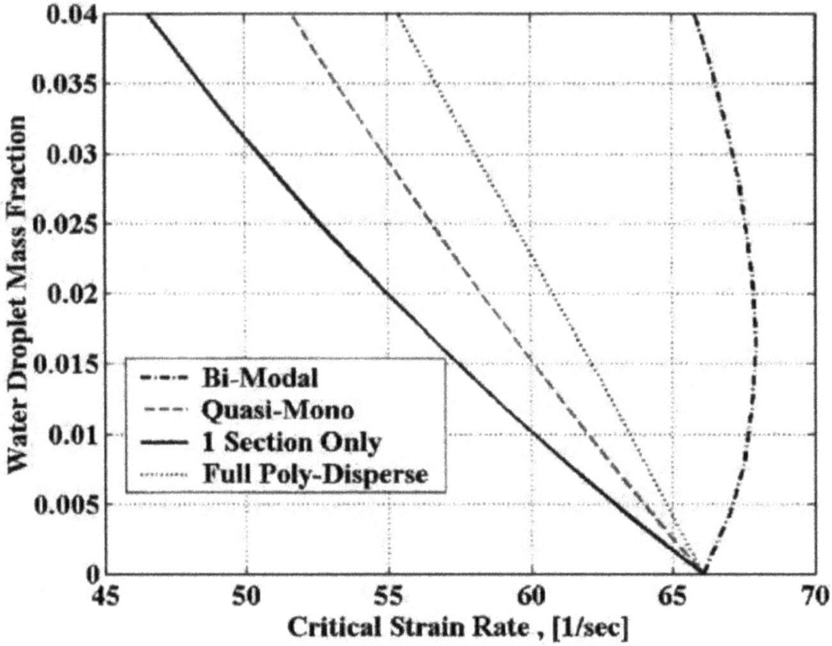

Fig. 15 Comparison of temperature profiles in opposed flow diffusion flames with and without various water sprays present: all sprays have same SMD [24]

oxidizer stream was gradually reduced to obtain the extinction condition. Numerical simulation results were in good agreement with the experimental result reported in Sasongko et al. [26] Approximately 3% deviation was observed from the experimental results of Sasongko et al. [26] in the extinction concentration. The maximum flame temperature and temperature contour of the flame with and without spray at different O_2 concentration are presented in Figs. 21 and 22 respectively.

Yoshida et al. [28] experimentally studied extinction of a counterflow methane/air diffusion flame using polydisperse fine water droplets. A piezoelectric atomizer is used to generate fine water spray. Droplet size distribution and velocities are obtained using Phase Doppler particle analyzer (PDPA). Figure 23 presents the variation of flame thickness with strain rate for different water spray mass loading. In Fig. 23 filled symbol indicate yellow luminous flame and hollow symbol indicate blue color flame. It is observed that flame thickness and flame color significantly change with the application of water spray. Figure 24 presents the velocity of the droplets at different locations for strain rate 80 s^{-1}. It is seen that smaller droplets start to decelerate according to the gas flow at far upstream of the flame front at y = 14.4 mm. Closer to the flame front at y = 8.5 mm, all the droplets almost move with the same velocity of the gas phase. However, at higher strain rate (230 s^{-1}), droplets velocities

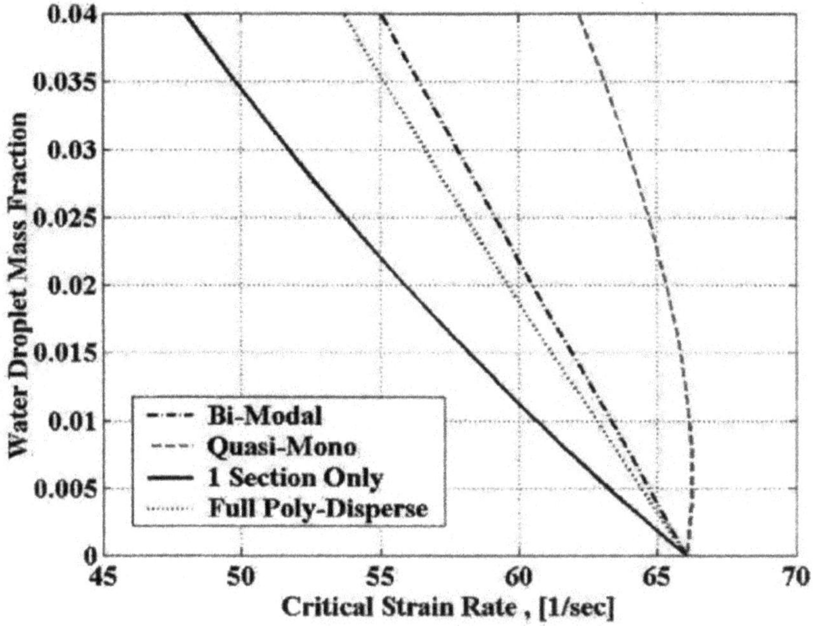

Fig. 16 Comparison of temperature profiles in opposed flow diffusion flames with and without various water sprays present: all sprays have same SMD [24]

are significantly different from gas velocity as high inertia force does not allow to establish equilibrium with the gas phase velocity (Fig. 25).

Lazzarini et al. [29] investigated the role of adding hydroxides of alkali metals to water to enhance the extinction rate. An upper limit of the concentration of alkali metal hydroxide allows one to combine the thermal fire suppression ability of water droplets with chemical suppression ability of alkali metals. Arias et al. [30] numerically investigated the extinction of counterflow non-premixed laminar and turbulent flames with water droplets. Through the analysis of turbulent flames, they demonstrated the role of parameters like flame curvature and local extinction.

Most of the studies referred above are performed with the opposing-jet burners. However there exist few more simple configurations other than counterflow burner for flame water spray interaction study like coflow burner, cup burner etc. Takahashi et al. [31] studied extinguishment of methane and polymethylmethacrylate (PMMA) flame using water spray in cylindrical burner configuration where cylindrical burner is inserted vertically downward into an upward coaxial air stream. The effect of gaseous agents (N_2 and CO_2) and water mist on flame structure and extinction is reported. Ndubizu et al. [32] studied water mist fire suppression mechanisms of a methane air diffusion flame using a modified Wolfhard-Parker burner setup. Liao et al. [33] studied interaction of fine water spray with liquid pool fires.

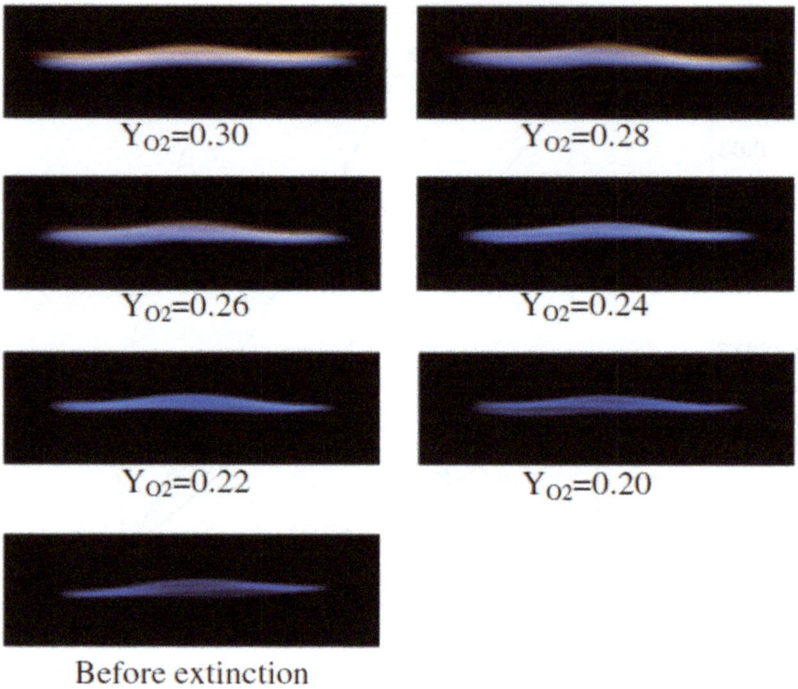

Fig. 17 Photographs of counterflow diffusion flame with water spray for different oxygen concentrations of the oxidizer Y $_{O2}$ (SMD = 97.6 lm, Y $_w$ = 0:022; a = 160 s^{-1}). [26]

5 Premixed Flame and Water Spray Interaction

5.1 Dynamics of Droplets

Chelliah [23] numerically investigated the effects of droplet size on flame inhibition for laminar methane-air premixed flames. Monodispersed droplet size of 10, 20 and 30 μm are chosen for numerical study. Figure 26 shows the variation of normalized droplet diameter for those initial droplet sizes in the flow field. It is observed that smallest droplets completely evaporate in the preheat zone of the flame. Whereas 30 μm droplets completely penetrate inside the flame and continue to evaporate in the post flame region. Gas velocity rapidly increases in front the flame due to thermal expansion. Droplets also start to accelerate due to drag. The response of droplets is presented in Fig. 26. A velocity lag is observed for larger droplets due to higher inertia. The thermal response is presented in Fig. 28. The difference between the droplet and gas phase temperature is minimum for 10 μm droplets. Mass source term due to evaporation for different droplet size is presented in Fig. 29. It is observed that 10 μm droplets completely in preheat zone and flame front. Whereas 30 μm droplets majorly evaporated in the post flame region (Figs. 27 and 30).

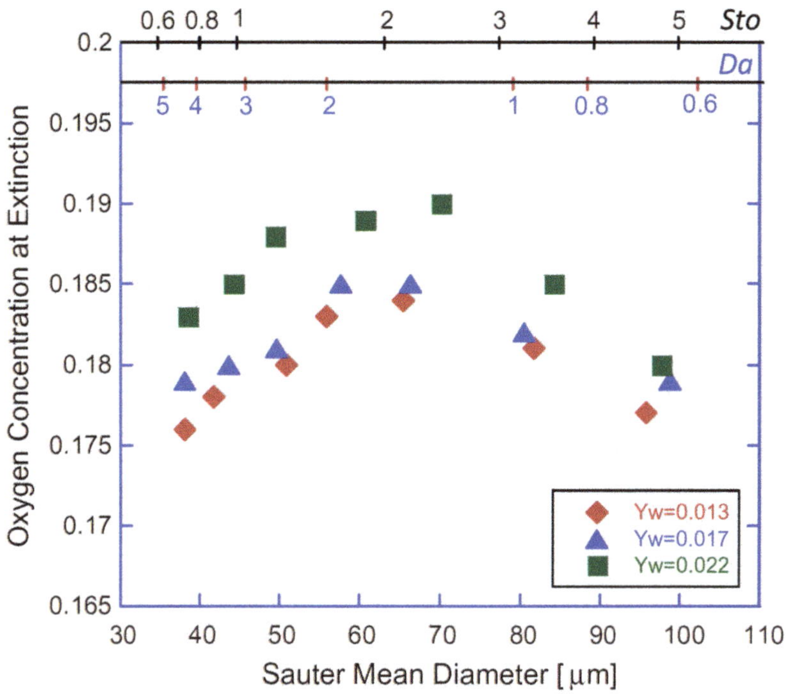

Fig. 18 Dependence of oxygen concentration at extinction on the Sauter mean droplet diameter for different water droplet mass fraction Y_w ($a = 160\ s^{-1}$) [26]

5.2 *Effect on Flame Speed*

Significant flame speed drop was observed due to with the application of water spray. Flame speed decreases with spray mass loading. The maximum decrement of flame speed was obtained for 10 μm droplets. Numerical results also very good agreement with the experimental results.

Yoshida et al. [34] have experimentally investigated the suppressing effect of fine water droplets in propane/air premixed flames in stagnation flow field. They have also found the dependence of laminar flame speed on the stretch rate which is in good agreement with the previously reported data. Laminar flame speed was observed to increase with increasing stretch in absence of any water droplets similar to the study of Law et al. [35] On the other hand laminar flame speed started reducing with increasing stretch rate in presence of fine water spray due to change in the mixture Lewis number and Markstein length (Fig. 31). However, it was not in agreement with the numerical study conducted by Yoshida et al. [34].

It was concluded that flame speed reduction due to mist accumulation at the stagnation plate is larger than the flame speed increment due to the flame stretch, hence providing a net decrease in the flame speed due to water mist addition. It was

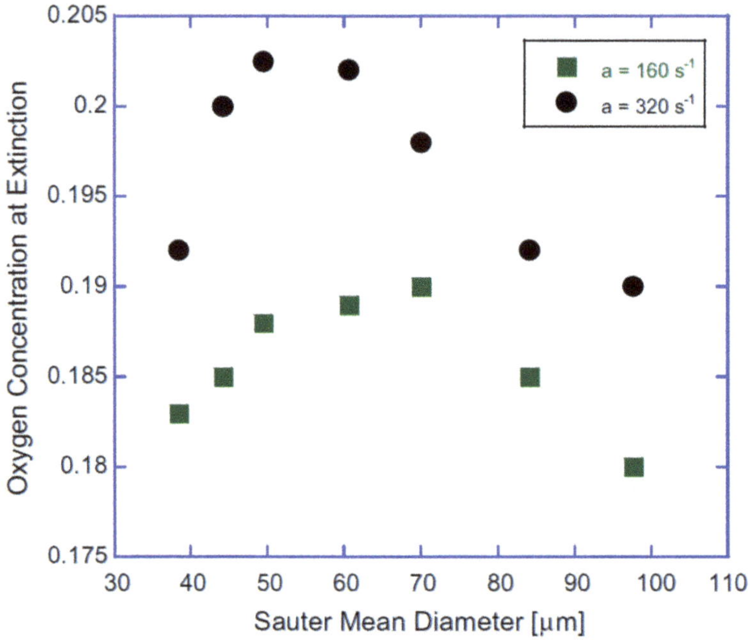

Fig. 19 Dependence of oxygen concentration at extinction on the Sauter mean droplet diameter for two different strain rates (Y $_w$ =0:022) [26]

also observed that for water vapor mole fraction of 0.2 or more, no flame can be established further.

There is a handful of literature present which have investigated the effect of water droplets as a fire suppressing agent using stagnation flow configuration. These experimental and numerical studies have revealed important characteristics such as the effect of water droplet size distribution, in case of poly dispersed water spray, on the laminar flame speed. Yang and Kee [36] have presented a computational model to describe two phase interactions between freely propagating premixed methane-air flame and mono-dispersed water droplets using commercial PREMIX code. The gas phase, which includes detailed chemistry, was modeled using adaptive Eulerian mesh whereas the discrete droplets were represented by the Lagrangian frame. It was observed that water droplets below the critical size (10 μm) is ineffective to further affect the extinction process of methane-air flame. The model also predicted turning-point extinction behavior for larger droplets.

Modak et al. [37] have studied numerically, using PREMIX code, the influence of fine water mist on the suppression of laminar freely propagating strain-free premixed methane-, propane-, hydrogen-air flames in atmospheric pressure. It was observed that smaller size droplets are more effective than larger droplets which are already concluded previously by many others experimentally and numerically. The critical small diameter limit was observed to be 10 μm (for methane-air and propane-air)

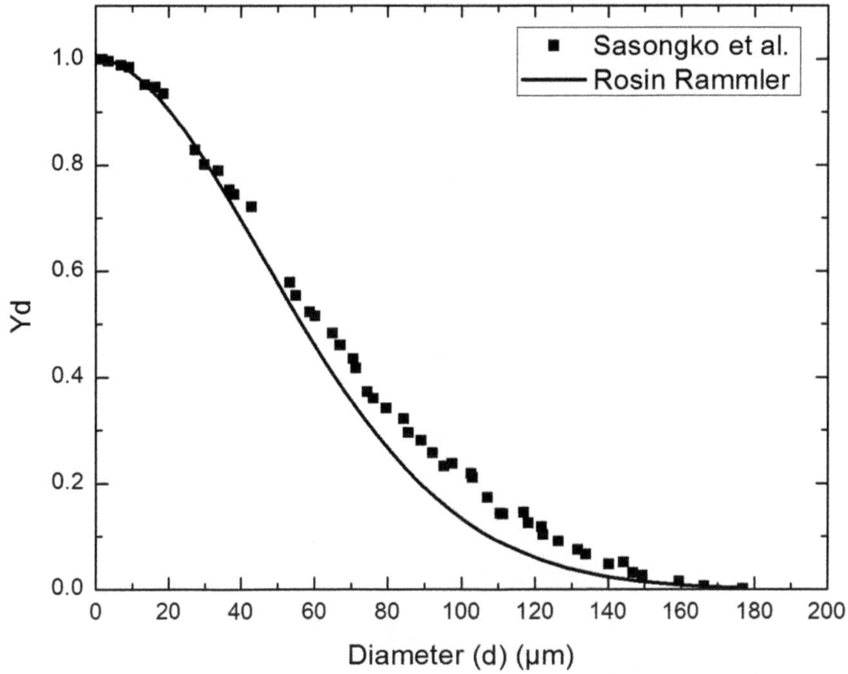

Fig. 20 Rosin Rammler curve fits to experimental droplet distributions

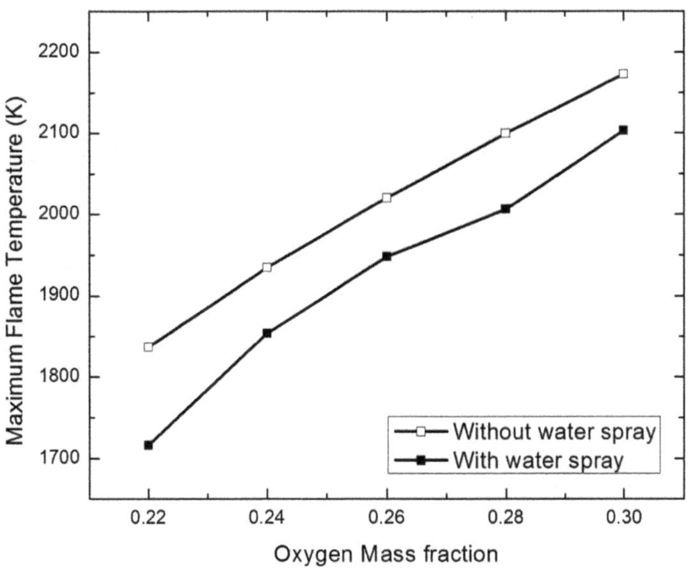

Fig. 21 Comparison of maximum flame temperature

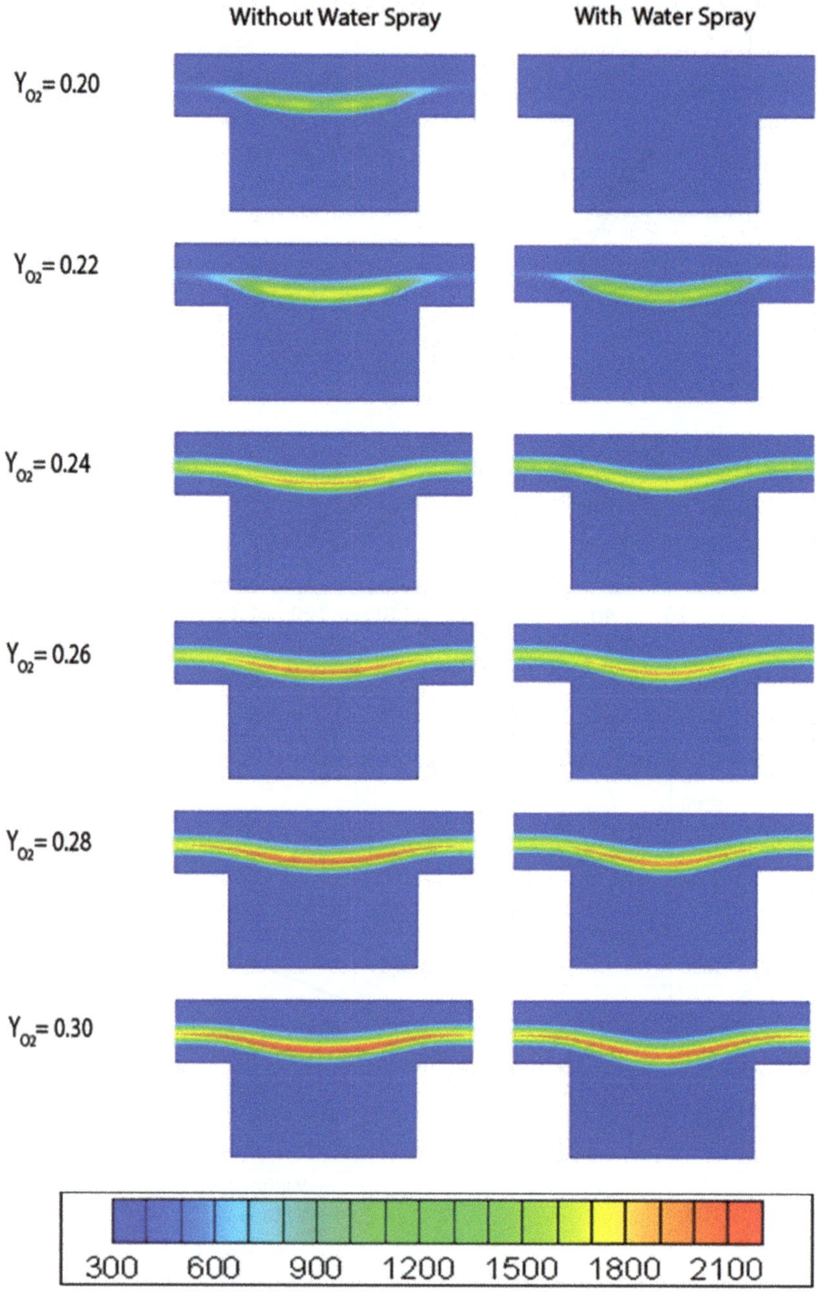

Fig. 22 Comparison of temperature contour

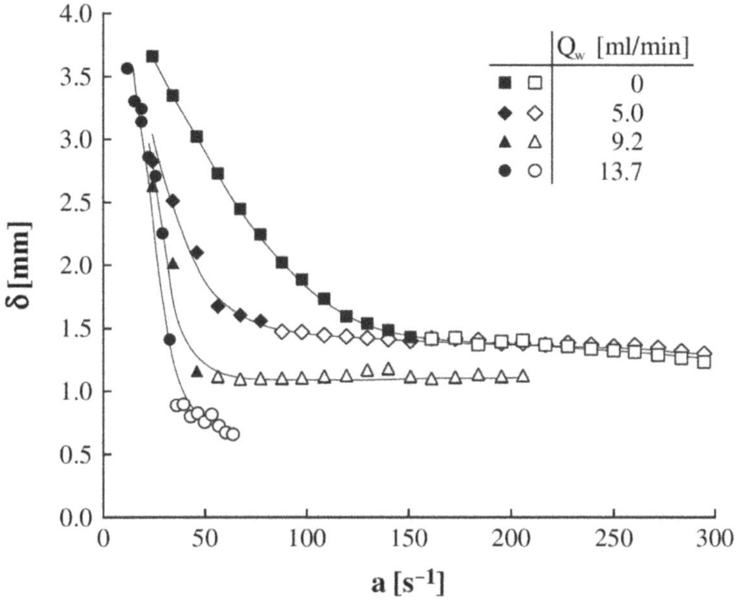

Fig. 23 Variation of flame thickness with strain rate [28]

and 2 μm (for hydrogen-air) below which there is no further influence on the droplet size which is in agreement with the analysis of Yang and Kee. 30 μm was denoted as turning-point extinction where burning velocity for methane-air and propane-air was observed to be half of the burning velocity without any water droplets.

Reduction of flammability region of hydrogen-air mixture by using cold fog nozzles and ultrasonic foggers was experimented by Jones et al. [38]. It was concluded that effect of water mist is strongly dependent on the droplet size distribution and the effect is more pronounced for very fine water mist diameters less than 10 μm. The experiment was conducted mainly for nuclear waste storage plant after decommissioning where significant hydrogen generation could occur.

Ingram et al. [39] have studied laminar flame speed for a premixed hydrogen-oxygen-nitrogen mixture using nozzle burner setup and have used water mist with NaOH additive. It was observed that above a critical concentration of water with NaOH additive (SMD ~4 μm), sudden significant reduction in burning velocity occurred. It is concluded that addition of NaOH helps to chemically inhibit the combustion in addition to the evaporative cooling effect due to the addition of water mist.

Joseph et al. [40] have presented a simple lumped parameter approach to study thermodynamic aspects of the interaction between water droplets and hydrogen-air flames in a closed container. The final pressure and temperature variation due to the addition of water can not only suppress the fire; sometimes the addition of water mist could over pressurize the vessel and could create turbulence effect which

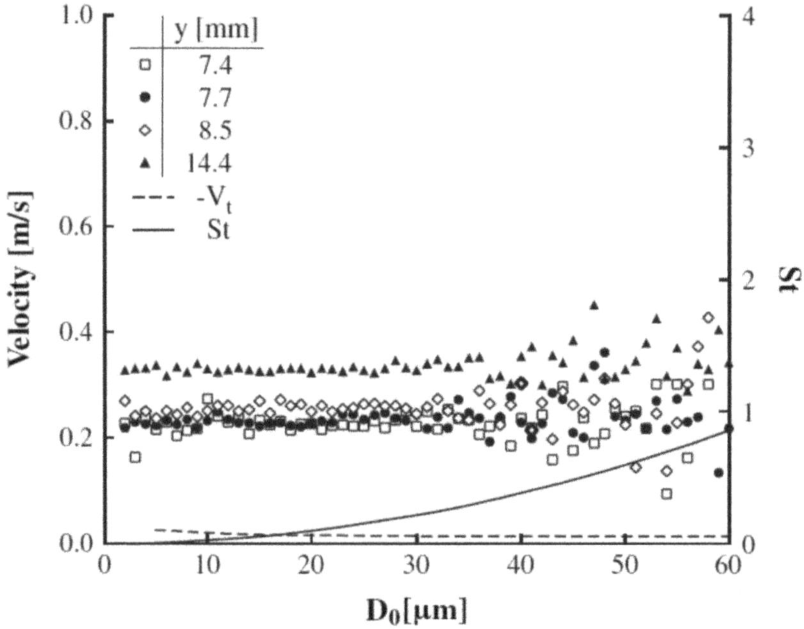

Fig. 24 Droplet velocity, Stokes number and thermophoretic velocity for $a = 80\ s^{-1}$ and $Q_w = 7.33$ ml/min [28]

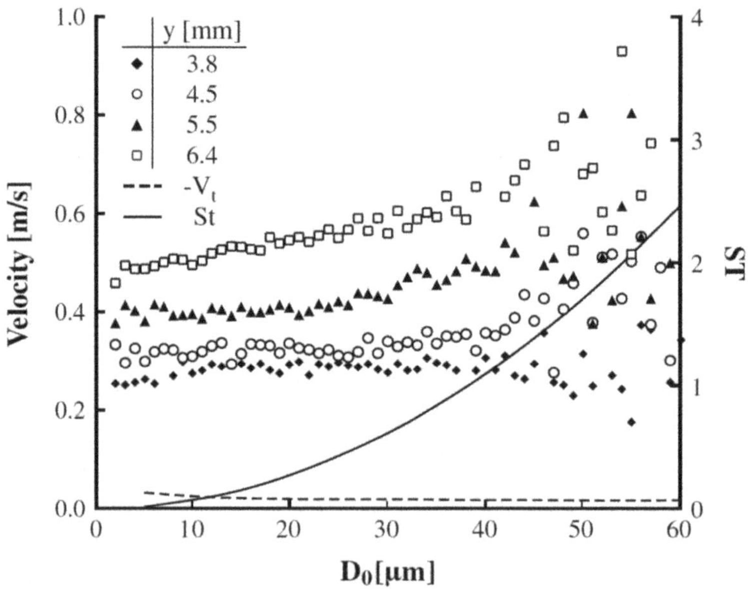

Fig. 25 Droplet velocity, Stokes number and thermophoretic velocity for $a = 230\ s^{-1}$ and $Q_w = 7.33$ ml/min [28]

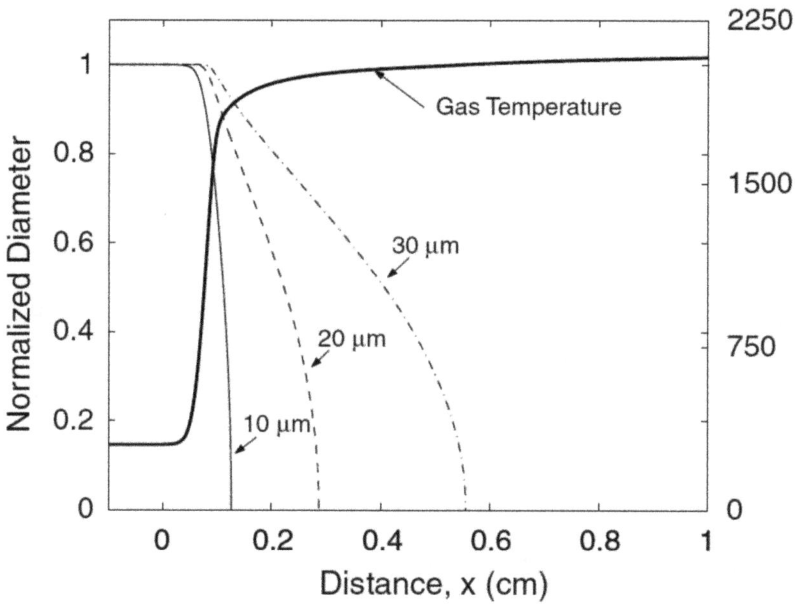

Fig. 26 Variation of normalized droplet diameter across the premixed flame structure, for different initial droplet sizes. Also shown is the gas-phase temperature versus distance [23]

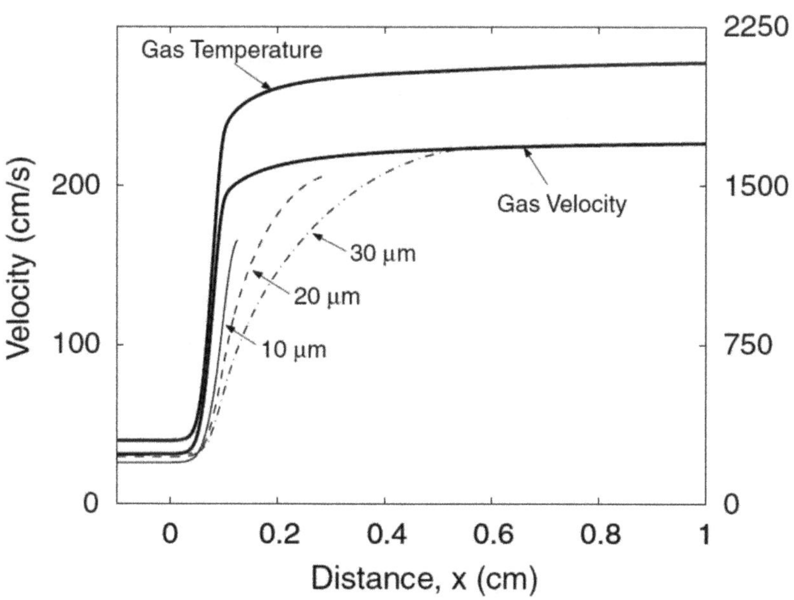

Fig. 27 Comparison of the gas and droplet velocity across the premixed flame structure, for different initial droplet sizes. Also shown is the gas-phase temperature versus distance [23]

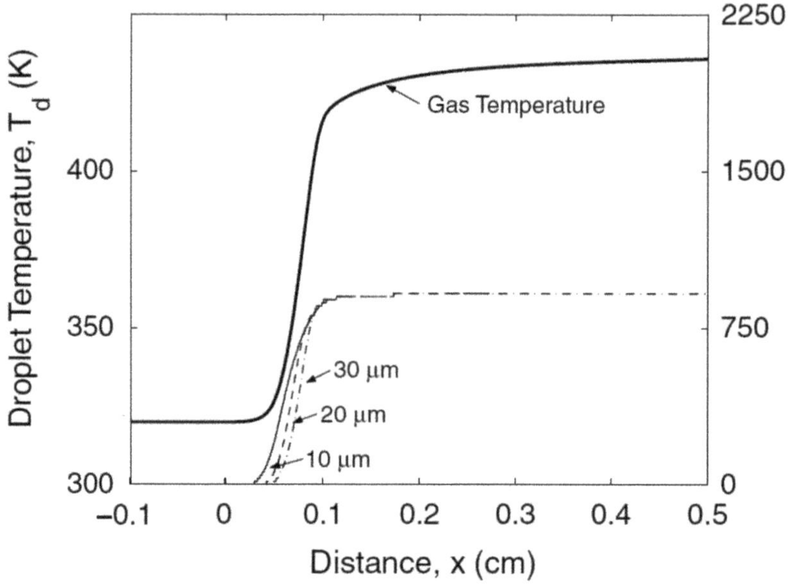

Fig. 28 Comparison of gas and droplet equilibrium temperature (T_{equil}) across the premixed flame structure, for different initial drop sizes [23]

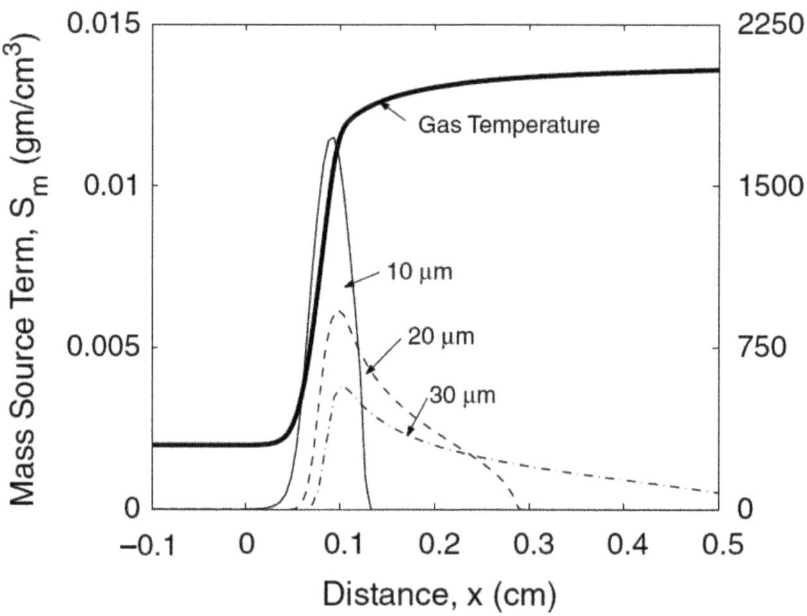

Fig. 29 The mass source term (S_m) across the premixed flame structure, for different initial droplet sizes and $Y_0 = 0.02$. Also shown is the gas-phase temperature profile [23]

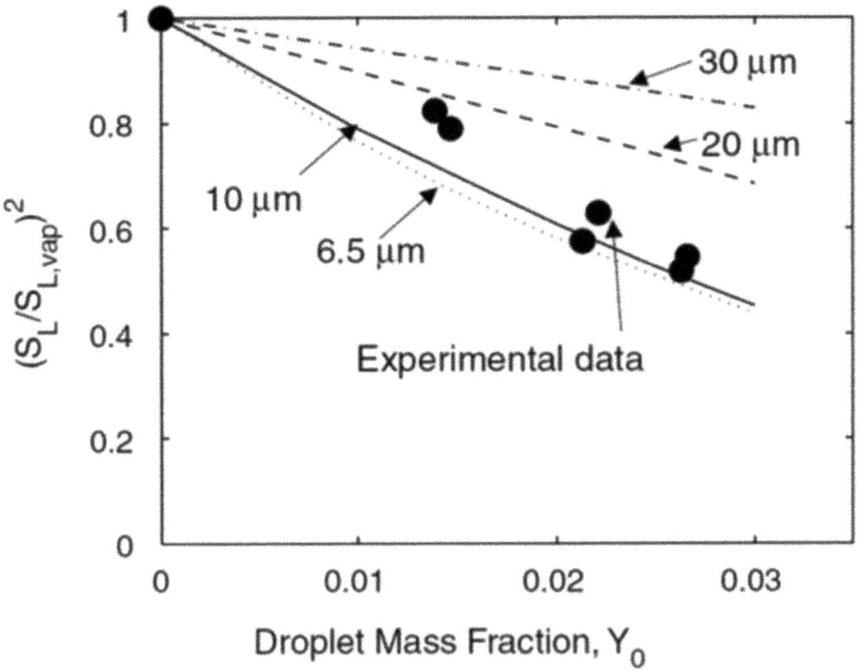

Fig. 30 Variation of the flame strength of a premixed flame versus droplet mass fraction, for different initial drop sizes [34]

would eventually enhance the flame propagation. The lumped parameter approach can only predict the final pressure and temperature in case of addition of water mist to the hydrogen-air flame. It was concluded that although liquid water acts as a heat sink, steam generated due to vaporization of liquid water in the reaction zone becomes important when hydrogen concentration attains certain critical value. The final pressure, in that case, becomes higher than the final pressure that would have attained without liquid water addition.

Cheikhravat et al. [41] have observed that in standard atmospheric pressure and temperature, the addition of water mist to the dry hydrogen/air mixture did not shift the lower flammability limit until a critical droplet density number was reached. The experiment was done in a spherical constant volume vessel and it was observed that for larger droplet size with Sauter mean diameter (SMD) in the range of 200–250 μm, effect of water spray on the flame speed is negligible. With water droplet diameter less than 10 μm, violence of explosion was mitigated due to reduction of flame speed; however opposite effect was observed for very lean hydrogen/air mixture (10% mole fraction of H_2) as the turbulence effect was enhanced by addition of droplets in this case and it helped to complete the combustion.

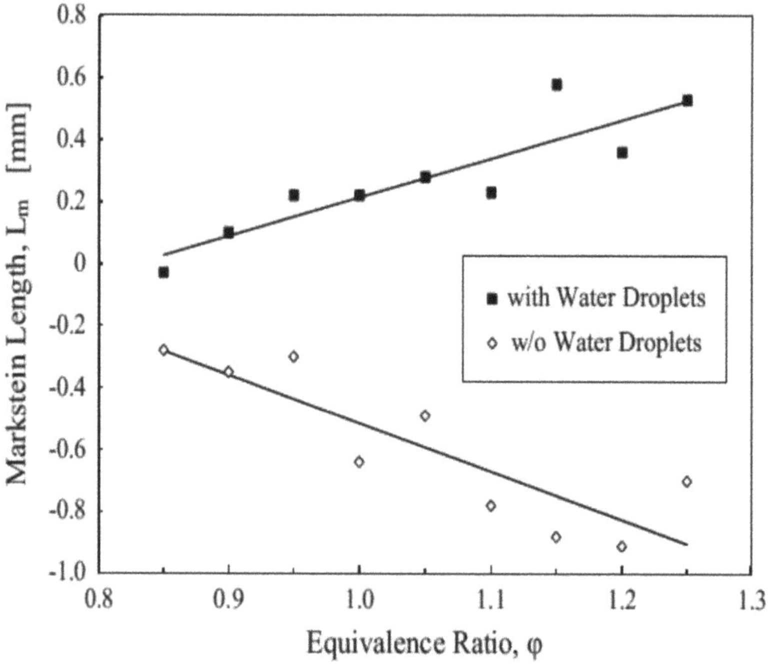

Fig. 31 Markstein length with and without water droplets [34]

5.3 *Flame Acceleration*

Munshi et al. [42] numerically studied effects of polydisperse water spray in hydrogen air premixed flame. A stagnation jet plate configuration was chosen for flame speed determination. Stabilized flames in different stretch conditions are now introduced with polydisperse water spray of different number mean diameters. Figure 32 depicts flame speed normalized with respect to the flame speed in absence of spray and flame temperature normalized with respect to flame temperature in absence of spray respectively with an initial flame stretch of 270.7 s^{-1}. For the stretch mentioned, it can be clearly observed to increase the flame speed and temperature for a certain range of spray mass loading. The only difference, in the two different stretch scenarios, is that the inlet velocity in case of lower stretch is 4 m/s whereas for the highest stretch it is 7 m/s. Due to increase in the velocity, flow Reynolds number is increased in case of higher inlet velocity condition giving rise to the turbulence in the flow field in presence of water spray. At this situation momentum and mass transfer take place in case of droplets interacting with the continuous phase, which eventually could lead to a turbulization of the flow field when two-way turbulence coupling is considered. This is the predominant physics which is significant in interpreting the phenomenon

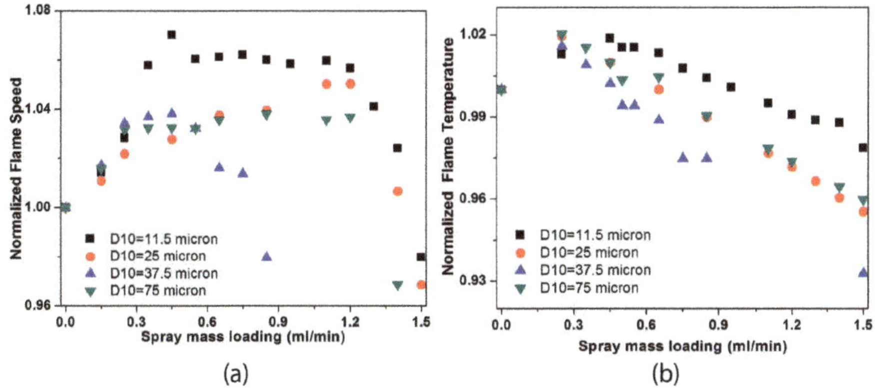

Fig. 32 Effect of spray on flame speed (**a**) and temperature (**b**) for stretch = 270.7 s^{-1}

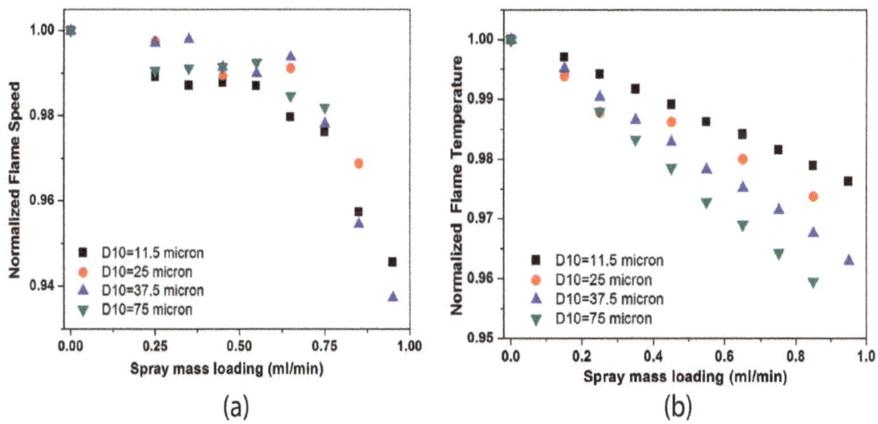

Fig. 33 Effect of spray on flame speed (**a**) and temperature (**b**) for stretch = 125.2 s^{-1}

depicted in Figs. 32 and 33. Figure 33 shows that at low stretch of 125.2 s^{-1} this flame acceleration is absent due to low gas phase velocity. This is due to the fact that, once the stretch is higher than a transition point, turbulization effect of spray surpasses the evaporative effect and instead of flame extinction due to heat loss by evaporation, flame accelerates due to turbulence.

6 Summary and Recommendations for Future Research

In this chapter, an overview of the interaction of flame and water spray was presented. Different simple laboratory scale flames were used to investigate the effects of water spray. Significant experimental and numerical works are summarized for both premixed and non-premixed flame. Counter flow burner was found to be majorly used configuration to study the effects of water spray in non-premixed flame. It is clearly understood that droplet size significantly affects the flame water spray interaction. Droplet inertia and evaporation are the two phenomena which govern the flame suppression. An optimum droplet diameter is found for which effectiveness of the water spray is maximum. Flame speed is influenced by the presence of water spray for the premixed flame. A decrement of flame speed observed due to evaporation heat loss of the droplets. It is also found that flame speed increases for hydrogen air premixed flame in presence of fine water droplets at high stretch rate. Turbulization effect which is caused by droplet-gas phase interaction is identified as a cause for flame acceleration.

References

1. G. Grant, J. Brenton, D. Drysdale, Fire suppression by water sprays. Prog. Energy Combust. Sci. **26**, 79–130 (2000)
2. R. Mugele, H.D. Evans, Droplet size distribution in sprays. Ind. Eng. Chem. **43**, 1317–1324 (1951)
3. S.R. Turns, *An Introduction to Combustion: Concepts and Applications*, 2nd edn. (The McGraw-Hill Companies Inc, USA, 2000)
4. H. Tsuji, Counterflow diffusion flames. Prog. Energy Combust. Sci. **8**, 93–119 (1982)
5. M.D. Smooke, R.A. Yetter, T.P. Parr, D.M. Hanson-Parr, M.A. Tanoff, M.B. Colket, R.J. Hall, Computational and experimental study of ammonium perchlorate/ethylene counterflow diffusion flames. Proc. Combust. Inst. **28**, 2013–2020 (2000)
6. M.D. Smooke, I.K. Puri, and K. Seshadri, A comparison between numerical calculations and experimental measurements of the structure of a counterflow diffusion flame burning diluted methane in diluted air, in *Twenty-First Symposium (International) on Combustion*, vol. 21, 1988, pp. 1783–1792
7. K. Seshadri, C. Trevino, M.D. Smooke, Analysis of the structure and mechanisms of extinction of a counterflow methanol-air diffusion flame. Combust. Flame **76**, 111–132 (1989)
8. G. Dixon-Lewis, T. David, P.H. Gaskell, S. Fukutani, H. Jinno, J.A. Miller, R.J. Kee, M.D. Smooke, N. Peters, E. Effelsberg, J. Warnatz, F. Behrendt, Calculation of the structure and extinction limit of a methane-air counterflow diffusion flame in the forward stagnation region of a porous cylinder, in *Twentieth Symposium (International) on Combustion Symposium (International) on Combustion*, vol. 20, 1985, pp. 1893–1904
9. G. Dixon-Lewis, V. Giovangigli, R.J. Kee, J.A. Miller, B. Rogg, M.D. Smooke, G. Stahl, J. Warnatz, Numerical modeling of the structure and properties of tubular strained laminar premixed flames. Am. Inst. Aeronaut. Astronaut. **131**, 125–144 (1991)
10. G. Balakrishnan, M.D. Smooke, F.A. Williams, A numerical investigation of extinction and ignition limits in laminar nonpremixed counterflowing hydrogen-air streams for both elementary and reduced chemistry. Combust. Flame **102**, 329–340 (1995)
11. G. Amantini, J.H. Frank, M.D. Smooke, A. Gomez, Computational and experimental study of steady axisymmetric non-premixed methane counterflow flames. Combust. Theory Model. **11**, 47–72 (2007)
12. G. Amantini, J.H. Frank, M.D. Smooke, A. Gomez, Computational and experimental study of standing methane edge flames in the two-dimensional axisymmetric counterflow geometry. Combust. Flame **147**, 133–149 (2006)

13. N. Peters, *Laminar flamelet concepts in turbulent combustion*, in *Symposium (International) on Combustion*, vol. 21, 1988, pp. 1231–1250
14. N. Peters, Laminar diffusion flamelet models in non-premixed turbulent combustion. Prog. Energy Combust. Sci. **10**, 319–339 (1984)
15. A. Lock, S.K. Aggarwal, I.K. Puri, Effect of fuel type on the extinction of fuel and air stream diluted partially premixed flames. Proc. Combust. Inst. **32**, 2583–2590 (2009)
16. J.S. Park, D.J. Hwang, J. Park, J.S. Kim, S. Kim, S.I. Keel, T.K. Kim, D.S. Noh, Edge flame instability in low-strain-rate counterflow diffusion flames. Combust. Flame **146**, 612–619 (2006)
17. A. Lock, S.K. Aggarwal, I.K. Puri, U. Hegde, Suppression of fuel and air stream diluted methane–air partially premixed flames in normal and microgravity. Fire Saf. J. **43**, 24–35 (2008)
18. A. Lock, A.M. Briones, S.K. Aggarwal, I.K. Puri, U. Hegde, Liftoff and extinction characteristics of fuel- and air-stream-diluted methane–air flames. Combust. Flame **149**, 340–352 (2007)
19. A.J. Lock, A.M. Briones, X. Qin, S.K. Aggarwal, I.K. Puri, U. Hegde, Liftoff characteristics of partially premixed flames under normal and microgravity conditions. Combust. Flame **143**, 159–173 (2005)
20. A.M. Lentati, and H.K. Chelliah, Physical, thermal, and chemical effects of fine-water droplets in extinguishing counterflow diffusion flames, in *Symposium (International) on Combustion*, vol. 27, 1998, pp. 2839–2846
21. E.J.P. Zegers, B.A. Williams, R.S. Sheinson, J.W. Fleming, Dynamics and suppression effectiveness of monodisperse water droplets in non-premixed counterflow flames. Proc. Combust. Inst. **28**, 2931–2937 (2000)
22. A.M. Lentati, H.K. Chelliah, Dynamics of water droplets in a counterflow field and their effect on flame extinction. Combust. Flame **115**, 158–179 (1998)
23. H.K. Chelliah, A.K. Lazzarini, P.C. Wanigarathne, G.T. Linteris, Inhibition of premixed and non-premixed flames with fine droplets of water and solutions. Proc. Combust. Inst. **29**, 369–376 (2002)
24. A. Dvorjetski, J.B. Greenberg, Theoretical analysis of polydisperse water spray extinction of opposed flow diffusion flames. Fire Saf. J. **39**, 309–326 (2004)
25. S. Pramanik, A. Mukhopadhyay, Numerical study of counterflow diffusion flame and water spray interaction. J. Therm. Sci Eng. Appl. **8**, 011–018 (2014)
26. M.N. Sasongko, M. Mikami, A. Dvorjetski, Extinction condition of counterflow diffusion flame with polydisperse water sprays. Proc. Combust. Inst. **33**, 2555–2562 (2011)
27. S. Sarkar, A. Mukhopadhyay, S. Sen, Numerical study of the effect of fine polydisperse water spray on laminar flame speed of hydrogen/air premixed flames stabilized in the stagnation flow field, in *ILASS-Asia 2016, 18th Annual Conference on Liquid Atomization and Spray Systems—Asia*, Chennai, India
28. A. Yoshida, R. Takasaki, K. Kashiwa, H. Naito, Y. Saso, Extinguishment of counterflow methane/air diffusion flame by polydisperse fine water droplets. Combust. Flame **160**, 1357–1363 (2013)
29. A.K. Lazzarini, R.H. Krauss, H.K. Chelliah, G.T. Linteris, Extinction conditions of non-premixed flames with fine droplets of water and water/NaOH solutions. Proc. Combust. Inst. **28**, 2939–2945 (2000)
30. P.G. Arias, H.G. Im, P. Narayanan, A. Trouvé, A computational study of non-premixed flame extinction by water spray. Proc. Combust. Inst. **33**, 2591–2597 (2011)
31. F. Takahashi, V.R. Katta, Extinguishment of diffusion flames around a cylinder in a coaxial air stream with dilution or water mist. Proc. Combust. Inst. **32**, 2615–2623 (2009)
32. C.C. Ndubizu, R. Ananth, P.A. Tatem, V. Motevalli, On water mist fire suppression mechanisms in a gaseous diffusion flame. Fire Saf. J. **31**, 253–276 (1998)
33. G. Liao, J. Liu, J. Qin, B. Yao, Experimental study on the interaction of fine water spray with liquid pool fires. J. Therm. Sci. **10**, 377 (2001)
34. A. Yoshida, T. Okawa, W. Ebina, H. Naito, Experimental and numerical investigation of flame speed retardation by water mist. Combust. Flame **162**, 1772–1777 (2015)

35. C.K. Law, P. Cho, M. Mizomoto, H. Yoshida, Flame curvature and preferential diffusion in the burning intensity of bunsen flames, in *Twenty-First Symposium (International) on Combustion*, vol. 21, 1988, pp. 1803–1809
36. W. Yang, R.J. Kee, The effect of monodispersed water mists on the structure, burning velocity, and extinction behavior of freely propagating, stoichiometric, premixed, methane-air flames. Combust. Flame **130**, 322–335 (2002)
37. A.U. Modak, A. Abbud-Madrid, J.-P. Delplanque, R.J. Kee, The effect of mono-dispersed water mist on the suppression of laminar premixed hydrogen–, methane–, and propane–air flames. Combust. Flame **144**, 103–111 (2006)
38. A. Jones, P.F. Nolan, Discussions on the use of fine water sprays or mists for fire suppression. J. Loss Prev. Process Ind. **8**, 17–22 (1995)
39. J.M. Ingram, A.F. Averill, P.N. Battersby, P.G. Holborn, P.F. Nolan, Suppression of hydrogen–oxygen–nitrogen explosions by fine water mist: Part 1 Burning velocity. Int. J. Hydrog. Energy **37**, 19250–19257 (2012)
40. C. Joseph-Auguste, H. Cheikhravat, N. Djebaïli-Chaumeix, E. Deri, On the use of spray systems: An example of R&D work in hydrogen safety for nuclear applications. Int. J. Hydrog. Energy **34**, 5970–5975 (2009)
41. H. Cheikhravat, J. Goulier, A. Bentaib, N. Meynet, N. Chaumeix, C.E. Paillard, Effects of water sprays on flame propagation in hydrogen/air/steam mixtures. Proc. Combust. Inst. **35**, 2715–2722 (2015)
42. J. Munshi, S. Sarkar, S. Sen, A. Mukhopadhyay, Numerical study of the effect of fine polydisperse water spray on laminar flame speed of hydrogen/air premixed flames stabi-lized in the stagnation flow field, in *ILASS-Asia 2016, 18th Annual Conference on Liquid Atomization and Spray Systems—Asia*, Chennai, India

Part III
Turbulent Combustion Modeling and Simulations

Quantification of External Enthalpy Controlled Combustion at Unity Damköhler Number

Fabian Hampp and Rune Peter Lindstedt

Abstract The use of external enthalpy support (e.g. via heat recirculation) can enable combustion beyond normal flammability limits and lead to significantly reduced emissions and fuel consumption. The present work quantifies the impact of such support on the combustion of lean ($\Phi = 0.6$) turbulent premixed DME/air flames with a Damköhler number around unity. The flames were aerodynamically stabilised against thermally equilibrated hot combustion products (HCP) in a back-to-burnt opposed jet configuration featuring fractal grid generated multi-scale turbulence ($Re \simeq 18{,}400$ and $Re_t > 370$). The bulk strain ($a_b = 750\ \mathrm{s}^{-1}$) was of the order of the extinction strain rate ($a_q = 600\ \mathrm{s}^{-1}$) of the corresponding laminar opposed twin flame with the mean turbulent strain ($a_I = 3200\ \mathrm{s}^{-1}$) significantly higher. The HCP temperature ($1600 \leq T_{HCP}(\mathrm{K}) \leq 1800$) was varied from close to the extinction point ($T_q \simeq 1570$ K) of the corresponding laminar twin flame to beyond the unstrained adiabatic flame temperature ($T_{ad} \simeq 1750$ K). The flames were characterised using simultaneous Mie scattering, OH-PLIF and PIV measurements and subjected to a multi-fluid analysis (i.e. reactants and combustion products, as well as mixing, weakly reacting and strongly reacting fluids). The study quantifies the (i) evolution of fluid state probabilities and (ii) interface statistics, (iii) unconditional and (iv) conditional velocity statistics, (v) conditional strain along fluid interfaces and (vi) scalar fluxes as a function of the external enthalpy support.

F. Hampp · R. P. Lindstedt (✉)
Department of Mechanical Engineering, Imperial College, London SW7 2AZ, UK
e-mail: p.lindstedt@imperial.ac.uk

F. Hampp
e-mail: f.hampp11@imperial.ac.uk

A. K. Runchal et al. (eds.), *Energy for Propulsion*, Green Energy and Technology,
https://doi.org/10.1007/978-981-10-7473-8_8

Nomenclature

Roman Letters

a	Rate of strain [s^{-1}].
$\overline{c}$	Reaction progress variable [–].
c	Progress variable; Instantaneous conditioning variable [–].
$\overline{cu}$	Scalar Flux [$m\ s^{-1}$].
D	Burner nozzle diameter [m].
Da	Conventional Damköhler number [–].
Da_{ign}	Turbulent auto–ignition Damköhler number [–].
$d_{p,x}$	Al_2O_3 particle diameter x% [m].
e	Planar rate of strain tensor [s^{-1}].
H	Burner nozzle separation [m].
I	OH signal intensity [–].
$I^{\ddagger}$	Reference signal intensity [–].
Ka	Conventional Karlovitz number [–].
Ka_{ign}	Auto-ignition Karlovitz number [–].
[k]	Theoretical concentration of species k [$mol\ m^{-3}$].
L_{η}	Kolmogorov length scale [m].
L_I	Integral length scale of turbulence [m].
N	Total number of images [–].
$\hat{\mathbf{n}}$	Unit vector of the iso-contour normal [–].
$\dot{Q}$	Heat release rate [$W\ m^{-3}$].
Re_t	Turbulent Reynolds number [–].
S_L	Laminar burning velocity [$m\ s^{-1}$].
$\hat{\mathbf{s}}$	Unit vector of the streamline tangent [–].
T	Temperature [K].
T_{ad}	Adiabatic flame temperature [K].
T_{ign}	Auto-ignition temperature [K].
T_{HCP}	Lower nozzle hot combustion product temperature [K].
U	Flow velocity [$m\ s^{-1}$].
$\overline{U}$	Mean unconditional axial velocity [$m\ s^{-1}$].
$\overline{U_{...}}$	Mean conditional axial velocity [$m\ s^{-1}$].
u	Velocity component [$m\ s^{-1}$].
$\sqrt{u'^2}$	Unconditional axial velocity fluctuation [$m\ s^{-1}$].
$\sqrt{u'^2_{...}}$	Conditional axial velocity fluctuation [$m\ s^{-1}$].
u_{rms}	Root mean square velocity fluctuation [$m\ s^{-1}$].
$\overline{U_s}$	Slip velocity [$m\ s^{-1}$].
$\overline{V}$	Mean unconditional radial velocity [$m\ s^{-1}$].
$\dot{V}$	Lower nozzle volumetric flow rate [$m^3\ s^{-1}$].
$\sqrt{v'^2}$	Unconditional radial velocity fluctuation [$m\ s^{-1}$].
$\sqrt{v'^2_{...}}$	Conditional radial velocity fluctuation [$m\ s^{-1}$].
X	Mole fraction [–].

x	Axial coordinate [m].
x_s	Distance from origin of first thermal alteration [m].
y	Radial coordinate [m].

Greek Letters

β	Fluid state material surface [–].
δ_f	Laminar fuel consumption layer thickness [m].
ε_r	Rate of dissipation within the reactants [m^2 s^{-3}].
Λ	Threshold value [–].
λ_B	Batchelor scale [m].
λ_D	Mean scalar dissipation layer thickness [m].
λ_{MF}	Multi-fluid spatial resolution [m].
λ_{PIV}	PIV spatial resolution [m].
ν_r	Reactants kinematic viscosity [m^2 s^{-1}].
ω	Planar vorticity tensor [s^{-1}].
Φ	Equivalence ratio [–].
τ_c	Chemical timescale [s].
τ_η	Kolmogorov timescale [s].
τ_{ign}	Auto-ignition delay time [s].
τ_I	Integral timescale of turbulence [s].
ξ	Blending fraction [$\%_{vol}$].

Sub-/Superscripts

0	Alignment at the origin; initial value.
‡	Reference value.
BTB	Back-to-burnt configuration.
b	Bulk flow motion.
d	Total.
FS	Fluid state.
HCP	Hot combustion products.
I	Integral scale; turbulent.
i, j	Pixel index.
k	Velocity component.
M	Mixing fluid iso-contour.
m	Mixing fluid.
n	Instantaneous image; normal.
NE	Nozzle exit.
p	Product fluid.
q	Extinction conditions.

R	Reactant fluid iso-contour.
r	Reactant fluid.
SR	Strongly reacting fluid iso-contour.
s	Strongly reacting (flamelet) fluid.
T	Dependency on HCP temperature.
t	Tangential.
WR	Weakly reacting fluid iso-contour.
w	Weakly reacting fluid.

1 Introduction

Low Damköhler number combustion has the potential to reduce NO_x emissions through stable fuel lean operation [1]. However, thermal support is typically required to sustain chemical reactions [2] with common enthalpy sources including exhaust gas recirculation (EGR) [3] and reactant preheating via heat exchangers [4]. Raising the initial temperature (T_0) via preheating reduces the chemical timescale [5]. The strongly fuel-dependent auto-ignition delay time may become sufficiently short at high T_0 to influence the combustion behaviour [6]. Several combustion concepts utilise the exhaust gas enthalpy to stabilise fuel lean combustion. Examples include flameless oxidation in gas turbine engines [7] and low NO_x burners [8]. Blending of high temperature (internal) EGR yields a complex competition of advanced chain branching, possibly supported by the presence of residual intermediates, and quenching due to the diluents (e.g. CO_2 and H_2O) [9]. The DNS data of Minamoto et al. [10] have shown that the broadening and distribution of chemically active zones is not solely dependent on turbulence, but is strongly affected by the dilution level and mixture reactivity. At low equivalence ratios, comparatively low levels of turbulence have been found sufficient to disturb and broaden the reaction zone [11]. In a related study, Zhou et al. [12] have shown a significant broadening of the CH layer that experiences a deeper penetration of CH_2O and OH [11, 13] with increasing turbulence intensity. Accordingly, flamelet-based bimodal descriptions that assume a negligible probability of encountering chemically active states become problematic [10].

Canonical configurations, e.g. the Sandia–Sydney piloted jet flames [14, 15] and the opposed jet configuration [16, 17], are frequently used to advance the fundamental understanding of combustion processes. Mastorakos et al. [2] investigated the removal of conventional flame extinction limits for lean premixed CH_4 flames using a back-to-burnt (BTB) opposed jet geometry. No extinction was observed for a burnt gas temperature >1550 K. An unstable region was detected in the range from 1450 to 1550 K with conventional extinction criteria valid for reduced temperatures. Goh [18] and Coriton et al. [19] extended the study by means of a wide range of Φ and increased turbulence levels. Related combustion regime transitions in lean premixed JP-10 (exotetrahydrodicyclopentadiene) flames were studied by Goh et al. [20] and compared to conventional twin flames approaching extinction [21]. Hampp et al. [22, 23] investigated burning mode transitions from close to the corrugated flamelet into the distributed reaction zone regime by varying the stoichiometry of lean premixed dimethyl ether (DME) flames at a constant turbulent Reynolds number ($Re_t \simeq 380$). A constant burnt gas state was used. Coriton et al. [24] have shown that lean counter-flowing combustion products promote stable operation compared to stoichiometric flames or hot inert gas due to a delayed radical pool depletion. The promotion of chain branching by intermediates has also been observed under moderate or intense low oxygen dilution (MILD) conditions by Abtahizadeh et al. [25].

The current study investigates the impact of the counter-flowing hot combustion products (HCP) on turbulent DME/air flames at a constant $Re_t \simeq 395$ and a close to unity Damköhler number ($Da \simeq 1.1$). DME was selected as it is regarded as an attractive alternative fuel [26, 27] with a comparatively well-established combustion chemistry [28]. The conditions mark the approximate transition from thin to distributed reaction zones in a conventional regime diagram [22]. The investigated HCP temperature range of $1600 \leq T_{HCP}(\text{K}) \leq 1800$ covers conditions from close to the extinction point ($T_q \simeq 1570$ K) of the corresponding laminar twin flame to beyond the unstrained adiabatic flame temperature ($T_{ad} \simeq 1750$ K). The flames were studied using simultaneous Mie scattering, OH-PLIF and PIV. The study quantifies the impact of the thermal support on burning modes via the (i) evolution of the probability of encountering multiple fluid states (i.e. reactants, combustion products, mixing, weakly reacting and strongly reacting fluids) and (ii) interface statistics, (iii) unconditional and (iv) conditional velocity statistics, (v) conditional strain on fluid state interfaces and (vi) scalar fluxes.

2 Experimental Set-up

The twin flame variant of the current opposed jet burner was developed by Geyer et al. [16]. The current revised back-to-burnt (BTB) configuration, schematically depicted in Fig. 1, is identical to that used by Hampp et al. [22, 23, 30, 31] with multi-scale turbulence [17, 32, 33] generated via a cross fractal grid (CFG) [18, 32]. Operation in the BTB mode [2, 19, 20, 22, 23] enables the stabilisation under fuel lean conditions. The premixed DME/air mixture was injected through the upper nozzle (UN) and stabilised against hot combustion products emerging from the lower nozzle (LN) as described below. The nozzle separation was set to one nozzle diameter ($D = 30$ mm).

2.1 *Upper Nozzle Flow Conditions*

Premixed DME/air ($\Phi = 0.60$, $T_0 = 320$ K) was injected through the UN at a constant bulk velocity ($U_b = 11$ m s^{-1}). The CFG, offering a blockage ratio of 65% with a maximum to minimum bar width ratio of 4, was installed 50 mm upstream of the UN exit and provided well-developed multi-scale turbulence with $Re_t \simeq 395$. The turbulent (τ_I) and chemical (τ_c) timescales of the UN reactant flow were maintained

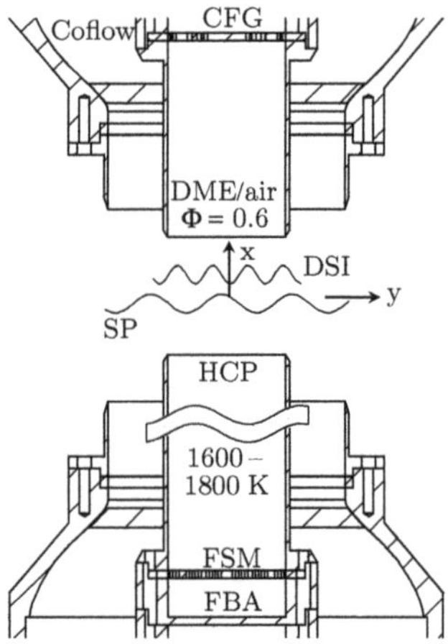

Fig. 1 Schematic of the experimental configuration. Unreacted premixed DME/air ($\Phi = 0.6$) is injected in the upper nozzle and stabilised by hot combustion products (HCP) from a stoichiometric H_2/CO_2/air flame emerging the lower nozzle. CFG—cross fractal grid, DSI—density segregation iso-contour, SP—stagnation plane, FBA—flashback arrestor, FSM—flame stabilisation mesh

constant and the investigated conditions occupy a nominally identical point in a conventional regime diagram with $Da \simeq 1.1$. The Re_t and τ_I were determined based on the integral length scale of turbulence ($L_I = 4.1$ mm) and the velocity fluctuations at the nozzle exit ($u_{rms} = 1.64$ m s^{-1}) measured using hot wire anemometry and particle image velocimetry (PIV), respectively. The calculated kinematic viscosity in the reactants was $\nu_r = 17.0 \times 10^{-6}$ m^2 s^{-1}. The Da number and τ_c, see Eq. 1, were estimated using a laminar burning velocity ($S_L = 0.21$ m s^{-1}) and a laminar flame thickness ($\delta_f = 0.22$ mm) [22] obtained from strained laminar opposed jet calculations ($a = 75$ s^{-1}) using the detailed chemistry of Park [28]. The flame thickness was based on the 5–95% fuel consumption thickness (i.e. inner thickness [34]).

$$\tau_I = \frac{L_I}{u_{rms}}, \qquad \tau_c = \frac{\delta_f}{S_L}, \qquad Re_t = \frac{L_I \cdot u_{rms}}{\nu_r}, \qquad Da = \frac{\tau_I}{\tau_c} \tag{1}$$

The extinction strain of the corresponding laminar flame ($a_q = 600$ s^{-1}), the bulk ($a_b = 2 \cdot U_b/H = 750$ s^{-1}) and mean turbulent strain ($a_I = 3200$ s^{-1}) were determined by Hampp and Lindstedt [22].

2.2 Lower Nozzle Flow Conditions

The stabilising hot combustion products (HCP) were produced from highly diluted stoichiometric premixed H_2 flames anchored on a flame stabilisation mesh (FSM) located 100 mm upstream of the LN exit. The FSM was optimised to preclude any

Table 1 Lower nozzle conditions with X being the mole fractions (the missing percentile accounts for air), $[OH]_T^{\ddagger}$ the equilibrium OH concentration at the nozzle exit, $[OH]_q$ the peak OH concentration at the twin flame extinction point, $\dot{V}$ the lower nozzle cold gas volumetric flow rate and U_{HCP} the HCP nozzle exit velocity.

Property	Unit	T_{HCP} [K]				
		1600	1650	1700	1750	1800
$X(H_2)$	–	0.218	0.230	0.238	0.244	0.249
$X(CO_2)$	–	0.263	0.225	0.197	0.175	0.159
$[OH]_T^{\ddagger} \times 10^{-3}$	mol m^{-3}	7.38	8.40	8.89	9.72	10.8
$[OH]_q \times 10^{-3}$	mol m^{-3}	30.9				
$\dot{V} \times 10^{-3}$	m^3 s^{-1}	3.11	3.11	3.10	3.08	3.06
U_{HCP}	m s^{-1}	22.7	23.4	24.1	24.6	25.1

flame instability and featured a blockage ratio of 62% with an aperture of 0.40 mm. A second finer mesh served as flashback arrestor (FBA). The HCP temperature at the nozzle exit was controlled from 1600 to 1800 K (peak-to-peak variation of 1%) by adjusting the CO_2 dilution rate from 15–25$\%_{vol}$ (prior combustion). The nozzle exit temperature was measured with a 50 μm R-type thermocouple. The lower limit corresponds approximately to the temperature at the twin flame extinction point ($T_q \simeq 1570$ K at $a_q = 600$ s^{-1}), and the upper limit exceeds the adiabatic flame temperature ($T_{ad} \simeq 1750$ K). The stagnation plane was positioned in the proximity of the burner centre by jet momentum matching with differences in HCP density compensated by modest adjustments of the cold gas bulk velocity as shown in Table 1.

2.3 *Diagnostic Set-up*

Simultaneous Mie scattering, PIV and hydroxyl planar laser-induced fluorescence (OH-PLIF) measurements were performed utilising the barium nitrate crystal technique of Kerl et al. [35] as detailed by Hampp et al. [22, 30]. The two overlaid light sheets (281.7 nm and 532 nm) exhibited a height of 1D and thicknesses <0.50 mm and <0.25 mm, respectively. The fluorescence signal was spectrally separated from the Mie scattering by a dichroic filter. Two interline-transfer CCD cameras (LaVision Imager Intense) were used with one connected to an intensified relay optics (LaVision IRO) unit to record the OH signal. The OH fluorescence was collected by a 105 mm ultraviolet lens (f/2.8) from LaVision and the Mie scattering with a Tokina AF 100 mm lens (f/2.8). Both lenses were equipped with bandpass filters featuring a transmissivity >85 % for the respective spectral range and an optical density >5 in order to block the laser lines. The PIV laser pulses were separated by 25 μs to minimise spurious vectors, and the OH-PLIF images were obtained from the first laser pulse. Each nozzle was seeded separately using aluminium oxide powder ($d_{p,90} = 1.7$ μm). Cross-correlation PIV (LaVision Davis 8.1) was performed with decreasing interrogation regions size (128 × 128 to 48 × 48 with a 75% overlap). The final pass was conducted using a high accuracy mode and shape adaptation of the weighted windows to incorporate the local flow field acceleration [36]. The determined velocity field consisted of 115 × 88 vectors providing a spacing of 0.30 mm and spatial resolution of 0.60 mm. A comprehensive uncertainty analysis and the error associated with 3D effects were presented by Hampp et al. [22, 23].

For each condition, 3000 statistically independent realisations (minimum temporal separation of τ_I) were recorded. Image pre-processing (i.e. alignment, data reduction, noise reduction, shot-to-shot intensity fluctuations and white image correction) was performed as described by Hampp et al. [22, 30]. The well-defined conditions close to the upper and lower nozzle exits were used to obtain calibration signals in predefined interrogation windows [22].

3 Burnt Gas State and the Impact on Flame Parameters

The stabilisation of low *Da* flows against hot combustion products removes conventional extinction criteria with chemical reactions sustained by the thermal support [2]. By contrast, high *Da* self-sustained flames detach from the stagnation plane and decouple from the external enthalpy source influence [23]. The integrated heat release ($\int \dot{Q}$) of self-sustained flames correlates well with the thermochemical state (e.g. peak radical concentrations and peak temperature) and has been shown to be close to configuration (twin or BTB) independent [22]. The peak OH concentration at the twin flame extinction point ($[\mathrm{OH}]_q = 30.9 \times 10^{-3}$ mol m^{-3}; see Table 1) can thus be used to approximately delineate self-sustained burning in the BTB configuration. The determined twin flame extinction point is independent of the HCP and is consequently constant in the current investigation. By contrast, the chemical activity of thermally supported burning with $\int \dot{Q}_{BTB} < \int \dot{Q}_q$ is influenced or governed by the HCP and dependent on the mixture composition and temperature of the external enthalpy source [9, 24, 25].

3.1 Burnt Gas State

The temperature of the HCP was controlled by means of CO_2 dilution prior combustion of the stoichiometric H_2 flames; see Sect. 2.2. The extended lower nozzle (100 mm) realised HCP in (close to) thermochemical equilibrium at the nozzle exit, which provided a well-defined experimental reference state [22]. In order to provide a reference point for comparisons with theoretical considerations, the latter can also be estimated using laminar flame calculations replicating the experimental conditions (i.e. mixture composition and residence time). The computations featured 660 locally refined cells providing a resolution of 7 μm in the reaction zone. The measured and computed temperatures were matched using an overall heat loss of 7.2–8.9% via a radiation correction [37]. The estimated OH concentrations ($[OH]_T^\ddagger$) at the nozzle exit range from 7.38 to 10.8×10^{-3} mol m^{-3}; see Table 1.

3.2 HCP Impact on Combustion Properties

The failure to establish self-sustained flames in the BTB opposed jet configuration results in turbulent mixing of the HCP with unreacted DME/air [23]. The blending of hot products with fresh reactants increases the initial temperature, reduces the auto-ignition delay times, alters burning properties (e.g. increased S_L and reduced δ_f) [9, 30] and thus eases ignition [38]. Under such conditions flamelet-like structures can coexist with pockets undergoing auto-ignition [39, 40]. Conventional definitions of the integral (τ_I), Kolmogorov (τ_η) and chemical (τ_c) timescales, associated with self-sustained burning, define the Da number (see Eq. 1) and the corresponding Karlovitz (Ka) number as shown in Eq. 2.

$$L_\eta = \left(\frac{\nu_r^3}{\varepsilon_r}\right)^{1/4}, \qquad \tau_\eta = \sqrt{\frac{\nu_r}{\varepsilon_r}}, \qquad Ka = \frac{\tau_c}{\tau_\eta} \tag{2}$$

The rate of dissipation (ε_r) in the reactants was obtained using the method of George and Hussein [41] for locally axisymmetric turbulence as described by Hampp and Lindstedt [22].

The potential influence of auto-ignition was estimated in terms of the associated delay times (τ_{ign}) as a function of the blending fraction (ξ) of reactants with HCP using shock tube calculations, see Fig. 2, where ξ is defined as nil in the reactants and unity in the HCP. With increasing T_{HCP} and ξ, the auto-ignition delay time decreases. The corresponding Damköhler and Karlovitz numbers based on the auto-ignition delay time (τ_{ign}) are defined below.

$$Da_{ign} = \frac{\tau_I}{\tau_{ign}}, \qquad Ka_{ign} = \frac{\tau_{ign}}{\tau_\eta} \tag{3}$$

At a given initial temperature, Da_{ign} and Ka_{ign} provide a measure of the likelihood of the mixture being subject to auto-ignition in fluid pockets with mixing timescales corresponding to the large- and small-scale turbulent motion. The blending fraction required to cause an auto-ignition delay time similar to the integral timescale of turbulence ($\tau_{ign} = \tau_I \rightarrow Da_{ign} \sim 1$) is reduced from $\xi = 0.70$ to 0.55 as T_{HCP} is increased from 1600 and 1800 K. The data are summarised in Table 2.

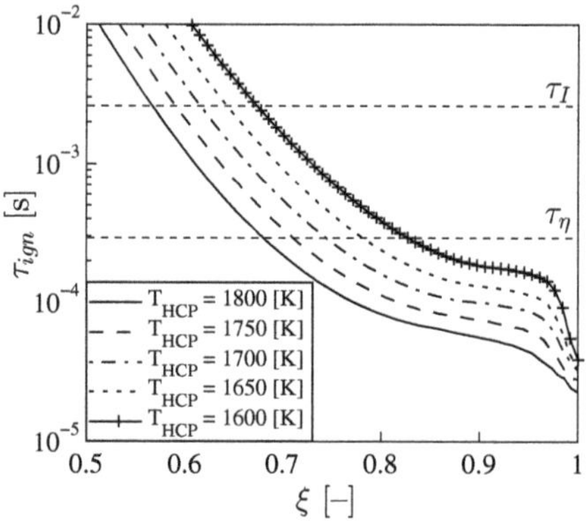

Fig. 2 Shock tube calculations to evaluate the auto-ignition delay time (τ_{ign}) as function of blending fraction ξ and T_{HCP}. The horizontal dashed lines show the integral (τ_I) and Kolmogorov timescales (τ_η)

Table 2 Summary of turbulent and chemical parameters used to derive the turbulent Reynolds, Damköhler and Karlovitz numbers for DME/air at $\Phi = 0.6$ at a low strain rate ($a = 75$ s^{-1}). The turbulence conditions were evaluated within the reactants. The auto-ignition delay times (τ_{ign}), Da_{ign} and Ka_{ign} were evaluated at the respective T_{HCP}

T_{HCP}	K	1800	1750	1700	1650	1600
Φ	–	0.60				
S_L	m s^{-1}	0.21				
δ_f	mm	0.46				
τ_c	ms	2.19				
τ_{ign}	μs	5.07	6.66	8.96	12.4	17.6
u_{rms}	m s^{-1}	1.63	1.67	1.65	1.60	1.64
L_I	mm	4.1	4.1	4.1	4.1	4.1
τ_I	ms	2.51	2.46	2.48	2.56	2.50
L_η	μm	75	72	73	72	73
τ_η	ms	0.33	0.31	0.31	0.30	0.31
a_q	s^{-1}	600				
ε_r	m^2 s^{-3}	158	178	172	183	176
$\nu_r \times 10^6$	m^2 s^{-1}	17.0				
Re_t	–	393	403	397	385	395
Da	–	1.15	1.12	1.13	1.17	1.14
Da_{ign}	–	495	369	277	206	142
Ka	–	6.67	7.09	6.96	7.19	7.05
Ka_{ign}	–	0.015	0.021	0.029	0.041	0.057

4 Multi-fluid Post-processing

Hampp and Lindstedt [22, 23] presented an extension to bimodal (i.e. reactants and products) statistical descriptions by incorporating chemically active fluid states such as mixing, weakly and strongly reacting. The methodology, briefly summarised below, was found particularly beneficial for low *Da* flows and was consequently adopted in the present study in order to delineate the impact of the external enthalpy source on the combustion process. The identified fluids states are defined as:

Reactants: Fresh reactants emerging from the UN that have not undergone any thermal alteration (i.e. no oxidation or mixing processes). The reactants were detected via a PIV tracer particle-based density segregation technique, e.g. [33, 42, 43], that is capable of detecting multiple and fragmented splines.

Mixing fluid: A fluid state that has been exposed to a thermal change prior the onset of OH producing chemical reactions (i.e. via mixing with HCP). The mixing fluid is detected by superposition of the Mie scattering and OH-PLIF images and identified as regions with low seeding density and no OH signal.

Strongly reacting fluid: Regions with a high OH signal intensity caused by self-sustained (e.g. flamelet) burning (see Sect. 4.1). Conventional aerothermochemistry conditions and extinction criteria apply [19, 44].

Weakly reacting fluid: A fluid state with modest levels of OH, e.g. ultra lean flames sustained by thermal support from an external enthalpy source and/or combustion products approaching equilibrium (see Sect. 4.1).

Hot combustion products: The HCP emerge the LN in close to chemical equilibrium and provide a well-defined reference state with constant OH concentration and signal intensity for a given temperature; see Table 1.

4.1 OH Containing Fluid States

Chemically active fluid states were delineated based on the OH signal intensity using two thresholds. The first is based on experimental data, and the second is linked to well-established combustion theories [22] as outlined below. The methodology uses a linear relation [22, 46] between the OH concentration and fluorescence intensity that was found sufficient for the conditions of interest (uncertainty $< 10\,\%$) to identify characteristic intensity bands [22, 30, 45]. The OH fluorescence signal intensity at the lower nozzle exit provides the well-defined reference state ($I_T^{\ddagger}$) that is used for calibration. The segregation of the HCP from chemically active fluid material that originates the DME combustion was based on the maximum measured OH signal for an isothermal case (upper nozzle $\Phi = 0.0$) [22, 30]. The limiting threshold was determined to $\Lambda_p = \lceil I/I_T^{\ddagger} \rceil = 2.0$ for all T_{HCP}. The excess OH was attributed to the oxidation of residual reactants in the HCP by the fresh UN air. The resulting iso-contour can be related to the gas mixing layer interface of Coriton et al. [19].

Self-sustained (strongly reacting) flames detach from the stagnation plane and decouple from the HCP [23]. However, high rates of strain may suppress conventional burning and result in thermally supported (weakly reacting) burning that is dominated by the HCP. The thermochemical state at the twin flame extinction point (e.g. $[OH]_q$; see Table 1) arguably provides a natural segregation of self-sustained from supported burning [22]. The strongly reacting fluid (i.e. self-sustained burning) is therefore assumed to be present in regions where the OH signal intensity exceeds the corresponding non-dimensional extinction threshold defined by Eq. 4, i.e. $I > \Lambda_{q(T)} \cdot I_T^{\ddagger}$ with $\Lambda_{q(1600\ K)} = 4.2$, $\Lambda_{q(1650\ K)} = 3.7$, $\Lambda_{q(1700\ K)} = 3.5$, $\Lambda_{q(1750\ K)} = 3.2$ and $\Lambda_{q(1800\ K)} = 2.9$. The corresponding $[OH]_T^{\ddagger}$ and $[OH]_q$ are listed in Table 1.

$$\Lambda_{q(T)} = \frac{[OH]_q}{[OH]_T^{\ddagger}} = \frac{I_q}{I_T^{\ddagger}} \quad \forall\, T \tag{4}$$

The weakly reacting fluid follows as $\Lambda_p < I/I_T^{\ddagger} < \Lambda_{q(T)}$ and can stem from (i) ignition events, (ii) decaying OH concentration in combustion products or (iii) chemically active material that is diluted by the HCP.

4.2 *Multi-fluid Fields and Spatial Resolution*

The identification of the individual fluid states from the instantaneous Mie scattering and OH-PLIF images and subsequent superposition leads to quinary multi-fluid fields. An example is shown in Fig. 3 for a flame with $\Phi = 0.6$ and $T_{HCP} = 1700$ K. The spatial resolution and uncertainty analysis of the planar PIV and multi-fluid images as well as the laminar flame thickness for DME/air at $\Phi = 0.6$ were determined by Hampp et al. [22, 23] and are summarised in Table 3. The integral length scale of turbulence is resolved with $L_I/\lambda_{PIV} \simeq 7$ and $L_I/\lambda_{MF} \simeq 16$.

5 Results and Discussion

The multi-fluid probability, interface and conditional velocity statistics were aligned at the first thermal alteration iso-contour (i.e. $x_s = 0$ and detected via the density segregation technique; see Fig. 3) to eliminate modest variations of the stagnation plane location. The multi-fluid probabilities (Sects. 5.1 and 5.2) and unconditional and conditional velocity statistics in Sects. 5.3 and 5.4 (besides the nozzle exit velocity profiles) were conditioned on the theoretical stagnation point streamline (SPS),

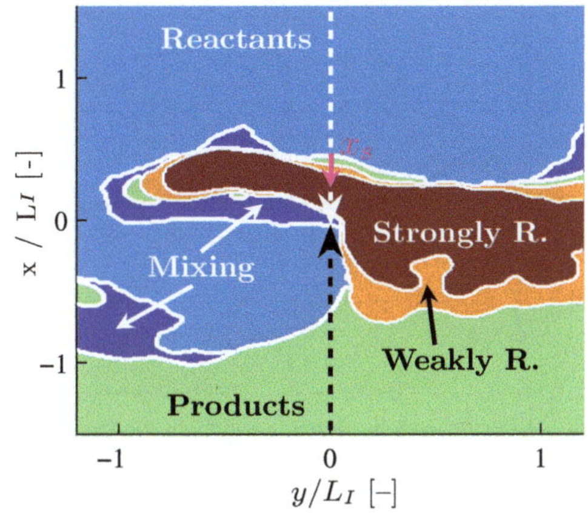

Fig. 3 Instantaneous quinary multi-fluid field for $T_{HCP} = 1700$ K with the vertical white/black arrows showing the theoretical stagnation point streamline (SPS). Interfaces are defined by the intersection of the SPS and material surfaces (white iso-contours). Reactants (light blue); Mixing (blue); Weakly reacting (orange); Strongly reacting (red); Products (green). The magenta arrow shows the x_s origin

Table 3 Physical and resolved length scales in µm, where λ_{PIV} and λ_{MF} are the PIV and multi-fluid resolutions, respectively. δ_f is the laminar flame thickness for a DME/air flame at $\Phi = 0.6$, λ_D is the mean scalar dissipation layer thickness [47] and λ_B the Batchelor scale [48]

Scales	Reactants	HCP
λ_{PIV}	598	
λ_{MF}	250	
δ_f	440	N.A.
λ_D	621 ± 18	N.A.
λ_B	86 ± 15	~370

i.e. $y = 0$ in Fig. 3. The strain analysis in Sect. 5.5 was condition on $y = 0 \pm 1/2$ L_I to include the radial movement of the stagnation point [32].

5.1 *Multi-fluid Statistics*

The reactant fluid, aligned at $x_s = 0$, inherently drops from unity to nil (see Fig. 4 top left) yet recurs independent of T_{HCP} at $x_s > 0$ with a peak probability ~5% due to turbulent transport [22]. The influence of the HCP enthalpy becomes apparent in the mixing fluid probability as shown in the top right of Fig. 4. The probability is reduced away from the origin due to the earlier onset of OH producing chemical reactions with increasing enthalpy support. For $T_{HCP} > T_{ad}$, a reduction in the mixing fluid peak

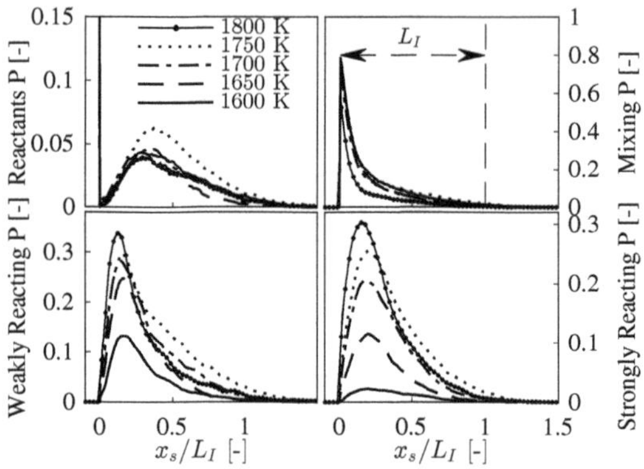

Fig. 4 Multi-fluid probability (P) for DME cases at $\Phi = 0.6$ with $T_{HCP} = 1600$–1800 K: Top left: Reactant fluid; Right: Mixing fluid; Bottom left: Weakly reacting fluid; Right: Strongly reacting (flamelet-like) fluid probability. The markers are drawn for identification purposes and do not represent the actual resolution

probability from 75 to 53 % is observed at $x_s = 0$ as OH producing chemical reactions are induced within a length scale below the multi-fluid resolution of $\simeq 250$ μm. The threefold increase of Da_{ign} $(= \tau_I/\tau_{ign}) = 140$–$500$ for $T_{HCP} = 1600$–1800 K results in a pronounced augmentation of the weakly reacting fluid peak probability from 13 to 35% as depicted in bottom left of Fig. 4. The strongly reacting fluid probability increases from 2% for $T_{HCP} = 1600$ K to approximately 30% for $T_{HCP} = 1800$ K. The combustion products make the balance for the missing percentiles. The spatial extent of all fluid probabilities is limited to approximately one integral length scale of turbulence (L_I) for all investigated conditions.

5.2 Multi-fluid Flow Structure

The interfaces encountered by traversing along the SPS through the quinary multi-fluid fields (see Fig. 5) are evaluated for three conditions: (i) the streamline tangent ($\hat{\mathbf{s}}$) is approximately aligned with the iso-contour normal ($\hat{\mathbf{n}}$) so that $\hat{\mathbf{s}} \cdot \hat{\mathbf{n}} > 0.05$, (ii) flow into the opposite direction with $\hat{\mathbf{s}} \cdot \hat{\mathbf{n}} < -0.05$ and (iii) tangential flow with $||\hat{\mathbf{s}} \cdot \hat{\mathbf{n}}|| < 0.05$ (i.e. 72–108°), where $\hat{\mathbf{s}}$ and $\hat{\mathbf{n}}$ are defined positive in flow direction and from reactants to products, respectively. A schematic illustrating the three flow scenarios is shown in Fig. 5a with an example depicted in Fig. 5b. The diagrams in Fig. 6 show major flow paths (i.e. >5%) for the $T_{HCP} = 1600$, 1700 and 1800 K cases. Reduced values of T_{HCP} result in a flow structure that highlights the need for extensive HCP blending to cause reaction onset and to sustain the chemical activity, i.e. primary fluxes through the mixing and weakly reacting fluid. The latter fluid state is predominately formed via fluxes from the HCP fluid due to smooth OH gradients [22] and longer auto-ignition delay times (see Fig. 2). By contrast, a high temperature burnt gas state promotes an earlier onset of OH producing chemical reactions and realises fluxes from the reactant fluid directly into strongly reacting

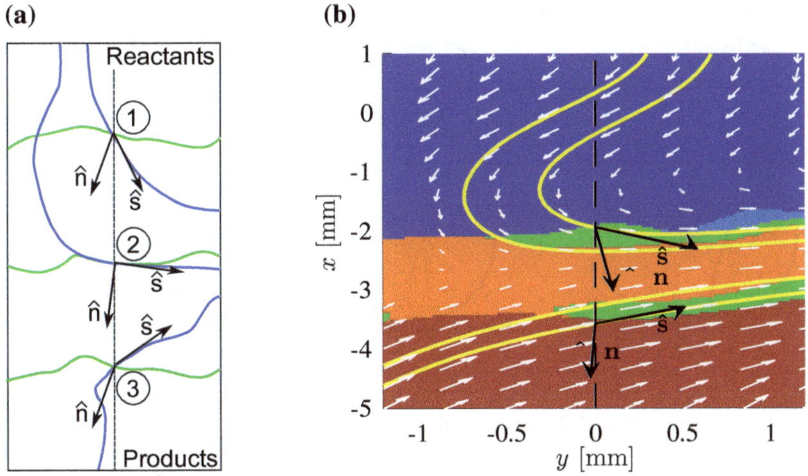

Fig. 5 **a** Schematic illustrating the three defined flow scenarios where blue lines represent streamlines and green lines iso-contours. The streamline tangent $\hat{\mathbf{s}}$ and the iso-contour normal $\hat{\mathbf{n}}$ are defined positive in flow direction and from reactants to products: (1) $\hat{\mathbf{s}} \cdot \hat{\mathbf{n}} > 0.05$, (2) $||\hat{\mathbf{s}} \cdot \hat{\mathbf{n}}|| < 0.05$ (i.e. 72–108°) and (3) $\hat{\mathbf{s}} \cdot \hat{\mathbf{n}} < -0.05$. **b** Multi-fluid image at $T_{HCP} = 1800$ K with streamlines (yellow) and PIV vectors overlaid: reactants (blue), mixing, (light blue), weakly (green), strongly reacting (orange), products (red). The vertical dashed line shows the theoretical SPS and the arrows the unit vectors of the iso-contour normal ($\hat{\mathbf{n}}$) and streamline tangent ($\hat{\mathbf{s}}$)

regions (e.g. a flamelet structure). However, the complex interaction of flame propagation, short auto-ignition delay times and high rates of strain cause fluxes from the reactants into all fluid states. The fluxes into the strongly reacting fluid via the weakly reacting and product fluid can be attributed to contact burning or small blending fractions. The weakly reacting fluid is still primarily formed via adjacent HCP layers indicating the strong need for thermal support for this fluid state.

5.3 Unconditional Velocity Statistics

The present work seeks to isolate the impact of the HCP temperature. The CO_2 concentration in the HCP stream was therefore the only parameter varied. This results in flow deviations/instabilities at the end points of $T_{HCP} = 1600$ and 1800 K, i.e. mean radial HCP velocities >1 m/s at the theoretical stagnation point streamline and increased axial velocity fluctuations. Consequently, the following analysis focuses on the intermediate cases with $T_{HCP} = 1650$, 1700 and 1750 K. The HCP conditions can be optimised for a wider temperature range at the expense of additional changes in experimental parameters. The nozzle exit flow directions of the reactants and HCP are defined as negative and positive, respectively; see Fig. 1.

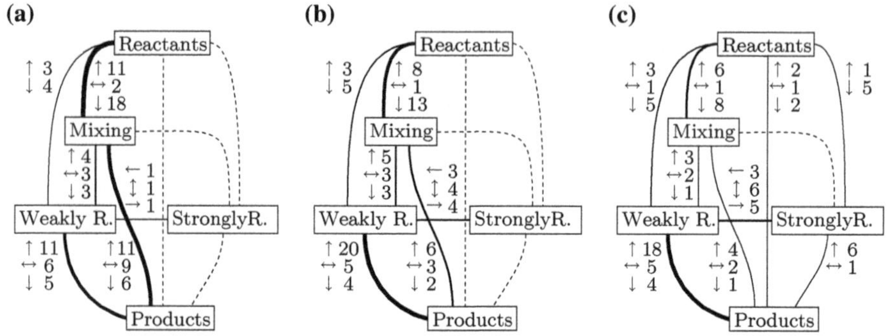

Fig. 6 Multi-fluid interface diagram for DME/air at $\Phi = 0.6$ and a supporting HCP temperature of: **a** $T_{HCP} = 1600$, **b** 1700 and **c** 1800 K. The weighted connections and values illustrate the number of interfaces in % and the arrows indicate the flow direction ($\updownarrow$ with $\leftrightarrow$ indicating near tangential flow). The total numbers of interfaces are 6700, 7900 and 7500 for $T_{HCP} = 1600$, 1700 and 1800 K

Unconditional velocity statistics were obtained 1 mm away from the upper nozzle exit, see Fig. 7a, and along the theoretical stagnation point streamline; see Fig. 7b. The former show nearly identical inlet velocity statistics for all T_{HCP} conditions. By contrast, with increasing T_{HCP} a modest reduction of the mean axial velocity along the stagnation point streamline is observed with a pronounced reduction in axial ($\sqrt{u'u'}/\mathrm{U}_b$) and radial ($\sqrt{v'v'}/\mathrm{U}_b$) velocity fluctuations. This can be attributed to the earlier onset of exothermic reactions and additional dilatation at higher temperature HCP blending and a resulting stronger impact on the flow field. Similar trends have been observed with increasing UN reactivity (i.e. Φ) by Goh et al. [20, 33] and Hampp and Lindstedt [22].

5.4 Conditional Velocity Statistics

Multi-fluid conditional velocity statistics [22] are used to clarify the influence of the HCP enthalpy as defined in Eq. 5.

$$
\begin{aligned}
\overline{U_{k,FS,i,j}} &= \frac{1}{N}\sum_{n=1}^{N} c_{FS,n,i,j}\cdot U_{k,n,i,j} \quad \forall\, i,\, j \\
(u'u')_{k,FS,i,j} &= \frac{1}{N}\sum_{n=1}^{N} c_{FS,n,i,j}\cdot \left(U_{k,n,i,j} - \overline{U_{k,FS,i,j}}\right)^2 \quad \forall\, i,\, j \\
C_{FS,i,j} &= \frac{1}{N}\sum_{n=1}^{N} c_{FS,n,i,j} \quad \forall\, i,\, j
\end{aligned}
\tag{5}
$$

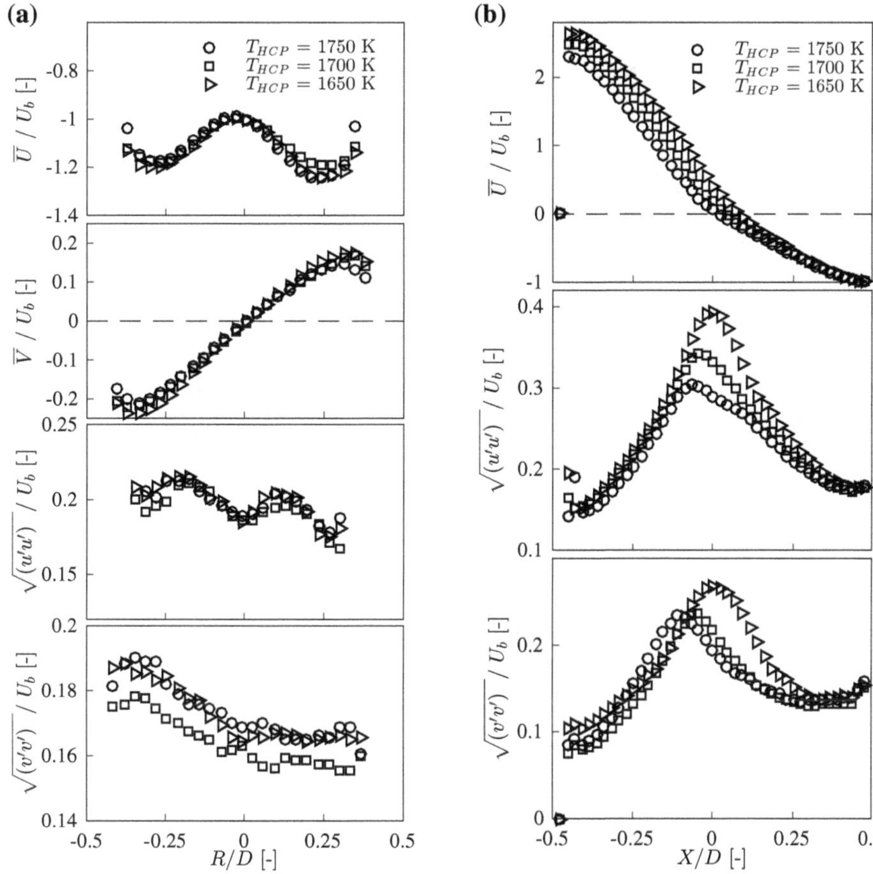

Fig. 7 Unconditional velocity statistics measured; **a** 1 mm of the UN exit: Mean axial velocity (top), mean radial velocity (2nd), axial velocity fluctuations (3rd) and radial velocity fluctuations (bottom) for varying HCP enthalpy; **b** along the stagnation point streamline for varying HCP enthalpy: Mean axial velocity (top row), axial velocity fluctuations (middle row) and radial velocity fluctuations (bottom row). Only every second data point is drawn to enhance the readability

In Eq. 5, $c_{FS,n}$ is the instantaneous (n) conditioning variable (i.e. unity within the individual fluid state (FS) and nil elsewhere), k the velocity component, N the total number of images (3000) and i and j the index variables. The sum of all fluid state progress variables (C_{FS}) is unity. The conditional velocity statistics were evaluated along the theoretical SPS and aligned at the instantaneous $x_s = 0$; see Fig. 3. A minimum of 75 vectors was used for the determination of conditional velocities. The conditional mean axial fluid state (FS) velocities ($\overline{U_{0,FS}}$) and axial ($\sqrt{u'u'_{0,FS}}$) and radial ($\sqrt{v'v'_{0,FS}}$) fluctuations were normalised by the respective reactant values measured 1 mm away from the upper nozzle exit; see Fig. 7a.

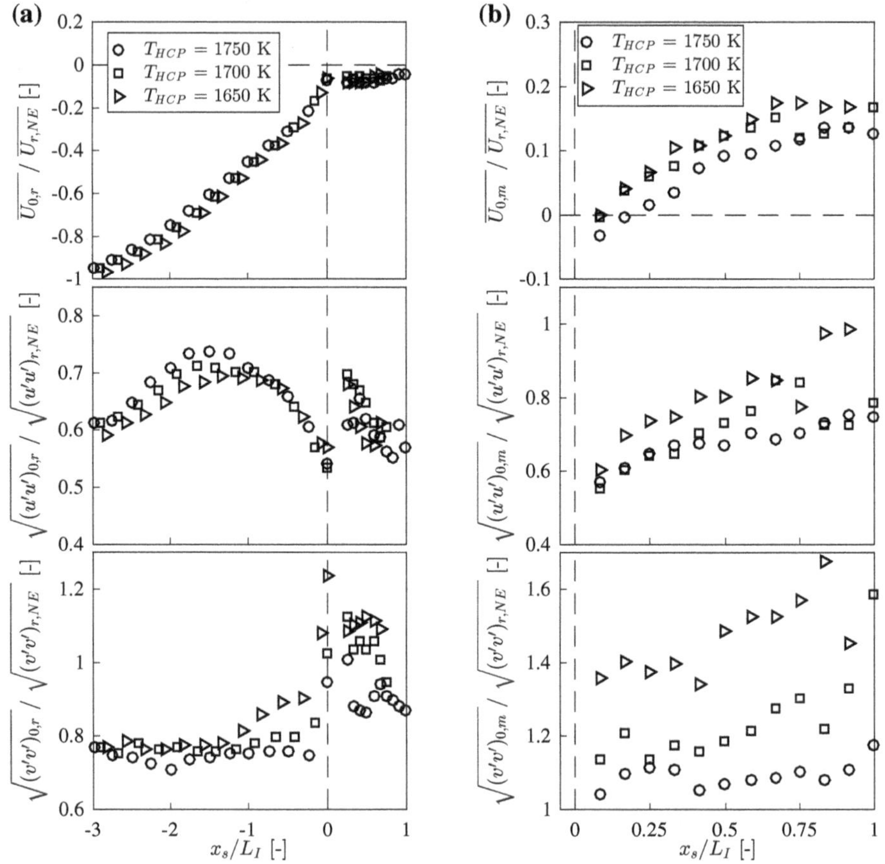

Fig. 8 Conditional mean axial **a** reactant and **b** mixing fluid velocity and fluctuations for varying HCP enthalpy along the stagnation point streamline aligned at the Mie scattering iso-contour (x_s = 0): Top—$\overline{U_{0,FS}}/\overline{U_{r,NE}}$; Middle—$\sqrt{(u'u')_{0,FS}}/\sqrt{(u'u')_{r,NE}}$; Bottom—$\sqrt{(v'v')_{0,FS}}/\sqrt{(v'v')_{r,NE}}$. At $x_s/L_I < 0$ only every third data point is plotted to enhance the readability. At $x_s/L_I > 0$, all data points are shown

5.4.1 Conditional Reactant Fluid Velocity

The conditional reactant fluid velocity is depicted in Fig. 8a. The reaction onset is anchored at a fixed mean reactant fluid velocity with $\overline{U_{0,r}} = -0.58 \pm 0.05$ m s^{-1} at $x_s = 0$. The axial velocity fluctuations are only just affected by the HCP enthalpy with the slight increase around $x_s/L_I \simeq -1.5$ consistent with modest differences in the mean velocity. The radial fluctuations show a tendency to increase close to the origin with decreasing T_{HCP} and a consistent increase in $\sqrt{v'v'_{0,R}}/\sqrt{v'v'_{r,NE}}$ is observed at $x_s > 0$. This can be attributed to a reaction onset that is increasingly governed by strong HCP blending.

5.4.2 Conditional Mixing Fluid Velocity

The impact of the HCP enthalpy on the thermally altered fluids is evident for the conditional mixing fluid velocity ($\overline{U_{0,m}}/\overline{U_{r,NE}}$). The fluid pockets become increasingly driven by the HCP momentum for reduced T_{HCP} as shown in Fig. 8b. The value of $\overline{U_{0,m}}/\overline{U_{r,NE}} < 0$ in the direct proximity of the origin (i.e. in line with the reactant flow) for $T_{HCP} = 1750$ K, while $\overline{U_{0,m}}/\overline{U_{r,NE}} \geq 0$ for $T_{HCP} \leq 1700$ K. Moreover, with decreasing T_{HCP} the axial and radial velocity fluctuations increase. This is consistent with the theoretical discussion on increased auto-ignition delay times with decreasing T_{HCP}, see Fig. 2, and the need for additional HCP blending with reactants prior the onset of OH producing reactions. Similar trends for decreasing mixture reactivity have been observed by Hampp and Lindstedt [22].

5.4.3 Conditional Weakly Reacting Fluid Velocity

The weakly reacting fluid velocity $\overline{U_{0,w}}/\overline{U_{r,NE}}$ is independent of the HCP enthalpy in the proximity of the origin as shown in Fig. 9a. The impact becomes increasingly evident for $x_s/L_I > 0.5$ leading to a separation of $\overline{U_{0,w}}/\overline{U_{r,NE}}$. The reduced gradients with higher T_{HCP} are caused by the more distinct dilatation due to the reduced HCP blending fraction required to initiate combustion (e.g. OH producing reactions). The values of $\sqrt{\overline{(u'u')_{0,w}}}/\sqrt{\overline{(u'u')_{r,NE}}}$ and $\sqrt{\overline{(v'v')_{0,w}}}/\sqrt{\overline{(v'v')_{r,NE}}}$ show a modest reduction with increasing T_{HCP} that is consistent with the changes in the mixture reactivity [22]. The weakly reacting axial velocity fluctuations show a modest decrease (∼10%) compared to the mixing fluid, while the radial component exhibits elevated (∼10%) fluctuations. This can be attributed to the heat release associated with the OH producing chemical reactions and the corresponding modestly enhanced dilatation of the fluid state.

5.4.4 Conditional Strongly Reacting Fluid Velocity

The conditional mean axial strongly reacting fluid velocity is insensitive to the thermal support as depicted in Fig. 9b. This is characteristic for self-sustained burning at constant Da and Re_t in the absence (or with vanishing levels) of HCP dilution. In the direct proximity of the origin, the $\overline{U_{0,s}}/\overline{U_{r,NE}}$ is consistently aligned with the reactant flow direction. Away from the origin, the strongly reacting fluid flow is increasingly governed by the momentum of the HCP stream, yet at significantly attenuated levels compared to the weakly reacting fluid. The axial ($\sqrt{\overline{(u'u')_{0,s}}}/\sqrt{\overline{(u'u')_{r,NE}}}$) and radial ($\sqrt{\overline{(v'v')_{0,s}}}/\sqrt{\overline{(v'v')_{r,NE}}}$) fluctuations show a distinct reduction with increasing HCP enthalpy that is more pronounced than the weakly reacting velocity fluctuations due to the impact of increased dilatation.

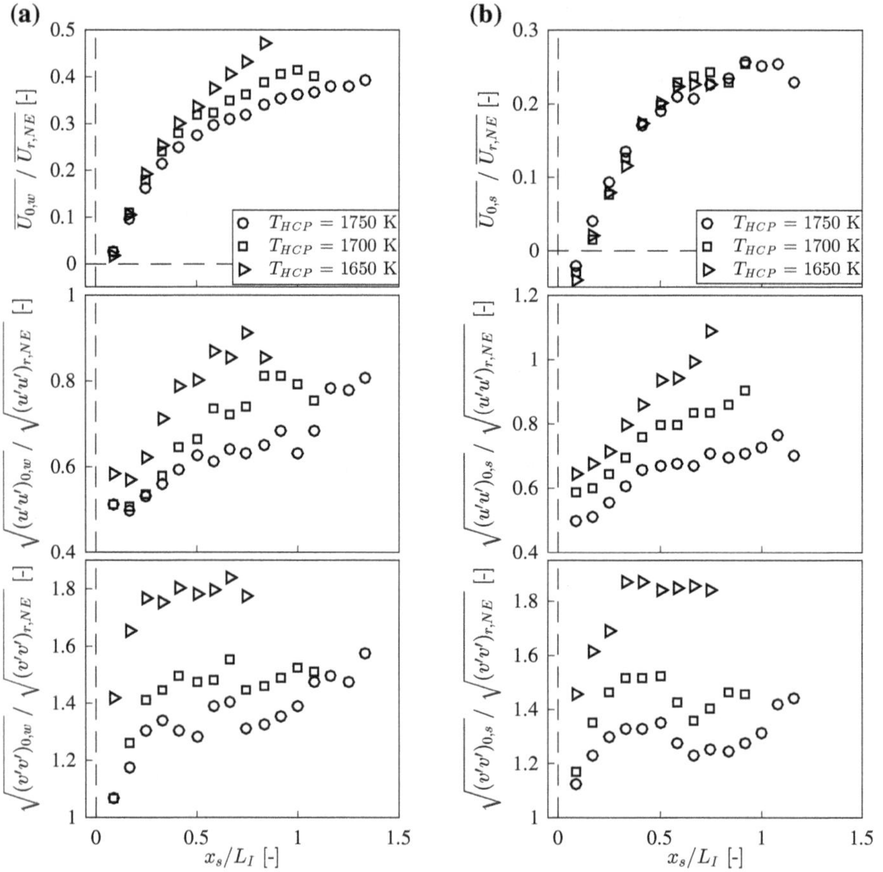

Fig. 9 Conditional mean axial **a** weakly and **b** strongly reacting fluid velocity and fluctuations for varying HCP enthalpy along the stagnation point streamline aligned at the Mie scattering iso-contour ($x_s = 0$): Top—$\overline{U_{0,FS}}/\overline{U_{r,NE}}$; Middle—$\sqrt{(u'u')_{0,FS}}/\sqrt{(u'u')_{r,NE}}$; Bottom—$\sqrt{(v'v')_{0,FS}}/\sqrt{(v'v')_{r,NE}}$

5.5 *Conditional Strain on Material Surfaces*

The in-plane velocity gradients were conditioned upon the fluid state material surfaces (β) [23]. The strain rate ($e_{ij} = 0.5(\partial u_i/\partial x_j + \partial u_j/\partial x_i)$) and vorticity ($\omega_{ij} = \partial u_i/\partial x_j - \partial u_j/\partial x_i$) were determined from the instantaneous planar PIV data and subsequently conditioned upon β. The normal (a_n), tangential (a_t) and total ($a_d = e_{\beta,11} + e_{\beta,22}$) strain as well as vorticity were determined for the cases T_{HCP} = 1650, 1700 and 1750 K within $\pm\, L_I/2$ radially away from the theoretical SPS to include the movement of the stagnation point [32].

Table 4 Mean and spread of the PDFs of the normal (a_n), tangential (a_t) and total (a_d) strain and vorticity (ω) conditioned upon the fluid state iso-contours (β) for T_{HCP} = 1650, 1700 and 1750 K. R—reactants; M—mixing; WR—weakly reacting; SR—strongly reacting; P—product fluid

T_{HCP}	β	Mean				Spread (rms)			
		a_n	a_t	a_d	ω	a_n	a_t	a_d	ω
1750 K	R	−1251	802	−504	2312	886	760	737	1351
1700 K	R	−1414	890	−643	2578	924	801	727	1391
1650 K	R	−1569	996	−733	2813	964	829	808	1442
1750 K	M	−1350	699	−401	2383	1060	763	826	1533
1700 K	M	−1611	767	−614	2758	1102	822	811	1581
1650 K	M	−1827	912	−735	3099	1177	897	925	1669
1750 K	WR	−1248	709	−521	2378	872	734	762	1331
1700 K	WR	−1450	737	−797	2690	928	858	786	1413
1650 K	WR	−1534	915	−717	2948	987	888	876	1512
1750 K	SR	−920	660	−276	1953	806	729	743	1278
1700 K	SR	−1138	668	−544	2296	861	810	767	1353
1650 K	SR	−1248	778	−479	2575	956	929	901	1522
1750 K	P	−1477	821	−518	2682	991	791	801	1452
1700 K	P	−1801	882	−889	3127	1029	870	815	1501
1650 K	P	−1915	1052	−849	3429	1125	923	901	1625

5.5.1 Strain Distribution on the Reactant Fluid Surface

The normal (a_n) and tangential (a_t) strain conditioned upon the reactant fluid iso-contour (R) are depicted in Fig. 10 along with the corresponding total rate of strain (a_d) and vorticity (ω). To recapitulate, the HCP from the LN exerted a vanishing effect on the conditional reactant fluid velocities. The $a_n|R$, however, shows a modest shift of the PDF towards reduced mean compressive values ($-1570 < a_n$ (s^{-1}) < -1250 for $1650 < T_{HCP}$(K) < 1750). The reduction can be attributed to increased flow acceleration of adjacent reactive fluid states and slight detachment of the reactant fluid iso-contour from the stagnation plane [23]. The latter further yields a reduced mean tangential ($1000 < a_t$ (s^{-1}) < 800) and total contracting ($-730 < a_d$ (s^{-1}) < -500) strain with reduced mean vorticity levels ($2810 < \omega$ (s^{-1}) < 2310). Moreover, the spread of all PDFs is reduced by ~10% with increasing T_{HCP}. This suggests a first thermal alteration that is increasingly caused by adjacent exothermic reactions with a distinct dilatation direction that is consistent with the observed interface statistics in Sect. 5.2 and the study presented by Hampp and Lindstedt [23].

The velocity gradients conditioned upon the mixing ($a_n|M$, $a_t|M$, $\omega|M$) and weakly reacting fluid surfaces ($a_n|WR$, $a_t|WR$, $\omega|WR$) exhibit similar trends to the reactant fluid and are thus not discussed separately. Values are listed in Table 4.

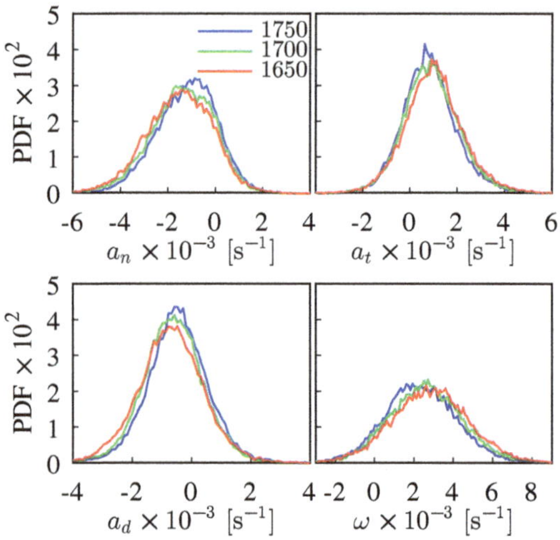

Fig. 10 PDF of the rate of strain and vorticity evaluated along the reactant fluid iso-contour: Top left: Normal strain; Right: Tangential strain; Bottom left: Total strain; Right: Vorticity. The legend entries refer to T_{HCP} [K]

5.5.2 Strain Distribution on the Strongly Reacting Fluid Surface

The rate of strain and vorticity conditioned upon the strongly reacting (SR) fluid material surface ($a_n|SR$, $a_t|SR$, $a_d|SR$, $\omega|SR$) are depicted in Fig. 11. For $T_{HCP} =$ 1750 K, the reaction onset is anchored in comparatively low compressive strain regions with a mean of $a_n = -920$ s^{-1} in contrast with -1250 s^{-1} for $T_{HCP} =$ 1650 K. The normal strain along the strongly reacting fluid material surface is ~30% lower than the rate of strain on the reactant and weakly reacting fluid iso-contour. The PDF($a_n|SR$) at $T_{HCP} = 1750$ K further shows an elevated skewness towards decreased compressive strain compared to lower HCP temperature as well as to fluid states with reduced reactivity. The merging of the strongly reacting material surface with the stagnation plane with decreasing T_{HCP} results in higher tangential strain ($660 < a_t$ (s^{-1}) < 780) as well as a_d. The corresponding Euclidean norm, defined in Eq. 6, increases from $1132 < |a_{SR}|$ (s^{-1}) < 1471 with decreasing T_{HCP}.

$$|a_{SR}| = \sqrt{\left(a^2_{n|SR} + a^2_{t|SR}\right)} \tag{6}$$

The extinction strain of the corresponding laminar flame was determined to $a_q = 600$ s^{-1} [22]. The increasing ratio from $0.41 < a_q/|a_{SR}| < 0.53$ with T_{HCP} further highlights the enhanced likelihood of establishing a self-sustained flame. The elevated skewness and reduced spread of the PDF(a_n) and PDF(a_t) with increasing reactivity was also observed by Hampp and Lindstedt [23] and Hartung et al. [49]. The lack of a preferential alignment of the flame normal with the extensive rate of strain at modest dilatation was also discussed by Chakraborty and Swaminathan [50]. A change of T_{HCP} from 1650 to 1750 K causes a 25% vorticity reduction, and the

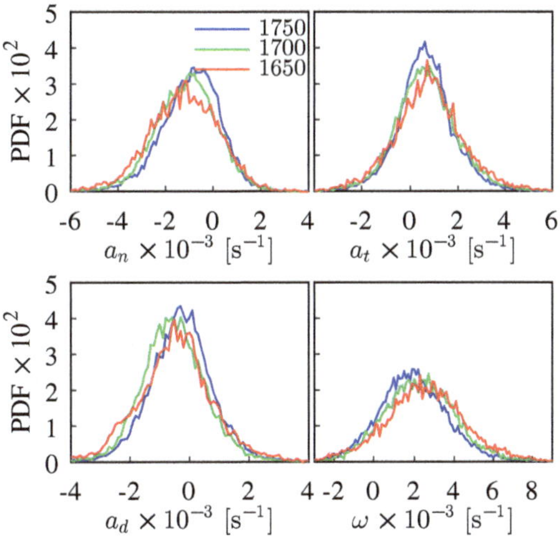

Fig. 11 PDF of the rate of strain and vorticity evaluated along the strongly reacting fluid iso-contour: Top left: Normal strain; Right: Tangential strain; Bottom left: Total strain; Right: Vorticity. The legend entries refer to T_{HCP} [K]

values are up to 15% lower than for the reactant or weakly reacting fluid material surface (see Table 4).

5.6 *Bimodal Flow Analysis*

Bimodal statistics [29] were obtained by combining the thermally altered fluid states (i.e. mixing, weakly reacting, strongly reacting and product fluid) as products, defining the reaction progress variable ($\overline{c}$) [23, 29]. This illustrates the impact of the HCP enthalpy on the turbulent flame brush and scalar transport. The reactant fluid velocity ($\overline{U_r}/U_b$) is only marginally affected by the HCP temperature (see Fig. 12). The observation is consistent with the multi-fluid analysis, which showed that reactant fluid statistics are independent of the HCP enthalpy (see Figs. 4 and 8a–9b). By contrast, the velocity of the combined product fluid ($\overline{U_p}/U_b$) shows a distinct impact of the HCP support on the flame brush for $\overline{c} > 0.25$ as increasing amounts of HCP are required to sustain chemical activity with decreasing T_{HCP}. This results in $\overline{U_p}/U_b$ being increasingly governed by the HCP stream momentum. By contrast, higher T_{HCP} values necessitate reduced HCP blending fractions, result in a more pronounced dilatation and, in turn, reduced slip velocities ($\overline{U_s}/U_b$). The increased dilatation further yields a noticeable reduction in the gradient scalar flux ($\overline{cu}$) as also shown in Fig. 12. However, a transition to counter-gradient transport is not observed. A similar impact on scalar transport was also observed by Goh et al. [18, 20] and Hampp and Lindstedt [23].

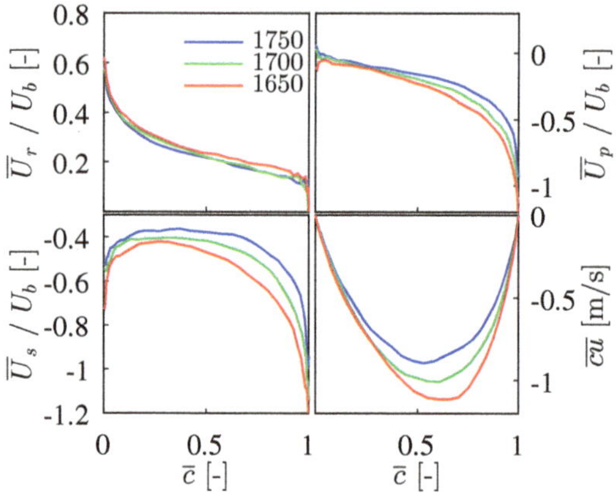

Fig. 12 Mean conditional reactant ($\overline{U}_r$, top left) and product ($\overline{U}_p$, top right) velocities in progress variable ($\overline{c}$) space along with the slip velocity ($\overline{U}_s$, bottom left) and scalar flux ($\overline{cu}$, bottom right) for varying HCP enthalpy. The legend entries refer to T_{HCP} [K]

6 Conclusions

The impact of the burnt gas state on the combustion of lean premixed DME/air flames in fractal grid generated multi-scale turbulence with constant $\Phi = 0.6$, $Re_t \simeq 395$, $Da \simeq 1.1$ and $Ka \simeq 7.0$ was investigated. The temperature of the hot combustion products, emerging the lower nozzle of a back-to-burnt opposed jet configuration, was varied ($1600 < T_{HCP}$(K) < 1800) by means of the CO_2 dilution level. The variation ranges from close to the extinction temperature ($T_q \simeq 1570$ K) of the corresponding twin flame configuration to beyond the unstrained adiabatic flame temperature ($T_{ad} \simeq 1750$ K). The burning mode transition was quantified by means of a multi-fluid analysis. With increasing T_{HCP}, combustion shifts progressively away from strong HCP dilution influence towards self-sustained burning that may be initialised by small HCP blending fractions or contact burning. By contrast, the strongly reacting fluid (representing self-sustained burning) nearly vanished for T_{HCP} of the order of the extinction temperature. The impact of turbulence–chemistry interaction on the flow field was evaluated in the range $1650 < T_{HCP}$(K) < 1750. With increasing HCP enthalpy a gradually stronger and more directed dilatation was observed that caused a reduction in the gradient scalar flux. Characteristic for self-sustained burning and the velocity statistics of the strongly reacting fluid were not affected by the HCP enthalpy. However, the strain distribution along the material surfaces indicated a gradual detachment of the reaction onset from the stagnation plane and adjacent flow acceleration with increasing T_{HCP}. By contrast, at reduced T_{HCP} the chemically active fluids were governed by the thermal support and characterised by high vorticity and rates of strain. The current data set is expected to challenge numerical models that aim to delineate combustion mode transitions.

Acknowledgements The authors would like to acknowledge the support of the AFOSR and EOARD under Grant FA9550-17-1-0021 and thank Dr. Chiping Li and Dr. Russ Cummings for encouraging the work. The US Government is authorised to reproduce and distribute reprints for

Governmental purpose notwithstanding any copyright notation thereon. The authors would also like to thank Dr. Robert Barlow for his support.

References

1. F.J. Weinberg, Combustion temperatures: the future? Nature **233**, 39–241 (1971)
2. E. Mastorakos, A.M.K.P. Taylor, J.H. Whitelaw, Extinction of turbulent counterflow flames with reactants diluted by hot products. Combust. Flame **102**, 101–114 (1995)
3. J.A. Wünning, J.G. Wünning, Flameless oxidation to reduce thermal NO-formation. Prog. Energy Combust. **23**, 81–94 (1997)
4. I.B. Özdemir, N. Peters, Characteristics of the reaction zone in a combustor operating at mild combustion. Exp. Fluids **30**, 683–695 (2001)
5. Z. Zhao, A. Kazakov, J. Li, F.L. Dryer, The initial temperature and N_2 dilution effect on the laminar flame speed of propane/air. Combust. Sci. Technol. **176**, 1705–1723 (2004)
6. H.J. Curran, S.L. Fischer, F.L. Dryer, Reaction kinetics of dimethylether. II: Low-temperature oxidation in flow reactors. Int. J. Chem. Kinet. **32**, 741–759 (2000)
7. R. Lückerath, W. Meier, M. Aigner, FLOX combustion at high pressure with different fuel compositions. J. Eng. Gas Turb. Power **130**, 011505 (2008)
8. T. Plessing, N. Peters, J.G. Wünning, Laseroptical investigation of highly preheated combustion with strong exhaust gas recirculation. Proc. Combust. Inst. **27**, 3197–3204 (1998)
9. J. Sidey, E. Mastorakos, R.L. Gordon, Simulations of autoignition and laminar premixed flames in methane/air mixtures diluted with hot products. Combust. Sci. Technol. **186**, 453–465 (2014)
10. Y. Minamoto, N. Swaminathan, R.S. Cant, T. Leung, Reaction zones and their structure in MILD combustion. Combust. Sci. Technol. **186**, 1075–1096 (2014)
11. B. Zhou, C. Brackmann, Q. Li, Z. Wang, P. Petersson, Z. Li, M. Aldén, X.-S. Bai, Distributed reactions in highly turbulent premixed methane/air flames. Part I. Flame structure characterization. Combust. Flame **162**, 2937–2953 (2015)
12. B. Zhou, C. Brackmann, Z. Li, M. Aldén, X.-S. Bai, Simultaneous multi-species and temperature visualization of premixed flames in the distributed reaction zone regime. Proc. Combust. Inst. **35**, 1409–1416 (2015)
13. B. Zhou, C. Brackmann, Z. Wang, Z. Li, M. Richter, M. Aldén, X.-S. Bai, Thin reaction zone and distributed reaction zone regimes in turbulent premixed methane/air flames: scalar distributions and correlations. Combust. Flame **175**, 220–236 (2016)
14. A.R. Masri, R.W. Dibble, R.S. Barlow, The structure of turbulent nonpremixed flames revealed by Raman-Rayleigh-LIF measurements. Prog. Energy Combust. Sci. **22**, 307–362 (1996)
15. R.S. Barlow, J.H. Frank, Effects of turbulence on species mass fractions in methane/air jet flames. Proc. Comb. Inst. **27**, 1087–1095 (1998)
16. D. Geyer, A. Kempf, A. Dreizler, J. Janicka, Turbulent opposed-jet flames: a critical benchmark experiment for combustion LES. Combust. Flame **143**, 524–548 (2005)
17. K.H.H. Goh, P. Geipel, R.P. Lindstedt, Lean premixed opposed jet flames in fractal grid generated multiscale turbulence. Combust. Flame **161**, 2419–2434 (2014)
18. Goh KHH (2013) Investigation of conditional statistics in premixed combustion and the transition to flameless oxidation in turbulent opposed jets. Ph.D. thesis, Imperial College. http://hdl.handle.net/10044/1/28073
19. B. Coriton, J.H. Frank, A. Gomez, Effects of strain rate, turbulence, reactant stoichiometry and heat losses on the interaction of turbulent premixed flames with stoichiometric counterflowing combustion products. Combust. Flame **160**, 2442–2456 (2013)
20. K.H.H. Goh, P. Geipel, F. Hampp, R.P. Lindstedt, Regime transition from premixed to flameless oxidation in turbulent JP-10 flames. Proc. Combust. Inst. **34**, 3311–3318 (2013)
21. K.H.H. Goh, P. Geipel, R.P. Lindstedt, Turbulent transport in premixed flames approaching extinction. Proc. Combust. Inst. **35**, 1469–1476 (2015)

22. F. Hampp, R.P. Lindstedt, Quantification of combustion regime transitions in premixed turbulent DME flames. Combust. Flame **182**, 248–268 (2017)
23. F. Hampp, R.P. Lindstedt, Strain distribution on material surfaces during combustion regime transitions. Proc. Combust. Inst. **36**, 1911–1918 (2017)
24. B. Coriton, M.D. Smooke, A. Gomez, Effect of the composition of the hot product stream in the quasi-steady extinction of strained premixed flames. Combust. Flame **157**, 2155–2164 (2000)
25. E. Abtahizadeh, J. van Oijen, P. de Goey, Numerical study of mild combustion with entrainment of burned gas into oxidizer and/or fuel streams. Combust. Flame **159**, 2155–2165 (2012)
26. T. Fleisch, C. McCarthy, A. Basu, C. Udovich, P. Charbonneau, W. Slodowske, S.-E. Mikkelsen, J. McCandless, A new clean diesel technology: demonstration of ULEV emissions on a Navistar diesel engine fueled with dimethyl ether, SAE, 950061 (1995)
27. S.C. Sorenson, S.E. Mikkelsen, Performance and emissions of a 0.273 liter direct injection diesel engine fuelled with neat dimethyl ether, SAE, 950064 (1995)
28. S.-W. Park, Detailed chemical kinetic model for oxygenated fuels. Ph.D. thesis, Imperial College. http://hdl.handle.net/10044/1/9599 (2012)
29. K.N.C. Bray, Laminar flamelets and the Bray, Moss, and Libby Model, in *Turbulent Premixed Flames*, ed. by N. Swaminathan, K.N.C. Bray (Cambridge University Press, 2001), pp. 41–60. ISBN: 978-0-521-76961-7
30. F. Hampp, Quantification of combustion regime transitions. Ph.D. thesis, Imperial College. http://hdl.handle.net/10044/1/32582 (2016)
31. F. Hampp, R.P. Lindstedt, Fractal grid generated turbulence—a bridge to practical combustion applications, in *Fractal Flow Design: How to Design Bespoke Turbulence and Why*. CISM International Centre for Mechanical Sciences 568, ed. by Y. Sakai, C. Vassilicos (Springer, 2016). https://doi.org/10.1007/978-3-319-33310-6_3
32. P. Geipel, K.H.H. Goh, R.P. Lindstedt, Fractal-generated turbulence in opposed jet flows. Flow Turbul. Combust. **85**, 397–419 (2010)
33. K.H.H. Goh, P. Geipel, F. Hampp, R.P. Lindstedt, Flames in fractal grid generated turbulence. Fluid Dyn. Res. **45**, 061403 (2013)
34. N. Peters, Kinetic foundation of thermal flame theory, in, Advances in *Combustion Science: In Honor of Y. B. Zel'dovich*. Progress in Astronautics and Aeronautics 173, ed. by W.A. Sirignano, A.G. Merzhanov, L. de Luca (AIAA, 1997), pp. 73–91
35. J. Kerl, T. Sponfeldner, F. Beyrau, An external Raman laser for combustion diagnostics. Combust. Flame **158**, 1905–1907 (2011)
36. B. Wieneke, K. Pfeiffer, Adaptive PIV with variable interrogation window size and shape, in *15th International Symposium on the Application of Laser Techniques to Fluid Mechanics* (2010). http://ltces.dem.ist.utl.pt/LXLASER/lxlaser2010/upload/1845_qkuqls_1.12.3.Full_1845.pdf
37. W.P. Jones, R.P. Lindstedt, The calculation of the structure of laminar counterflow diffusion flames using a global reaction mechanism. Combust. Sci. Technol. **61**, 31–49 (1988)
38. J. Sidey, E. Mastorakos, Visualization of MILD combustion from jets in cross-flow. Proc. Combust. Inst. **35**, 3537–3545 (2015)
39. Y. Minamoto, N. Swaminathan, Scalar gradient behaviour in MILD combustion. Combust. Flame **161**, 1063–1075 (2014)
40. Y. Minamoto, N. Swaminathan, S.R. Cant, T. Leung, Morphological and statistical features of reaction zones in MILD and premixed combustion. Combust. Flame **161**, 2801–2814 (2014)
41. W.K. George, H.J. Hussein, Locally axisymmetric turbulence. J. Fluid Mech. **233**, 1–23 (1991)
42. I.G. Shepherd, R.K. Cheng, P.J. Goix, The spatial scalar structure of premixed turbulent stagnation point flames. Proc. Combust. Inst. **23**, 781–787 (1991)
43. A.M. Steinberg, J.F. Driscoll, S.L. Ceccio, Measurements of turbulent premixed flame dynamics using cinema stereoscopic PIV. Exp. Fluids **44**, 985–999 (2008)
44. B. Böhm, C. Heeger, I. Boxx, W. Meier, A. Dreizler, Time-resolved conditional flow field statistics in extinguishing turbulent opposed jet flames using simultaneous highspeed PIV/OH-PLIF. Proc. Combust. Inst. **32**, 1647–1654 (2009)

45. B.E. Battles, R.K. Hanson, Laser-induced fluorescence measurements of NO and OH mole fraction in fuel-lean, high-pressure (1–10 atm) methane flames: fluorescence modeling and experimental validation. J. Quant. Spectrosc. Radiat. Transf. **54**, 521–537 (1995)
46. S. Krishna, R.V. Ravikrishna, Quantitative OH planar laser induced fluorescence diagnostics of syngas and methane combustion in a cavity combustor. Combust. Sci. Technol. **187**, 1661–1682 (2015)
47. K.A. Buch, W.J.A. Dahm, Experimental study of the fine-scale structure of conserved scalar mixing in turbulent shear flows. Part 2. Sc = 1. J. Fluid Mech. **364**, 1–29 (1998)
48. G.K. Batchelor, Small-scale variation of convected quantities like temperature in turbulent fluid. Part 1. General discussion and the case of small conductivity. J. Fluid Mech. **5**, 113–133 (1959)
49. G. Hartung, J. Hult, C.F. Kaminski, J.W. Rogerson, N. Swaminathan, Effect of heat release on turbulence and scalar-turbulence interaction in premixed combustion. Phys. Fluids **20**, 035110 (2008)
50. N. Chakraborty, N. Swaminathan, Influence of the Damköhler number on turbulence-scalar interaction in premixed flames. I. Physical insight. Phys. Fluids **19**, 045103 (2007)

A Dynamic LES Model for Turbulent Reactive Flow with Parallel Adaptive Finite Elements

Jiajia Waters, David B. Carrington, Xiuling Wang and Darrell W. Pepper

Abstract An adaptive finite element method (FEM) is used for the solution of turbulent reactive flows in 3-D utilizing parallel methods for fluid dynamic and combustion modeling associated with engines. A dynamic LES method permits transition from laminar to turbulent flow without the assumptions usually required for turbulent sublayers near wall area. This capability is ideal for engine configurations where there is no equilibrium in the turbulent wall layers and the flow is not always turbulent and often in transition. The developed adaptive FEM flow solver uses "h" adaptation to provide for grid refinement. The FEM solver has been optimized for parallel processing employing the message passing interface (MPI) for clusters and high-performance computers.

Keywords Dynamic LES · Turbulent reactive flow · Mesh adaptation · Parallel implicit solver · Finite element · FEM · CFD · Computational fluid dynamics

Nomenclature

~	Designates a Favre-averaged variable
–	Designates a grid-filtered variable
c	Sound speed (m/s)
C_p	Specific heat capacity at constant P (J/kg.K)
C_{vm}	Vreman fixed SGS eddy viscosity coefficient
C_{DVMG}	Vreman dynamic SGS eddy viscosity coefficient

J. Waters · D. B. Carrington
Los Alamos National Laboratory, T-3 Division, Los Alamos, USA

X. Wang
Purdue University Northwest, Hammond, IN 46323, USA

D. W. Pepper (✉)
NCACM, Department of Mechanical Engineering, University of Nevada
Las Vegas, Las Vegas, USA
e-mail: darrell.pepper@unlv.edu

A. K. Runchal et al. (eds.), *Energy for Propulsion*, Green Energy and Technology,
https://doi.org/10.1007/978-981-10-7473-8_9

D_j	Diffusion coefficient of the *j*th species (m^2/s)
D_k	Turbulent diffusion coefficient (m^2/s)
E	Total internal energy (J/kg)
$f_{k,j}$	Body forces (N/m^3)
f_{drop}	Body forces related to particulate or droplets in flow (N/m^3)
H_j	Enthalpy of species j (J)
H_{oj}	Enthalpy of formation (J)
P	Pressure (Pa)
Pr	Molecular Prandtl number
Pr_{sgs}	SGS eddy Prandtl number
Pr_{DVMG}	Vreman dynamic SGS eddy Prandtl number
Q_j	Subtest-scale heat flux vector
q_i	Heat flux vector
Re	Reynolds number
$\tilde{S}_{ij}$	Strain rate tensor $\left(\frac{N}{m^2}, kg/m\,s^2\right)$
Sc	Schmidt number
Sct	Subgrid-scale turbulent Schmidt number
T	Temperature (K)
T_{ij}	Subgrid test-scale stress tensor
t_{ij}	Grid-scale (resolved scale) shear stress $\left(\frac{N}{m^2}, kg/m\,s^2\right)$
u_i	Velocity component (m/s)
$\Upsilon_j f_j$	Body force term for the jth component
$\dot{w}^j_{chem}$	Chemical reaction
$\dot{w}^j_{spray}$	Spray evaporation

Greek Symbols

∂t	Discrete time step size (s)
κ	Coefficient of thermal conductivity (W/m K)
ρ	Density (kg/m^3)
Υ^j	Mass fraction (jth species) $\left(\frac{\rho^j}{\rho}\right)$
τ_{ij}	Subgrid-scale stress tensor
μ	Fluid viscosity (Pa s)
μ_{sgs}	Turbulent eddy viscosity

1 Introduction

A number of challenges found in combustion modeling are related to developing methods that handle the numerical requirements of coupled physical and chemical processes. These requirements related to engines are pronounced, with the highly transient three-dimensional (3-D) dynamics ranging over flow regimes with evaporating fuel sprays, the ignition of gases, and the subsequent chemical reactions. In addition, heat transfer and species transport modeling are critical to the success of the models accuracy. When dealing with multiple phases in the liquid sprays, the process will get even more complicated to track the dynamically evolving interfaces. Our efforts in developing accurate methods and models to calculate such flows in engine cylinders, including the effects of turbulence, improve the current Los Alamos National Laboratory's KIVA [1] suite of codes with the creation of an hp-adaptive FEM all flow regime solve (KIVA-hpFE), where h is associated with mesh adaptation and p is the order of the basis function for FEM. Among the adaptation families, the hp-adaptive FEM is one of the best mesh-based algorithms. The computational mesh is automatically refined and unrefined based on the change of flow features; the shape function order is dynamically controlled by the computational error. In this study, we only focus on h-adaptation (mesh refinement).

The most widely used approach for modeling turbulent flow, Reynolds-averaged Navier–Stokes (RANS), is implemented using the two-equation k-ω modeled by Wilcox [2]. Time averaging methods produce a mean value for turbulent variables, kinetic energy, dissipation rate, and turbulent viscosity and do not capture flow structures in unsteady turbulence. The two-equation models generally use a law-of-the-wall relation for emulating the turbulent boundary layer, but this law is not appropriate at varying times during the cycle of an engine. For example, the flow is nearly always perpendicular to the surface near the piston.

A large eddy simulation (LES) model provides a method for solution of unsteady turbulence intensity, varying turbulent viscosity, and dynamic flow structures. In many modern combustion devices, the LES method is more widely adopted to study combustion [3–7]. A LES model solves the spatially averaged Navier–Stokes equations using a grid-based filtering process. In LES, the larger eddies are directly resolved at the grid resolution and eddies smaller than the grid are modeled. Using a dynamic Vreman approach [8] as implemented in the finite element method (FEM), guarantees vanishing subgrid-scale (SGS) dissipation for various laminar shear flows and thus there is no need to use any wall-damping functions in simulating boundary layer flows. These features of this type of LES system are well suited for wall-bounded shear flows [9]. Hence, the dynamic Vreman LES is capable of modeling various flow regimes, laminar, transitional, and turbulent flows simultaneously. More details of the LES implementation can be found in Waters et al. [10].

In this study, Vreman LES modeling, with the use of adaptive FEM and parallel implicit solution methods, is examined in more detail. LES models, in comparison to the RANS models, require finer density meshes because the modeled flow size depends on the filter size decided by the mesh size. Unstructured CFD algorithms used to model engine combustion typically require large computing resources, typically provided through parallel computer systems. By linking together hundreds and thousands of individual processors (or nodes), parallel computer systems deliver significant enhancements in computational memory, storage, and overall computing speed. In this study, a message passing interface (MPI) is employed to model 3-D turbulent flow in engines. Jimack [11] describes a similar procedure using MPI for

FEM. The parallel system in the KIVA-hpFE code also tracks the motion of spray particle parcels (droplets) from one position to another and from one processor to another [12]. Krylov solvers are used with matrix-vector multiplication and dot product operations for shared element updating and convergence of shared nodes [13].

2 Governing Equations

2.1 *Turbulent Flow Modeling with Multi-species*

The grid-filtered and Favre-filtered continuity, momentum, energy, and species equations governing the process of large-scale eddies are expressed as

$$\frac{\partial \bar{\rho}}{\partial t} + \frac{\partial(\bar{\rho}\tilde{u}_i)}{\partial x_i} \tag{1}$$

$$\frac{\partial(\bar{\rho}\tilde{u}_i)}{\partial t} + \frac{\partial(\bar{\rho}\tilde{u}_i\tilde{u}_j)}{\partial x_j} = \frac{\partial \tilde{t}_{ji}}{\partial x_j} - \frac{\partial \bar{p}}{\partial x_i} + \frac{\partial \tau_{ji}}{\partial x_j} + \bar{f}_{drop} + \bar{\rho} \sum_{k=1}^{NumSpecies} \tilde{\Upsilon}_k f_{k,j} \tag{2}$$

where $\tilde{t}_{ij}$ is the stress tensor evaluated using the Stoke's hypothesis as

$$\tilde{t}_{ij} = \mu\left(\frac{\partial \tilde{u}_i}{\partial x_j} + \frac{\partial \tilde{u}_j}{\partial x_i}\right) - \frac{2}{3}\mu\frac{\partial \tilde{u}_k}{\partial x_k}\delta_{ij} \tag{3}$$

$$\begin{aligned}\frac{\partial \tilde{E}}{\partial t} &= -\frac{\partial}{\partial x_i}\left(\tilde{E}\tilde{u}_i + p\tilde{u}_i\right) + \frac{\partial}{\partial x_i}\kappa\frac{\partial \tilde{T}}{\partial x_i} - \frac{\partial(C_p q_i)}{\partial x_i} \\ &+ \frac{\partial}{\partial x_i}(t_{ij} + \tau_{ij}) + \frac{\partial}{\partial x_i}\left(\bar{\rho}\sum_{j=1}^{NumSpecies} \overline{H}_K(D_K + \frac{\mu_{sgs}}{Sc_t})\frac{\partial \tilde{\Upsilon}_K}{\partial x_i}\right) \\ &+ \bar{\rho}\sum_{j=1}^{NumSpecies} \tilde{\Upsilon}_j f_j(x_i)\cdot \tilde{u}_i - \sum_{k=1}^{NumSpecies} H_{o,k} w_k\end{aligned} \tag{4}$$

$$\begin{aligned}\frac{\partial \bar{\rho}\tilde{\Upsilon}_j}{\partial t} &= -\frac{\partial}{\partial x_i}\left(\bar{\rho}\tilde{u}_i\tilde{\Upsilon}_j\right) + \frac{\partial}{\partial x_i}\bar{\rho}\left[\left(D_{j,N} + \frac{\mu_{sgs}}{Sc_t}\right)\frac{\partial \tilde{\Upsilon}_j}{x_i}\right] + \bar{\rho}\tilde{\Upsilon}_j f_j(x_i) \\ &+ \dot{w}^j_{chem} + \dot{w}^j_{spray}\end{aligned} \tag{5}$$

The SGS stress tensor τ_{ij} and SGS heat flux vector q_i in Eqs. (2) and (4) are defined, respectively, as

$$\tau_{ij} - \frac{1}{3}\tau_{kk}\delta_{ij} = -2\mu_{sgs}\left(\tilde{S}_{ij} - \frac{1}{3}\tilde{S}_{kk}\delta_{ij}\right) \tag{6}$$

$$q_j = -\frac{\mu_{sgs}}{Pr_{sgs}}\frac{\partial \tilde{T}}{\partial x_j} \tag{7}$$

where μ_{sgs} is the SGS viscosity, Pr_{sgs} is the SGS Prandtl number, and $\tilde{S}_{ij} = \frac{1}{2}\left(\frac{\partial \tilde{u}_i}{\partial x_j} + \frac{\partial \tilde{u}_j}{\partial x_i}\right)$ is the strain rate tensor. Here ~ is a Favre-filtered variable obtained from filtering its grid-filtered component. In this work, the box or top hat filter is applied for the grid-filtered component.

2.2 *Mass Conserving Projection Method for Compressible Flow*

In order to create a fractional split method, an initial guess for specific momentum is advanced in time, utilizing the expression

$$\Delta U_i^* = -\Delta t \times M_u^{-1}[A_u U_i + K_{\tau u} U_i - F_u]^n \tag{8}$$

where $U_i = \rho u_i$, M_u is the mass matrix, A_u is the advection matrix, $K_{\tau u}$ is the diffusion matrix, and F_u is the source term. The projection method is presented here with solving momentum explicitly, but the same algorithm is used when our governing equations are solved implicitly.

$$\Delta U_i^* = U_i^* - U_i^n \tag{9}$$

The corrected momentum is determined from the estimated momentum and the pressure gradient, given by

$$U^{n+1} - U^* = -\Delta t \frac{\partial P'}{\partial x_i} \tag{10}$$

Changes of density or pressure are determined from solving an implicit pressure/density Poisson equation created as a result of conservation of mass. This leads to the value for P' as shown in the following continuity solution process:

Mass conservation:

$$\frac{\partial \rho}{\partial t} = -\frac{\partial(\rho u_i)}{\partial x_i} = -\frac{\partial U_i}{\partial x_i} \tag{11}$$

Time advancement in discrete terms for continuity is

$$\frac{\rho^{n+1} - \rho^n}{\Delta t} = -\frac{\partial U_i'}{\partial x_i} \tag{12}$$

where

$$U' = \theta_1 U^{n+1} + (1 - \theta_1) U^n \tag{13}$$

with $U = \rho u$.

We define $P' = \theta_2 P^{n+1} + (1 - \theta_2) * P^n$ and $\Delta P = P^{n+1} - P^n$. The final specific momentum as obtained with the explicit corrector defined previously by

$$\Delta U_i = U^{n+1} - U^n = \Delta U^* - \Delta t \frac{\partial P'}{\partial x_i} = \Delta U^* - \Delta t \left(\theta_2 \frac{\partial \Delta P}{\partial x_i} + \frac{\partial P^n}{\partial x_i} \right) \tag{14}$$

The finite element form in matrix form is given by

$$\{\Delta \mathbf{U}_i\} = \{\Delta \mathbf{U}^*\} - \Delta t \left[\mathbf{M}_u^{-1}\right] (\theta_2 [\mathbf{G}] \{\Delta p_i\} [\mathbf{G}] \{p_i^n\}) \tag{15}$$

The final specific momentum is obtained using the corrector,

$$\{\mathbf{U}_i^{n+1}\} = \{\Delta \mathbf{U}_i\} + \{\mathbf{U}_i^n\} \tag{16}$$

The mass velocity or momentum is solved and velocity is extracted

$$u^{n+1} = U^{n+1} / \rho^{n+1} \tag{17}$$

Density is recovered from the equation of state. The speed of sound, c, is calculated by

$$c = \sqrt{\gamma R T}. \tag{18}$$

where R is the specific gas constant, for air 287 J/kg K, or it can be determined by a mass-averaging process.

When it is incompressible flow, we use artificial compressibility β for c.

$$\beta = \max(\in, u_{conv}, u_{diff}) \tag{19}$$

where $\in$ is a small constant to ensure β in Eq. (19) is not approaching zero. $u_{conv} = |U| = \sqrt{u_i u_i}$ and $u_{diff} = \frac{\nu}{h}$, where h is the element size and ν is the kinematic viscosity. The time step for the artificial compressibility method may become limited as $\beta \to \infty$. Therefore, pseudo time-stepping is needed in the incompressible region.

2.3 Dynamic Vreman SGS LES Model

Vreman [9] developed a fixed model coefficient SGS model where the SGS viscosity is determined as

$$\mu_{sgs} = \bar{\rho} C_{vm} \Pi^g \tag{20}$$

in which

$$\Pi^{g}=\sqrt{\frac{B_{\beta}^{g}}{\alpha_{ij}\alpha_{ij}}} \tag{21}$$

$$\alpha_{ij}=\frac{\partial\tilde{u}_j}{\partial x_i} \tag{22}$$

$$\beta_{ij}=\sum_{m=1}^{3}\Delta_m^2\alpha_{mi}\alpha_{mj} \tag{23}$$

$$B_{\beta}^{g}=\beta_{11}\beta_{22}-\beta_{12}^2+\beta_{11}\beta_{33}-\beta_{13}^2+\beta_{22}\beta_{33}-\beta_{23}^2 \tag{24}$$

where $C_{vm}=0.07$ and the SGS Prandtl number $\mathrm{Pr_{sgs}}=0.4$.

If $\alpha_{ij}\alpha_{ij}$ or B_{β}^{g} is approaching zero, then $\Pi^{g}=0$. This is equivalent to running laminar flow.

It is not realistic to assume the coefficient to be constant throughout the simulation. The development of the dynamic subgrid-scale model (DSGS) model captures significant progress in the subgrid-scale modeling of non-equilibrium flows. The DSGS model coefficient is calculated from the energy of the smallest resolved scale (grid size), instead of setting a priori parameters. The DSGS is obtained by two-filter processes: in the first filter, we used the grid size Δ, where the filtered expressions are given by (1)–(4). In this portion, the SGS Reynolds stress is included. By adding a test filter $\hat{\Delta}=2\Delta$ to the filtering Eqs. (1)–(4) leads to the subtest-scale stress tensor $\mathrm{T_{ij}}$ and subtest-scale heat flux vector $\mathrm{Q_j}$:

$$T_{ij}-\frac{1}{3}T_{kk}\delta_{ij}=-2\mu_{sgs}\left(\hat{\tilde{S}}_{ij}-\frac{1}{3}\hat{\tilde{S}}_{kk}\delta_{ij}\right) \tag{25}$$

and

$$Q_j=-\frac{\mu_{sgs}}{Pr_{sgs}}\frac{\partial\hat{\tilde{T}}}{\partial x_j}, \tag{26}$$

We define $\mu_{sgs}=\bar{\rho}C_{DVMG}\Pi^{t}$ and $Pr_{sgs}=Pr_{DVMG}$. Under the aid of the Germano identity [14] and a least-squares error minimization technique of Lilly [15], the coefficients C_{DVMG} and Pr_{DVMG} are calculated as

$$C_{DVMG}=\frac{\langle L_{ij}M_{ij}\rangle_V}{\langle M_{ij}M_{ij}\rangle_V} \tag{27}$$

and

$$Pr_{DVMG}=\frac{\left\langle M_j^{\theta}M_j^{\theta}\right\rangle_V}{\left\langle L_j^{\theta}M_j^{\theta}\right\rangle_V}, \tag{28}$$

More details regarding the filtering system of this dynamic LES method is provided in Water et al. [10].

3 Adaption Methodology

The computational mesh is automatically refined and unrefined based on the change of flow features. The smooth flow region usually associates with small computational error, and the fast-changing flow region usually associated with large computational error, as discussed in Waters et al. [10]. Mesh adaptation starts with an initial coarse mesh. A local element refinement indicator is defined to determine if a local refinement for an element is needed, i.e.,

$$\xi_i = \frac{\|e\|_i}{\bar{e}_{avg}} \tag{29}$$

when $\xi_i > 1$ the element is refined; when $\xi_i < 1$ the element is unrefined.

The local relative percentage error of any single element is $\|e\|_i$ while the average element error is defined as $\bar{e}_{avg}$. The gradient of the momentum in each element is used to calculate the error estimate, e.g., $e = \nabla U$. There are different levels of refinement. When an element is refined once, this is labeled a level one refinement, refining again produces a level two for refinement, and so on. The adaptation process is illustrated in Fig. 1. All physical nodes are vertexes of the quadrilaterals. Virtual nodes are only added physically for visualization. More details can be found in Wang and Pepper [16].

4 Implicit Solution Method

Developing an implicit solution scheme allows for a larger time step size and maintaining the accuracy of the overall system of model equations. In this system, the diffusive terms such as turbulence and other stresses, are moved to the left-hand side of the equation and integrated into a matrix equation. The advection and source terms are for the load vector. The method is developed by starting with the momentum Eq. (2) without the pressure gradient term to get the intermediate velocity u_i^*. For simplicity, we drop all of the superscripts and source terms.

$$\rho^n u_i^* - \Delta t \times \left(\frac{\partial t_{ij}^*}{\partial x_j} - \frac{\partial \tau_{ij}^*}{\partial x_j} \right) = \rho^n u_i^n - \Delta t \times \frac{\partial (\rho^n u_i^n u_j^n)}{\partial x_j} \tag{30}$$

where $t_{ij}^* = \mu\left(\frac{\partial u_i^*}{\partial x_j} + \frac{\partial u_j^*}{\partial x_i}\right) - \frac{2}{3}\mu \frac{\partial u_k^*}{\partial x_k}\delta_{ij}$ and $\tau_{ij} - \frac{1}{3}\tau_{kk}\delta_{ij} = -2\mu_{sgs}\left(\tilde{S}_{ij} - \frac{1}{3}\tilde{S}_{kk}\delta_{ij}\right)$, where the strain rate tensor is given by $S_{ij}^* = \frac{1}{2}\left(\frac{\partial u_i^*}{\partial x_j} + \frac{\partial u_j^*}{\partial x_i}\right)$.

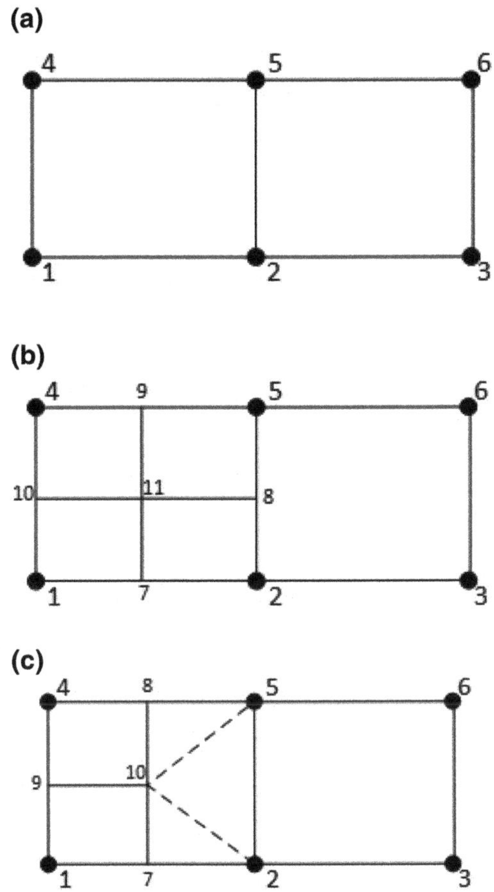

Fig. 1 Adaptation process

Using the mass conserving projection method described previously results in u_i^{n+1} by multiplying u_i^* by ρ^n to form U_i^*, hence $\Delta U_i^* = U_i^* - U_i^n$, as usual in the semi-implicit projection.

After determining the pressure as stated earlier, the specific internal energy is solved again in implicit form:

$$E^{n+1} - \Delta t \times \frac{\partial}{\partial x_i}\left(\frac{\kappa}{C_v}\frac{\partial E^{n+1}}{\partial x_i}\right) - \Delta t \times \frac{\partial}{\partial x_i}\left(\frac{C_p \mu_{sgs}}{\mathrm{Pr}_{sgs}\, C_v}\frac{\partial E^{n+1}}{\partial x_i}\right)$$
$$= E^n - \Delta t \times \frac{\partial}{\partial x_i}\left(E^n u_i^{n+1} + P^{n+1} u_i^{n+1}\right) + \Delta t \times \frac{\partial}{\partial x_i}\left(t_{ij}^{n+1} + \tau_{ij}^{n+1}\right) \tag{31}$$

Here $E^{n+1} = \rho^n e^{n+1}$ and e^{n+1} is the internal energy, and we can get temperature T^{n+1} with $T^{n+1} = \frac{e^{n+1}}{C_v}$.

The solution to the species transport equations follows in similar manner:

$$\rho^n \Upsilon_j^{n+1} - \Delta t \times \frac{\partial}{\partial x_i} \rho^n \left[\left(D_{j,N} + \frac{\mu_{sgs}}{Sc_t} \right) \frac{\partial \Upsilon_j^{n+1}}{\partial x_i} \right] = \rho^n \Upsilon_j^n - \Delta t \times \frac{\partial}{\partial x_i} (\rho^n u_i^{n+1} \Upsilon_j^n) \tag{32}$$

For compressible flow, the sound speed is given by $c = \sqrt{\gamma RT}$, where γ and R are aggregated properties determined from the different species mass concentrations. The aggregation process is described in Carrington [17]. The implicit system is not fully implicit because we want to keep the matrix as symmetric as possible for large-scale parallel calculation to reduce computational time, especially for preconditioning.

5 Parallel Solver System

Choice of either a semi-explicit or an implicit solver system can be selected, both of which use the message passing interface (MPI) paradigm for parallel communication; truly a requirement for 3-D simulations and LES. The linear equation system uses the preconditioned conjugate gradient (PCG) solver package having various Krylov solvers. The user supplies matrix-vector multiplication and dot product operations. Our in situ preconditioning methods use SSOR or SOR, since no off-block information is required during the distributed solution of the matrix equations that might require a block Jacobi scheme. The global system is solved by a Beam-Warming method, checking for global convergence, and evoking the additive Schwartz preconditioning system. Communication for preconditioning requires the matrix-vector multiple and vector dot product; both requiring collocations, that is, a gathering and reduction processes facilitated by MPI. An array utilizing an array of globally shared nodes stored on mother processor. The Beam-Warming iteration process provides for the global convergence of the equation system as developed and described by Carrington [18]. For the semi-explicit process, the PCG linear equation solver is used for the pressure; for the implicit method, the linear equation solver is used for all the equations. Subcylcing of the species transport equations is recommended since there are many replications for each species and can be easily threaded onto processing units by nesting in OpenMP calls, usually this only requires one explicit step, however. Details regarding this MPI implementation for FEM and its efficiency are given in Waters and Carrington [19], showing nearly 30 times speed-up (strong scaling test) over the serial implementation of the overall CFD solver, KIVA-hpFE and is shown to be super-linear.

Essentially, the super-linear behavior of the system results from the local integration of each element and the additive property of integrands. A section of the mesh used for the FEM discretization is shown in Fig. 3. Only those elements dissected are considered overlapped. Some information about the material properties and primitive variables required for integration on an overlapping element must be gathered. This

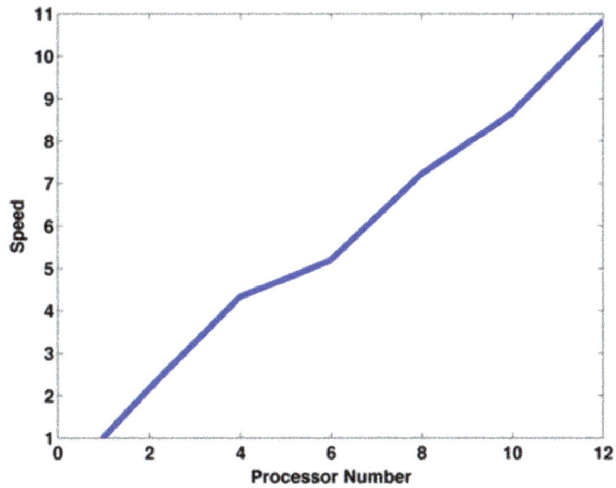

Fig. 2 Speed up improvement

is a simple message passing gather, as performed by a Message Passing Interface (MPI). Jimack and Touheed [20] give a brief introduction about using MPI for FEM. Only the material and primitive variables changing in time need to be part of the communication. Values at each node are determined by integration over elements, on which this node resides. The integration over an overlapped element, shown in Fig. 2, requires gathering values whenever a node is off processor. Therefore, mostly all information that is required for each processor is as follows:

1. Which elements are overlapped, and placed into a 1-D array of shared element s having dimension of number of shared elements.
2. How many nodes should be received from each processors—this is a 1-D array with the dimension of number of processors.
3. How many nodes should be sent to the other processors—this is 1-D array with the dimension of number of processors.
4. Which nodes should be sent to which processor—this is an 2-D array of shared nodes between domains in the global domain, called global_sharednode.
5. What ordering of nodes and elements is present for the global numbering, a mapping from local to global numbering and is called NEWORDER and NEWORDER_E.

In the explicit case, the solution is merely updated at new time value at n + 1 and no matrix equation needs to be solved except for the pressure Poisson equation (as described in the next paragraph on implicit procedures). Elements that are needed to be integrated are elements residing on this processor and the shared elements found in the shared element array. We only update nodes on each processor when the value of the array NEWORDER is not zero and is less than the number of nodes on that processor. After all nodes on that processor are integrated or updated that information is passed into the shared nodes array and distributed to their corresponding processors (e.g. in Fig. 3, nodes 2, 5, and 8 will be passed to processor (PE) 1

Fig. 3 Overlapped nodes

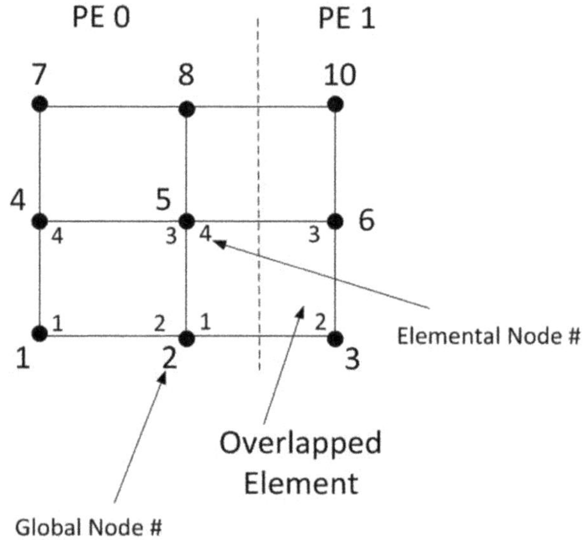

and 3, 6 and 10 need to be passed to PE 0). This information is found in the array global_sharednode and the process occurs via MPI using an all gather command for vector arrays (MPI_Allgatherv). The procedure is applicable to the momentum predictor and corrector steps, along with all transport equations, provided both the viscous and convection terms are stated explicitly.

For the implicit solution processes for pressure or any of the implicit matrices produced, the PCG linear equation solver package of Krylov solvers with user-supplied matrix-vector multiplication and dot product operations is incorporated. In addition, the PCG package provides for user-developed equation preconditioning and overall equation convergence. In this case, a Beam-Warming method is used for the solution to the equations using an additive Schwartz preconditioning system. Only the communication for preconditioning requires the matrix-vector multiplication and dot product routines in KIVA-hpFE. These systems require the collocation, i.e., gathering (GATHERV) and reducing (ALLREDUCE) processes from MPI and employing the 1-D array of shared nodes in the global domain called global_sharednode. The Beam-Warming iteration process provides for the global convergence of the equation system.

The MPI code is implemented in Fortran and has been tested for 3-D flow over a cylinder with 50,720 elements and 57,508 nodes. Figure 2 shows the super-linear speed improvement with MPI incorporated into the semi-implicit algorithm.

6 Numerical Results

6.1 Implicit Solver for 3-D Flow Over a Cylinder

Turbulent flow over a circular cylinder for Reynolds numbers 1.2e+05 (determined by the cylinder diameter and inlet velocity) is investigated here using the dynamic Vreman LES with implicit solver. For simplicity, we assume all species consist of air. The dimension of the domain is

$$x \in [-2.75\,\text{m}, 2.75\,\text{m}],\, y \in [-1.375\,\text{m}, 1.375\,\text{m}],\, z \in [-0.275\,\text{m}, 0.275\,\text{m}].$$

Inlet velocity U = 9.0 m/s, V = 0.0 m/s, and W = 0.0 m/s, and the cylinder diameter is 0.2 m. Here we want to show the efficiency of our implicit solver; therefore, we used a mesh with 114 K nodes and 104 K elements. The flow domain is decomposed into eight subdomains which are assigned to eight processing elements. The domain decomposition is done by ParMETIS, which is an MPI-based parallel library that implements a variety of algorithms for partitioning unstructured graphs, meshes, and for computing fill-reducing orderings of sparse matrices [21]. The domain decomposition is shown in Fig. 4a and the simulation mesh set up is as Fig. 4b. Since no reactions are occurring for the problems described here, the source terms in the governing equations are zero. We use free outflow and no slip as boundary conditions on all walls, and expect 3-D effects from the side walls (along Z direction). The wake starts in a symmetric fashion, as seen in Fig. 5a, then the symmetry breaks down: one eddy becomes larger, as shown in Fig. 5b. Secondary eddies form downstream where the boundary layer separates from the cylinder surface. This is the onset of the eddy separation process, as shown in Fig. 3c. Asymmetry continues to develop in higher Re flows, and eddies are shed from the cylinder, producing the familiar flow pattern of periodic vortex shedding. At later times, the vortices in the wake generate large recirculation zones on the leeward side of the cylinder and develop into the oscillatory von Karman vortex street, as shown in Fig. 5d, e. These results agree with Mustto and Bodstein [22]. Figure 5 shows the instantaneous velocity contour plot at meridional slice of Z = 0 for different time where the vortex shedding is well captured. The Reynolds number is close to the critical Reynolds number and under the influence of 3-D side well effect, we experience some turbulence in the downstream after the cylinder and vortex shedding starts at a further distance compared to small Reynolds number, e.g., Re = 1000. Figure 6 shows the turbulent viscosity and local Mach number. In order to compare the pressure coefficient with the experimental data from Merrick and Bitsuamlak [23], we have the same set up as the experiments shown in [23]. The comparison is shown in Fig. 7. The detachment of the boundary layer flow from the cylinder occurs at $\theta = 85°$, which is approximately 5° different from that described in the literature where $\theta = 80°$ [24]. This difference may be a result of the 3-D side-wall effects (Taylor–Gortler vortex) in the narrow domain with no-slip walls. Overall, results match the experimental values for the coefficient of pressure, shown in Fig. 7. The implicit solver allows for a time step

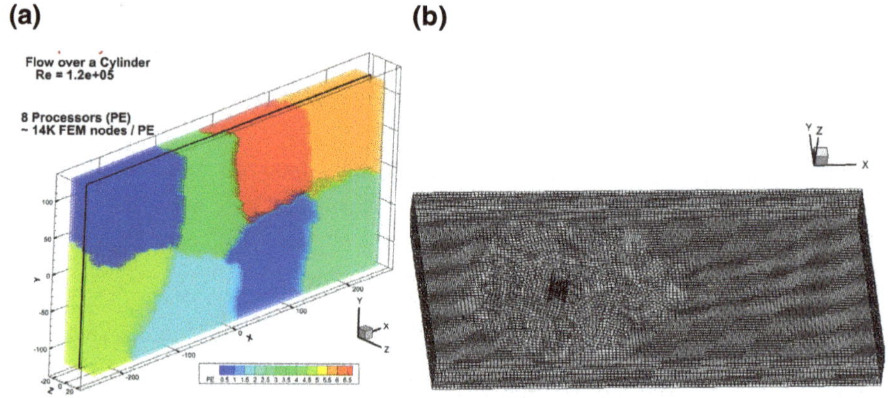

Fig. 4 **a** Simulation domain decomposition onto eight processors (PE is the number of the processor) and **b** mesh set up

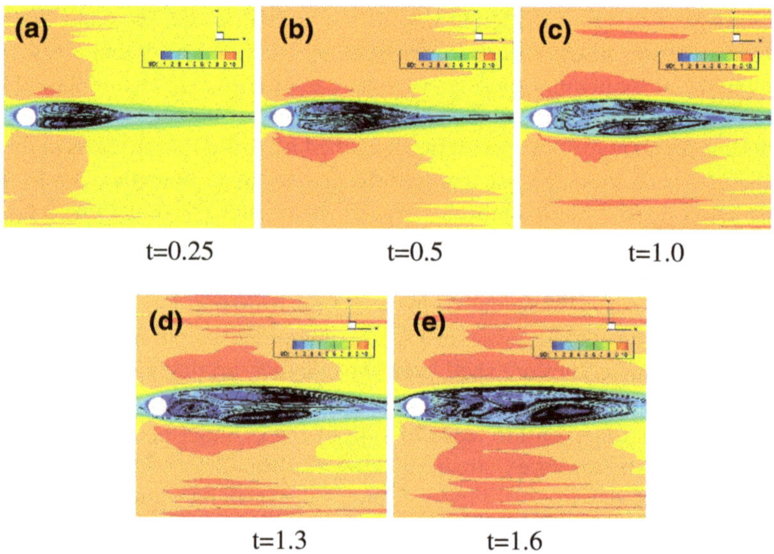

Fig. 5 Instantaneous velocity streamlines at different time (seconds) of the simulation, Re $= 1.2 \times 10^5$ and SD is the velocity magnitude

size of $\Delta t = 1.64 \times 10^{-5}$ while $\Delta t = 1.21 \times 10^{-7}$ is needed for the explicit solver, running this lower than stability limits require. The implicit solver requires more computational effort, taking 2.62 s for each time step compared to only 0.25 s for the explicit solver. However, the implicit solution scheme produces a shorter wall-clock time per calculation by a factor of 10.

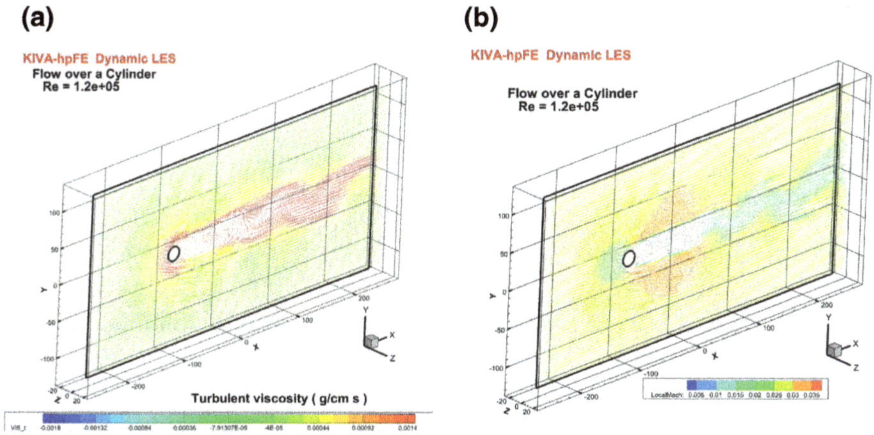

Fig. 6 **a** Turbulent viscosity and **b** local Mach number

Fig. 7 Pressure coefficient comparison with experiment for Re = 1.2×10^5

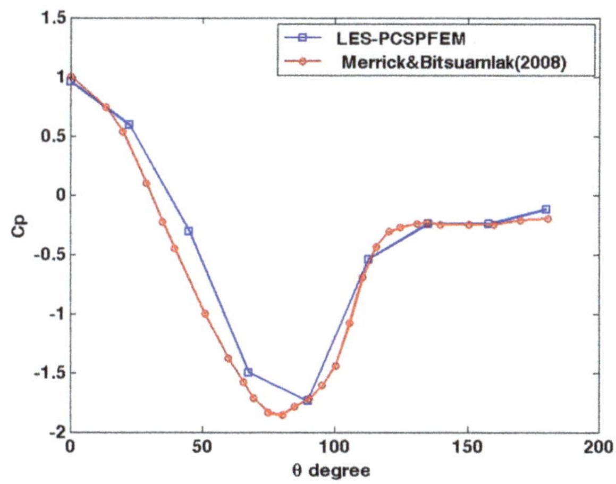

6.2 *3-D Flow Over a Backward-Facing Step Using Adaptive FEM and LES*

For this test case, 3-D flow over a backward-facing step is examined using the h-adaptive system and LES. The Reynolds number for this test case is 28,000 (determined by the step height) using air as the fluid at 296 K° with Prandtl number (Pr) of 0.71. The step height (H) is 0.025 m. The outflow boundary is modeled with a zero-gradient applied for momentum, which is located 30 step height downstream from the expansion. No-slip boundary conditions are applied on the walls except for the Z direction, which receives a periodic boundary condition to remove 3-D effects from the side walls. The 2-D version of the model is discussed in Waters et al. [10]. Here, we present the comparison of LES and RANS model with the help of the h-

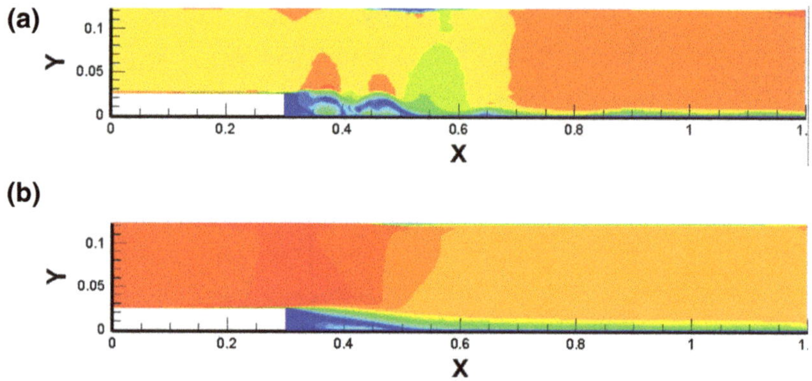

Fig. 8 2-D instantaneous flow magnitude with adaptation by LES (**a**) and RANS, (**b**) at 24 s

adaptation in Fig. 8 where LES in Fig. 8a is better at capturing the instantaneous flow than RANS in Fig. 8b where only the average results are shown. Figure 9 is showing the h-adaptation and the heat flux is well simulated here. In this work, we adapted the mesh twice (level 2 adaptation). 3-D simulation is demonstrated in this study. The initial mesh starts with a coarse mesh having 6,955 nodes and 4,976 elements as shown in Fig. 10a. LES requires a resolved grid in order to evaluate the boundary layer properly so with the use of mesh refinement algorithm, h-adaptation, the grid density is altered automatically in the region of interest. The refinement is limited to this region since the solution is performed on 1 processor, restricting the amount of memory available for more cells and greater refinement. In order to capture the recirculation after the step in the downstream, the initial grid is adapted twice and the final grid has 18,930 nodes and 16,316 elements shown in Fig. 10b. Figure 10c shows the adaptation at a meridional slice at Z = 0, where the recirculation region has been h-adapted to two levels. Taking the time-averaged velocity allows for the measurement of the mean recirculation length behind the step. Figure 11a shows the reattachment region ranges from −0.25 to −0.07 m, with a secondary eddy being well captured in the corner. The reattachment length is therefore 7.6H, agreeing with the experimental value ~7.2H and previous work by Waters and Carrington [19]. Figure 11b shows the instantaneous flow at t = 0.6 s, where fully turbulent flow is shown behind the step.

7 Conclusions

A parallel implicit solver with a dynamic 3-D LES model has been embedded in KIVA-hpFE. Flow over a cylinder with Re = 1.2×10^5 is tested and the numerical results compared with experimental data. The vortex shedding is shown to be accurately captured in the 3-D simulation. Comparing with the use of an explicit solver,

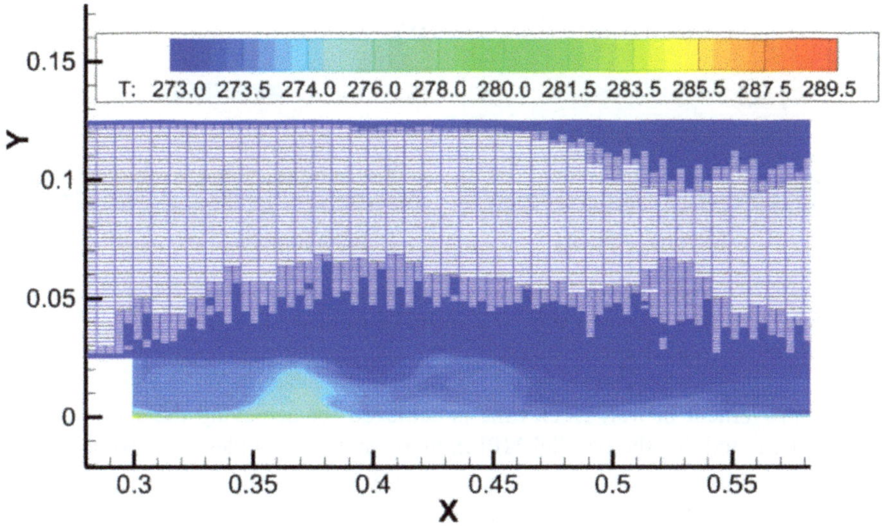

Fig. 9 Shown the adaptation up to level 2 adaptation with temperature contour

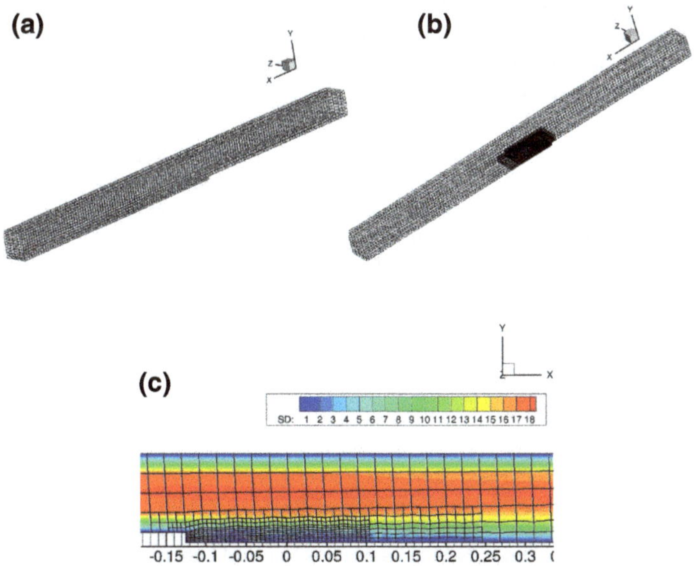

Fig. 10 Initial mesh (**a**), final mesh after adaptation (**b**), and (**c**) slice mesh adaptation at z = 0

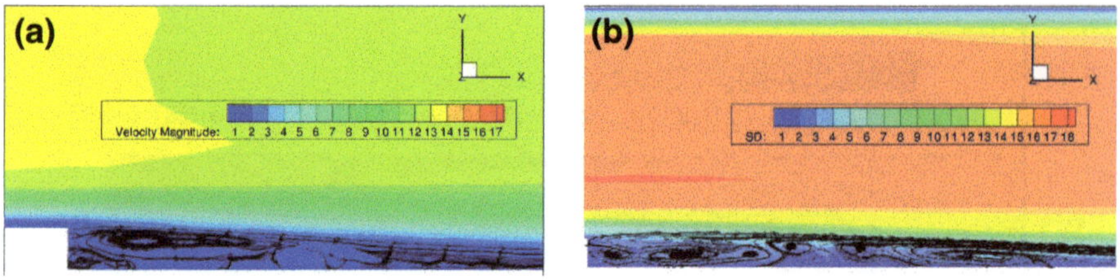

Average velocity — Instantaneous flow at t=0.6s

Fig. 11 Average velocity streamline and velocity magnitude are the average velocity magnitude, instantaneous flow at t = 0.6 s (**a**) average velocity (**b**) instantaneous flow at t = 0.6 s

a 10x improvement in wall-clock time is achieved for achieving the same results. It should be noted that the parallel MPI implementation provides for ~30x speed-up over serial solution. Overall, the implicit method, when combined with parallelization, essentially yields a 300x speed-up over the explicit serial system. The adaptive, dynamic LES model was also applied to 3-D flow over a backward-facing step and achieved very good results compared with studies reported in the literature with just two levels of grid refinement. Efforts are underway to expand the system to include a parallel hp-adaptive scheme. The hp-adaptive FEM strategy follows a three-step strategy using posterior error estimators based on the L_2 norm and is expected to achieve exponential convergence rates.

Acknowledgements The DOE's Office of Energy Efficiency and Renewable Energy (EERE) Advanced Combustion Program (Gurpreet Singh and Leo Breton) is supporting this effort. Los Alamos National Laboratory, an affirmative action/equal opportunity employer is operated by the Los Alamos National Security, LLC for the National Nuclear Security Administration of the US Department of Energy (DOE) under contract DE-AC52-06NA25396. Los Alamos National Laboratory strongly supports academic freedom and a researcher's right to publish; as an institution, however, the Laboratory does not endorse the viewpoint of a publication or guarantee its technical correctness.

References

1. A.A. Amsden, P.J. O'Rourke, T.D. Butler, KIVA-II, a computer program for chemically reactive flows with sprays. Los Alamos, N.M.: Los Alamos National Laboratory Scientific Report, LA-11560-MS (1989)
2. D.C. Wilcox, *Turbulence Modeling for CFD*, 3rd edn. (DCW Industries Inc, La Canada CA, 2006)
3. P.E. Desjardins, S.H. Frankel, Two dimensional Large Eddy Simulation of soot formation in the near field of a strongly radiating non-premixed acetylene-air jet flame. Combust. Flame **119**, 121–133 (1999)
4. O. Colin, F. Ducros, D. Veynante, T. Poinsot, A thickened flame model for large eddy simulations of turbulent premixed combustion. Phys. Fluids **12**, 1843–1863 (2000)
5. C. Angelberger, F. Egolfopoulos, D. Veynante, Large Eddy Simulations of chemical and acoustic effects on combustion instabilities. Flow Turbul. Combust. **65**, 205–222 (2000)

6. H. Pitsch, Duchamp, L. de la Geneste, Large Eddy simulation of premixed turbulent combustion using a level-set approach, in *Proceedings of the Combustion Institute*,vol. 29 (2002) (in press)
7. C.D. Pierce, P. Moin, Progress-variable approach for large eddy simulation of non-premixed turbulent combustion. J. Fluid Mech. **504**, 73–97 (2004)
8. G. E Lau, G.H. Yeoh, V. Timchenko, J.A. Reizes, Application of dynamic global-coefficient subgrid-scale models to turbulent natural convection in an enclosed tall cavity. Phys. Fluids (1994-present) **24**, 094105 (2012)
9. A.W. Vreman, An eddy-viscosity subgrid-scale model for turbulent shear flow: algebraic theory and applications. Phys. Fluids **16**, 3670–3678 (2004)
10. J. Waters, D.B. Carington, D.W. Pepper, An adaptive finite element method with dynamic LES for turbulent reactive flows. Comput. Therm. Sci. Int. J. **2940–2503**(8–1), 57–71 (2016)
11. P.K. Jimack, N. Touheed, Developing parallel finite element software using MPI. HPC Comput. Mech. 15–38 (2000)
12. J. Waters, D.B. Carrington, Modeling turbulent reactive flow in internal combustion engines with an LES in a semi-implicit/explicit finite element projection method, *in Proceedings of the ASME 2016 Internal Combustion Fall Technical Conference*, ICEF2016, 9–12 Oct 2016, Greenville, SC, USA (accepted)
13. W.D. Joubert, G.F. Carey, PCG: a software package for the iterative solution of linear systems on scalar, vector and parallel computers, in *Society for Industrial and Applied Mathematics (SIAM) Proceedings of the 6th SIAM Conference on Parallel Processing for Scientific Computing*, 22–24 Mar, Norfolk, Virginia (1993)
14. M. Germano, U. Piomelli, P. Moin, W.H. Cabotm, A dynamic subgrid-scale eddy viscosity model. Phys. Fluids A **3**(7), 1760–1765 (1991)
15. D.K. Lilly, A proposed modification of the Germano subgrid-scale closure method. Phys. Fluids A **4**(3), 633–635 (1992)
16. X. Wang, D.W. Pepper, Application of an hp-adaptive FEM for solving thermal flow problems. AIAA J. Thermophys. Heat Transf. **21**(1), 190–198 (2007)
17. D.B. Carrington, X. Wang, D.W. Pepper, A Predictor-Corrector Split projection method for turbulent reactive flow. Comput. Therm. Sci. **5**(4), 333–353 (2013)
18. D.B. Carrington, A parallel first-order spherical harmonics (P1) matrix-free method for radiative transport, in *Numerical Heat Transfer, Part B: Fundamentals*, vol. 53 (Taylor and Francis, 2008), pp. 1–21
19. J. Waters, D.B. Carrington, A parallel large eddy simulation in a finite element projection method for all flow regimes. Numer. Heat Transf. Part A Appl. **70**(2), 117–131 (2016)
20. P.K. Jimack, N. Touheed, Developing parallel finite element software using MPI. High Perform. Comput. Comput. Mech. 15–38 (2000)
21. G. Karypis, V. Kumar, MeTis: Unstructured Graph Partitioning and Sparse Matrix Ordering System, Version 4.0. 2009, http://www.cs.umn.edu/~metis, University of Minnesota, Minneapolis, MN
22. A.A. Mustto, G.C.R. Bodstein, Subgrid-scale modeling of turbulent flow around circular cylinder by mesh-free vortex method. Eng. Appl. Comput. Fluid Mech. **5**(2), 259–275 (2011)
23. R. Merrick, G. Bitsuamlak, Control of flow around a circular cylinder by the use of surface roughness: a computational and experimental approach, http://www.ihrc.fiu.edu/wpcontent/uploads/2014/03/MerrickandBitsuamlak_FlowAroundCircularCylinders.pdf. Accessed 2008
24. T. Kawamura, T. Nakao, M. Takahashi, M. Hayashi, K. Murayama, N. Gotoh, Synchronized vibrations of a circular cylinder in cross flow at supercritical reynolds numbers. ASME. J. Press. Vessel Technol. **125**(1), 97–108 (2003)

Fundamentals of Rate-Controlled Constrained-Equilibrium Method

Fatemeh Hadi, Guangying Yu and Hameed Metghalchi

Abstract The Rate-Controlled Constrained-Equilibrium (RCCE) dimension reduction methodology models complex reacting systems within acceptable accuracy with a number of constraints N_c, much smaller than the number of species N_s, in the corresponding Detailed Kinetics Model (DKM). It describes the time evolution of chemical kinetics systems using a sequence of constrained-equilibrium states specified by the chosen constraints. The comprehensive chemical composition at each constrained-equilibrium state is determined by maximizing entropy (or minimizing Gibbs free energies) given the instantaneous values of the constraints. RCCE guarantees final equilibrium concentrations since Lagrange multipliers of all non-elemental constraints will be zero at final state. In this chapter, RCCE fundamentals, constraint and constraint potential representations, methods of initializing constraint potentials (non-dimensional Lagrange multipliers) as well as a brief discussion of RCCE constraint selection are presented. To show its accuracy against DKM, RCCE method is applied to H_2/O_2 and CH_4/O_2 zero-dimensional, constant energy/volume combustion over a wide range of initial conditions. The results show that both mixture results are in excellent agreement with the DKM predictions.

1 Introduction

Describing chemical reactions in complex multidimensional flows requires implementation of an efficient computational method in the computational fluid dynamics (CFD) code. In such flows, directly integrating Detailed Kinetics Model (DKM)

F. Hadi (✉)
Department of Mechanical and Manufacturing Engineering,
Tennessee State University, Nashville, TN 37212, USA
e-mail: fhadi@tnstate.edu

G. Yu · H. Metghalchi
Mechanical and Industrial Engineering Department, Northeastern University,
Boston, MA 02115, USA
e-mail: yu.g@husky.neu.edu

H. Metghalchi
e-mail: metghalchi@coe.neu.edu

A. K. Runchal et al. (eds.), *Energy for Propulsion*, Green Energy and Technology,
https://doi.org/10.1007/978-981-10-7473-8_10

involving a large number of species and elementary chemical reactions with a vast diversity of chemical time scales is computationally very expensive and most probably not practical at all because a transport equation with a chemical source term for each species in the underlying DKM must now be included. To avoid solving a relatively large set of stiff ordinary differential equations (ODEs) in conjunction with the complex fluid flow, dimension reduction techniques are currently widely used.

Dimension reduction methods are based on the notion that compositions in reactive systems evolve by many time scales. This, in turn, causes the species compositions to exist on, or in the close vicinity of a low-dimensional slow invariant manifold (SIM) in the composition space [50]. As an approximation to the actual SIM, in dimension reduction techniques [2–4, 9, 10, 12, 13, 19, 25–28, 31, 32, 34–36, 39, 40, 52–54, 60, 62, 63], a low-dimensional manifold which is explicitly or implicitly identified represents the reactive system. In the reduced presentation, a reactive system is modeled by transporting and solving only a smaller number of reduced variables such as major species. Some of the dimension reduction techniques are quasi-steady-state approximation (QSSA) [5, 60], partial equilibrium approximation [53], intrinsic low-dimensional manifold (ILDM) [40], trajectory-generated low-dimensional manifolds (TGLDMs) [45], computational singular perturbation (CSP) [34], G-scheme [64], the relaxation redistribution method (RRM) [9], the equation-free approach [8], and invariant constrained-equilibrium-edge pre-image curve (ICE-PIC) [55, 56].

Among dimension reduction techniques, the Rate-Controlled Constrained-Equilibrium (RCCE) is specially attractive due to its foundation in the second law of thermodynamics. The RCCE is originally initiated by Keck [31, 32] and further developed by Metghalchi and coworkers [2–4, 19, 25–27, 35, 42–44, 63]. RCCE method [2–4, 19, 25–27, 35, 63] is based on the fact that relatively slow reactions can be separated from relatively fast equilibrating kinetic mechanisms in the underlying DKM. Slow reactions control the evolution of the system toward the final equilibrium while fast mechanisms equilibrate the system locally. For a chosen level of accuracy in a particular problem with a set of initial conditions, a threshold time scale exists that divides the chemical mechanism into fast and slow reactions. The slow reactions control the non-equilibrium dynamics in that they effectively identify a sequence of constrained-equilibrium states on a low-dimensional constrained-equilibrium manifold (CEM) in the composition space. It is conjectured that for the acceptable degree of approximation, the kinetics take place in the CEM which is an estimation of the actual SIM controlled by kinetics in a reactive system. The rate-controlling reactions, in general, are slow because they have to go through one or more "bottlenecks" as called in Ref. [1]. As an example to these "bottlenecks," the three-body reactions are considered slow reactions because three-body collision rates are small. Consequently, the corresponding slowly varying time-dependent constraint is the total number of moles, which cannot change without three-body reactions [1, 27]. Due to the importance of this constraint, it is considered a "universal" constraint in Ref. [27]. In RCCE, the whole chemical composition is reconstructed by maximizing (or minimizing) the appropriate thermodynamic properties, e.g., entropy (or Gibbs free energies), with respect to instantaneous values of the constraints. In RCCE description, these constraints which are the reduced composition variables are transported and solved. For hydrocarbon fuels, the number of reduced variables and thus the number of equations to solve can be substantially smaller than the number of species in the underlying DKM. Apart from methods based on chemical model reduction, there also exists chemistry tabulation methods, e.g., in situ adaptive tabulation (ISAT) [38, 46].

Following the developments, as referenced above, the RCCE method has been used in many studies. These include development of ISAT-RCCE [61], in which the RCCE reduced variables are tabulated by ISAT. The ISAT-RCCE is applied to a constant pressure adiabatic pairwise mixing stirred reactor (PMSR) in which mixing and reaction are solved for in separate fractional steps. In a following study [62], the ISAT-RCCE methodology is further advanced by considering an alternative projection based on the close parallel inertial manifold and simulations are performed of an adiabatic, isobaric plug-flow reactor (PFR) using rate equations for constraint potentials. Subsequently, three different representations of RCCE for computing the reaction mapping in the ISAT-RCCE approach are presented in Ref. [22]. In Refs. [28–30], the RCCE method is used in differential algebraic equations (DAEs) and constraint potential form. These formulations are employed to model homogeneous and pairwise mixed stirred reactors as well as laminar premixed and non-premixed flames. The RCCE formulation is coupled with the transport equations, including those for RCCE constraints. An operator-splitting approach is employed in which transport of kinetic constraints is considered in a separate step; subsequently, the reaction step is carried out. In this step, using a system of algebraic equations the constraints are first transformed into constraint potentials, which are then advanced in time. The RCCE formulation, used in Refs. [28–30], is coupled with artificial neural network (ANN) and probability density function (PDF) schemes in Reynolds-Averaged Navier–Stokes (RANS) to simulate turbulent non-premixed flames [7]. RCCE has also been applied to large eddy simulation (LES) to model turbulent flames via two approaches: the representation used in Refs. [28–30] coupled with conditional moment closure [37, 41] and the ISAT-RCCE along with PDF methods [20, 21]. Recently, combined dimension reduction and tabulation is used to simulate a turbulent lean-premixed propane using LES-PDF method [33]. Reference [15] demonstrates the efficiency and accuracy of the RCCE method as well as its capacity to accurately predict reacting systems with different mixing intensities. The importance of this issue is because mixing can influence the constrained-equilibrium state and thus effectiveness of constraints representing chemical kinetics. It is thus essential to study the capacity of RCCE to predict the coupling between chemical reaction and fluid motion at different levels of mixing. Reference [51] investigates different possible constraints to be used with the RCCE method for reacting compressible flows and the predicted performance of the constraints before running a full RCCE simulation. A very important issue in RCCE is selection of constraints. This challenging issue has been addressed in limited investigations. In previous contributions [25, 27], the C_1 hydrocarbon fuel oxidation process is studied and a very successful set of structurally valid constraints are developed based on studying oxidation pathways. Algorithms for automated selection of constraints include that developed by Yousefian in Ref. [65] as well as other methods based on level of importance to identify single species constraints by determining the species governed by fast/slow mechanisms [59]; greedy algorithm [23, 24] to select single species constraints by cyclic direct integration of chemical kinetics; degree of disequilibrium (DoD) of chemical reactions introduced in Refs. [1, 26] and time scale analysis [17, 18, 57].

This chapter is organized as follows. RCCE fundamentals, its constraint potential and constraint representations are presented in Sects. 2, 2.1 and 2.2, respectively. Methods of initializing constraint potentials based on constraints to constraint potentials conversion and calculating chemical potentials of major species are discussed in Sect. 2.3. A brief discussion on RCCE constraint selection based on studying oxidation paths and DoD of chemical reactions is presented in Sect. 3. The results of RCCE application to constant energy/volume combustion of hydrogen and methane for different sets of initial conditions are provided in Sect. 4 along with a summary in Sect. 5.

2 Rate-Controlled Constrained-Equilibrium Principles

The evolution of a reacting system that is macroscopically homogeneous and involves N_s chemical species is governed by

$$\frac{d\phi_\beta}{dt} = \mathscr{M}_\beta(t) + S_\beta(t) \qquad (\beta = 1, 2, \ldots, N_s + 1) \tag{1}$$

where compositional variables $[\phi_1, \phi_2, \ldots, \phi_{N_s+1}]$ constitute the full representation of the reacting system. These variables are $\phi_\beta \equiv Y_\beta$, for $\beta = 1, \ldots, N_s$ and $\phi_{N_s+1} \equiv h$, where Y_β and h denote mass fraction of chemical species and specific enthalpy of the mixture, respectively. The right-hand side (RHS) includes the effects of micro-mixing $\mathscr{M}_\beta$ and chemical reaction S_β. A detailed description of the effects of the mixing on the RCCE predictions is presented in [15]. For the micro-mixing term, we employ the linear mean square estimation (LMSE) model [11], also known as interchange by exchange with the mean (IEM) closure [6],

$$\mathscr{M}_\beta(t) = -\frac{1}{\tau_{mix}}\left(\phi_\beta - \overline{\phi}_\beta\right) \tag{2}$$

where τ_{mix} denotes the mixing time scale and $\overline{\phi}_\beta$ is the mean value of composition variables. The reaction source term can be calculated by the following equation

$$S_\beta(t) = \frac{\dot{\omega}_\beta M_\beta}{\rho} \quad (\beta = 1, \ldots, N_s) \tag{3}$$

where $\dot{\omega}_\beta$ and M_β are the net production rate and molar mass of species β, respectively, and ρ is the density of the gas mixture. In low Mach number flows, we assume

$S_{N_s+1} \approx 0$ for the specific enthalpy. $\dot{\omega}_\beta$ can be determined by considering the chemical reaction mechanism which is represented by

$$\sum_{\beta=1}^{N_s} \nu'_{\beta\gamma} Z_\beta \rightleftharpoons \sum_{\beta=1}^{N_s} \nu''_{\beta\gamma} Z_\beta \quad (\gamma = 1, \ldots, N_r) \tag{4}$$

where $\boldsymbol{Z}$ is the species vector, $\boldsymbol{\nu}'$ and $\boldsymbol{\nu}''$ are $N_s \times N_r$ matrices of stoichiometric coefficients of reactants and products, respectively, and N_r is the number of reactions. The following equation gives the corresponding rate equations for the species

$$\dot{\omega}_\beta = \sum_{\gamma=1}^{N_r} \nu_{\beta\gamma} \Omega_\gamma = \sum_{\gamma=1}^{N_r} (\nu''_{\beta\gamma} - \nu'_{\beta\gamma}) \Omega_\gamma \tag{5}$$

where $\nu_{\beta\gamma}$ and Ω_γ are the net stoichiometric coefficient of species β in reaction γ and the net production rate of reaction γ which is defined as

$$\Omega_\gamma = k_\gamma^f \prod_{\beta=1}^{N_s} [Z_\beta]^{\nu'_{\beta\gamma}} - k_\gamma^b \prod_{\beta=1}^{N_s} [Z_\beta]^{\nu''_{\beta\gamma}} \tag{6}$$

The first and second terms on the RHS constitute forward and backward rate of progress of reaction γ, respectively. k_γ^f and k_γ^b are the forward and backward reaction rate constant for reaction γ, respectively. $[Z_\beta]$ denotes molar concentration of species Z_β. The RCCE method provides a reduced representation of the reacting system using N_c slowly varying scalars which impose constraints on the evolution of the system. The constraints are assumed to be linear combinations of the concentrations of the gas mixture species [32]

$$C'_\alpha = \sum_{\beta=1}^{N_s} A'_{\alpha\beta} \frac{Y_\beta}{M_\beta} \qquad (\alpha = 1, 2, \ldots, N_c) \tag{7}$$

where N_c and C'_α denote the number and the value of constraints. $A'_{\alpha\beta}$, a predefined transformation matrix here, identifies in each constraint, the contribution of different chemical species. Following previous work [27], the constant coefficient matrix is transformed to a new form to diagonalize the square sub-matrix relating the constraints and the major species. This modification has shown to improve robustness of calculations [27]. The transformed constraint vector and the species mass fractions are related similar to Eq. (7),

$$C_\alpha = \sum_{\beta=1}^{N_s} A_{\alpha\beta} \frac{Y_\beta}{M_\beta} \tag{8}$$

where C_α and $A_{\alpha\beta}$ are transformed constraint vector and coefficient matrix, respectively. These are, henceforth, called constraint vector and coefficient matrix for simplicity. The reduced representation $[C_1, C_2, \ldots, C_{N_c}, h]$ is governed by rate equations for constraints and specific enthalpy. The former is obtained by transforming Eq. (1) according to Eq. (8),

$$\frac{dC_\alpha}{dt} = \sum_{\beta=1}^{N_s} \frac{A_{\alpha\beta}}{M_\beta} \left(\mathscr{M}_\beta(t) + S_\beta(t)\right) \tag{9}$$

In RCCE, Eq. (1) is replaced by Eq. (9) which is advantageous since Eq. (9) includes fewer ordinary differential equations (ODEs) with reduced numerical stiffness. The rapidly varying (unconstrained) species are assumed to be instantaneously in a constrained-equilibrium state, at which the composition of chemical species are determined. The constrained-equilibrium state is specified using the method of undetermined Lagrange multipliers together with maximizing the entropy or minimizing an appropriate free energy with respect to the instantaneous values of constraints [27]. We consider the summation of entropy and constraints

$$L = s + \sum_{\alpha=1}^{N_c} \lambda_\alpha \left(C_\alpha - \sum_{\theta=1}^{N_s} A_{\alpha\theta} \frac{Y_\theta}{M_\theta} \right) \tag{10}$$

where $s \equiv s(p, h, Y_1, \ldots, Y_{N_s})$ is the specific entropy of the mixture, p is pressure, and λ_α denotes the Lagrange multiplier corresponding to constraint α. To optimize this relation with respect to mass fraction, we set

$$\frac{\partial L}{\partial Y_\beta} = 0 \qquad (\beta = 1, \ldots, N_s) \tag{11}$$

We, thus, have

$$\frac{\partial s}{\partial Y_\beta} - \sum_{\alpha=1}^{N_c} \lambda_\alpha \sum_{\theta=1}^{N_s} A_{\alpha\theta} \frac{1}{M_\theta} \frac{\partial Y_\theta}{\partial Y_\beta} = 0 \tag{12}$$

For a multicomponent system at constant specific enthalpy and pressure, the derivative of specific entropy with respect to mass fraction of species β can be written as

$$\frac{\partial s}{\partial Y_\beta} = -\frac{\mathring{\mu}_\beta}{M_\beta T} \tag{13}$$

Substituting Eq. (13) into Eq. (12) yields

$$-\frac{\mathring{\mu}_\beta}{M_\beta T} - \frac{1}{M_\beta} \sum_{\alpha=1}^{N_c} A_{\alpha\beta} \lambda_\alpha = 0 \tag{14}$$

where $\mathring{\mu}_\beta$ is the chemical potential per unit mole of species β. We divide both sides of this equation by universal gas constant R_u. The normalized Lagrange multiplier conjugate to each constraint, called the "constraint potential," is defined as

$$\Lambda_\alpha = \frac{\lambda_\alpha}{R_u} \tag{15}$$

Equation (14) can thus be written as

$$\frac{\mathring{\mu}_\beta}{R_u T} = -\sum_{\alpha=1}^{N_c} A_{\alpha\beta}\Lambda_\alpha \tag{16}$$

where the left-hand side (LHS) can be expanded according to the relationship between the chemical potential and the Gibbs free energy of the pure substance per unit mole of species β $\mathring{g}_\beta$

$$\mathring{\mu}_\beta = \mathring{g}_\beta + R_u T \ln X_\beta \tag{17}$$

The Gibbs free energy $\mathring{g}_\beta \equiv \mathring{g}_\beta(T, p)$ is related to that at the standard pressure p_0 as

$$\mathring{g}_\beta = \mathring{g}_\beta^0 + R_u T \ln \frac{p}{p_0} \tag{18}$$

where $\mathring{g}_\beta^0 \equiv \mathring{g}_\beta(T, p_0)$. Substituting Eqs. (17, 18) into Eq. (16) gives

$$\frac{1}{R_u T}\left[\mathring{g}_\beta^0 + R_u T \ln \frac{pX_\beta}{p_0}\right] = -\sum_{\alpha=1}^{N_c} A_{\alpha\beta}\Lambda_\alpha \tag{19}$$

Solving this equation for X_β gives

$$X_\beta^{RCCE} = \frac{p_0}{p}\exp\left(-\frac{\mathring{g}_\beta^0}{R_u T}\right)\exp\left(-\sum_{\alpha=1}^{N_c} A_{\alpha\beta}\Lambda_\alpha\right) \tag{20}$$

where X_α^{RCCE} is the mole fraction evaluated at the RCCE state of the system. Equation (20) can also be written in terms of mass fractions as

$$Y_\beta^{RCCE} = \frac{M_\beta}{\rho}\frac{p_0}{R_u T}\exp\left(-\frac{\mathring{g}_\beta^0}{R_u T}\right)\exp\left(-\sum_{\alpha=1}^{N_c} A_{\alpha\beta}\Lambda_\alpha\right) \tag{21}$$

where Y_β^{RCCE} is the mass fractions at the constrained-equilibrium state.

2.1 Rate Equations for the Constraint Potentials

The constraint potential representation of RCCE is obtained following a procedure similar to Refs. [25–27]. This representation of RCCE involves transforming Eq. (9) into differential equations for constraint potentials. In previous studies [25–27], similar equations are obtained for temperature and constraint potentials for known energy and volume. Here, constraint potential formulation with constraint potentials along with temperature and density as independent variables for given enthalpy and pressure, following Refs. [15, 16], is presented. By differentiating Eq. (8) with respect to time, we obtain

$$\frac{dC_\alpha}{dt} = \sum_{\beta=1}^{N_s} A_{\alpha\beta} \frac{1}{M_\beta}\frac{dY_\beta}{dt} = \sum_{\beta=1}^{N_s} A_{\alpha\beta} \frac{Y_\beta}{M_\beta}\frac{d\ln Y_\beta}{dt} \tag{22}$$

The RHS is expressed by taking derivative of Eq. (21) with respect to time as

$$\frac{d\ln Y_\beta}{dt} = -\frac{1}{\rho}\frac{d\rho}{dt} + \frac{1}{R_u T^2}\frac{dT}{dt}\left[\mathring{g}_\beta^0 - T\frac{d\mathring{g}_\beta^0}{dT} - R_u T\right] - \sum_{\alpha=1}^{N_c} A_{\alpha\beta}\frac{d\Lambda_\alpha}{dt} \tag{23}$$

The second term on the RHS can be simplified to $\frac{\mathring{e}_\beta}{R_u T^2}\frac{dT}{dt}$ where $\mathring{e}_\beta$ denotes the internal energy per unit mole of species β. For a pure chemical species, governed by ideal gas equation of state, we have

$$\mathring{g}_\beta^0 = \mathring{h}_\beta - T\mathring{s}_\beta^0 = \mathring{e}_\beta + R_u T - T\mathring{s}_\beta^0 \tag{24}$$

In this equation, $\mathring{h}_\beta$ and $\mathring{s}_\beta^0$ designate the enthalpy per unit mole of species β and the entropy of the pure substance at pressure p_0 per unit mole of species β, respectively. According to Gibbs fundamental equation, for a constant pressure and chemical composition process we have

$$\mathring{s}_\beta = -\left(\frac{\partial \mathring{g}_\beta}{\partial T}\right)_{p,n_1,n_2,\ldots,n_{N_s}} \tag{25}$$

where n_θ and $\mathring{s}_\beta$ denote the number of moles of species θ and the entropy of the pure substance per unit mole of species β, respectively. For a pure substance at pressure p_0, this relation along with Eq. (24) gives

$$\mathring{e}_\beta = \mathring{g}_\beta^0 - T\frac{d\mathring{g}_\beta^0}{dT} - R_u T \tag{26}$$

which relates the entropy to the Gibbs free energy and the temperature of the system. Therefore, the time derivative of constraints is obtained by substituting Eq. (23) into Eq. (22) as

$$\frac{dC_\alpha}{dt} = -\left(\sum_{\beta=1}^{N_s} A_{\alpha\beta}\frac{Y_\beta}{M_\beta}\right)\frac{1}{\rho}\frac{d\rho}{dt} + \left(\sum_{\beta=1}^{N_s} A_{\alpha\beta}\frac{Y_\beta}{M_\beta}\frac{\mathring{e}_\beta}{R_u T}\right)\frac{1}{T}\frac{dT}{dt} - \sum_{\theta=1}^{N_c}\sum_{\beta=1}^{N_s} A_{\alpha\beta}A_{\theta\beta}\frac{Y_\beta}{M_\beta}\frac{d\Lambda_\theta}{dt} \tag{27}$$

To include the energy balance, we take time derivative of the specific enthalpy $h = \sum_{\beta=1}^{N_s} h_\beta Y_\beta$ to obtain

$$\frac{dh}{dt} = c_p\frac{dT}{dt} + \sum_{\beta=1}^{N_s} h_\beta Y_\beta \frac{d\ln Y_\beta}{dt} \tag{28}$$

where c_p is the specific heat capacity of the mixture at constant pressure and h_β denotes specific enthalpy of species β. Employing Eq. (23) results in

$$\frac{dh}{dt} = -h\frac{1}{\rho}\frac{d\rho}{dt} + \left[c_p T + \sum_{\beta=1}^{N_s} h_\beta Y_\beta \frac{\mathring{e}_\beta}{R_u T}\right]\frac{1}{T}\frac{dT}{dt} - \sum_{\theta=1}^{N_c}\sum_{\beta=1}^{N_s} A_{\theta\beta}h_\beta Y_\beta\frac{d\Lambda_\theta}{dt} \tag{29}$$

The set of equations corresponding to species concentrations and energy are completed by the ideal gas equation of state

$$p = \rho R_u T \sum_{\beta=1}^{N_s}\frac{Y_\beta}{M_\beta} \tag{30}$$

Taking derivative of Eq. (30) with respect to time gives

$$\frac{1}{p}\frac{dp}{dt} = \frac{1}{\rho}\frac{d\rho}{dt} + \frac{1}{T}\frac{dT}{dt} + M\sum_{\beta=1}^{N_s}\frac{Y_\beta}{M_\beta}\frac{d\ln Y_\beta}{dt} \tag{31}$$

where $M = \left(\sum_{\beta=1}^{N_s}\frac{Y_\beta}{M_\beta}\right)^{-1}$ is the average molar mass of the mixture. Using Eq. (23), we have

$$\frac{1}{p}\frac{dp}{dt} = \left(M\sum_{\beta=1}^{N_s}\frac{Y_\beta}{M_\beta}\frac{\mathring{e}_\beta}{R_u T} + 1\right)\frac{1}{T}\frac{dT}{dt} - M\sum_{\theta=1}^{N_c}\sum_{\beta=1}^{N_s} A_{\theta\beta}\frac{Y_\beta}{M_\beta}\frac{d\Lambda_\theta}{dt} \tag{32}$$

Equations (27, 29, 32) constitute the set of $N_c + 2$ equations describing the time variation of N_c constraint potentials, temperature, and density. The final form of these equations are written as [15, 16]

$$-C_\alpha \frac{1}{\rho}\frac{d\rho}{dt} + Q_{c\alpha}\frac{1}{T}\frac{dT}{dt} + \sum_{\theta=1}^{N_C} R_{c_{\alpha\theta}} \frac{d\Lambda_\theta}{dt} = \sum_{\beta=1}^{N_S} \frac{A_{\alpha\beta}}{M_\beta}\left(\mathscr{M}_\beta(t) + S_\beta(t)\right) \tag{33a}$$

$$-h\frac{1}{\rho}\frac{d\rho}{dt} + Q_h\frac{1}{T}\frac{dT}{dt} + \sum_{\theta=1}^{N_C} R_{h_\theta}\frac{d\Lambda_\theta}{dt} = \frac{dh}{dt} \tag{33b}$$

$$Q_p\frac{1}{T}\frac{dT}{dt} + \sum_{\theta=1}^{N_C} R_{p_\theta}\frac{d\Lambda_\theta}{dt} = \frac{1}{p}\frac{dp}{dt} \tag{33c}$$

where the coefficients are

$$\begin{aligned}
Q_{c_\alpha} &= \sum_{\beta=1}^{N_s} A_{\alpha\beta}\frac{Y_\beta}{M_\beta}\frac{\mathring{e}_\beta}{R_u T} \; ; \; R_{c_{\alpha\theta}} = -\sum_{\beta=1}^{N_s} A_{\alpha\beta}A_{\theta\beta}\frac{Y_\beta}{M_\beta} \\
Q_h &= c_p T + \sum_{\beta=1}^{N_s} h_\beta Y_\beta \frac{\mathring{e}_\beta}{R_u T} \; ; \; R_{h_\theta} = -\sum_{\beta=1}^{N_s} A_{\theta\beta}h_\beta Y_\beta \\
Q_p &= M\sum_{\beta=1}^{N_s}\frac{Y_\beta}{M_\beta}\frac{\mathring{e}_\beta}{R_u T} + 1 \; ; \; R_{p_\theta} = -MC_\theta
\end{aligned} \tag{34}$$

For adiabatic, isobaric system, we can assume $\frac{dh}{dt} \approx 0$ and $\frac{1}{p}\frac{dp}{dt} \approx 0$ on the RHS of Eq. (33b) and Eq. (33c), respectively. For cases where variation of internal energy is known, Eq. (33b) can be replaced with

$$-e\frac{1}{\rho}\frac{d\rho}{dt} + Q_e\frac{1}{T}\frac{dT}{dt} + \sum_{\theta=1}^{N_c} R_{e_\theta}\frac{d\Lambda_\theta}{dt} = \frac{de}{dt} \tag{35}$$

where

$$Q_e = c_v T + \sum_{\beta=1}^{N_s}\frac{\mathring{e}_\beta^2}{R_u T}\frac{Y_\beta}{M_\beta} \; ; \; R_{e_\theta} = -\sum_{\beta=1}^{N_s} A_{\theta\beta}\mathring{e}_\beta\frac{Y_\beta}{M_\beta} \tag{36}$$

In these equations, e and c_v are internal energy per unit mass of the mixture and specific heat capacity of the mixture at constant volume, respectively. It is also noticed that since chemical reaction conserves the atomic elements, for the elemental constraints, chemical reaction source term is zero; i.e., it does not appear on the RHS of Eq. (33a). Given an initial thermodynamic state $(\rho, T, \Lambda_1, \ldots, \Lambda_{N_c})$, Eq. (56) is integrated to obtain $T(t)$, $\rho(t)$ and $\Lambda_\theta(t)$ $(\theta = 1, \ldots, N_c)$. These variables provide the mass fraction of all the species present in the chemical mechanism according to Eq. (21). An advantage of RCCE equations as presented in Eq. (56) is that the effects of mixing and chemical reaction can be accounted for simultaneously during numerical integration.

2.2 Rate Equations for the Constraints

Rate equations of constraints Eq. (9), in the constraint representation of RCCE, are solved rather than the rate equations for constraint potentials. To obtain constraint potentials, mass fractions, temperature, and density at the constrained-equilibrium state, the following system of equations must be solved simultaneously

$$C_\alpha = \sum_{\beta=1}^{N_s} A_{\alpha\beta} \frac{Y_\beta}{M_\beta} \tag{37a}$$

$$Y_\beta = \frac{M_\beta}{\rho} \frac{p_0}{R_u T} \exp\left(-\frac{\mathring{g}_\beta^0}{R_u T}\right) \exp\left(-\sum_{\alpha=1}^{N_c} A_{\alpha\beta} \Lambda_\alpha\right) \tag{37b}$$

$$h = \sum_{\beta=1}^{N_s} h_\beta(T) Y_\beta \tag{37c}$$

$$1 = \sum_{\beta=1}^{N_s} Y_\beta \tag{37d}$$

Normally Newton's method is considered to find constrained-equilibrium composition, as performed in STANJAN [58] equilibrium code. In Newton's procedure, first, an initial guess is made for Λ_α. Afterward, Λ_α is advanced iteratively until it satisfies Eq. (37) while $\mathring{g}_\beta^0$ remains constant. However, this method is liable to suffer from singularity difficulties and sometimes the step size in Λ_α cannot be stably calculated. The substitute code proposed here, CEQ [47] based on Gibbs function continuation method, is principally created to overcome this problem based on the observation that for arbitrary $d\mathring{g}_\beta^0$, a corresponding value for $d\Lambda_\alpha$ can always be stably determined [49]. As a result of this observation, an analogous set of ODEs in pseudo-time, ξ, is established to represent the constraint potential equations, Eq. (37). The solution to these equations is obtained by integrating this set of equations in pseudo-time. The numerical procedure involves making infinitesimal changes in $d\Lambda_\alpha$ and $d\mathring{g}_\beta^0$ and evaluating the corresponding variations in dY_β and dC_α. First, the pseudo-Gibbs free energy $\overline{\mathring{g}_\beta}(\xi)$ and constraint potentials $\overline{\Lambda_\alpha}(\xi)$ as functions of ξ are defined. Subsequently, the pseudo-Gibbs free energy is varied linearly from a specified initial condition at $\xi = 0$ to the accurate value at $\xi = 1$. $\overline{\Lambda_\alpha}(\xi)$ is determined by integrating the set of ODEs from a defined initial condition $\overline{\Lambda_\alpha}(0)$. The constraint potential values at $\xi = 1$, $\overline{\Lambda_\alpha}(1)$, are solutions to the constraint potential equations (37). We refer to Refs. [48, 49] for detailed description of the method.

The numerical procedure in the constraint representation of RCCE involves integrating the transport equations of constraints, Eq. (9) in time, using ODE solvers. Constraints are integrated from time t to $t + \Delta t$ in each reaction time step. The ODE solver breaks each time step into several sub-time steps. In each smaller sub-time step, constrains, enthalpy, and pressure are given as input to the constrained-equilibrium code to determine mass fractions, temperature, and density at the constrained-equilibrium state. Subsequently, these variables are then utilized to calculate the source term on the RHS of Eq. (9), which is then used to advance the constraint values. These calculations are repeated until time $t + \Delta t$ is reached. Unlike the constraint potential approach, this formulation of RCCE does not need matrix inversion.

In Ref. [16], the RCCE method is applied in the simulation of methane/oxygen combustion in an adiabatic, isobaric well-stirred reactor to compare the accuracy and efficiency of the two RCCE formulations: constraint potential and constraint. The two representations of RCCE show different numerical performance because the two approaches involve quite different mathematical operations; former requires matrix inversion, while the latter involves constrained-equilibrium calculations. This dissimilarity in numerical solution methodology produces different number of mathematical operations as well as dissimilar reduction in stiffness of the kinetics. Reference [16] demonstrates that both constraint potential and constraint methods provide very accurate representation of the kinetics. Also, it shows that while the constraint implementation exhibits less numerical stiffness, the constraint potential representation results in more saving in computation time [16].

2.3 *Initializing the Constraint Potentials*

The constraint potentials have finite initial values when reacting systems are in a constrained-equilibrium state. Nonetheless, for reacting systems that are in non-equilibrium states with one or more species having zero concentrations, Eqs. (16) and (21) show that one or more constraint potentials must be infinite. This condition is often encountered in ignition delay time calculations because initially mass fractions of all species except the fuel and oxidizer are often assumed zero and the system is far from equilibrium. This difficulty can be overcome by assigning infinitesimal partial pressures to species that have zero concentrations to obtain constraint potentials with finite values [27].

Once small partial pressures are assigned to zero concentration species, given the initial values of specific enthalpy h, pressure p, and constraint vector C_α for $\alpha = 1, 2, \ldots, N_c$, the following algebraic equations can be solved simultaneously to determine species mass fractions, temperature, density, and constraint potentials:

$$C_\alpha = \sum_{\beta=1}^{N_s} A_{\alpha\beta} \frac{Y_\beta}{M_\beta} \tag{38a}$$

$$\frac{\mathring{\mu}_\beta}{R_u T} = -\sum_{\alpha=1}^{N_c} A_{\alpha\beta} \Lambda_\alpha \tag{38b}$$

$$h = \sum_{\beta=1}^{N_s} h_\beta(T) Y_\beta \tag{38c}$$

$$p = \rho R T \tag{38d}$$

where $R = \frac{R_u}{M}$ is the gas constant. For cases where variation of specific internal energy e is known, Eq. (38c) should be replaced with $e = \sum_{\beta=1}^{N_s} e_\beta(T) Y_\beta$ in which e_β is the internal energy per unit mass of species β. It is noted that Eq. (38) is essentially performing the same function as Eq. (37). Also, the original transformation matrix can also be used in Eq. (38) which results in untransformed constraint potentials. Constrained-equilibrium state calculations for gas-phase mixtures can be carried out using readily available programs such as NASA CEA [14], STANJAN [58], and CEQ [47] equilibrium codes. Although the first two codes were originally created to determine the final chemical equilibrium state of a system subject to conservation of the elemental constraints, they can also be employed for evaluating the constrained-equilibrium states of reacting systems subject to general constraints which are linear combination of the chemical species concentrations. This can be achieved by modifying their inputs to enable them to accept general constraints. Previous publications have shown that that these codes might result in excessive computation times [3, 4, 63] or be unreliable due to stability problems [49]. Thus, the CEQ code which is a Fortran library to compute equilibrium compositions using Gibbs function continuation method [47, 48] serves better for chemical constrained-equilibrium calculations as discussed in Sect. 2.2. This method is shown to be able to specify the chemical equilibrium state for all well-posed constrained and unconstrained problems [49].

There are special applications where a simpler approach leads to basically the same initial RCCE conditions and constraint potentials. For example, due to the choices of fast and slow species, all ignition delay simulation cases described in Ref. [27] fall in this category. In the constraint potential initialization approach discussed in Ref. [27], the species mass fraction vector Y_β, $\beta = 1, \ldots, N_s$ is reorganized in such a way that the first N_c species are the major species. A good initial selection for a major species corresponding to a constraint, in the group of species included in the constraint, is the species with the minimum standard Gibbs free energy. Therefore, the constraint vector can be decomposed as follows

$$C_\alpha = \sum_{\beta=1}^{N_c} A_{1\alpha\beta} \frac{Y_{1\beta}}{M_\beta} + \sum_{\beta=Nc+1}^{N_s} A_{2\alpha\beta} \frac{Y_{2\beta}}{M_\beta} \tag{39}$$

where $A_{1\alpha\beta}$ gives the contribution of the major species, $Y_{1\beta}$, to the constraint, C_α, and $A_{2\alpha\beta}$ gives the contribution of $Y_{2\beta}$ to C_α. The corresponding decomposition of Eq. (16) using the entropic chemical potentials

$$\hat{\mathring{\mu}}_\beta = -\frac{\mathring{\mu}_\beta}{R_u T} \qquad (\beta = 1, 2, \ldots, N_s) \tag{40}$$

is as follows

$$\hat{\mathring{\boldsymbol{\mu}}} = \begin{bmatrix} \hat{\mathring{\boldsymbol{\mu}}}_1 \\ \hat{\mathring{\boldsymbol{\mu}}}_2 \end{bmatrix} = \boldsymbol{A}^T \boldsymbol{\Lambda} = \begin{bmatrix} \boldsymbol{A}_1^T \\ \boldsymbol{A}_2^T \end{bmatrix} \boldsymbol{\Lambda} \tag{41}$$

where $\hat{\mathring{\boldsymbol{\mu}}}_1$ and $\hat{\mathring{\boldsymbol{\mu}}}_2$ are the entropic chemical potential vectors corresponding to major and minor species, respectively. Constraint potentials initial values can now be determined by assuming $\hat{\mathring{\boldsymbol{\mu}}}_1(0) \gg \hat{\mathring{\boldsymbol{\mu}}}_2(0)$. This gives

$$\boldsymbol{\Lambda}(0) = (\boldsymbol{A}_1^T)^{-1} \hat{\mathring{\boldsymbol{\mu}}}_1(0) \tag{42}$$

It can be seen from Eq. (42) that since $\boldsymbol{A}_1$ is a square, diagonalized matrix, in the decomposed representation, the initial values for the constraint potentials are simply the initial entropic chemical potentials of the major species chosen as surrogates. To ensure the initial values determined for the constraint potentials, by this approach, are correct, the assumption of small initial partial pressures for the minor species must be checked. If the assumption is not satisfied, an alternative selection for the major species may solve the problem.

3 Selection of RCCE Constraints

The accurate selection of RCCE constraints is crucial in the successful application of the RCCE method to represent reacting systems. RCCE constraints, in general, must satisfy the following requirements: (1) They must be linearly independent combinations of the chemical species concentrations; (2) they must be able to keep the reacting system in the specified initial state; (3) they must restrain the system from producing products from reactants or intermediates by global reactions; and (4) they must determine other properties such as entropy and energy for the acceptable degree of approximation.

3.1 Constraint Selection Based on Studying Oxidation Paths

In this section, a summary of the constraint selection procedure based on evaluating oxidation pathways detailed in Ref. [27] is presented. The reactions considered here are only chemical reactions because the rates of ionization and nuclear reactions are negligible in the pressure and temperature range of interest. The fixed constraints are the number of moles of elemental hydrogen, oxygen, carbon, nitrogen, etc., designated by EH, EO, EC, EN, etc.

The slowest mechanisms controlling the evolution of the reacting system are three-body recombination/dissociation reactions, under these conditions. These reactions are slow in the exothermic direction because three-body collisions are much less frequent than two-body collisions, and in the endothermic direction because they require high activation energies. Thus, they force rate-controlling constraints on the number of moles, M, of the mixture. Other slow reactions are reactions that form and break valence bonds. These reactions are considered slow in the endothermic and exothermic directions because of required high activation energies and the small radical concentrations, respectively. They impose slowly varying constraints on the number of free valence bonds, FV, of the gas.

The third group of slow reactions are the reactions that break OO bonds. These reactions impose slowly varying time-dependent constraints on the number of moles of free oxygen, FO, defined as any oxygen atom not directly connected to another oxygen atom. The formation of the major products of hydrocarbon combustion such as H_2O, CO_2, and CO requires an increase in FO.

Two additional rate-controlling constraints considered in Ref. [27] are $OHO \equiv OH + O$ and $DCO \equiv HCO + CO$. These constraints slightly improve RCCE predictions under some circumstances. The OHO constraint is enforced as a result of the combined action of the fast reaction $RH + O = OH + R$ and the relatively slow reaction $RH + OH \rightleftharpoons H_2O + R$. The DCO constraint is a result of the slow spin-forbidden reaction $CO + HO_2 \rightleftharpoons CO_2 + OH$ and the fast reaction $HCO + O_2 \rightleftharpoons CO + HO_2$.

For reacting systems made of the three elements H, O, and C, the eight constraints consisting of elemental constraints, total number of moles, free valence and free oxygen, OHO, and DCO are considered "universal" constraints because they are independent of the initial hydrocarbon fuel. These constraints coupled with the following equilibrium reactions are capable of determining the constrained-equilibrium composition of the eleven major hydrocarbon oxidation products H, O, OH, HO_2, H_2, O_2, H_2O, H_2O_2, HCO, CO, and CO_2.

$$\begin{aligned}
&H_2 + O = OH + H\\
&H_2 + HOO = H_2O_2 + H\\
&HCO + O_2 = CO + HO_2
\end{aligned}$$

In combustion of hydrocarbon fuels with one carbon element, to improve the agreement between DKM and RCCE, four additional constraints which depend on the specific fuel are required. These constraints are on the fuel, fuel radicals, alkylperoxides, and alcohol plus formaldehyde designated by FU, FR, APO, and ALCD, respectively. The first which is essential to hold the reacting system in its initial state is enforced by slow hydrogen abstraction reactions $FU + O_2 \equiv FR + HO_2$ as well as recombination/dissociation reactions $AB + M \equiv A + B + M$.

The second constraint is required to restrain the equilibration of forbidden exothermic global reactions such as $CH_3 + 2O_2 + 2H_2O = CO_2 + 2H_2O_2 + H_2 + H$. The third APO $\equiv CH_3OOH + CH_3OO + CH_2OOH$ is imposed by combined effect of rate-limiting reactions which convert APO to hydroperoxides and fast reactions which equilibrate the chemical species of APO. The last constraint ALCD $\equiv$ $CH_3OH + CH_3O + CH_2OH + CH_2O$ is imposed by slow reactions which form or remove ALCD as well as fast reactions which equilibrate the species of ALCD.

3.2 *Constraint Derivation Based on Degree of Disequilibrium of Chemical Reactions*

In this section, a summary of the constraint selection procedure which is focused on the behavior of the degrees of disequilibrium of the individual chemical reactions introduced in Refs. [1, 26] is presented. This methodology consists of a straightforward algebraic analysis of simulation results of the underlying DKM. A variation of this methodology computes an approximate singular value decomposition of the actual degrees of disequilibrium (ASVDADD) [1] to find a low-dimensional subspace in the DoD space approximating the actual DoD traces with a given accuracy. In the following, a brief summary of the ASVDADD algorithm is given.

1—In the matrix of stoichiometric coefficients $\nu_{\beta\gamma}$ ($\beta = 1, \ldots, N_s, \gamma = 1, \ldots, N_r$), identify a set of r linearly independent $\boldsymbol{\nu}_{\gamma_k}$ ($k = 1, \ldots, r$) columns. The rank of the matrix of stoichiometric coefficients equals the difference between the number of species and the number of chemical elements because the DKM must contain the maximum number of linearly independent reactions. This implies that the maximum number of derived constraints is also equal to this rank. This ensures that when thermodynamic equilibrium is reached, mole densities of chemical elements are the only constraints remained.

2—Using these reactions as the basis vectors of the $\boldsymbol{\nu}$ ($N_s \times N_r$) matrix, we derive $\hat{\mathring{\boldsymbol{\mu}}}_p$, the projection of non-dimensionalized vector of chemical potentials $\mathring{\boldsymbol{\mu}}$ into the column space of $\boldsymbol{\nu}$. The DoD of reaction γ is related to chemical potentials and the matrix of stoichiometric coefficients as follows

$$\phi_\gamma = -\frac{1}{R_u T} \sum_{\beta=1}^{N_s} \boldsymbol{\nu}_{\beta\gamma} \hat{\mathring{\boldsymbol{\mu}}}_\beta \qquad (\gamma = 1, 2, \ldots, N_r) \tag{43}$$

Equation (43) indicates that the DoD of any chemical reaction γ is a linear combination of the stoichiometric coefficients spanned by the γth column of the stoichiometric matrix $\boldsymbol{\nu}$, with $-\frac{\mathring{\mu}_\beta}{R_u T}$ as the coefficients of the linear combination. This means that DoDs of reactions are dependent if the corresponding reactions are linearly dependent themselves. Using the definition of entropic chemical potential, Eq. (43) can be rewritten as

$$\phi_\gamma = \sum_{\beta=1}^{N_s} \boldsymbol{\nu}_{\beta\gamma} \hat{\mathring{\boldsymbol{\mu}}}_\beta = \langle \hat{\mathring{\boldsymbol{\mu}}} \mid \boldsymbol{\nu}_\gamma \rangle \tag{44}$$

We designate the r-dimensional column space of matrix $\boldsymbol{\nu}$ and the N_{el}-dimensional left null space of $\boldsymbol{\nu}$ which is extended by the elemental constraints by span($\boldsymbol{\nu}_\gamma$) and coker($\boldsymbol{\nu}$), respectively. N_{el} denotes the number of chemical elements. These two spaces are orthogonal components of the N_s-dimensional real vector space and are often referred to as the "reactive subspace" and the "inhert subspace," respectively. We can decompose the vector $\hat{\mathring{\boldsymbol{\mu}}}$ by using these subspaces

$$\hat{\mathring{\boldsymbol{\mu}}} = \hat{\mathring{\boldsymbol{\mu}}}_p + \hat{\mathring{\boldsymbol{\mu}}}_\perp \tag{45}$$

where for shorthand, we introduced

$$\hat{\mathring{\boldsymbol{\mu}}}_p = \hat{\mathring{\boldsymbol{\mu}}}_{span(\boldsymbol{\nu}_\gamma)} \qquad \hat{\mathring{\boldsymbol{\mu}}}_\perp = \hat{\mathring{\boldsymbol{\mu}}}_{coker(\boldsymbol{\nu})} \tag{46}$$

We call $\hat{\mathring{\boldsymbol{\mu}}}_p$ the overall degree of disequilibrium (ODoD) vector. We rewrite Eq. (44) as following because the $\hat{\mathring{\boldsymbol{\mu}}}_\perp$ component of $\hat{\mathring{\boldsymbol{\mu}}}$ is normal to the set of r linearly independent columns of the stoichiometric matrix $\boldsymbol{\nu}_\gamma$,

$$\phi_\gamma = \langle \hat{\mathring{\boldsymbol{\mu}}}_p \mid \boldsymbol{\nu}_\gamma \rangle \tag{47}$$

Assuming $\hat{\mathring{\boldsymbol{\mu}}}_p$ is a linear combination of the γth columns, we can write

$$\hat{\mathring{\boldsymbol{\mu}}}_p = \sum_{k=1}^{r} \boldsymbol{\nu}_{\gamma_k} \alpha_k \tag{48}$$

Using Eq. (47) α_ks can be obtained using

$$\alpha_k = \sum_{k'=1}^{r} \phi_{k'} W_{k'k} \tag{49}$$

where $\boldsymbol{W} = \boldsymbol{M}^{-1}$ and in return $\boldsymbol{M}$ is a $r \times r$ matrix whose components are calculated using $M_{kk'} = \langle \boldsymbol{\nu}_{\gamma_k} \mid \boldsymbol{\nu}_{\gamma_{k'}} \rangle$. Matrix $\boldsymbol{M}$ is not singular because of the linear independence of the basis vectors $\boldsymbol{\nu}_{\gamma_k}$. We rewrite Eq. (48) using Eq. (49)

$$\hat{\boldsymbol{\mu}}_p = \sum_{k=1}^{r} \phi_k \sum_{k'=1}^{r} W_{kk'} \boldsymbol{\nu}_{\gamma k'} \tag{50}$$

where ϕ_k is the DoD of the kth linearly independent stoichiometric vector $\boldsymbol{\nu}_{\gamma_k}$. This gives us the ODoD vector.

3—Use DKM-derived values of $\hat{\boldsymbol{\mu}}_p$ at m discrete points and store the data in the $N_s \times m$ matrix $\hat{\boldsymbol{\mu}}_p(z_m)$, where z stands for either x, t or $(\boldsymbol{x}, t)$, depending on the problem.

4—Determine the Singular Value Decomposition (SVD) of $\hat{\boldsymbol{\mu}}_p(z_m)$:

$$\hat{\boldsymbol{\mu}}_p(z_m) = \boldsymbol{U}\boldsymbol{\Sigma}\boldsymbol{V}^T \tag{51}$$

$\boldsymbol{U}$ is a $N_s \times N_s$ rank r orthonormal matrix, $\boldsymbol{V}$ is a $m \times m$ rank r orthonormal matrix, and $\boldsymbol{\Sigma}$ is a $N_s \times m$ diagonal matrix having square roots of the r $\hat{\boldsymbol{\mu}}_p\hat{\boldsymbol{\mu}}_p^T$ and $\hat{\boldsymbol{\mu}}_p^T\hat{\boldsymbol{\mu}}_p$ positive eigenvalues σ_r placed on its first r diagonal elements. Note that only the first r columns of $\boldsymbol{U}$ matrix are in the column space of $\hat{\boldsymbol{\mu}}_p(z_m)$, and only the first r rows of $\boldsymbol{V}^T$ matrix (or the first r columns of $\boldsymbol{V}$ matrix) are in the row space of $\hat{\boldsymbol{\mu}}_p(z_m)$. This means Eq. (51) can be replaced by:

$$\hat{\boldsymbol{\mu}}_p(z_m) = \boldsymbol{U}_r\boldsymbol{\Sigma}_r\boldsymbol{V}_r^T \tag{52}$$

The subscript "r" means reduced dimension. $\boldsymbol{U}_r$, $\boldsymbol{V}_r^T$, and $\boldsymbol{\Sigma}_r$ are $N_s \times r$, $r \times m$, and diagonal $r \times r$ matrices, respectively. Now let $\boldsymbol{A}_p^T = \boldsymbol{U}_r$ and $\boldsymbol{\Lambda}_p = \boldsymbol{\Sigma}_r\boldsymbol{V}_r^T$, then Eq. (52) becomes

$$\hat{\boldsymbol{\mu}}_{p_\beta}(z_m) = \sum_{k=1}^{r} A_{p_{k\beta}} \Lambda_{p_k}(z_m) \tag{53}$$

Comparing this with Eq. (16) indicates that $\boldsymbol{A}_p$, which is a $r \times N_s$ matrix, defines the coefficient matrix for computing constraint vector C_{p_k} $(k = 1, \ldots, r)$. In *decreasing order of importance*, the rows of $\boldsymbol{C}_p$ gives the entire spectrum of RCCE constraints.

$$C_{p_k} = \sum_{\beta=1}^{N_s} A_{p_{k\beta}} \frac{Y_\beta}{M_\beta} \tag{54}$$

5—Add elemental constraints to the derived set of constraints Eq. (54) to get:

$$\boldsymbol{A} = \begin{bmatrix} \boldsymbol{A}_p \\ \boldsymbol{A}_{el} \end{bmatrix} \qquad C_\alpha = \sum_{\beta=1}^{N_s} A_{\alpha\beta} \frac{Y_\beta}{M_\beta} \qquad (\alpha = 1, \ldots, N_c) \tag{55}$$

$\boldsymbol{A}_{el}$ is the $N_{el} \times N_s$ constraint coefficient matrix of chemical elements. Note that at this point, the N_s species have been replaced by $N_c = r + N_{el} = N_s$ constraints that are the linear combination of these species, which means simulating the system using these N_s constraints will result in the exact DKM solution. The number of constraints can be decreased by abandoning the constraint C_{p_k} for $k \geq n$ if $\sigma_{n-1} \gg \sigma_n$, according to this model. The effectiveness of the procedure is exhibited in Ref. [1] for several simulations of increasing difficulty including: (1) a one-dimensional expansion of the products of the hydrogen/oxygen combustion in a supersonic nozzle (8 species, 24 reactions); (2) methane/oxygen combustion in the same supersonic nozzle (29 species, 133 reactions) and (3) ignition delay simulation of homogeneous methane and oxygen mixture.

4 Results

The RCCE method has been applied to model H_2/O_2 and CH_4/O_2 combustion at constant energy and volume by solving Eqs. (56a) and (56b) together with equation of state. These equations are obtained from Eqs. (33a) and (35) by making no mixing and constant energy and volume assumptions:

$$Q_{c_\alpha}\frac{1}{T}\frac{dT}{dt} + \sum_{\theta=1}^{N_c} R_{c_{\alpha\theta}}\frac{d\Lambda_\theta}{dt} = \sum_{\beta=1}^{N_s}\frac{A_{\alpha\beta}}{M_\beta}S_\beta(t) \quad (\alpha = 1, \ldots, N_c) \tag{56a}$$

$$Q_e\frac{1}{T}\frac{dT}{dt} + \sum_{\theta=1}^{N_c} R_{e_\theta}\frac{d\Lambda_\theta}{dt} = 0 \tag{56b}$$

where the coefficients are repeated for convenience

$$\begin{aligned} Q_{c_\alpha} &= \sum_{\beta=1}^{N_s} A_{\alpha\beta}\frac{Y_\beta}{M_\beta}\frac{\mathring{e}_\beta}{R_u T} \; ; \quad R_{c_{\alpha\theta}} = -\sum_{\beta=1}^{N_s} A_{\alpha\beta}A_{\theta\beta}\frac{Y_\beta}{M_\beta} \\ Q_e &= c_v T + \sum_{\beta=1}^{N_s}\frac{\mathring{e}_\beta^2}{R_u T}\frac{Y_\beta}{M_\beta} \; ; \quad R_{e_\theta} = -\sum_{\beta=1}^{N_s} A_{\theta\beta}\mathring{e}_\beta\frac{Y_\beta}{M_\beta} \end{aligned} \tag{57}$$

4.1 H_2/O_2 Combustion System

In this section, we have applied RCCE using the constraints identified by Janbozorgi and Metghalchi [26] to simulate hydrogen/oxygen combustion consisting of 8 species

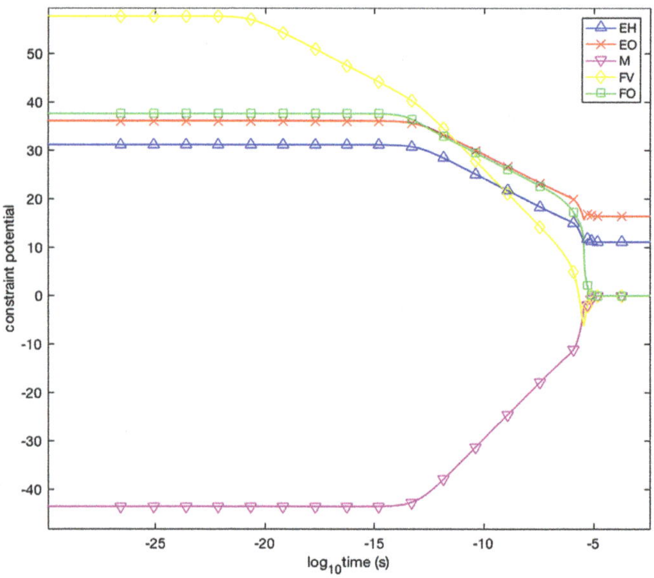

Fig. 1 Constraint potential profiles of stoichiometric H_2 and O_2 combustion at 1500 K and 1 atm

and 24 reactions. The constraints are listed in Table 1. Figure 1 shows the change of (original) constraint potentials versus time at initial temperature of 1500 K, pressure of 1 atm and equivalence ratio equal to 1. As shown in the figure, the constraint potentials of elements remain nonzero at equilibrium while the constraint potentials of the rest of the constraints vanish. This shows RCCE leads to the right equilibrium state. Figure 2 shows the temperature profile of H/O reaction system with different initial temperatures. All simulations are carried out at initial pressure of 1 atm and equivalence ratio equal to 1. It is shown that RCCE results are in a good agreement with those of DKM with a wide range of initial temperatures. Figure 3 shows the temperature profile of H/O reaction system with different initial pressures. All simulations are carried out at initial temperature of 1500 K and equivalence ratio equal to 1. As it is shown in Fig. 3, RCCE slightly underpredicts the ignition delay time; however, RCCE predictions of ignition delay times for these mixtures are within 1–5% of those obtained from DKM. Figures 4 and 5 show concentration predictions in the stoichiometric hydrogen and oxygen combustion at initial temperature of 1500 K and pressure of 1 atm. It is shown that RCCE predictions of major species concentrations are in good agreement with those of DKM.

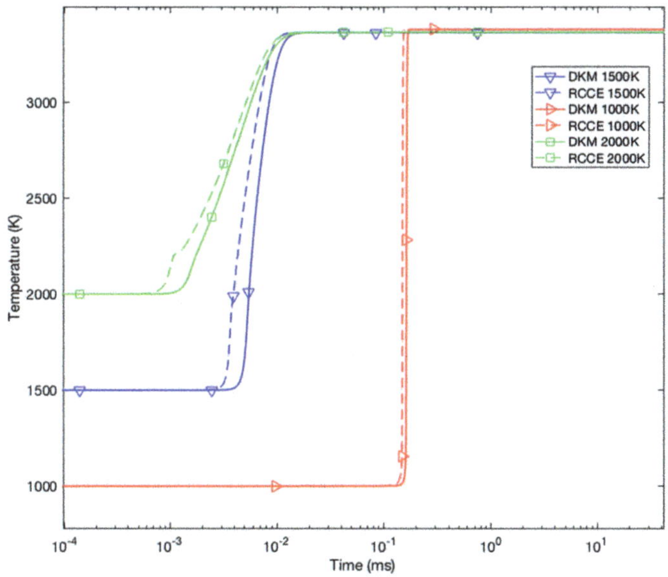

Fig. 2 RCCE and DKM temperature predictions of stoichiometric H_2 and O_2 combustion at initial pressure of 1 atm, and equivalence ratio equal to 1. Solid and dashed lines represent DKM and RCCE results, respectively

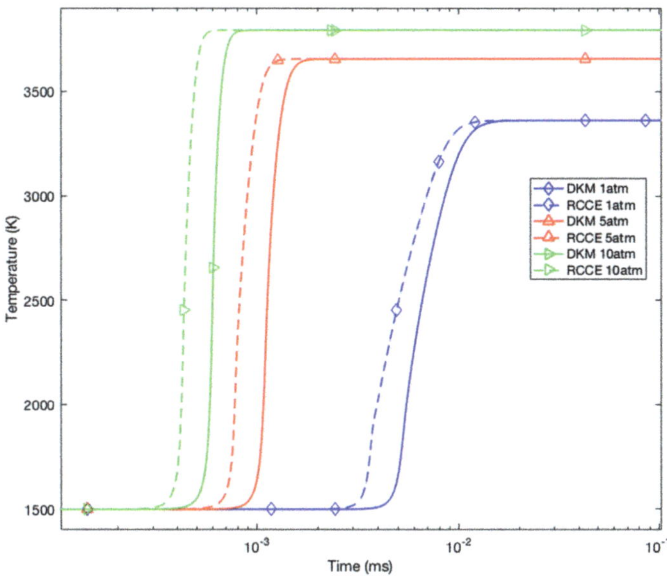

Fig. 3 RCCE and DKM temperature predictions of stoichiometric H_2 and O_2 combustion at initial temperature of 1500 K and equivalence ratio equal to 1. Solid and dashed lines represent DKM and RCCE results, respectively

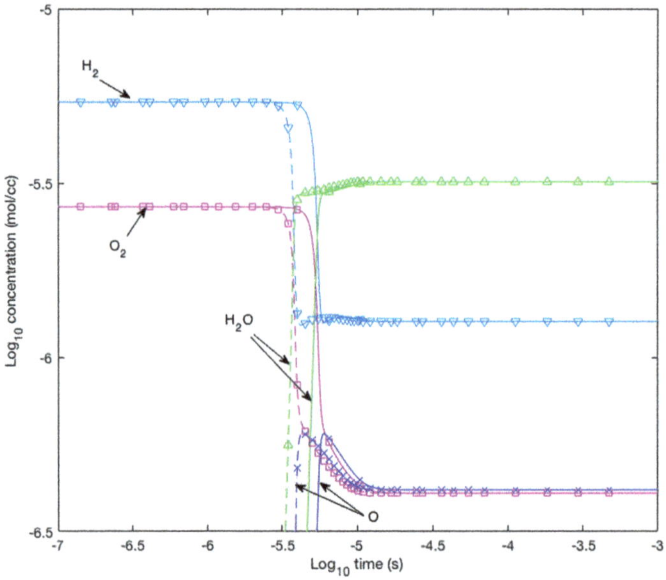

Fig. 4 RCCE and DKM concentration predictions of stoichiometric H_2 and O_2 combustion at 1500 K and 1 atm. Solid and dashed lines represent DKM and RCCE results, respectively

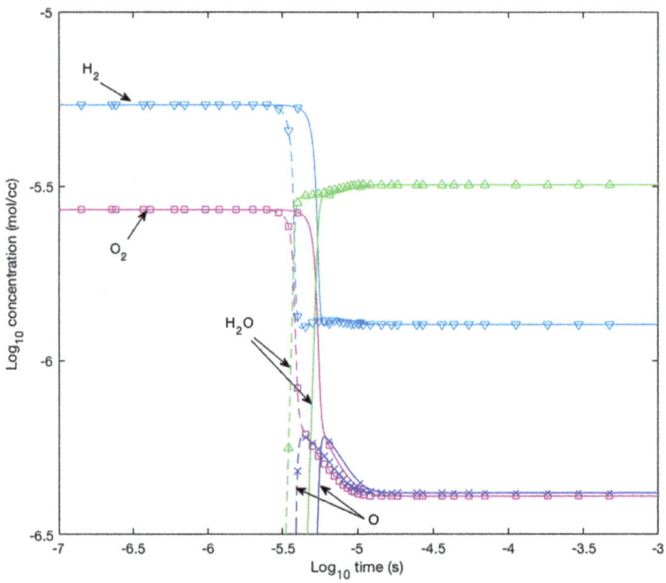

Fig. 5 RCCE and DKM concentration predictions of stoichiometric H_2 and O_2 combustion at 1500 K and 1 atm. Solid and dashed lines represent DKM and RCCE results, respectively

Table 1 Definition of constraints for hydrogen combustion in oxygen

Constraints	Definition
H	Elemental hydrogen
O	Elemental oxygen
M	Total number of moles
FV	Moles of free valance (any unpaired valence electron)
FO	Moles of free oxygen (any oxygen not directly attached to another oxygen)

Table 2 Definition of constraints for methane combustion in oxygen

Constraints	Definition
H	Elemental hydrogen
O	Elemental oxygen
C	Elemental carbon
M	Total number of moles
FV	Moles of free valance (any unpaired valence electron)
FO	Moles of free oxygen (any oxygen not directly attached to another oxygen)
OHO	Moles of water radicals OH + O
DCO	Moles of HCO + CO
FU	Moles of fuel molecule (CH_4 in the case of methane)
FR	Moles of fuel radical(s) (CH_3 in the case of methane)
APO	Moles of AlkyPerOxydes $CH_3OOH + CH_3OO + CH_2OOH$
ALCD	Moles of Alcohols+Aldehydes ($CH_3OH + CH_3O + CH_2OH + CH_2O$)

4.2 CH_4/O_2 Combustion System

In this section, the RCCE method is applied to predict the combustion of CH_4/O_2 reacting system using the constraints identified by Janbozorgi et al. [27]. A mechanism which contains 29 species and 133 reactions [27] is employed for the simulations. The CH_4/O_2 reacting system is represented by the 12 constraints listed in Table 2. Figure 6 shows the (original) constraint potential variation versus time at initial temperature of 900 K, pressure of 100 atm and equivalence ratio of 1. Consistent with the previous case, all constraint potentials vanish at equilibrium except for those of elements (EH, EO, and EC), which indicates that RCCE predicts equilibrium accurately. It can be seen that for different initial temperatures and initial pressures (Figs. 7 and 8), RCCE shows a very good agreement with DKM. For lean and rich CH_4/O_2 mixtures, the RCCE predictions are within 1–5% of those obtained from DKM (Fig. 9). Figures 10 and 11 illustrate the change of concentrations for major species for stoichiometric CH_4 combustion at initial temperature and pressure of 900 K and 10 atm, respectively. It is clearly shown that RCCE method shows a

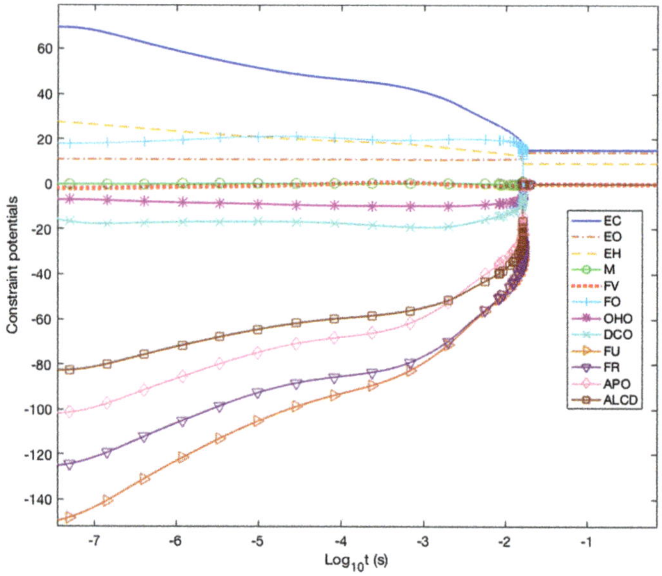

Fig. 6 RCCE constraint potentials predictions of CH_4 and O_2 stoichiometric combustion at initial temperature of 900 K and pressure of 100 atm

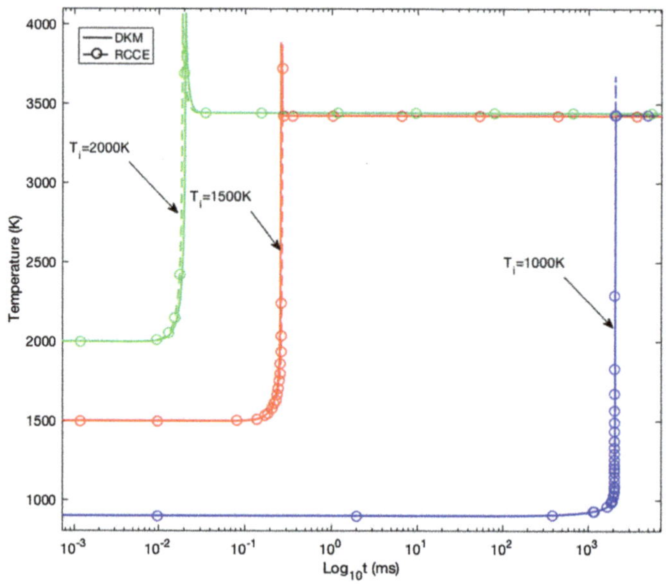

Fig. 7 Comparison of temperature predictions in RCCE and DKM simulations of CH_4 and O_2 stoichiometric combustion with initial temperatures ranging from 900 to 2000 K. All cases have initial pressure of 1 atm and equivalence ratio of 1

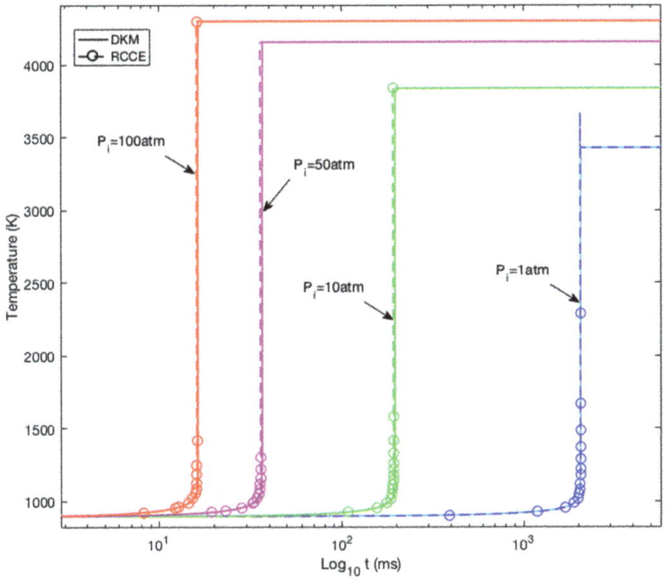

Fig. 8 Comparison of temperature predictions in RCCE and DKM simulations of CH_4 and O_2 stoichiometric combustion with a wide range of initial pressures. Initial temperature is 900 K and equivalence ratio is equal to 1 for all cases

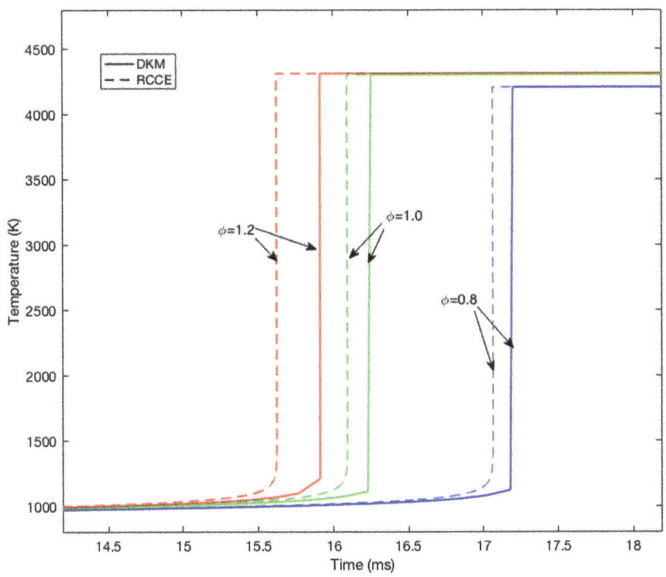

Fig. 9 Comparison of temperature predictions in RCCE and DKM simulations of CH_4 and O_2 stoichiometric combustion with different equivalence ratios. For all simulations, initial temperature is 900 K and initial pressure is 100 atm

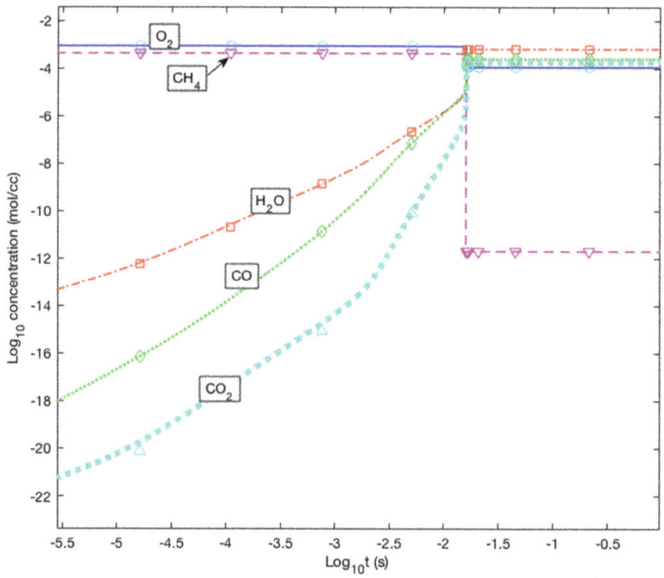

Fig. 10 Comparison of major species concentration in RCCE and DKM predictions of CH_4 and O_2 stoichiometric combustion at 900 K and 1 atm. Lines and marks denote RCCE and DKM results, respectively

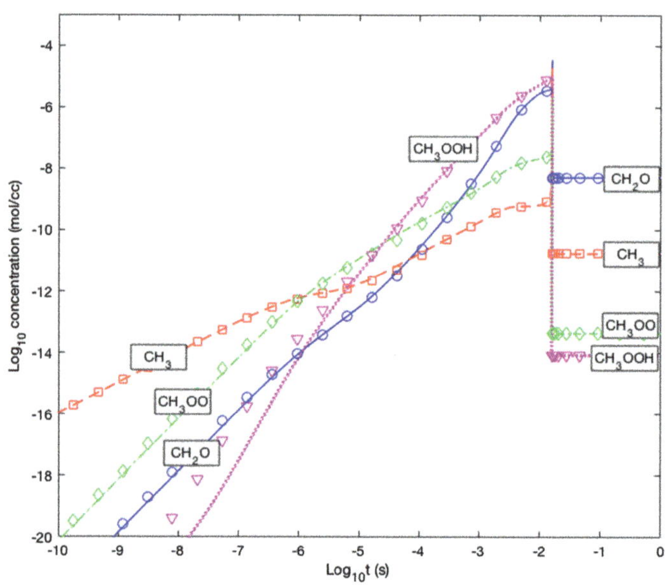

Fig. 11 Comparison of major species concentration in RCCE and DKM predictions of CH_4 and O_2 stoichiometric combustion at 900 K and 1 atm. Lines and marks denote RCCE and DKM results, respectively

perfect agreement with DKM method. As shown in these figures, RCCE method provides accurate results with much smaller number of differential equations.

5 Summary

In this chapter, a summary of Rate-Controlled Constrained-Equilibrium (RCCE) principles is first presented. Then, rate equations in constraint and constraint potential representations of RCCE are obtained. These two formulations are mathematically equivalent but they follow different numerical procedures. Next, two approaches based on calculating chemical potentials of major species and constraint to constraint potential conversion for initializing constraint potentials in the constraint potential formulation are presented. Finally, constraint selection based on kinetics knowledge and degree of disequilibrium (DoD) analysis of chemical reactions in an underlying DKM calculation is briefly discussed. RCCE calculations of hydrogen and methane combustion in oxygen over a wide range of initial temperatures, pressures, and equivalence ratios have been made in a constant energy/volume reactor. RCCE calculations have been performed using 5 and 12 constraints, and excellent agreement with Detailed Kinetics Model (DKM) calculations using 8 and 29 species has been obtained in hydrogen and methane cases, respectively.

Acknowledgements This paper was made possible by NPRP award 7-252-2-113 from the Qatar National Research Fund (a member of The Qatar Foundation). The statements made herein are solely the responsibility of the authors. The authors would like to express their appreciation to Dr. Vreg Yousefian for very helpful discussions regarding constraint potentials initialization and RCCE constraint selection.

References

1. G.P. Beretta, M. Janbozorgi, H. Metghalchi, Degree of disequilibrium analysis for automatic selection of kinetic constraints in the Rate-Controlled Constrained-Equilibrium method. Combust. Flame **168**, 342–364 (2016)
2. G.P. Beretta, J.C. Keck, M. Janbozorgi, H. Metghalchi, The Rate-Controlled Constrained-Equilibrium approach to far-from-local-equilibrium thermodynamics. Entropy **14**, 92–130 (2012)
3. P. Bishnu, D. Hamiroune, M. Metghalchi, Development of constrained equilibrium codes and their applications in nonequilibrium thermodynamics. J. Energy Res. Technol. **123**(3), 214–220 (2001)
4. P. Bishnu, D. Hamiroune, M. Metghalchi, J.C. Keck, Constrained-equilibrium calculations for chemical systems subject to generalized linear constraints using the NASA and STANJAN equilibrium programs. Combust. Theor. Model. **1**(3), 295–312 (1997)
5. M. Bodenstein, S.C. Lind, Geschwingigkeitder bildung des bromwasserstoffs aus seinen elementen. Z. Phys. Chem **57**, 168 (1906)
6. R. Borghi, Turbulent combustion modeling. Prog. Energy Combust. Sci. **14**, 245–292 (1988)

7. A.K. Chatzopoulos, S. Rigopoulos, A chemistry tabulation approach via Rate-Controlled Constrained Equilibrium (RCCE) and Artificial Neural Networks (ANNs), with application to turbulent non-premixed $CH_4/H_2/N_2$ flames. Proc. Combust. Inst. **34**(1), 1465–1473 (2013)
8. E. Chiavazzo, C. Gear, C. Dsilva, N. Rabin, I. Kevrekidis, Reduced models in chemical kinetics via nonlinear data-mining. Processes **2**, 112–140 (2014)
9. E. Chiavazzo, I. Karlin, Adaptive simplification of complex multiscale systems. Phys. Rev. E **83**, 036,706 (2011)
10. A. Doostan, A. Validi, G. Iaccarino, Non-intrusive low-rank separated approximation of high-dimensional stochastic models. Comput. Methods Appl. Mech. Engrg. **263**, 42–55 (2013)
11. C. Dopazo, E.E. O'Brien, An approach to the autoignition of a turbulent mixture. Acta Astronaut. **1**(9–10), 1239–1266 (1974)
12. S. Fraser, The steady state and equilibrium approximations: a geometrical picture. J. Chem. Phys. **88**, 4732–4738 (1988)
13. A. Gorban, I. Karlin, Method of invariant manifold for chemical kinetics. Chem. Eng. Sci. **58**, 4751–4768 (2003)
14. S. Gordon, B.J. McBride, Computer program for calculation of complex chemical equilibrium compositions and applications. NASA Reference Publication 1311, National Aeronautics and Space Administration (1994)
15. F. Hadi, M. Janbozorgi, M.R.H. Sheikhi, H. Metghalchi, A study of interactions between mixing and chemical reaction using the Rate-Controlled Constrained-Equilibrium method. J. Non-Equilib. Thermody. **41**(4), 257–278 (2016)
16. F. Hadi, M.R.H. Sheikhi, A comparison of constraint and constraint potential forms of the Rate-Controlled Constrained-Equilibrium method. J. Energy Res. Technol. **138**(2), 022,202 (2015)
17. F. Hadi, V. Yousefian, M.R.H. Sheikhi, H. Metghalchi, A study of the RCCE constraint potential formulation incorporating a constraint selection algorithm, in *Proceedings of the 2016 ESSCI Spring Meeting*, The Combustion Institute, Princeton, NJ (2016)
18. F. Hadi, V. Yousefian, M.R.H. Sheikhi, H. Metghalchi, Time scale analysis for rate-controlled constrained-equilibrium constraint selection, in *Proceeding of the 10th U.S. National Combustion Meeting of the Combustion Institute*, The Combustion Institute, College Park, Maryland (2017)
19. D. Hamiroune, P. Bishnu, M. Metghalchi, J.C. Keck, Controlled constrained equilibrium method using constraint potentials. Combust. Theory Model. **2**(1), 81–94 (1998)
20. V. Hiremath, S.R. Lantz, H. Wang, S.B. Pope, Computationally-efficient and scalable parallel implementation of chemistry in simulations of turbulent combustion. Combust. Flame **159**(10), 3096–3109 (2012)
21. V. Hiremath, S.R. Lantz, H. Wang, S.B. Pope, Large-scale parallel simulations of turbulent combustion using combined dimension reduction and tabulation of chemistry. Proc. Combust. Inst. **34**(1), 205–215 (2013)
22. V. Hiremath, S.B. Pope, A study of the rate-controlled constrained-equilibrium dimension reduction method and its different implementations. Combust. Theory Model. **17**(2), 260–293 (2013)
23. V. Hiremath, Z. Ren, S.B. Pope, A greedy algorithm for species selection in dimension reduction of combustion chemistry. Combust. Theory Model. **14**(5), 619–652 (2010)
24. V. Hiremath, Z. Ren, S.B. Pope, Combined dimension reduction and tabulation strategy using ISAT-RCCE-GALI for the efficient implementation of combustion chemistry. Combust. Flame **158**(11), 2113–2127 (2011)
25. M. Janbozorgi, H. Metghalchi, Rate-Controlled Constrained-Equilibrium theory applied to the expansion of combustion products in the power stroke of an internal combustion engine. Int. J. Thermodyn. **12**(1), 44–50 (2009)
26. M. Janbozorgi, H. Metghalchi, Rate-controlled constrained-equilibrium modeling of H/O reacting nozzle flow. J. Propuls. Power **28**(4), 677–684 (2012)
27. M. Janbozorgi, S. Ugarte, H. Metghalchi, J. Keck, Combustion modelling of mono-carbon fuels using the rate-controlled constrained-equilibrium method. Combust. Flame **156**(10), 187–1885 (2009)

28. W.P. Jones, S. Rigopolous, Rate controlled constrained equilibrium: formulation and application of nonpremixed laminar flames. Combust. Flame **142**, 223–234 (2005)
29. W.P. Jones, S. Rigopolous, Reduction of comprehensive chemistry via constraint potentials. Proc. Combust. Inst. **30**(1), 1325–1331 (2005)
30. W.P. Jones, S. Rigopolous, Reduced chemistry for hydrogen and methanol premixed flames via RCCE. Combust. Theory Model. **11**, 755–780 (2007)
31. J.C. Keck, Rate-controlled constrained-equilibrium theory of chemical reactions in complex systems. Prog. Energy Combust. Sci. **16**(2), 125–154 (1990)
32. J.C. Keck, D. Gillespie, Rate-controlled partial-equilibrium method for treating reacting gas mixtures. Combust. Flame **17**(2), 237–241 (1971)
33. J. Kim, S.B. Pope, Effects of combined dimension reduction and tabulation on the simulations of a turbulent premixed flame using a large-eddy simulation/probability density function method. Combust. Theory Model. **18**(3), 388–413 (2014)
34. S.H. Lam, D.A. Goussis, The CSP method for simplifying kinetics. Int. J. Chem. Kinet. **26**, 461–486 (1994)
35. R. Law, M. Metghalchi, J.C. Keck, Rate-controlled constrained equilibrium calculations of ignition delay times in hydrogen-oxygen mixtures. Proc. Combust. Inst. **22**, 1705–1713 (1988)
36. D. Lebiedz, Computing minimal entropy production trajectories: an approach to model reduction in chemical kinetics. J. Chem. Phys. **120**, 6890–6897 (2004)
37. T. Løvås, S. Navarro-Martinez, S. Rigopoulos, On adaptively reduced chemistry in large eddy simulations. Proc. Combust. Inst. **33**(1), 133–1346 (2011)
38. L. Lu, S.B. Pope, An improved algorithm for in situ adaptive tabulation. J. Comput. Phys. **228**, 361–386 (2009)
39. T. Lu, Y. Ju, C. Law, Complex CSP for chemistry reduction and analysis. Combust. Flame **126**, 1445–1455 (2001)
40. U. Maas, S.B. Pope, Simplifying chemical kinetics: intrinsic low-dimensional manifolds in composition space. Combust. Flame **88**(3–4), 239–264 (1992)
41. S. Navarro-Martinez, S. Rigopoulos, Large eddy simulation of a turbulent lifted flame using conditional moment closure and rate-controlled constrained equilibrium. Flow Turbul. Combust. **87**(2), 407–423 (2011)
42. G. Nicolas, M. Janbozorgi, H. Metghalchi, Constrained-equilibrium modeling of methane oxidation in air. J. Energy Res. Technol. **136**(3), 032,205-1–032,205-7 (2014)
43. G. Nicolas, H. Metghalchi, Comparison between RCCE and shock tube ignition delay times at low temperatures. J. Energy Res. Technol. **137**(6), 062,203-1–062,203-4 (2015)
44. G. Nicolas, H. Metghalchi, Development of the rate-controlled constrained-equilibrium method for modeling of ethanol combustion. J. Energy Res. Technol. **138**, 022,205–1 (2016)
45. S. Pope, U. Mass, Simplifying chemical kinetics: trajectory-generated low-dimensional manifolds. Technical Report FDA 93-11, Cornell University, Ithaca, NY (1993)
46. S.B. Pope, Mean field equations in PDF particle methods for turbulent reactive flows. Technical Report FDA 97-06, Cornell University, Ithaca, NY (1997)
47. S.B. Pope, CEQ: A Fortran library to compute equilibrium compositions using Gibbs function continuation. http://eccentric.mae.cornell.edu/~pope/CEQ (2003)
48. S.B. Pope, The computation of constrained and unconstrained equilibrium compositions of ideal gas mixtures using Gibbs function contimuation. Cornell University Report FDA 03–02, Cornell University (2003)
49. S.B. Pope, Gibbs function continuation for the stable computation of chemical equilibrium. Combust. Flame **139**(3), 222–226 (2004)
50. S.B. Pope, Small scales, many species and the manifold challenges of turbulent combustion. Proc. Combust. Inst. **34**(1), 1–31 (2013)
51. J. Rabinovitch, G. Blanquart, Rate-controlled constrained equilibrium for nozzle and shock flows. J. Propuls. Power **33**, 776–792 (2017)
52. V. Raman, H. Pitsch, Large-eddy simulation of a bluff-body-stabilized non-premixed flame using a recursive filter-refinement procedure. Combust. Flame **142**, 329–347 (2005)

53. M. Rein, The partial-equilibrium approximation in reacting flows. Phys. Fluids A **4**, 873–886 (1992)
54. Z. Ren, S. Pope, The geometry of reaction trajectories and attracting manifolds in composition space. Combust. Theory Model. **10**, 361–388 (2006)
55. Z. Ren, S.B. Pope, A. Vladimirsky, J.M.J. Guckenheimer, The invariant constrained equilibrium edge preimage curve method for the dimension reduction of chemical kinetics. J. Chem. Phys. **124**, 114,111 (2006)
56. Z. Ren, S.B. Pope, A. Vladimirsky, J.M.J. Guckenheimer, Application of the ICE-PIC method for the dimension reduction of chemical kinetics coupled with transport. Proc. Combust. Inst. **31**, 473–481 (2007)
57. Z. Rena, Z. Lu, Y. Gao, T. Lu, L. Hou, A kinetics-based method for constraint selection in rate-controlled constrained equilibrium. Combust. Theory Model. **21**, 159–182 (2017)
58. W.C. Reynolds, The element potential method for chemical equilibrium analysis: implementation in the interactive program STANJAN. Stanford University Report ME 270 HO 7, Stanford University (1986)
59. S. Rigopoulos, T. Løvås, A LOI-RCCE methodology for reducing chemical kinetics, with application to laminar premixed flames. Proc. Combust. Inst. **32**, 569–576 (2009)
60. M.D. Smooke (ed.), *Reduced Kinetic Mechanisms and Asymptotic Approximations for Methane-Air Flames*, vol. 384. Lecture Notes in Physics (Springer, Berlin, Germany, 1991)
61. Q. Tang, S.B. Pope, Implementation of combustion chemistry by in situ adaptive tabulation of rate-controlled constrained equilibrium manifolds. Proc. Combust. Inst. **29**(1), 1411–1417 (2002)
62. Q. Tang, S.B. Pope, A more accurate projection in the rate controlled constrained equilibrium method for dimension reduction of combustion chemistry. Combust. Theory Model. **8**, 255–279 (2004)
63. S. Ugarte, S. Gao, H. Metghalchi, Application of maximum entropy principle in the analysis of a non-equilibrium chemically reacting mixture. Int. J. Thermodyn. **8**(1), 43–53 (2005)
64. M. Valorani, S. Paolucci, The G-Scheme: a frame work for multi-scale adaptive model reduction. J. Comput. Phys **228**, 4665–4701 (2009)
65. V. Yousefian, A rate controlled constrained equilibrium thermochemistry algorithm for complex reacting systems. Combust. Flame **115**, 66–80 (1998)

Direct Numerical Simulation Study of Lean Hydrogen/Air Premixed Combustion

Rohit Saini, Ashoke De and S. Gokulakrishnan

Abstract Turbulence-chemistry interaction is known to play a vital role in changing the characteristics of a flame surface. It changes evolution, propagation, annihilation, local extinction characteristics of the flame front. This study seeks to understand how turbulence interaction affects flame surface geometry and propagation of turbulent premixed H_2/Air flames in a three-dimensional configuration. 3D Direct Numerical Simulation (DNS) study of premixed turbulent H_2/Air flames has been carried out using an inflow–outflow configuration at moderate Reynolds number (Re) with a fairly detailed chemistry. The simulations are conducted at different parametric conditions in conjunction with differential diffusion (non-uniform Lewis number) effects. The topology of the flame surface is interpreted in terms of its propagation and statistics. Statistics related to the flame surface area and the correlations between the curvature and the gradient of temperature are obtained from the computed fields. It is found that the displacement speed increases with the negative mean curvature, while it correlates well for high turbulent cases and scattered for low turbulent cases. It is also observed that the diffusion effects become more dominant for deciding the flame structure when the mean flow is lower (low Re case). Further, the unsteady effects of tangential strain rate, curvature on flame propagation, and heat release rate are also investigated. Later, effects of prominent species and radicals are described to correlate the production of the maximum heat release rate in the lower temperature regions.

Keywords Direct numerical simulation · Premixed flames · Curvature
Strain rate · Displacement speed · Vorticity budget

R. Saini · A. De (✉) · S. Gokulakrishnan
Department of Aerospace Engineering,
Indian Institute of Technology Kanpur, Kanpur 208016, India
e-mail: ashoke@iitk.ac.in

A. K. Runchal et al. (eds.), *Energy for Propulsion*, Green Energy and Technology,
https://doi.org/10.1007/978-981-10-7473-8_11

1 Background

Turbulent flame characteristics such as flame surface evolution, propagation, annihilation, local extinction/re-ignition are affected by the interaction of flame front with various parameters such as turbulence, molecular transport, and energy transport and strongly coupled with chemistry. Experimental studies of turbulent-chemistry interaction are extremely rare due to practical difficulties associated with conducting such an intricate study. However, a number of studies have been made on the interaction of low-intensity turbulence with flame fronts. The development of faster processors and larger storage devices has rekindled interest in DNS of turbulent reactive flows. Studies of turbulent combustion with detailed chemistry and multi-component diffusion in the three-dimensional domain using DNS have been of great interests of many researchers [1].

Typically, turbulent premixed flames are locally stretched and wrinkled due to the mutual interaction between turbulence and chemical reactions and the flame front propagation is often represented by the scalar iso-surface [2]. The change in flame surface (a measure of a flame stretch) is primarily due to the underlying turbulence of the flow, which in turn impact the reaction rate [3]. On the other hand, stretching plays a key role in flame stabilization, flame speed determination, flammability limits, flame-front instability, concentration, and temperature. Another important parameter in premixed flames is Lewis number (Le), which influence the flame structure (stretch, strain etc.) and stability of the flame front [4]. The interaction of flame front with turbulence, in turn, the flame-vortex interaction, has been numerically studied by many researchers in the past. The studies include the studies by Poinsot et al. [5], Kadowaki and Hasegaw [6], Sankaran and Menon [7, 8], Najm and Wyckoff [9], Lipatnikov and Chomiak [10, 11], Markstein [12], Clavin and Joulin [13] and Matalon and Matkowsky [14]. Further, the effects of strain rate and curvature on the local displacement speed were also investigated while varying the degree of Lewis [15] and Damkohler numbers [16].

While looking at the flame characteristics under the turbulent condition, i.e. stretch, strain, curvature, displacement speed etc., numerous studies have been carried out on hydrogen-air premixed flames in two and three-dimensional domains. For example, Baum et al. [17] studied premixed H2/O2/N2 flames in two-dimensional turbulence with detailed chemistry and checked the validity of the flamelet assumption for premixed turbulent flames. Wang et al. [18] studied swirling hydrogen-air premixed flames at different equivalence ratios and investigated the effect of equivalence ratio on preferential diffusion effects and studied the role of small-scale eddies in interactions between turbulence and the flame front. Hamlington et al. [19] studied interactions between turbulence and flames in premixed reacting flows in a three-dimensional unconfined domain and examined the effect of turbulence intensity on flame orientation and thickness of preheat and reaction zones. Chen and Im [20] investigated the stretch effects on burning velocity of turbulent H_2/Air premixed flames. They found that the burning velocity was strongly correlated with the curvature for moderate turbulent intensities, while the displacement speed (S_d), in general, was found to be negatively correlated with curvature. Also, they mentioned that the reduction in displacement speed is sometimes as high as 80% for large strain and positive curvatures. The effect of preferential diffusion in hydrogen enriched flames have been explored by several researchers, e.g. Jiang et al. [21], Dinesh et al. [22, 23], and Han et al. [24] by analyzing the flame compositional structures and predominant intermediate radicals such as H, OH, HO_2, etc., which was primarily due to highly diffusive H_2 molecules and H radicals. More recently, Urankara et al. [25] used a flame-particle tracking approach to investigate the hydrogen-air premixed flames with unity Lewis number. They have also analyzed the flame characteristics in the context of strain, curvature and displacement speed for different parametric conditions. Along the similar line, the present study attempts to capture the effect of turbulence on the flame surface by observing the changes in mean curvature, displacement speed and species profiles using 3D DNS simulations for the turbulent premixed Hydrogen-air flames which fall in the thin-reaction zone in the regime diagram. The simulations are conducted at different parametric conditions with a detailed Hydrogen-Air reaction mechanism in conjunction with differential diffusion (non-uniform Lewis number) effects. An analytical expression derived for temperature iso-surface is considered for the calculation of displacement speed. The contribution by conduction, mass diffusion, heat release rate and temperature gradient to the displacement speed are estimated. Moreover, the contribution of flow unsteadiness on the flame stretch is captured by analyzing the tangential strain rate, and finally, the correlations are established between species mass fractions and temperature for different turbulent conditions.

2 Numerical Details

Direct Numerical Simulation of initially planar premixed flame that gets wrinkled by oncoming isotropic turbulence is carried out. The present computations involve an inflow–outflow type configuration characterized by Navier-Stokes Characteristic Boundary Conditions (NSCBC) along the direction of the mean flow and periodic boundary conditions otherwise. The simulations are carried out at three different levels of turbulence intensities, which resulted in different $Re\tau$, Da, and Ka conditions. The simulation parameters that varied between cases: Kolmogorov length η, $\tau\eta$, mean flow velocity $< U >$ and rms of fluctuating flow velocity u_{rms} at the inlet face, could be found in Table 1, while the parameters, which are kept same for all the simulations, are provided in Table 2.

Before performing the actual turbulent premixed flame simulations, a one-dimensional H_2-air planar premixed flame is computed which serves as an initial condition to the 3D DNS simulations. Standalone computations for non-reacting, isothermal, homogenous isotropic turbulence (NRIHIT) with all the relevant species are performed until the statistical stationary state is reached. For a given case, the resulting velocity field of one of these computations is fed through the inlet boundary with an imposed mean velocity $< U >$, while the inlet temperature has been set at

Table 1 List of simulation parameters

	Case A	Case B	Case C	Case D	Case E
< U >(cm/s)	700	700	700	500	1000
< urms > (cm/s)	503	557	832.5	557	557
urms/S_L	2.71	3.01	4.5	3.01	3.01
Re_τ	162	178	266	178	178
Da	0.682	0.616	0.412	0.616	0.616
Ka	18.66	21.66	39.59	21.66	21.66
η (μm)	14.75	13.75	10.17	13.75	13.75

Table 2 Common parameters for all simulations

Laminar flame speed (S_L) (cm/s)	185
Flame thickness (μm)	361
Pressure (atm)	1
Inlet temperature (K)	310
Equivalence ration ϕ	0.81
Domain size (cm^3)	$1.2 \times 0.4 \times 0.4$
Grid size	$400 \times 120 \times 120 = 5.76$ million

310 K (preheating) with an inlet equivalence ratio $\phi = 0.81$. The H_2-Air chemical reaction is represented using a reduced mechanism by Li et al. [26] involving 9 species and 19 reactions. The computational domain is a rectangular box and it is shown in Fig. 1a. Various parameters, such as divergence of heat flux, heat release rate, reaction rate, curvature, and strain rate, are calculated for a given scalar iso-surface. A progress variable is defined based on the temperature and it is used to find the flame surface. Iso-surface of C = 0.5 is identified as flame surface [27].

$$c = \frac{T_{local} - T_{unburnt}}{T_{burnt} - T_{unburnt}} \tag{1}$$

Figure 1b shows the parametric conditions for the five cases on a Borghi regime diagram, indicating that all cases lie in the thin reaction zones regime.

The simulations are carried out using the open source Pencil code [28, 29], which solves the continuity, momentum, energy, and species transport equations in fully compressible formulation on a structured 3D Cartesian grid. The spatial discretization is obtained by a sixth-order explicit finite difference scheme, while a third-order Runge–Kutta scheme is invoked for time integration. The transport properties are calculated using the commonly used mixture averaged formulation. The detailed description of evaluations of these properties and other relevant parameters are provided in [28, 29].

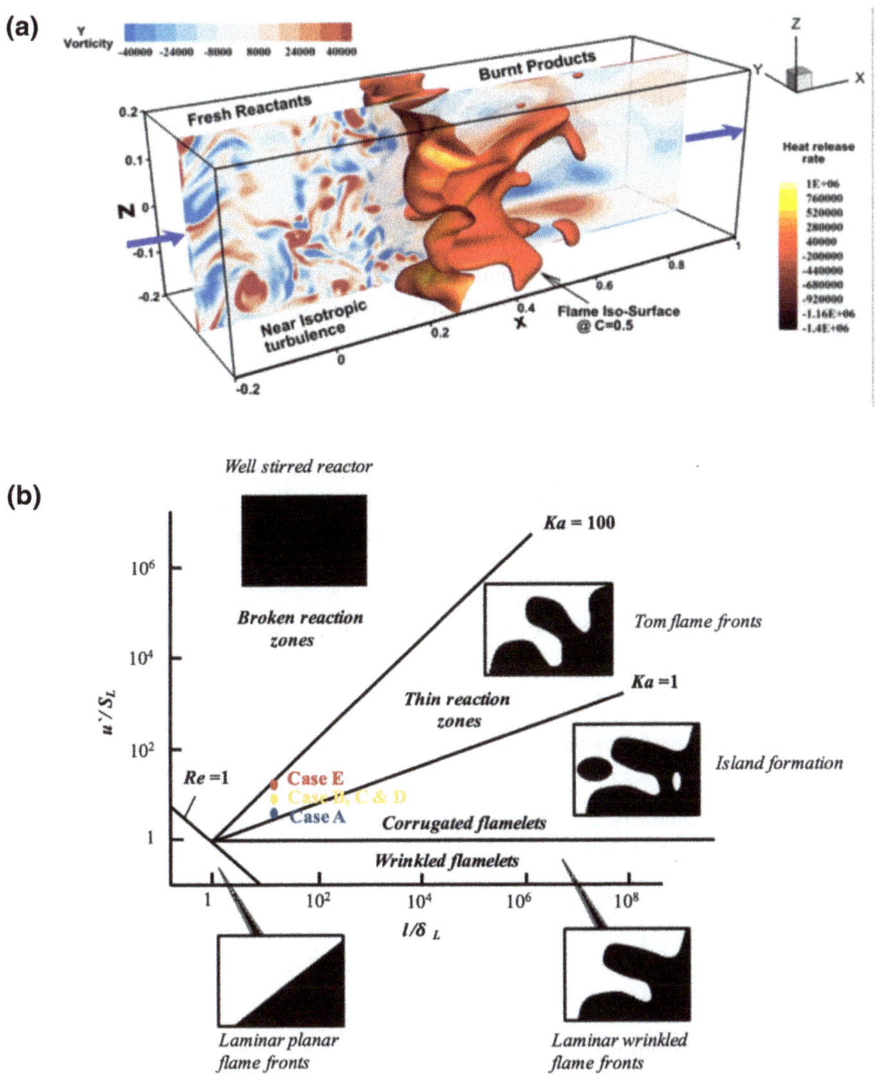

Fig. 1 **a** Schematic of the computational domain showing the contour of the Y vorticity and the distorted flame colored with heat release rate (mean flow is in the positivex-direction whereas flame propagates in the negative direction), **b** Broghi diagram showing the regimes of the simulated cases

2.1 Flame Stretch

In the present study, the flame stretch is defined by the fractional rate of change of a flame surface element A [27] as:

$$\kappa = \frac{1}{A}\frac{dA}{dt} \tag{2}$$

Stretch (κ) is a measure of fractional change in area due to flow non-uniformities, strain and curvature effects of the flame surface, and expressed as [27, 30]:

$$\kappa = -\vec{n}\vec{n} : \nabla\vec{u} + \nabla \cdot \vec{u} + s_d(\nabla \cdot \vec{n}) \tag{3}$$

where, $\vec{n}$ is the unit vector normal to the flame surface pointing towards the fresh gasses and $\vec{u}$ is the flow velocity, S_d is the displacement speed. The operator ($\vec{n}\vec{n} : \nabla$) represents the gradient operator normal to the flame surface and the first two terms contribute to the strain effects. Further, the stretch can be expressed in terms of tangential strain as:

$$\kappa = \nabla_t \cdot \vec{u} + s_d(\nabla_t \cdot \vec{n}) \tag{4}$$

2.2 Displacement Speed (S_d)

It is an important quantity affecting the flame stretch and varies depending on the choice of iso-values. Using the temperature iso-surface the equation for S_d is represented by:

$$S_d = \frac{1}{\rho C_p |\nabla T|}\left[\nabla \cdot (\lambda \nabla T) + \rho \nabla T \cdot \sum_{\alpha}(D_\alpha C_{p,\alpha} \nabla Y_\alpha) - \sum_{\alpha} h_\alpha \dot{\omega}_\alpha\right] \tag{5}$$

where the first term within the brackets corresponds to the heat conduction, the second term is the contribution due to diffusion, and the third term corresponds to heat release rate contributing to the displacement speed.

The numerical approach used for solving the turbulence-chemistry interactions, radiation modeling and soot models are described in the following section.

3 Results and Discussion

In this section, we present the results of all the simulated cases taken at $5F_L$, where the flame time (F_L) is defined as the ratio of the flame thickness to the flame speed.

We primarily concentrate on the annihilation and propagation part as it changes the texture of a flame surface due to turbulence-chemistry interaction. The different Lewis numbers of the species and intermediate radicals in H_2/air flames makes the difference in the local equivalence ratio and in places where the flame surface is convex to the reactants. It substantially alters the flame surface behavior. Hence propagation and annihilation process can be better explained using the mean curvature. The reported plots represent the iso-surface of progress variable at value $C=0.5$ colored by mean curvature taken for all the cases at $5F_L$. The sign convention of the mean curvature is followed as negative if the flame surface is concave towards the reactants and assigned positive if the flame surface is convex towards the reactants. It is observed from the iso-surfaces that case A and case E are weakly wrinkled and case B, C, and D are moderately wrinkled. From this, we can say that the flame wrinkling strongly depends on the turbulent intensity, while the value of the mean curvature increases with the rms of the fluctuating velocity. Regions highlighted in the blue color are prone to the local annihilation process.

3.1 Mean Curvature

Figure 2 depicts the plots of mean curvature for all the simulated cases at $5F_L$. The mean curvature of case A, as shown in Fig. 2a, is the maximum at convex surfaces, i.e. with positive curvature whereas the minimum is observed at the concave surfaces (negative curvature). It is found that the case A has fewer curvature extremes compared to the case B and case C, which can be primarily attributed to the less turbulent intensity and lower *Ka* number. For the case C, the flame surface is being pushed into the burnt side and creates larger concave surface; whereas in case D, the flame has moved towards the inlet due to the effect of preferential diffusion. The positive values dominating over the iso-surface confirms this phenomenon. Whereas for case E, the iso-surface is weakly wrinkled and displaced towards burnt side because of the high mean velocity.

3.2 Displacement Speed

Displacement Speed distribution on the flame surface is exhibited in Fig. 3. Previous studies noted that the displacement speed sharply increases nearing the annihilation process. In the present work also, the value of the displacement speed is found to be high in places where the flame surface is concave to the reactants for all the cases. This is consistent with the available literature. The region highlighted in red color, i.e. with negative curvatures, moves faster than the other portion of the flame surface and on the verge of separating from the flame surface. Under different turbulence conditions, the curvature of the flame surface changes and it affects the displacement speed through modified transport. From the plots, it can be noticed that for higher

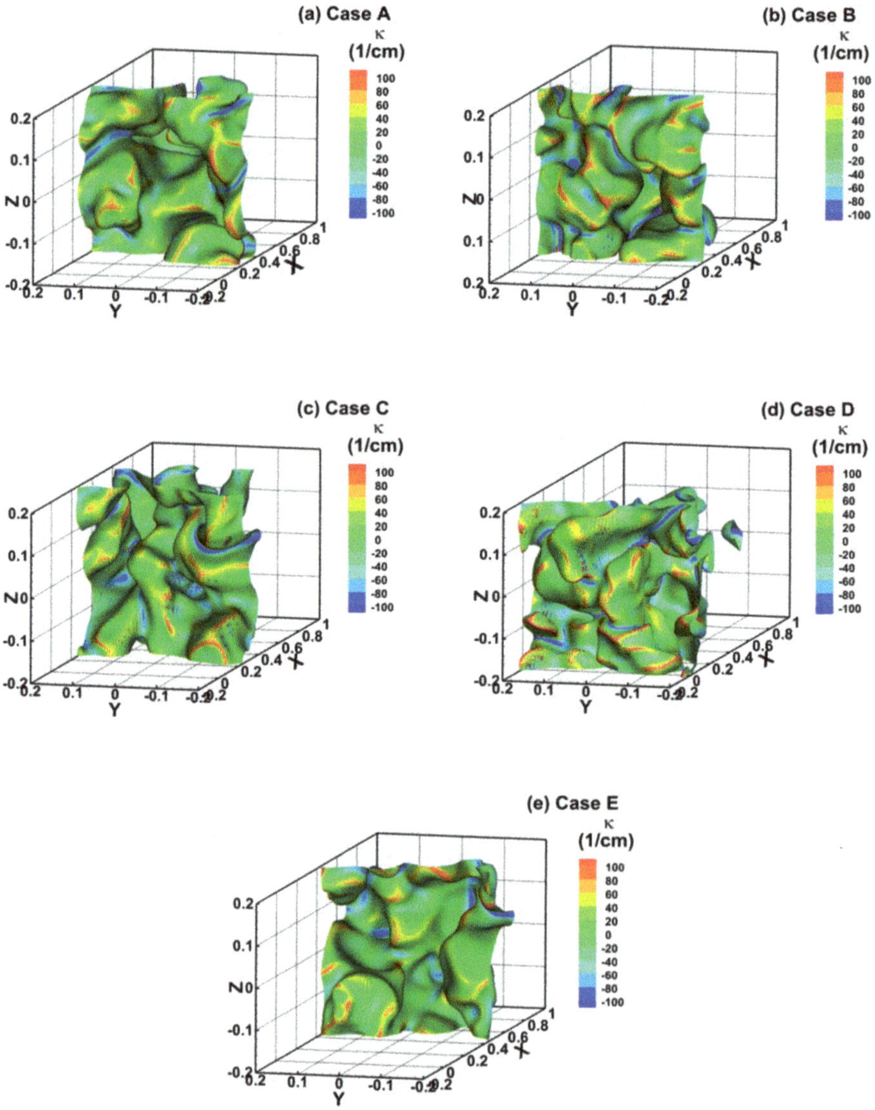

Fig. 2 Mean curvature at iso-surfaces conditioned at C = 0.5 for all the cases

turbulent intensity case, i.e. case D, most of the concave surfaces of the flame reaches highest displacement speed and is in the annihilation regime.

To identify the interdependency and the contribution by mean curvature to the displacement speed for different turbulent conditions, scatters plots are presented in Fig. 4a–e. Despite some scatter, the strong negative correlation between the displacement speed and mean curvature can be confirmed for all the cases. From case

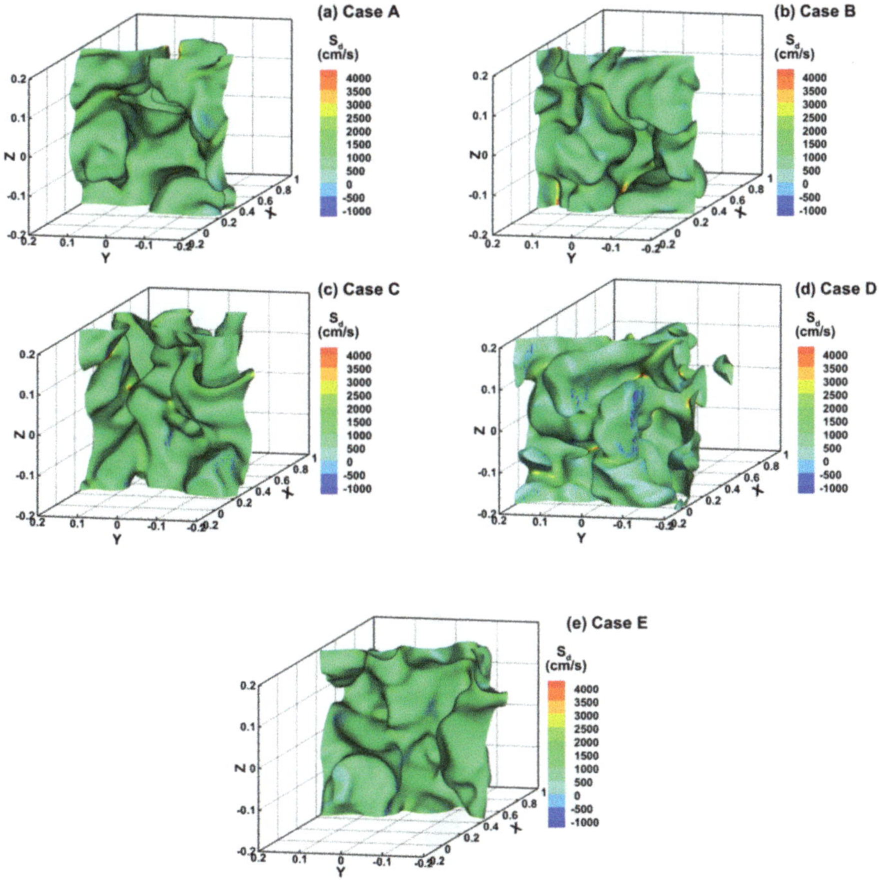

Fig. 3 Displacement flame speed at iso-surfaces conditioned at C = 0.5 for all the cases

A to case C, the correlation is scattered for the lower turbulence level and strongly correlated with higher turbulence level. While comparing the same with the cases A, B and D, it is evident that the displacement speed takes various values around zero mean curvature; in turn, it implies that the flame surface at some places is not wrinkled and still remains planar acting as a planar wrinkled flame. Particularly in the case of C, the dominance of negative curvature towards higher the displacement speed is observed (Fig. 4c), thereby making it more vulnerable to local annihilation in the post flame zone. Due to high turbulence intensity in case D, more dissipation of the scatter plot can be noted in the Fig. 4d.

To further elaborate the impact of the magnitude of temperature gradient on S_d, scatter plots are depicted in Fig. 4f–j. The magnitude of the temperature gradient is mainly affected by the mean curvature. When the mean curvature increases, it brings two temperature iso-surfaces close to each other, therefore, reducing the gradient of

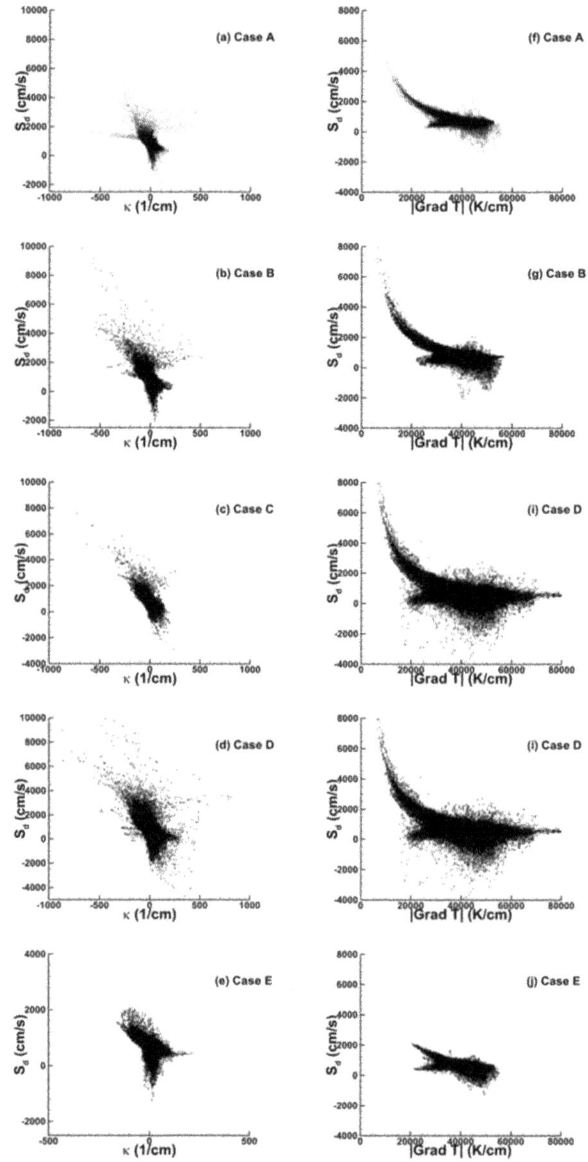

Fig. 4 Scatter plot of the displacement flame speed versus mean curvature (left column) and the gradient of the temperature (right column)

the temperature. As observed from the scatter plots of all the cases, there exist a unique correlation between S_d and $|\nabla T|$. It is negatively correlated, as the local flame surface having minimum $|\nabla T|$ experiencing higher S_d. For increasing *Ka* number cases, i.e. A, B and C are shown in Fig. 4, the regions belonging to lower $|\nabla T|$ are getting broadened, contributing more towards the higher flame displacement speed. This clearly depicts that the iso-surfaces in the vicinity of each other are propagating faster, making themselves more prone to the protrusion inside the post flame region. For higher intensity case, i.e. D, lower |∇T| is getting enormously broadened acquiring higher flame displacement speeds whereas for lower intensity case, case E, the band is not marginally scattered, apparently due to the lower turbulence in the system; hence S_d has not altered much. Furthermore, to quantify the correlation of displacement speed for different cases studied herein, we have carried out the term by term (Eq. 5) analysis in the following subsections.

3.2.1 Term 1

This term (first term of Eq. 4) tells about the heat conduction term behavior over the flame surface. From the plots of Fig. 5a–e, it is clear that this has higher values in the negative curvature region. And, this can be imputed from the reactants in the negative curvature region, surrounded by the temperature iso-surface which reduces the gradient of the temperature and leads to a reduction in the heat flux, thereby increasing the divergence of heat flux. From the iso-surface plots shown in Fig. 5, it is found that this term follows the curvature behavior similar to the displacement flame speed, thereby contributing more to the displacement speed in the concave regions. Further alike displacement speed, higher turbulence intensity case D shown in Fig. 5d, depicts prominent effect with maximum value on the negative curvatures, i.e. concave regions to the reactants. Whereas, a marginal effect of the increasing *Ka* number is observed on the maximum values between Fig. 5b, c.

3.2.2 Term 2

This term defining the mass diffusion contribution to the thermal transport, does have similar values in all the simulated cases and iso-surfaces as shown in Fig. 5f–j. Even though the magnitude of the gradient of the temperature is varying with respect to the turbulence as will be seen later, the contribution of this term is found to be negligible. This is primarily because the reduction in the magnitude of the gradient of temperature that is compensated by the gradient of mass fraction. Moreover, the contribution of the term 2 is nearly two orders of magnitude smaller than the term 1 and term 3 in terms of absolute magnitude and this behavior is consistent with the results of Altantzis et al. [31], pointing out a contribution by this term to S_d is virtually negligible.

3.2.3 Term 3

The change in gradient of local species concentration and the temperature is decided by the curvature. Because of the non-unity Lewis number, the system has different heat and mass diffusion rates. These combined effects of curvature and non-unity Lewis number affect the total enthalpy. The effect of term 3 of displacement speed

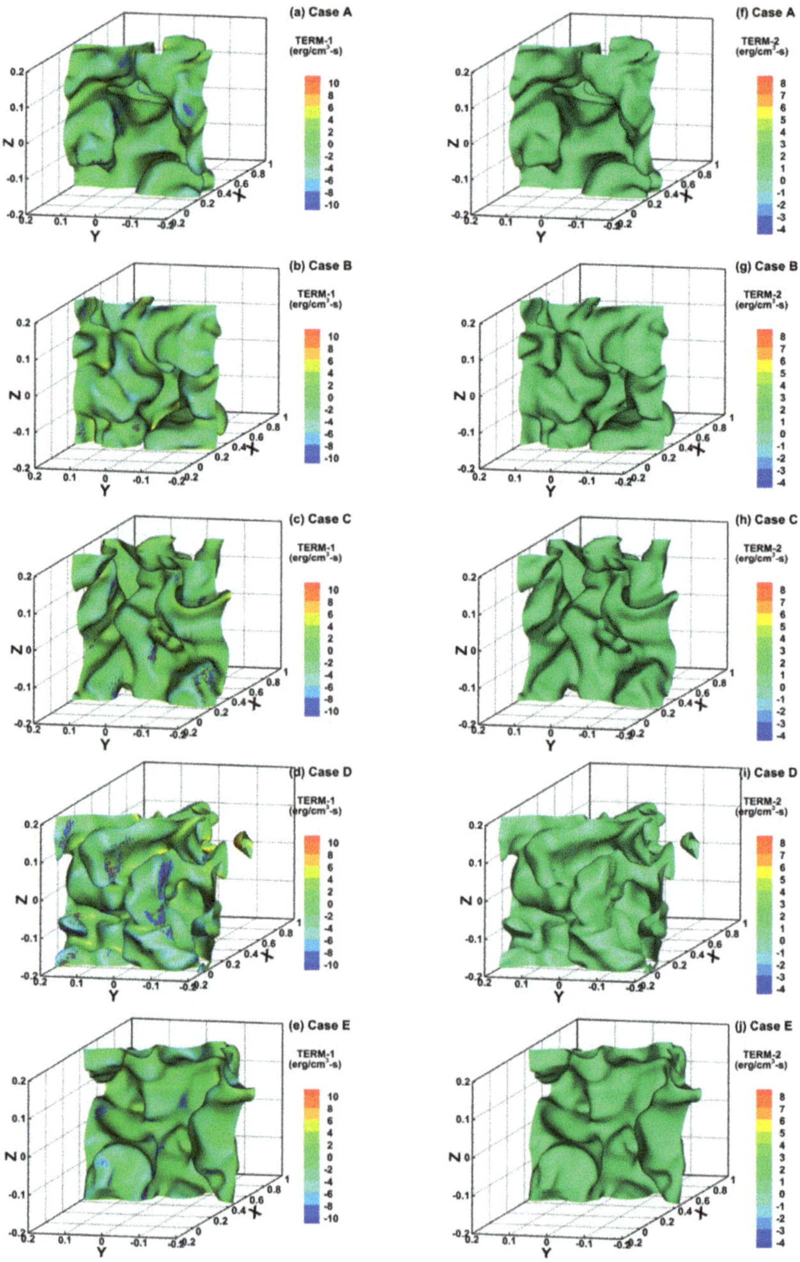

Fig. 5 Flame sheet colored with term 1 (left column) and term 2 (right column) of Eq. 4 i.e. measure of heat conduction and the mass diffusion, respectively, at iso-surfaces conditioned at C = 0.5 for all the cases

expression that is, heat release rate is identified by analyzing the Fig. 6. The maximum heat release rate is present in the negative curvatures and is consistent with all the five cases, whereas minimum value is present on the sharp convex tip on the reactant side. This, in particular, due to the concentrated thermal energy in the concave regions which further enhances the combustion process in the region resulting in intensified heat release rate. Focusing of the heat ahead of itself in the flame surface lead to the faster propagation of the flame locally, therefore inflation in the corresponding displacement speed justifies the positive correlation. Increasing *Ka* number enhances the wrinkling of the flame and is stated in the earlier sections but the effect on heat release rate is minimal in currently chosen *Ka* number regime, which can be seen in Fig. 6a–c. Interestingly, the minimum and maximum values of the heat release rate are intensified locally with the increased turbulence intensity and can be seen in Fig. 6d. Whereas, the mild effect is observed in the case of lower turbulence intensity case, i.e. case E as shown in Fig. 6e.

3.3 *Vorticity and Tangential Strain Rate Distribution*

To study the contribution of the flow on the flame stretch, vorticity has been calculated in the flow field. Studying wrinkling effect on flame using vorticity has been carried out previously. The Q-criterion analysis has been performed in the computational domain (not shown here) and it is inferred that the all vortical structures created by turbulence are accurately identified by the Q-criterion and similar structures are also present in z-vorticity as well. Hence, z-vorticity and tangential strain contours are used for enlightening the flame stretch and propagation and annihilation associated with it, as depicted in Fig. 7.

The previous study conducted on flame-vortex interaction by sending a pair of counter-rotating vortices to interact with a laminar flame concludes that the flame is being stretched in the direction along the line that passes through the midpoint of the two vortex cores. The similar effect can be noticed in the successive times by following the counter rotating pair near the flame which is highlighted in the black box. The red line designates the flame surface location in the corresponding slice of the domain. The counter-rotating vortices exert positive stretch on the flame surface leading to the formation of negative curvatures; whereas, the negative stretch due to similar vortices pair causes positive curvature. However, the surface area increases in both the cases, particularly due to free wrinkling in the latter case and this negative correlation is confirmed by the DNS predictions of Chakraborty and Cant [32]. The concave bulge ahead of the reactants side protruding inside the reaction zone is certainly due to the positive stretch created by counter-rotating vertical pairs inducing the negative curvatures. However, the vortex which is highlighted in the dark red box visually has no pair. But still, it wrinkles the flame surface primarily due to the local momentum imbalance of the flow. At $F_L = 5.02$, it can be seen that the vortex structure, previously marked/highlighted in rectangular bounded region (@$F_L = 5.00$ in Fig. 7) before, starts to wrinkle the flame at the same time the

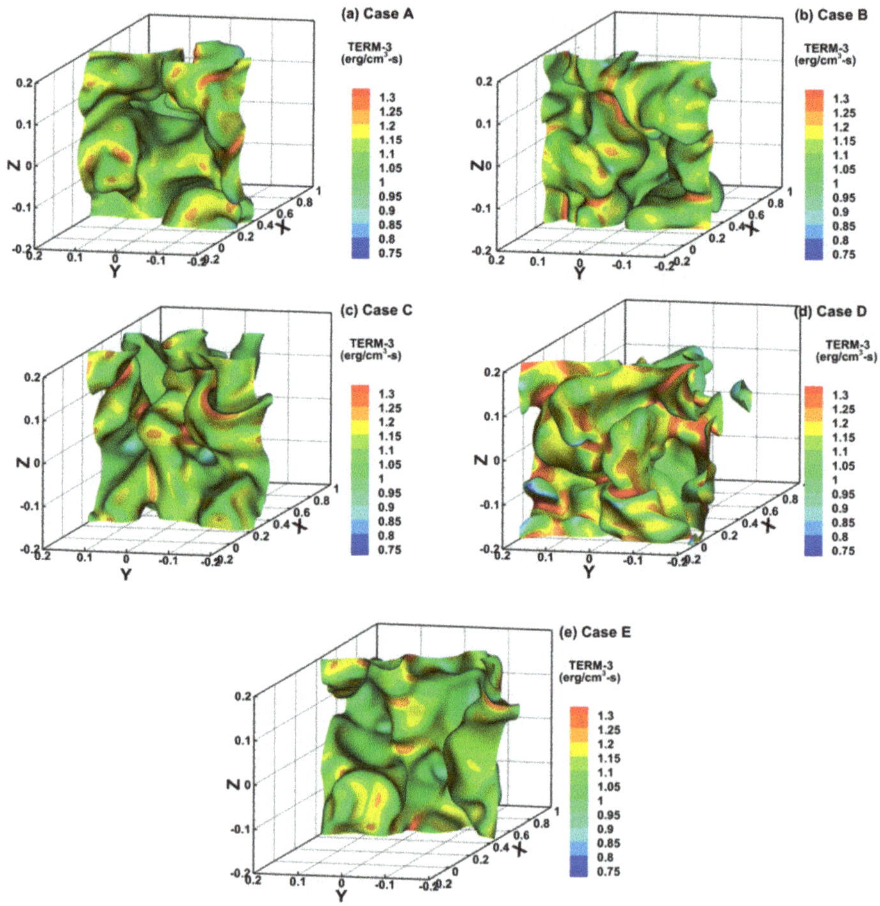

Fig. 6 Term 3 of Eq. 4, i.e. measure of the heat release rate at iso-surfaces conditioned at $C = 0.5$ for all the cases

flame separates from the main iso-surface due to local annihilation. Nonetheless, the influence of the vortex structure in the dark red box is not that prominent as compared to one leading in the extreme negative curvatures.

To further understand the change in the surface area due to vortical pairs, the effect of the tangential strain is also described and shown in Fig. 7. It is observed from the vorticity contours that the regions between the vortices are responsible for the change in the flame surface area. Hence, the strain rate field that is present in the flow should be also responsible for the change in the flame surface area. The strain rate field also serves as a description for the flame turbulence interaction. Since tangential strain has to be found for 3D, because it is equally contributed by Z direction gradients, it is calculated from the 3D domain and plotted on the same slice for which vorticity study has been previously conducted. It can be clearly seen that maximum contribution to

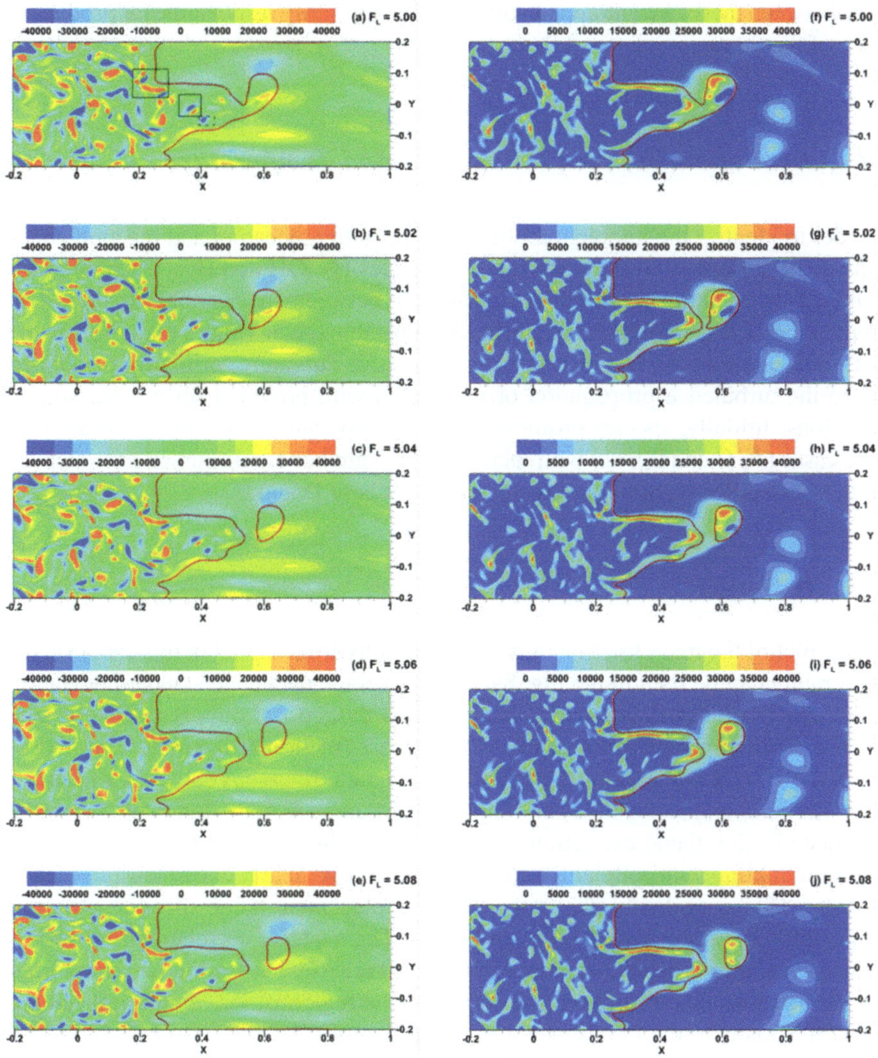

Fig. 7 Contour of Z vorticity (left column) and tangential strain (right column) at mid plane for the case A at different flame times

the tangential strain rate is gained through the positively valued vortices, i.e., these are primarily responsible for the alterations in the flame surface area in the concave regions. While traveling in the downstream regions with F_L, from 5.04 to 5.08, the tangential strain rate increases to the maximum at the negative curvatures and separates the vortical pair on the flame surface into the post flame zone, causing the segregation of the tangential strain in this zone. It is particularly due to the high shear between these two entrapped vortical pairs, primarily responsible for producing bulge of the flame into the post flame zone.

3.4 Correlation Between the Species and Temperature

In order to identify the interaction of turbulence with preheating and reaction zones of flame, scatter plots of mass fractions of species H_2, OH and HO_2 are plotted for all the cases. A unique correlation can be seen between species mass fraction and temperature. These intermediate radicals and unstable species play a major role in the heat release rate in the negative as well as positive curvature, primarily developed due to the turbulence propagation or mitigation with provided several parametric conditions. Initially, species profiles with the temperature dependence of the first three cases (A, B and C), i.e. with increasing *Ka* number are discussed and later the effect of increased fluctuations are described for case D and E.

3.4.1 H_2 Versus T

Scatter plot of the mass fraction of the molecular hydrogen (fuel in this case) versus temperature can be seen in the Fig. 8a–e. The average mean and standard deviation of the scatter plot is shown using the solid and dashed red line, respectively. The contour in the rectangular box represents the mass fraction of the molecular hydrogen in the xy-mid plane of the computational domain. In case A shown in Fig. 8a, at lower *Ka* number, the negative curvatures of the flame are quite dominant due to the presence of thick flame extinction zones (as highlighted in the blue box) and substantial wrinkling with cellular burning structures, particularly due to low turbulence intensity. This is evidently due to the lower displacement flame speed resulting in the positive curvatures (can be seen in Fig. 4). The negative correlation is observed with molecular hydrogen concentration and heat release, i.e. a minimum mass fraction of the molecular hydrogen is observed in the region of maximum heat release rate corresponds to the highly positive curvatures of the flame. These findings are concurrent with the simulations performed by Aspden et al. [33]. In the case of B, flame curvatures are perturbed due to higher turbulence intensity and can be noticed in Fig. 8b. The negative curvatures can be obtained by protruding into the post flame zone as long and narrow tubular structures, which can be particularly due to the fast convection of the radicals in the reaction zone resulting into large gradients. At further higher *Ka* number, i.e. case C, higher turbulent intensity mitigates the tendency for the thermo-diffusive instability to extinguish the flame or flame extinction in the regions of the negative curvatures. Though, the thickening of the reaction zone due to small scale stirring induced by higher turbulent intensity is observed. Interestingly, distribution of the scatter plot is nearly symmetrical around the mean line and is getting broadened while transitioning from case A to C.

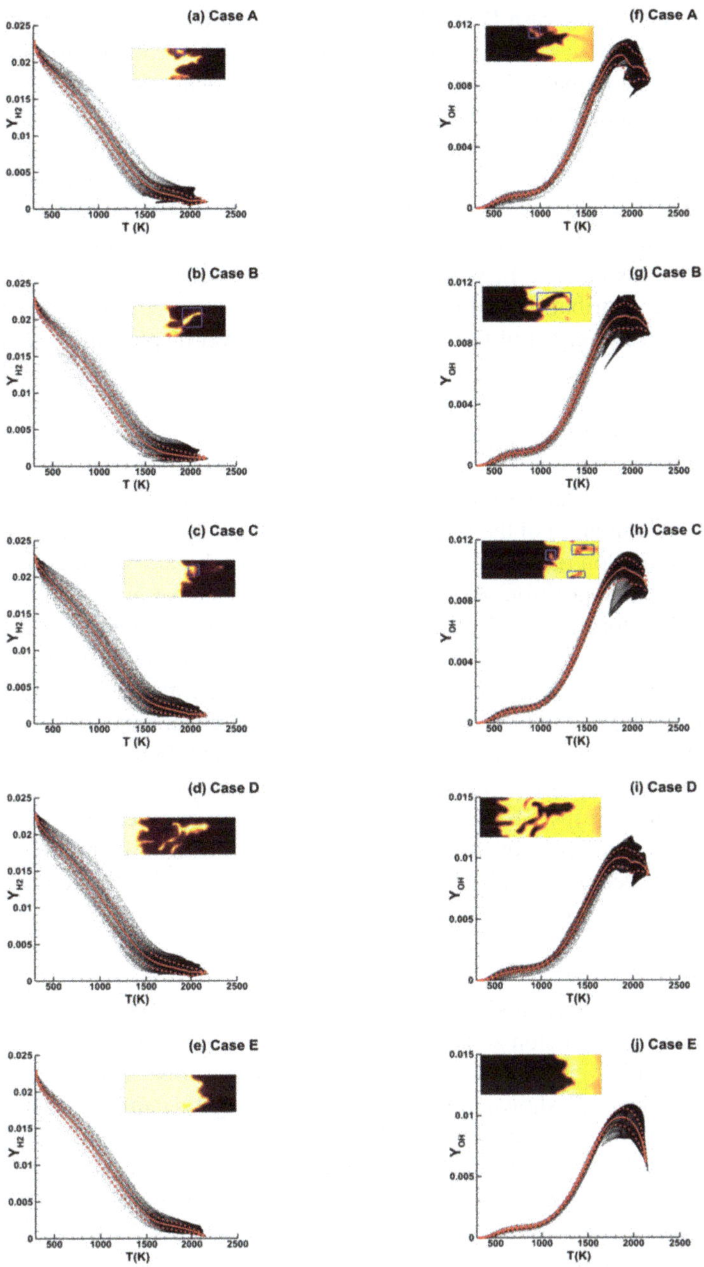

Fig. 8 Scatter plot of the mass fraction of H_2 molecule and OH radicals versus temperature, for all the cases. Solid and dashed red line denotes the average mean and the standard deviation of the scatter distribution corresponding to the subsequent temperature

As stated earlier for case D and E, the mean velocities remain constant whereas turbulence intensity is varied in the descending order by reducing the fluctuating component of the global velocity component. This lightens up the impact of turbulence intensity on the flame-turbulence interactions along with constant *Ka* number. In the case of D, highly broadened distribution can be observed around the mean line and thickened reaction as well as post flame zone seems to be plausible. Due to higher turbulent intensity, unburnt fuel mass fraction pockets are protruded inside the reaction zone leading to the number of local flame extinction zones causing annihilation. Whereas in Case E, with lower turbulence intensity as shown in Fig. 8e, thinning of the distribution and the positive curvatures can be found which can be the result of a lesser effect of turbulent stirring and lower displacement speed.

3.4.2 OH Versus T

Formation and dissociation of the OH radicals correspond to the reaction rate propagation after the initiation of the flame-turbulence interactions and correlates robustly with H_2 consumption rate. This is particularly used in laboratory flames as a marker to signify the flame position. At lower *Ka* number, as shown in Fig. 8f, large scale wrinkling in the form of highly negative curvatures is observed with anticipation of the local extinction in the reaction and these higher curvatures bulge towards the post flame zone. This can be dedicated to the dominance of thermo-diffusive instabilities causing extinction of the fame in the regions of the preheat zone, primarily due to the low turbulence intensity. Most of the heat release rate (HRR*) is observed in the preheat zone; which may be referred to as $T < 900$ K and is shown in Fig. 9. Carlesson et al. [34] defined the increased radical's concentration as a source of the heat release source in the low-temperature zone. This exothermic behavior is dominated by the reaction R9 defined in Li et al. [26] chemical mechanism and is written below:

$$H + O_2 + M \rightarrow HO_2 + M \tag{6}$$

where M denotes the third body species. This reaction compensates the consumption of H radicals in the reaction zone corresponding to the high-temperature zone, with the formation of HO_2 in the low temperature, i.e. preheat zone. The reaction rates of HO_2 and H radicals are shown in Fig. 10. For increasing *Ka* number cases, i.e. cases A–C, the trough of the reaction rate of H radicals denoting the consumption rate in preheat zone, is thickened in case A whereas standard deviation in the case C reduces marginally. This happens owing to the reduction in effective diffusion time due to high turbulent convective transport in higher *Ka* number case. Whereas, in cases D and E, a substantial difference in the crest and trough of the H and HO_2 radicals can be noticed in Fig. 10c, d. This clearly depicts the sensitivity of the turbulent intensity in the case D, acting as a stirrer during flame-turbulence interactions.

The above-mentioned observation explicates the major contribution of H radicals towards heat release rate as compared to the hydroperoxyl (HO_2) intermediates. With increased *Ka* number in case B, the thinning of the OH radical's mass fraction as

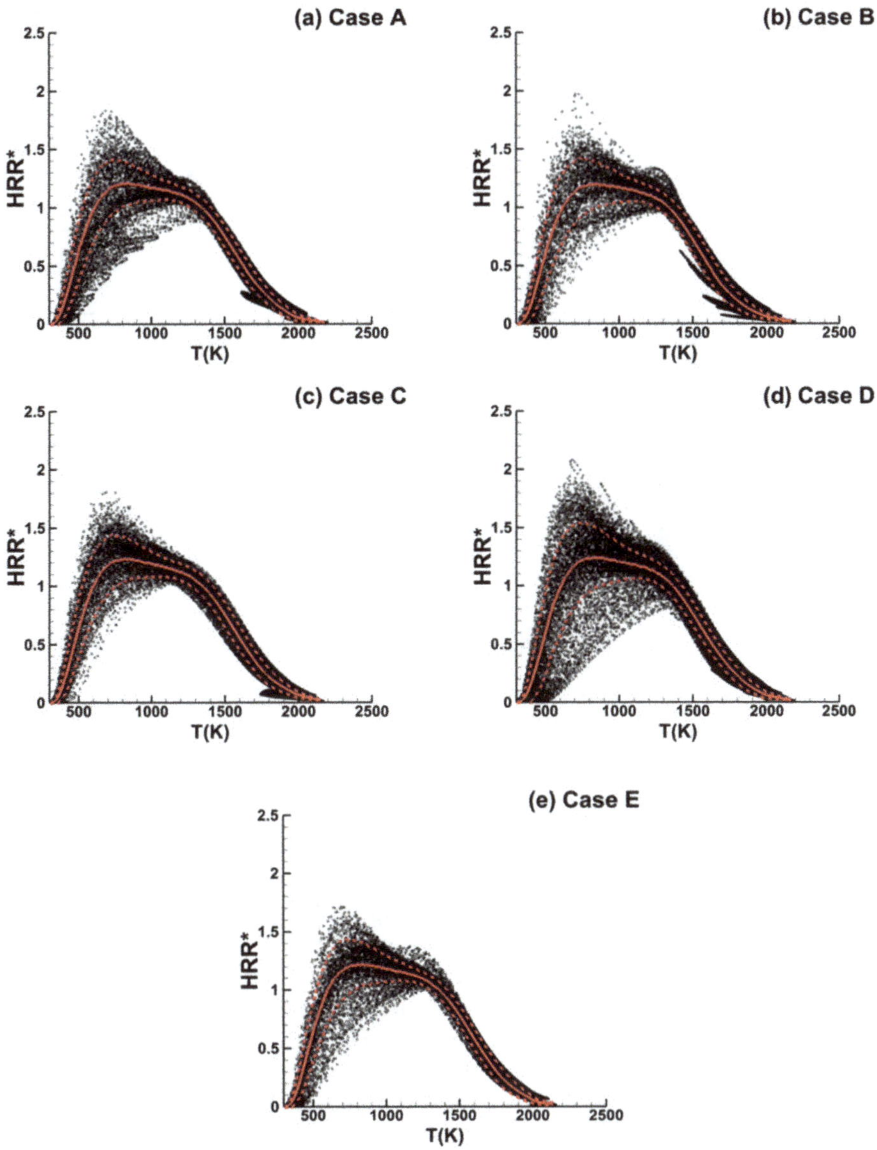

Fig. 9 Scatter plot distribution of the heat release rate (J/s) versus temperature. * designates normalization by 10e+10, for better understanding

well as the heat release rate concentration is observed. Increasing *Ka* number leads to the inflation of the small-scale turbulent structures which excites the local extinction zones in the reaction zone as well as in the post flame regime. This local extinction

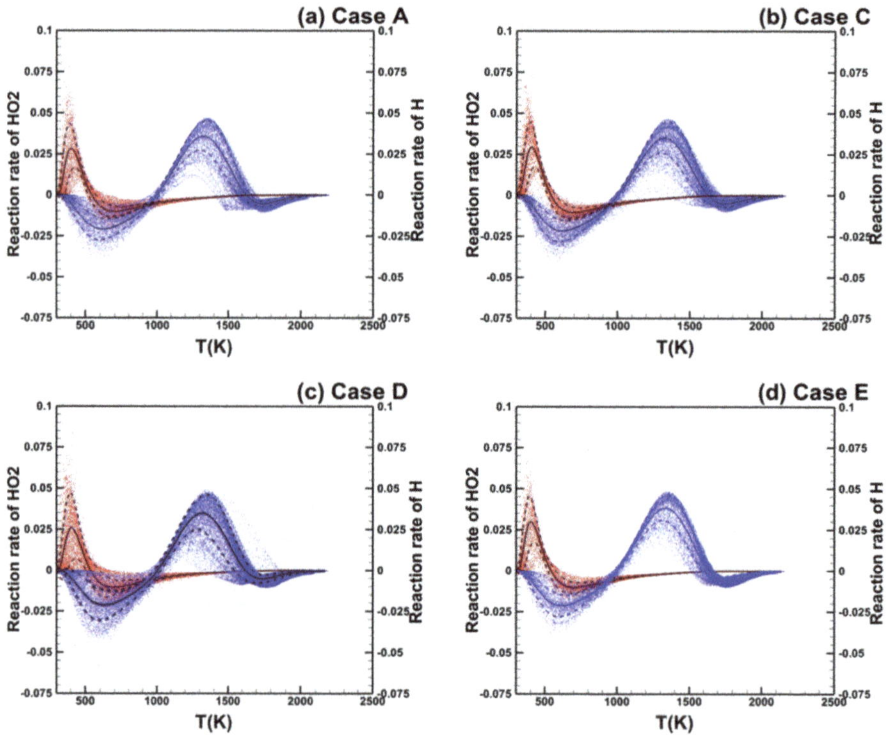

Fig. 10 Scatter plot distribution of the reaction rate of HO_2 and H radicals versus temperature. Red colored distribution corresponds to reaction rate of HO_2 and blue colored distribution corresponds to the H radicals

process shows the dominance behavior with higher *Ka* number in case C, with a higher concentration of heat release rate in the preheat zone.

Contrarily, in the case of D, more protrusive local extinction zones are observed certainly due to the higher level of the turbulence intensity, i.e. ~1.5 times higher compared to case B. On comparing Case B and D, one should note that the flame sensitivity towards turbulent intensity is more pronounced as compared to the increasing *Ka* number. This can be evidently seen from the augmentation and thickening of the heat release rates at a lower temperature (T< 900 K: may be referred to as preheat zone) shown in Fig. 9. This sensitivity analysis can be affirmed from case E, where thinning of the heat release rates under reduced turbulence intensity of nearly halved compared to case D.

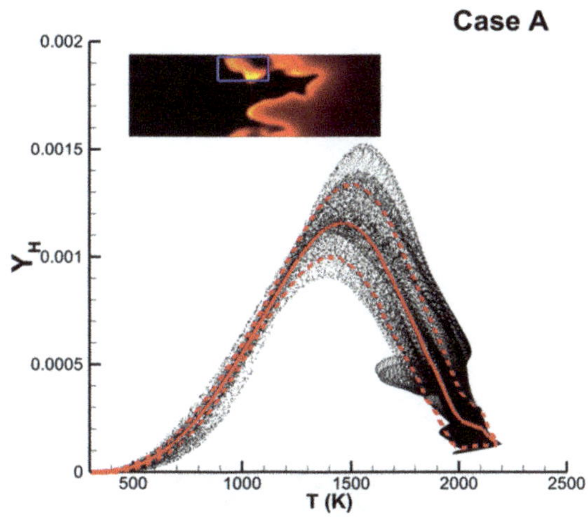

Fig. 11 Scatter plot distribution of H mass fraction versus temperature for case A

3.4.3 HO_2 Versus T

The HO_2 radicals are active centers of the hydrogen combustion reaction due to its primary role as a precursor of auto-ignition and the secondary role as an active center of propagating the chain reaction. Due to the exothermic behavior of the reaction (R9), maximum heat release rate corresponds to the peak HO_2 mass fractions in the low-temperature zone. Aspden et al. [33] and Carlesson et al. [34] observed identically pronounced effects correspond to the reaction (R9). Moreover, a maximum mean mass fraction of H radicals higher by an order as compared to the maximum mean mass fraction of HO_2 as exhibited in Fig. 11, depicting consumption of H radical among the primary role in the heat release rate. As the *Ka* number increases, i.e. in case of B and C, the contraction of the band of the heat release rate can be observed. Whereas, with higher turbulent intensity in case D, the peak heat release rate changes by ~1.1 times as compared to the case B. Interestingly, enhancing turbulence intensity results into the broadening of the distribution of the scatter points. This is associated to the enhanced volumetric burning rate in the lower temperature zones, i.e. preheat zone and more broadened HO_2 mass fraction, as depicted in Fig. 12. However, mean distribution remains insensitive towards these parametric changes. Enhanced turbulent intensity results into more prominent negative structures in the reaction zone. Whereas in the case of E, reduction in the turbulent intensity leads to the thinning of the distribution of the HO_2 mass fraction as well as a corresponding reduction in the heat release rates by a factor of ~1.20. This clearly depicts that the increasing turbulent intensity is more sensitive towards increment in the heat release rates as compared to increasing *Ka* number. The study clearly differentiates the mild sensitivity of the first three flames (Case A–C) towards Karlovitz number (*Ka*) and strong sensitivity of the turbulence intensity on comparing the case D and E.

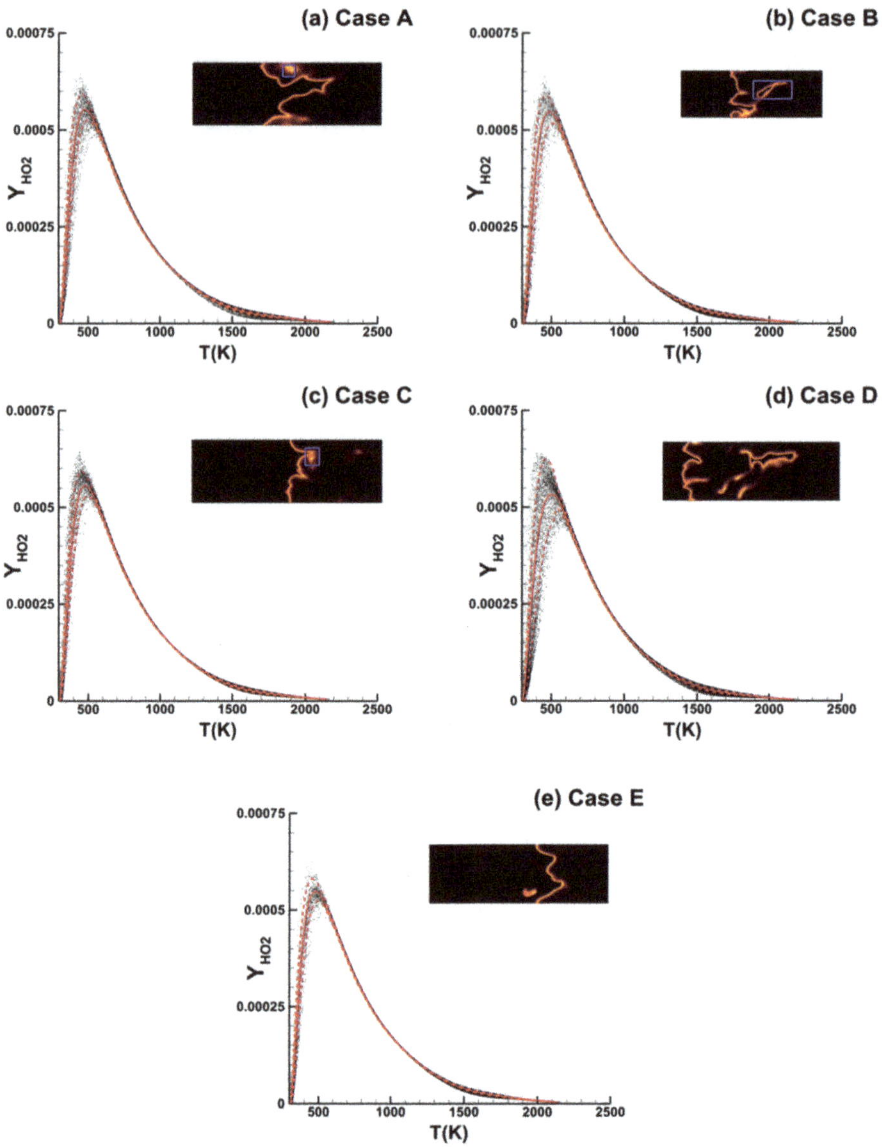

Fig. 12 Scatter plot distribution of the HO_2 mass fraction versus temperature for all the cases

4 Conclusion

Turbulence-flame interaction of moderately lean H_2/Air premixed flame is studied using Direct Numerical Simulation approach. Fairly detailed chemistry in conjunction with differential diffusion is invoked in the present study. The flame surface is interpreted in terms of its propagation and annihilation in the positive and the negative curvatures. Results are oriented towards determining the contribution of the heat conduction, mass diffusion and heat release rate, in the calculation of the displacement flame speed. The simulations clearly define the pronounced effect of the turbulent intensity over the Karlovitz number.

The results clearly elucidate the mean curvature and displacement flame speed exhibit negative correlation, i.e. maximum mean curvature regions hold the minimum displacement flame speed, particularly due to the annihilation of the flame particles in the preheat zone. Further, displacement flame speed is seen in negative correlation with a gradient of temperature as well. In the computation of displacement flame speed, a significant contribution of the heat conduction and heat release rates is observed in the negative curvatures corresponding to the maximum displacement flame speed. Whereas, mass diffusion term doesn't alter the combined equation due to a lesser order of magnitude as compared to the other two. Further, sharp intensive structures are found after increasing the turbulent intensity whereas cellular burning with curved tips is observed while augmenting the Karlovitz number. The pronounced effect of the vorticity and tangential strain caused by vortical pairs in the negative curvatures is found to be the major source of the propagation of the displacement flame speed.

A noticeable effect of species and radical's concentration and associated reaction rates have been studied comprehensively. The thickening of the scatter distribution and substantial condensation of the scatter particles of the H_2, HO_2 and OH species are observed in the case of higher Karlovtiz number and higher turbulent intensity as well. The heat release rate is found to be controlled by the production of the lower temperature controlled HO_2 species corresponding to the consumption of H radicals. Whereas, the production of H radicals appear to be dominating in the high-temperature zones only, thereby contributing least to the production of the heat release rates. Nonetheless, consumption of H radicals is found to be one order of magnitude higher than the consumption of HO_2 radicals, causing the heat release rate to be maximum in the lower temperature controlled region. Present simulation clearly ascertains the strong sensitivity of the turbulent intensity over the mild intensity of the Karlovitz number in the flame-turbulence interactions.

Acknowledgements The authors would like to acknowledge the IITK computer center (www.iitk.ac.in/cc) for providing support to perform the computation work, data analysis, and article preparation.

References

1. N. Babkovskaia, N.E. Haugen, A. Brandenburg, A high-order public domain code for direct numerical simulations of turbulent combustion. J. Comput. Phys. **230**(1), 1–2 (2011)
2. H.G. Im, P.G. Arias, S. Chaudhuri, H.A. Uranakara, Direct numerical simulations of statistically stationary turbulent premixed flames. Combust. Sci. Technol. **188**(8), 1182–1198 (2016)

3. C.K. Law, Dynamics of stretched flames. in *Symposium (international) on Combustion*, vol. 22, no. 1, (1989), pp. 1381–1402
4. P. Clavin, Dynamic behavior of premixed flame fronts in laminar and turbulent flows. Prog. Energy Combust. **11**(1), 1–59 (1985)
5. T. Poinsot, D. Veynante, S. Candel, Diagrams of premixed turbulent combustion based on direct simulation. in *Symposium (international) on Combustion*, vol. 23, no. 1, (1991), pp. 613–619
6. S. Kadowaki, T. Hasegawa, Numerical simulation of dynamics of premixed flames: flame instability and vortex–flame interaction. Prog. Energy Combust. **31**(3), 193–241 (2005)
7. V. Sankaran, S. Menon, Structure of premixed turbulent flames in the thin-reaction-zones regime. Proc. Combust. Inst. **28**(1), 203–209 (2000)
8. V. Sankaran, S. Menon, Subgrid combustion modeling of 3-D premixed flames in the thin-reaction-zone regime. Proc. Combust. Inst. **30**(1), 575–582 (2005)
9. H.N. Najm, P.S. Wyckoff, Premixed flame response to unsteady strain rate and curvature. Combust. Flame **110**(1–2), 92IN595-4IN6112 (1997)
10. A.N. Lipatnikov, J. Chomiak, Molecular transport effects on turbulent flame propagation and structure. Prog. Energy Combust. **31**(1), 1–73 (2005)
11. A.N. Lipatnikov, J. Chomiak, Effects of premixed flames on turbulence and turbulent scalar transport. Prog. Energy Combust. **36**(1), 1–02 (2010)
12. G.H. Markstein (ed.), *Nonsteady Flame Propagation*: *AGARDograph* (Elsevier, 2014)
13. P. Clavin, G. Joulin, Premixed flames in large scale and high intensity turbulent flow. J. Phys. Lett Paris **44**(1), 1–2 (1983)
14. M. Matalon, B.J. Matkowsky, Flames as gasdynamic discontinuities. J. Fluid Mech. **124**, 239–259 (1982)
15. N. Chakraborty, R.S. Cant, Influence of Lewis number on curvature effects in turbulent premixed flame propagation in the thin reaction zones regime. Phys. Fluids **17**(10), 105105 (2005)
16. N. Chakraborty, N. Swaminathan, Influence of the Damköhler number on turbulence-scalar interaction in premixed flames. I. Physical insight. Phys. Fluids **19**(4), 045103 (2007)
17. M. Baum, T.J. Poinsot, D.C. Haworth, N. Darabiha, Direct numerical simulation of H 2/O 2/N 2 flames with complex chemistry in two-dimensional turbulent flows. J. Fluid Mech. **281**, 1–32 (1994)
18. H. Wang, K. Luo, S. Lu, J. Fan, Direct numerical simulation and analysis of a hydrogen/air swirling premixed flame in a micro combustor. Int. J. Hydrog. Energy **36**(21), 13838–13849 (2011)
19. P.E. Hamlington, A.Y. Poludnenko, E.S. Oran, Interactions between turbulence and flames in premixed reacting flows. Phys. Fluids **23**(12), 125111 (2011)
20. J.B. Chen, H.G. Im, Stretch effects on the burning velocity of turbulent premixed hydrogen/air flames. Proc. Combust. Inst. **28**(1), 211–218 (2000)
21. J. Jiang, X. Jiang, M. Zhu, A computational study of preferential diffusion and scalar transport in nonpremixed hydrogen-air flames. Int. J. Hydrog. Energy **40**(45), 15709–15722 (2015)
22. K.R. Dinesh, X. Jiang, J.A. Van Oijen, R.J. Bastiaans, L.P. De Goey, Hydrogen-enriched nonpremixed jet flames: effects of preferential diffusion. Int. J. Hydrog. Energy **38**(11), 4848–4863 (2013)
23. K.R. Dinesh, H. Shalaby, K.H. Luo, J.A. van Oijen, D. Thévenin, High hydrogen content syngas fuel burning in lean premixed spherical flames at elevated pressures: effects of preferential diffusion. Int. J. Hydrog. Energy **41**(40), 18231–18249 (2016)
24. C. Han, D.O. Lignell, E.R. Hawkes, J.H. Chen, H. Wang, Examination of the effect of differential molecular diffusion in DNS of turbulent non-premixed flames. Int. J. Hydrog. Energy (2017)
25. H.A. Uranakara, S. Chaudhuri, H.L. Dave, P.G. Arias, H.G. Im, A flame particle tracking analysis of turbulence-chemistry interaction in hydrogen-air premixed flames. Combust. Flame **163**, 220–240 (2016)
26. J. Li, Z. Zhao, A. Kazakov, F.L. Dryer, An updated comprehensive kinetic model of hydrogen combustion. Int. J. Chem. Kinet. **36**(10), 566–575 (2004)
27. T. Poinsot, D. Veynante, *Theoretical and Numerical Combustion*. (RT Edwards, Inc., 2005)

28. http://pencil-code.nordita.org (2016)
29. N. Babkovskaia, N.E. Haugen, A. Brandenburg, A high-order public domain code for direct numerical simulations of turbulent combustion. J. Comput. Phys. **230**(1), 1–2 (2011)
30. S.M. Candel, T.J. Poinsot, Flame stretch and the balance equation for the flame area. Combust. Sci. Technol. **70**(1–3), 1–5 (1990)
31. C. Altantzis, C.E. Frouzakis, A.G. Tomboulides, K. Boulouchos, Direct numerical simulation of circular expanding premixed flames in a lean quiescent hydrogen-air mixture: phenomenology and detailed flame front analysis. Combust. Flame **162**(2), 331–344 (2015)
32. N. Chakraborty, R.S. Cant, Effects of strain rate and curvature on surface density function transport in turbulent premixed flames in the thin reaction zones regime. Phys. Fluids **17**(6), 065108 (2005)
33. A.J. Aspden, M.S. Day, J.B. Bell, Turbulence-chemistry interaction in lean premixed hydrogen combustion. Proc. Combust. Inst. **35**(2), 1321–1329 (2015)
34. H. Carlsson, R. Yu, X.S. Bai, Direct numerical simulation of lean premixed CH4/air and H2/air flames at high Karlovitz numbers. Int. J. Hydrog. Energy **39**(35), 20216–20232 (2014)

Investigation of the Role of Chemical Kinetics in Controlling Stabilization Mechanism of the Turbulent Lifted Jet Flame Using Multi-flamelet Generated Manifold Approach

Rohit Saini, Ashoke De, Venu Aggarwal and Rakesh Yadav

Abstract The study reports on the numerical investigation of lifted turbulent jet flames with H_2/N_2 fuel issuing into a vitiated coflow. The hot vitiated co-flow containing oxygen as well as combustion products stabilize the lifted turbulent flame by providing an autoignition source. A 2D axisymmetric formulation has been used for the predictions of the flow field, while multidimensional Flamelet Generated Manifold (multi-FGM) approach has been used for turbulence-chemistry interactions in conjunction with RANS approach. The chemical kinetics in H_2-O_2 combustion is followed by (Mueller et al, Int J Chem Kinet 31: 113–125, 1999 [1]) and (Li et al, Int J Chem Kinet 36(10): 566–575, 2004 [2]) mechanisms and the difference in chemical kinetics is analyzed (in terms of auto-ignition distance) using one-dimensional calculations. The major difference between the two mechanisms is the value of rate constants contributing towards the source of the autoignition and the corresponding enthalpy of formation of OH radicals. The lift-off height is determined from the axial distance (from the burner exit) at which the auto-ignition occurs, and is located through local concentration of OH radical equivalent to 2×10^{-4}. In order to understand the impact of chemical kinetics on the autoignition, speeding up (Set A) and delaying (Set B) auto-ignition controlled reaction rates are augmented and corresponding changes in lift-off height are observed. Hereafter, the comprehensive chemical kinetics sensitivity analysis is carried out in understanding the underlying behavior of HO_2 radicals as autoignition precursor and OH radicals as reaction rate

R. Saini · A. De (✉)
Department of Aerospace Engineering,
Indian Institute of Technology Kanpur, Kanpur 208016, India
e-mail: ashoke@iitk.ac.in

V. Aggarwal
Department of Mechanical Engineering, Indian Institute of Technology Gandhinagar,
Gandhinagar 382355, India

R. Yadav
ANSYS Inc, 5930 Cornerstone Court West, Suite 230, San Diego, CA 92121, USA

A. K. Runchal et al. (eds.), *Energy for Propulsion*, Green Energy and Technology,
https://doi.org/10.1007/978-981-10-7473-8_12

determinant. It is found that some specific reaction is most sensitive to auto-ignition and plays a vital role in lift-off height predictions. The results obtained in the current study elucidates that the flame is largely controlled by chemical kinetics.

Keywords Role of chemical kinetics · Lifted flame · Multi-FGM approach

1 Background

The fundamental mechanisms controlling the flame stabilization and extinction of lifted flames has been a subject of great deal of study since past many decades [3–17]. The primary characteristics of the lifted flames such as lift-off height, blow-out velocity etc., can only be predicted reliably through numerical tools once the mechanism governing flame stabilization is understood completely. Although the stabilization mechanism for the laminar lifted flames is well understood [3, 4], the relevant stabilizing mechanism behind turbulent lifted flame has not yet been understood completely [5]. The theories that have been proposed regarding the stabilization of turbulent lifted flame in the coflow can be categorized into four types: (1) premixed flame propagation theory [6] (2) triple flames stabilized [7–9] (3) non-premixed laminar flamelet [10–12] (4) large-scale turbulent structures [13–17]. Lyons [6] proposed flame stabilization through premixed flame propagation theory, which assumes that the fuel and oxidizer streams get completely premixed at the base of the flame and flames get stabilized through a balance of local mean velocity and turbulent flame speed at a location where mixture becomes stoichiometric. Muniz and Mungal [7] also came up with a similar conclusion that flame stabilizes at a region where the local incoming velocity of the reactants is near the burning velocity of laminar premixed flame. They also reported that leading-edge flame exhibits triple flame like behavior which was further supported by the works of Vervisch et al. [8] and Veynante et al. [9]. This led to the propagation of triple flame at the flame base as possible flame stabilization phenomenon. Peter and Williams [10] questioned the basic concept of premixed flame propagation theory, arguing that fuel and air are not sufficiently premixed at the stabilization base. The turbulent flame was instead seen as an ensemble of fluctuating laminar diffusion flamelets and that the flame was said to be stabilized through local quenching of these flamelets where the mean scalar dissipation rate was above the critical level. The lift-off height was thus seen as being controlled through quenching of laminar diffusion flamelets. Pitts [13] and Broadwell et al. [14] proposed that the processes responsible for flame stabilization are contained inside large scale turbulent structures (LSTS) formed as a result of the turbulent motion of incoming fuel jet stream. These structures were seen dominating the mixing of fuel and ambient air in mixing regions where stabilization of flame occurs. Schefer et al. [15] through their experimental findings suggested a model that includes both the large scale turbulent structures and premixed turbulent flame propagation. They also refuted Peter and Williams's [10] theory of flame stabilization through local quenching of diffusion flamelets, by reporting that the scalar dissipation rate at the flame base was far less than the quenching value at the flame stabilization points. Instead, the interaction between large scale turbulent structures and local stoichiometry was seen as the primary factor controlling the flame stability. Kelman and Masri [16] confirmed the findings of Schefer et al. [15] and further proposed that it is the entrainment of ambient co-flowing air into the base of flame zone by the large turbulent vortices in the fuel jet, which premixes the fresh mixture just below the flame zone. The flame is thus stabilized through premixed flame propagation. However, Tacke et al. [17] through their PLIF images showed that the stabilization point was located in the lean mixture and was insensitive to the change in stoichiometric mixture fraction of the fuel investigated. The experimental results were therefore inconsistent with the predictions of both the premixed flame propagation theory and laminar diffusion flamelet theory. Instead, the predictions by the concept of large scale turbulent structures were in correspondence with experimental results, thereby making it the only possible flame stabilization mechanism for turbulent jet flames in the ambient coflow air.

However, for lifted jet flames in the vitiated co-flow [18], the subject of present work, an autoignition has been proposed as a possible flame stabilization mechanism [19]. This flame introduced by Cabra and Myhrvold [18] is a lifted turbulent H_2/N_2 jet flame issuing into a co-flow of lean combustion products of premixed H_2/air flame. Previously, the chemical kinetics of H2/air combustion has been extensively represented through reaction mechanisms given by Li et al. [2], Mueller et al. [1], O'Conaire et al. [20] and GRI-H_2 mechanism [21]. Many numerical simulations have been performed on this burner to develop and validate new turbulent combustion models and to understand the mechanism behind flame stabilization. Turbulent modeling has usually been done using either LES [22] or RANS [23] approach. Among the combustion models, CMC (Conditional Moment Closure) [23–27], EDC (Eddy Dissipation Concept) [28] and PDF (Probability Density Functions) [18, 29–31] have been used to model the turbulence-chemistry interactions. Cabra and Myhrvold [18] through their experimental measurements and PDF calculations concluded that the flame was stabilized by turbulent premixed flame propagation enhanced by small scale recirculation and mixing of hot products into reactants and the subsequent rapid ignition of the mixture. In another attempt to investigate whether flame stabilization was actually controlled by auto-ignition dominated by chemical kinetics or premixed flame propagation controlled by mixing processes, Marsi et al. [29] employed composition PDF combustion model along with RANS k-ε turbulent method. Mueller et al. [1] and GRI2.1-H_2 [21] mechanisms were used to incorporate detailed chemical kinetics. It was observed that the lift-off height and hence the stabilization point was more sensitive to reaction rates in the chemical kinetic mechanism of H_2/O_2 combustion than to the mixing constants from the combustion model. The flame was thus found to be largely controlled by chemical kinetics rather than mixing processes. Cao et al. [30] used joint velocity turbulence frequency composition PDF method to study the Cabra and Myhrvold [18] flame. Two different chemical kinetics models (Li et al. [2] and Mueller et al. [1]) and three different mixing models were used to visualize its effect on the flame stabilization. Differences in lift-off height due to different mixing models were smaller than the differences due to different chemistry mechanisms, thus indicating chemistry controlled auto-ignition was the relevant mechanism behind flame stabilization. Gordon et al. [31], using numerical indicators, suggested auto-ignition as the possible flame stabilization mechanism for lifted flames at different co-flowing temperature. One of the indicators was based on the budget of convection, diffusion and production of key radical species such as OH and the other indicator treated HO_2 radical as auto-ignition precursor. Stankovic et al. [32] used unsteady laminar flamelet approach on Cabra H_2 flame [18], to study the sensitivity of different chemical reactions on auto-ignition delay. They concluded that for Li et al. [2] and O'Conaire et al. [20] reaction mechanisms, ignition delay times were greatly sensitive to some chemical reactions. Direct Numerical Simulations (DNS) [33–35] have also been used to simulate turbulent lifted flames and study the flame stabilization phenomenon. It was concluded from these DNS studies that the flame is stabilized by a competition of auto-ignition and large scale turbulent structures.

In view of the above-mentioned research findings, it is clear that an auto-ignition has been accepted as the most important stabilizing mechanism for lifted turbulent flames in vitiated co-flow. Also, OH has been recognized as the key radical in controlling the reaction rates and hence auto-ignition and HO_2 as the precursor to auto-ignition phenomenon. However, the role of individual reactions in reaction mechanism in the production of OH and HO_2 has not yet been explored within the framework of Multi-regime Flamelet Generated Manifold combustion modeling approach [36–40]. Multi-FGM is a generalized version of Flamelet type turbulent combustion models [41–44], which is able to incorporate detailed finite-rate chemistry only for a particular type of combustion regime, i.e. either premixed [42] or non-premixed combustion [44]. However, for flows where combustion occurs in multiple or mixed regimes, like partially premixed combustion, the pure flamelet based models (such as only diffusion or premixed) fail to model the combustion accurately. Multi-FGM uses a flame index to distinguish between premixed and non- premixed combustion regimes. While many modifications of the flame index have been formulated [39, 43, 45], the one used in the present work, is taken from literature [41, 46] which is constructed by examining whether gradients of fuel and oxidizer align. For premixed flames, the reactants approach from the same direction and their scalar product is positive while in the case of diffusion flames, the fuel and oxidizer approach the flame zone from different directions and have a negative scalar product. A detailed review can be found in the recent article [45], which elaborates on FGM and its applications from pure diffusion flame to stretched premixed flames.

In the present work, a comprehensive chemical kinetics analysis of stabilization of Cabra H_2 flame [18] is performed using Multi-FGM combustion modeling approach, coupled with standard k-ε RANS turbulent model. Two different chemical kinetics reaction mechanisms of H_2/O_2 combustion given by, Mueller et al. [1] and Li et al. [2] are used and investigated for their sensitivity towards auto-ignition. The aim of this work is to understand the sensitivity of different chemical reactions occurring inside the mixing zone towards auto-ignition, and identifying the reactions which are most sensitive to auto-ignition and hence in determining the lift-off height of the flame.

2 Multi-regime Flamelet Manifold Formulation

The Cabra H_2 flame [18] flame investigated in the current work is a partially premixed flame exhibiting the characteristics of both premixed and diffusion flames. To account for both the regimes, manifolds are generated using both laminar diffusion and premixed flamelets. The computed flamelets are obtained with the assumption of unity Lewis number (ratio of thermal diffusion to mass diffusion) for all the species. FGM model describes the thermo-chemical state as a function of mixture fraction and reaction progress variable. The reaction progress variable is defined as a linear combination of species mass fraction with constant weight as given by Eq. (1),

$$Y_c = \sum_k a_k \left(Y_k - Y_k^u\right) \tag{1}$$

where Y_c is the reaction progress variable, k is the species index, Y_k is the mass fraction of species k, superscript u denotes the unburned reactant and a_k is the weight of the kth species. The weights should be prescribed so that the reaction progress variable increases monotonically through the flame. They are typically taken as zero for reactants and unity for product species. For the present case of H_2 combustion, $a_{H_2O} = 1$ and rest all a_k are taken to be zero. A normalized reaction progress variable c defined by Eq. (2) is used, which varies from 0 at the unburned mixture to 1 in the burned zone. It is used for optimizing the performance of the PDF tables.

$$c = \frac{\sum_k a_k \left(Y_k - Y_k^u\right)}{\sum_k a_k \left(Y_k^{eq} - Y_k^u\right)} = \frac{Y_c}{Y_c^{eq}} \tag{2}$$

The Multi-regime FGM based modeling of the turbulent flame is based on the following four steps:

(i) Generation of Premixed flamelet database using FGM chemistry reduction technique.
(ii) Generation of Diffusion flamelet database using FGM chemistry reduction technique.
(iii) Calculation of PDF tables using both the flamelets and invoking them to the solver.
(iv) Using Flame Index to account for the contribution of both premixed and diffusion regimes to the final solution.

2.1 *Diffusion Flamelet Generation*

Diffusion flamelet manifold is constructed using 1-D adiabatic laminar counter flow diffusion flames. The laminar diffusion flames are transferred from physical space to mixture fraction space by solving N species mass fraction Eq. (3) and a temperature flamelet equation Eq. (4). Differential diffusion is neglected in these equations.

$$\rho \frac{\partial Y_k}{\partial t} = \frac{1}{2} \rho \chi \frac{\partial^2 Y_k}{\partial Z^2} + \dot{\omega}_k \tag{3}$$

$$\rho \frac{\partial T}{\partial t} = \frac{1}{2} \rho \chi_Z \frac{\partial^2 T}{\partial Z^2} - \frac{1}{C_p} \sum_k h_k \dot{\omega}_k + \frac{\rho \chi_Z}{2C_p} \left(\frac{\partial C_p}{\partial Z} + \sum_k C_{p,k} \frac{\partial Y_k}{\partial Z} \right) \frac{\partial T}{\partial Z} \tag{4}$$

The notation is as follow: Y_k, T, ρ, and Z are the kth species mass fraction, temperature, density and mixture fraction respectively. $C_{p,k}$ and C_p are the kth species specific heat and mixture-averaged specific heat, respectively. $\dot{\omega}_k$ and h_k are the kth species reaction rate, and specific enthalpy respectively. χ_Z is the mixture fraction scalar dissipation rate which is modelled across the flamelet as:

$$\chi_Z(Z) = \chi_{\max}^{sto} \exp\left(-2\left(erfc^{-1}(2Z)\right)^2\right) \tag{5}$$

where χ_{max}^{sto} is the maximum scalar dissipation rate which is at stoichiometric mixture fraction and $erfc^{-1}$ is the inverse complimentary error function. Steady diffusion flamelets are generated over a range of scalar dissipation rates by starting from a very small value of χ_{max}^{sto} and increasing this in increments until the flamelet extinguishes. As the strain rate increases, the flamelet chemistry departs further from chemical equilibrium and progress variable (c) decreases from unity towards the extinction reaction $c_{extinction}$. The FGM between $c_{extinction}$, and the unburnt state $c = 0$, is determined from the thermochemical states of the final, extinguishing diffusion flamelet.

2.2 Premixed Flamelet Generation

The premixed flamelets are generated using 1D unstrained laminar premixed flame configuration. The flamelets are first solved in physical space and are then transformed to reaction-progress variable space, using the flamelet equations for species mass fractions and temperature.

$$\rho \frac{\partial Y_k}{\partial t} + \frac{\partial Y_k}{\partial c} \dot{\omega}_c = \rho \chi_c \frac{\partial^2 Y_k}{\partial c^2} + \dot{\omega}_k \tag{6}$$

$$\rho \frac{\partial T}{\partial t} + \frac{\partial T}{\partial c} \dot{\omega}_c = \rho \chi_c \frac{\partial^2 T}{\partial c^2} - \frac{1}{C_p} \sum_k h_k \dot{\omega}_k + \frac{\rho \chi_c}{C_p} \left(\frac{\partial C_p}{\partial c} + \sum_k C_{p.k} \frac{\partial Y_k}{\partial c} \right) \frac{\partial T}{\partial c} \tag{7}$$

Similarly, the progress variable scalar dissipation rate χ_c is modelled as:

$$\chi_c(c) = \chi_{\max}^{sto} \exp\left(-2\left(erfc^{-1}(2c)\right)^2\right) \tag{8}$$

where χ_{max}^{sto} is the maximum scalar dissipation rate which is at stoichiometric mixture fraction. For partially premixed combustion, premixed laminar flamelets are generated over a range of mixture fractions and the solution beyond the flammability zone is interpolated between the last burned flamelet and the unburned mixture. Hence, away from the stoichiometric condition, the scalar dissipation rate of reaction progress variable at any mixture fraction (Z) is modelled as:

$$\chi_c(Z,c) = \chi_{\max}^{sto} \exp\left(-2\left(erfc^{-1}\left(\frac{Z}{Z_{sto}}\right)\right)^2\right) \exp\left(-2\left(erfc^{-1}(2c)\right)^2\right) \quad (9)$$

where the parameters remain same as defined earlier.

2.3 FGM Closure Using Presumed PDF Approach

The flamelet library created, has all the thermochemical quantities which are required to simulate a flame (like density, temperature, chemical source term) parameterized in terms of mixture fraction (Z) and reaction progress variable (c). After storing FGM in a database, it is linked to a standard CFD code which not only solves the continuity and momentum equations but also the transport equations for an un-normalized progress variable (PV) as it does not have any additional term which otherwise will be required for partially premixed combustion [46, 47]. Further, it is to be noted that the normalized progress variable is not well defined at the oxidizer boundaries. In the present study, Eq. (10) is solved for the Reynolds averaged un-normalized progress variable, $\widetilde{Y}_c$:

$$\frac{D}{Dt}\left(\bar{\rho}\tilde{Y}_c\right) = \frac{\partial}{\partial x_i}\left(\bar{\rho} D_{eff}\frac{\partial \tilde{Y}_c}{\partial x_i}\right) + \overline{\dot{\omega}}_{Y_c} \quad (10)$$

The mean source term $\overline{\dot{\omega}}_{Y_c}$ is modeled from the laminar finite rate of progress variable obtained from the flamelets. The turbulence-chemistry is then accounted for by integrating the thermochemical quantities in the 2-D table with a presumed joint *PDF* for mixture fraction and reaction progress variable, and calculate the average property as:

$$\widetilde{\phi} = \iint \phi(c, Z)P(c, Z)dcdZ \quad (11)$$

where, $\phi = \rho, T, C_p, Y_k$ or $\dot{\omega}_{Yc}$ and $\widetilde{\phi} = \tilde{T}, \tilde{Y}$ are Favre-filtered quantities in physical space.

2.4 Flame Index to Account for Multi-regime of Combustion

The Cabra H_2 flame [18], which is considered in the present work is a lifted partially premixed flame, therefore combustion at any location could be either premixed or non-premixed. This is identified by a marker called Takeno flame index, a normalized value of which is given by Eq. (12):

$$\alpha = 0.5\left(1 + \frac{\nabla Y_F \cdot \nabla Y_O}{|\nabla Y_F \cdot \nabla Y_O|}\right) \tag{12}$$

Here the subscript F and O refers to Fuel and Oxidizer respectively. A value of 0 for flame index means a pure diffusion flame while a value of 1 means a premixed flame. The reaction progress source term and mixture properties at any location inside the combustor are linearly interpolated between the two states, based on the flame index as follows:

$$\varphi_{mixture} = \alpha\varphi_{Premixed} + (1 - \alpha)\varphi_{Diffusion} \tag{13}$$

The flame index can take any value between 0 and 1, based on the combustion regime. Now for regions such as shear layers and recirculation zones, the value of flame index may vary drastically in adjacent computational cells in the domain. This sudden variation in the flame index may lead to issues of numerical stability due to fluctuations in mixture properties which are computed based on this flame index. For maintaining numerical stability, a volume-weighted smoothing is performed after obtaining the raw values of the flame index, ensuring, a smoothened transition from a diffusion to premixed flame and vice versa. The flame index is defined only in the flame zone where both fuel and oxidizer are present, whereas combustion characteristics in the post flame zone are treated same as that of the premixed regime.

2.5 Computational Details

The results presented in this paper are obtained through numerical computations performed using ANSYS Fluent [45], where we have invoked our user-defined function (UDF) to incorporate the multi-FGM feature. A schematic of the experimental setup of turbulent H_2/N_2 lifted flame, studied in the present study is shown in Fig. 1. The inlet boundary conditions for the lifted hydrogen flame as given by Cabra and Myhrvold [18] are mentioned in Table 1. The governing equations are solved in the rectangular domain which extends from 127D in the axial direction to 54D in the radial direction, where D is the diameter of the central jet equal to 4.57 mm. SIMPLE Algorithm is used for pressure-velocity coupling and 2nd order scheme is employed to discretize the diffusive and convective terms. Grid-independence study is demonstrated using two non-uniform grids: (I) a finer grid having 330 $\times$ 160 cells (II) a

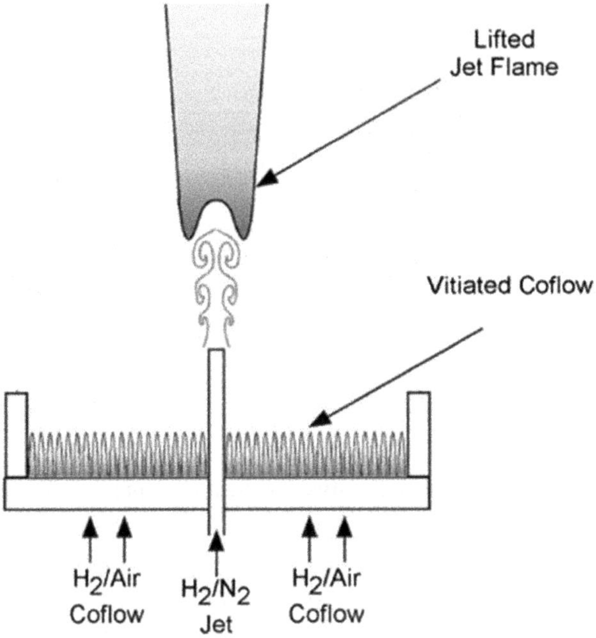

Fig. 1 Two-dimensional schematic of Cabra H_2 flame [18]

relatively coarse grid consisting of 220 × 106 cells in the radial and axial directions, respectively. The grid is non-uniform with higher cell density towards the fuel inlet. Figure 2 shows overlapping axial profiles of mean temperature and mixture fraction using all the grids. Therefore, the grid of 220 × 106 cells is accepted and used for all the further numerical computations. The laminar flamelet solution is obtained using 64 × 64 grid points in mean mixture fraction and progress variable space coordinates, where the variance of mixture fraction and progress variable are both resolved using 64 grid points. The turbulent nature of the flame is modeled through standard two equation RANS k-ε model. The chemical kinetics of H_2/air combustion is depicted by reaction mechanisms given by Li et al. [2] and Mueller et al. [1]. Mueller mechanism consists of 9 species, participating in 21 reversible reaction steps, whereas Li mechanism consists of 11 species participating in 26 reactions. The OH contours shown in the results section are clipped with a minimum value—of OH mass fraction equal to 2×10^{-4}, which is considered as a marker for the onset of auto-ignition, i.e. lift-off height [31].

Table 1 Inlet conditions for Cabra H_2 jet flame [18]

	Jet	Co-flow
Velocity (m/s)	107	3.5
Temperature (K)	305	1045
Turbulent kinetic energy (m^2/s^2)	1.0	0.0726
Dissipation rate (m^3/s^3)	1.0	19.56
Y_{H2}	0.02325	0
Y_{O2}	0	0.1709
Y_{H2O}	0	0.0645
Y_{N2}	0.97675	0.7646

3 Results and Discussion

The results of the chemical kinetics analysis of the turbulent lifted flame are presented in the following sequence:

I. Effect of Li et al. [2] and Mueller et al. [1] chemical mechanisms using one-dimensional calculation using CHEMKIN [48].
II. Turbulence modeling correction using different combinations of turbulence constants.
III. The sensitivity of the Lift-off height on the reaction rates by distinguishing the most dominant set of reactions, responsible for controlling the autoignition.

3.1 Effect of Chemical Mechanism

As chemical reactions play an important role in the calculation of the random spots leading to the lifted flame stabilization in the numerical computation, in this section we solely focus on the effect of the two different chemical mechanisms (Li et al. [2] and Mueller et al. [1]) using an idealized model problem. CHEMKIN [48] is used to simulate the one-dimensional channel, preferentially a plug-flow reactor, operating under a lean but reactive mixture issuing from one end at a velocity of 100 ms^{-1}. The composition of the fluid flow entering the channel is as follows: mixture fraction $(Z)=0.05$, temperature $(T)=1003.03$ K, and mass fractions of species are $Y_{H2}=0.00117129$, $Y_{O2}=0.162326$, $Y_{H2O}=0.0613191$ and rest of the mass fraction corresponds to the nitrogen (obtained from measurements of [18]). The mixture lies along the mixture line formed by the jet and pilot fluid considered in the present paper.

Figure 3 shows the variations of the autoignition distance in the channel using two different chemical mechanisms. Autoignition distance is determined using the quantification of the maximum differential temperature (dT/dx) along the channel setup. Using Mueller et al. [1] mechanism, the mixture auto-ignites at an axial dis-

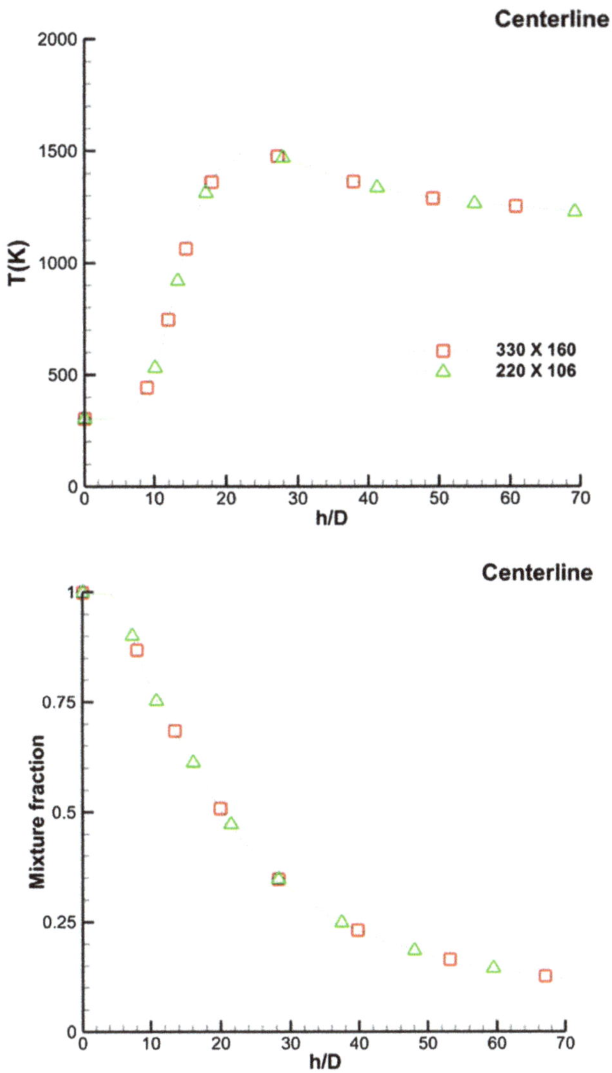

Fig. 2 Grid independence study for centerline axial plot of temperature and mean mixture fraction: □: 330 × 160 mesh, △: 220 × 106 mesh

tance x = 190 mm, for inlet condition of T = 1003.03 K. For this case, the maximum peak temperature obtained is 1116.58 K.

These results are consistent and in agreement with the turbulent flame studies considered by Masri et al. [29]. Further, an increment of the temperature equivalent to 5 K is followed up till the coflow temperature, considered in the present study, i.e. 1045 K. The increase in the temperature clearly shows the decrement in the ignition

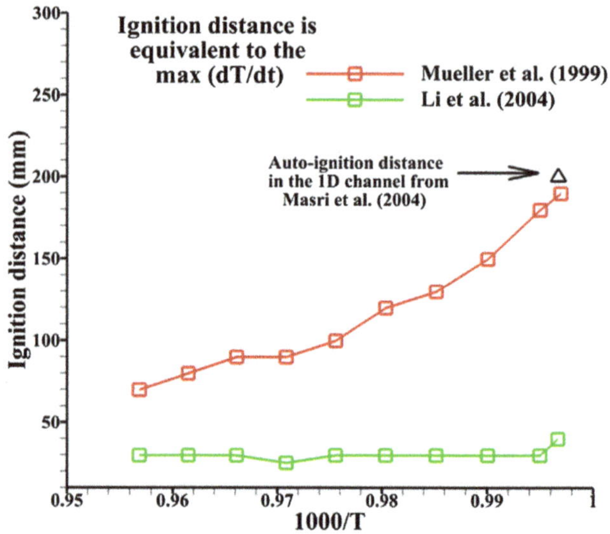

Fig. 3 Variation of ignition distance (mm) with 1000/T (K) in a one-dimensional channel behaving as plug flow reactor

distance using Mueller mechanism [1], whereas stagnation condition is observed in the ignition distance with increment in the case of Li mechanism [2]. These calculations clearly signify the importance of the chemical kinetics in controlling the autoignition region and results using Mueller mechanism [1] correctly the captures the shift in the autoignition distance which concerns chemistry at relatively low temperature, particularly in the order of 1000 K or less. Hence, the Mueller mechanism [1] is considered for further study in the current computations.

3.2 Turbulence Modeling Correction

The Standard RANS k-ε model used to model the turbulent nature of the lifted flame, is modified to account for the representation of the experimentally determined Cabra H_2 flame [18]. Numerical simulations are carried out with two different modifications of Standard k-ε model. In the first modification, $C_{\varepsilon 2} = 1.8$ (Case A) is used, and in the other $C_{\varepsilon 1} = 1.6$ (Case B) is applied. Usually, the standard k-ε model is well not known for not being able to predict the round-jet accurately, that's why some modifications are made to account for the round-jet corrections. The results for these modified constants along with Standard k-ε model and Reynold's Stress Model (RSM) with default constants are compared with the experimental data. Figure 4 depicts both the centerline and radial distribution of temperature, mean mixture fraction, H_2O, and OH mass fractions. It can be seen from Fig. 4 that the centerline axial profile for all the reactive scalars, is best captured by Case A modification to the standard k-ε model, as compared to Case B and RSM model as well. The axial distribution of temperature and H_2O mass fraction is under-predicted whereas the mean mixture fraction is over-predicted by Case B modification is due to the underprediction of the production term, indicating that the spread of jet is more in Case B. Jet spreading, shown by the radial profiles at $x/D = 14$, is also well represented by case A. However, the peak mass fraction of OH radicals is not well captured in the reaction zone. Case A predicts a reaction zone close to the jet exit as opposed to the experimentally determined mixing zone. Consequently, at this location, the radial profile for temperature, OH, and the H_2O mass fraction is over predicted. While looking at the OH mass fraction profiles, all the considered turbulence models appear to underpredict the lift-off height. The greater jet spread for Case B as seen in Fig. 4, causes the mixing zone to be pushed further downstream resulting in increased lift-off for Case B. Hence, Case B shows its inability to accurately capture the axial and radial profile makes its unsuitable for the further analysis. On the other hand, Case A is able to predict the flame behavior more accurately and is therefore chosen for further analysis in this study.

3.3 Distinguishing the Dominance of Reaction Sets in Controlling Auto-Ignition

Mueller mechanism, combined with modified k-ε turbulence model ($C_{\varepsilon 2} = 1.8$) as per discussed in the previous sub section, is used to study the impact of different chemical reactions on stabilization of the Cabra H_2 flame [18]. Table 2 lists down the two most dominant set of reactions in Mueller mechanism towards auto-ignition, as reported by Masri et al. [29]. On increasing the reaction rate constants, Set A reactions are shown to speed up the auto-ignition, whereas Set B reactions cause a delay in auto-ignition. The reactions in Set A are concerned with controlling the generation of the OH radical pool which is identified as the reaction rate determinant [31] and thus playing a prominent role in the onset of auto-ignition. Set B reactions describe the formation and consumption of HO_2, which plays the role of an auto-ignition precursor [31]. The reaction rate constants of both the set of reactions (Set A and Set B) are now altered to study the sensitivity of the two sets of reactions towards auto-ignition. The modified cases considered for Mueller reaction mechanism with respect to Set A and Set B reaction rates (A–F) and the corresponding lift-off height observed are tabulated in Table 3. The comparison of their OH contours, axial and radial profiles are shown in Figs. 5, 6 and 7, respectively. The following points are worth noting with regard to the comparison of OH contours in Fig. 5.

- The increase in the lift-off height on doubling the reaction rate of Set B reactions as shown in Fig. 5(i) is consistent with the notion of Set B delaying the auto-ignition. The increase in the reaction rate of Set B reactions, increase the consumption of HO_2 radicals, thereby pushing the mixing zone further downstream and delaying the auto-ignition.

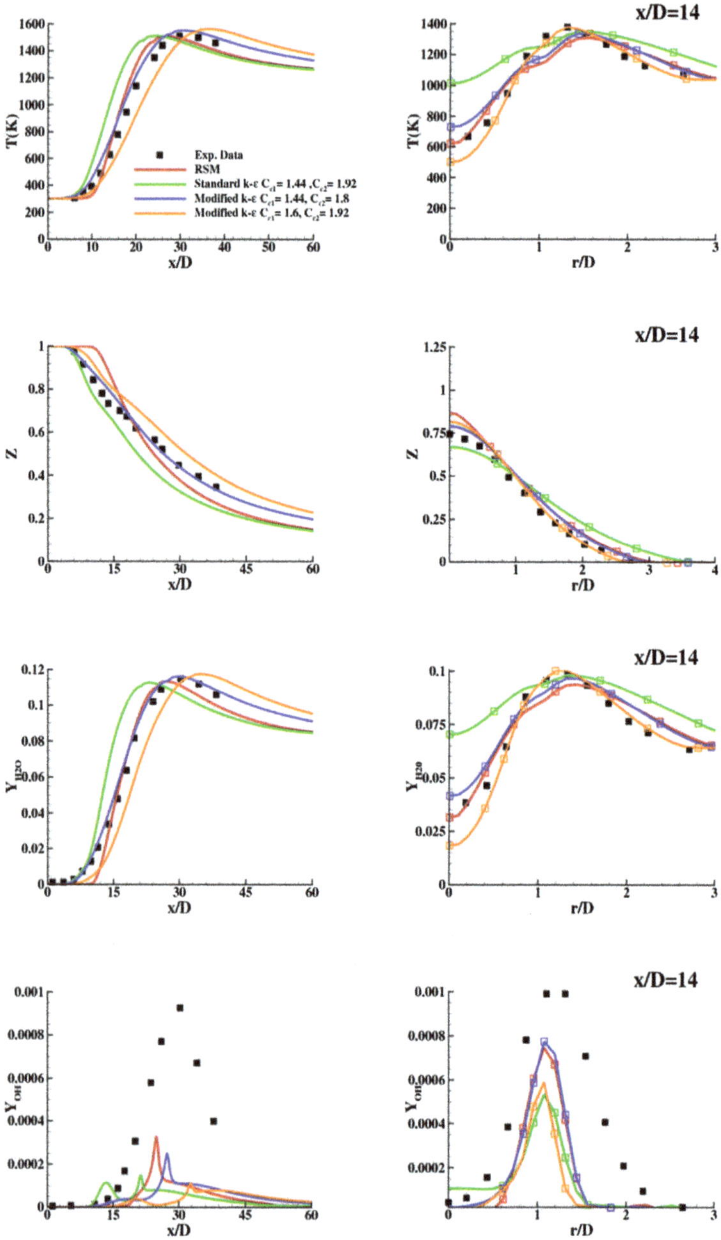

Fig. 4 Effect of different turbulence model parameters on the centerline and radial distribution of *T*, *Z*, Y_{OH} and Y_{H2O}

Table 2 Reactions contained in the Set-A and Set-B

Set-A (Increasing the reaction rate speeds up auto-ignition)	A1: $O_2 + H \leftrightarrow O + OH$ A2: $O + H_2 \leftrightarrow H + OH$ A3: $H_2 + OH \leftrightarrow H_2O + H$
Set-B (Increasing the reaction rate delays auto-ignition)	B1: $HO_2 + OH \leftrightarrow H_2O + O_2$ B2: $HO_2 + O \leftrightarrow OH + O_2$ B3: $HO_2 + H \leftrightarrow H_2 + O_2$

Table 3 Description of the reaction rates for Set-A and Set-B with their correspondingly calculated lift-off heights

Case	Set A	Set B	Lift-off height (h/D)
A	Unchanged	Unchanged	4.29
B	Doubled	Unchanged	4.20
C	Unchanged	Doubled	7.81
D	Halved	Unchanged	8.58
E	Halved	Doubled	9.02
F	One third	Doubled	25.008

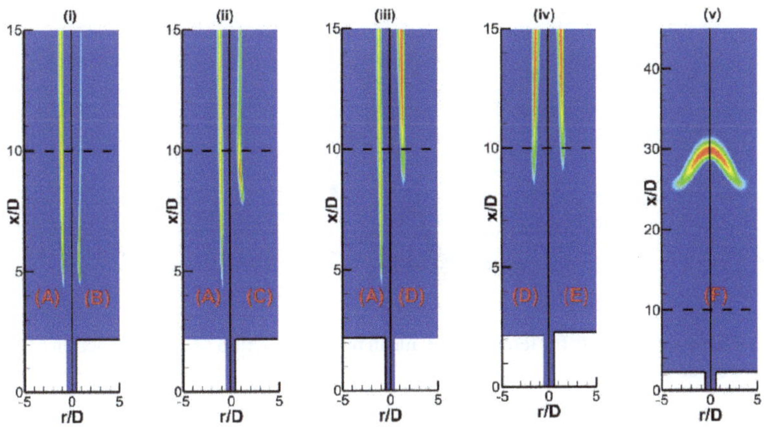

Fig. 5 OH contour comparisons for cases (A–F)

- The lift-off height in Fig. 5(ii) is not much affected on doubling the reaction rates of Set A. However, in Fig. 5(iii), on reducing the Set A reaction rates by a factor of half, the lift-off height is increased by a factor of two. Masri et al. [29] doubled the reaction rate of A1 in Set A and observed a decrease in lift-off height by a factor of two. Thus the role of Set A reactions in speeding up auto-ignition is verified.
- The increase in lift-off height on halving Set A reaction rate is larger than the increase observed on doubling the Set B reactions rates. This signifies the difference in sensitivity of Set A and Set B reaction rates towards flame stabilization.

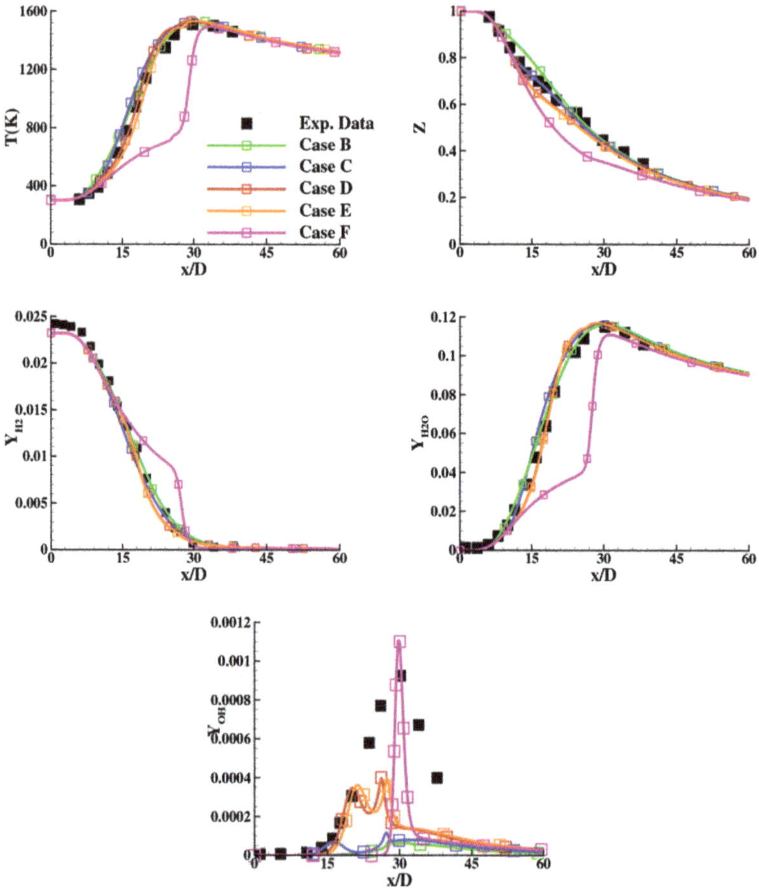

Fig. 6 Centerline axial plots for temperature, mean mixture fraction (Z), mass fraction of H_2, H_2O and OH for cases B–F

- The slight increase in lift-off height from 8.58D (in Case of D) to 9D (in Case of E) as shown in Fig. 5(iv), hints towards the lesser dominance of Set B reactions.
- The lift-off height, in Fig. 5(v), increases by a factor of 2.7 on further reducing the Set A reaction rates from half (case D) to one-third (case E). This marked increase in lift-off height clearly shows that the Set A reactions are more sensitive than Set B reactions towards auto-ignition and thus in determining the lift-off height.

Figure 6 shows the centerline axial profile of temperature, mean mixture fraction, the mass fraction of H_2, H_2O and OH for all the modified mechanisms (B–F), compared with the experimental data. The results simulated by all the cases except case F are in good agreement with the experimental data. However, the peak value of OH mass fraction in Fig. 6(v) is under predicted by all the mechanisms except case F, which over predicts the peak value.

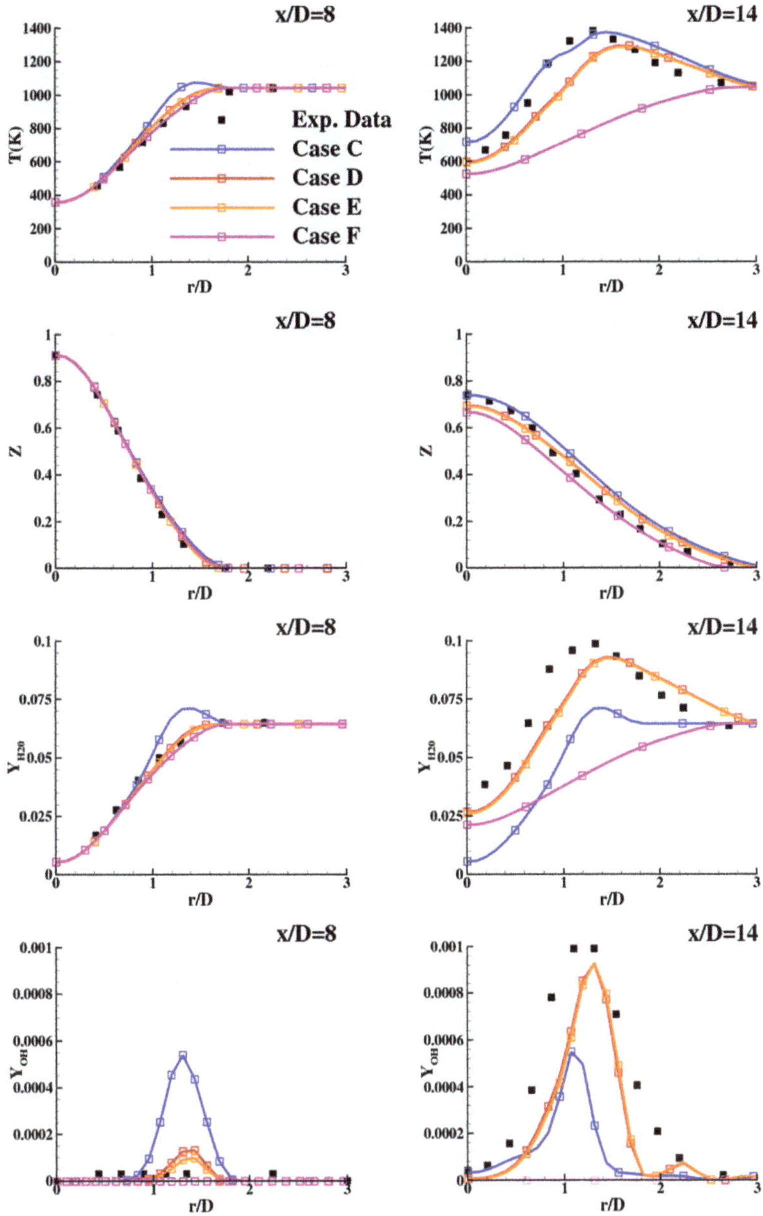

Fig. 7 Radial plots for temperature, mean mixture fraction (Z), mass fraction of OH and H_2O for cases C–F at $X/D=8$ and $X/D=14$

Figure 7, shows the radial profile of temperature, mean mixture fraction, the mass fraction of H_2O and OH at axial locations of $x/D=8$ and $x/D=14$. The radial profile

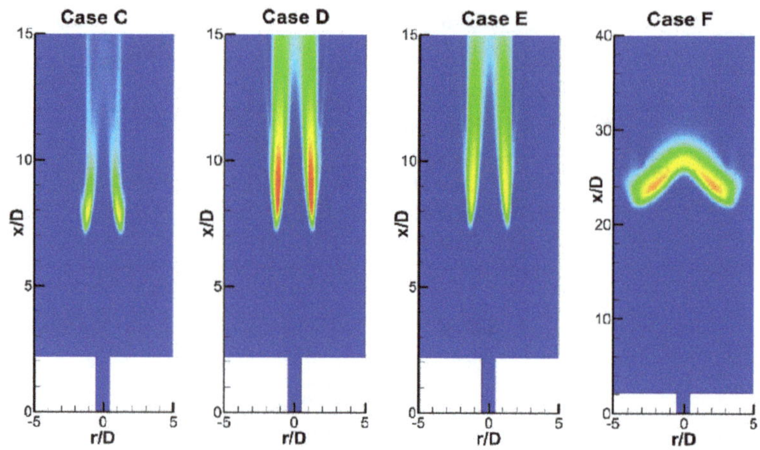

Fig. 8 Distribution of the HO_2 contours in two-dimensional domain for cases C–F

for all the quantities, predicted by case D and case E is in close agreement with the experimental values. The peak OH value at $x/D = 14$ (which is the experimentally determined reaction zone), captured by mechanism D and E, is very close to experimental value.

These results show that chemical kinetics is important in simulating the experimental flame structure. Some of the modified mechanisms (C & D) are able to predict the reactive scalars and the lift-off height better than the unmodified Mueller Mechanism. An optimum chemical mechanism to model auto-ignition of H_2 is therefore very important in simulating such reaction rate dependent controlled auto-ignition phenomenon in vitiated flame. The role of HO_2 as auto-ignition precursor is observed in Fig. 8. The contours for the cases (B–F) show a build-up of HO_2 radical just prior to flame stabilization zone, where the concentration of OH is minimum. Thus, it is inferred that the peak value of HO_2 mass fraction plays a prominent role in determining the extent of mixing zone and thereby controlling the auto-ignition location. The peak mass fraction of OH and HO_2 radicals for the considered cases (C–F) are depicted in Fig. 9. It can be seen from the bar graph that while the peak mass fraction of HO_2 does not vary much on altering the Set B reaction rates, the peak mass fraction of OH does change considerably on modifying Set A reaction rates. Thus asserting that Set A reactions are more sensitive in the control and production of OH radicals than Set B reactions are towards HO_2 radical. This further yields to the earlier conclusion that auto-ignition is more sensitive towards chemical kinetics rates of Set A reactions than Set B reactions. The lift-off height is thus seen to be more controlled by OH radicals which are dependent on Set A reactions rather than HO_2 radicals which are Set B dependent.

Fig. 9 Bar graph showing the peak mass fractions of OH and HO_2 for cases C–F (i–iv)

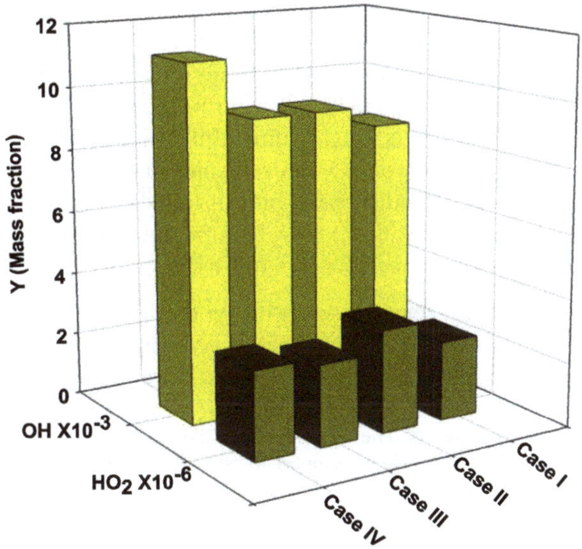

4 Conclusion

In the present work, Multi-FGM combustion modeling approach is used to analyze the sensitivity of chemical kinetic aspects of lifted turbulent Cabra H_2 flame [18]. Reaction mechanisms for H_2/O_2 combustion given by Li et al. [2] and Mueller et al. [1] are analyzed using one-dimensional computations representing channel flow as a plug-flow reactor. A modified k-ε RANS turbulence model is used to closely predict the experimental data. In light of the results presented in the previous section, following conclusions are made:

1. The Cabra H_2 flame [18] is found to be greatly dependent on the choice of the reaction mechanism, with Li and Mueller over-predicting and under-predicting the lift-off height, respectively. The lift-off height by Mueller mechanism is, however, found to be closer to the experimental value and was therefore used for further analysis.
2. Spreading of the turbulent fuel jet is corrected by modifying the turbulence constants, where $C_{\varepsilon 1} = 1.44$ and $C_{\varepsilon 2} = 1.8$ show better axial predictions along the centerline of the burner. However, peak location of the mass fraction of OH radicals is not well captured in the reaction zone.
3. A sensitivity study on Set A and Set B reactions illustrated, lift-off height to be more sensitive towards Set A reactions than Set B reactions.
4. The build-up of HO_2 radical prior to auto-ignition shows the vital role it plays as an auto-ignition precursor. It is thus inferred that the peak value of HO_2 plays an important role in determining the lift-off height, by signifying the amount of

mixing taking place in the pre-ignition zone leading to the auto-ignition phenomenon.

5. The dominance of Set A reactions (controlling OH radicals) over Set B reactions (controlling HO_2 radicals), is further depicted by the enhanced peak mass fraction of OH, which is roughly three times higher than a peak HO_2 mass fraction.
6. The peak value of HO_2 radical is almost consistent on altering Set B reaction rates, whereas OH radical peak value is found to vary on tuning the Set A reaction rates, suggesting the flame lift-off to be auto-ignition controlled rather than mixing control.

The highly sensitive variation in the lift-off height using ranges of the coflow temperatures and velocities are observed and increment in the coflow velocity leads to the increase in the lift-off height at lower coflow temperature.

Acknowledgements The authors would like to acknowledge the IITK computer center (www.iitk.ac.in/cc) for providing support to perform the computation work, data analysis, and article preparation.

References

1. M.A. Mueller, T.J. Kim, R.A. Yetter, F.L. Dryer, Flow reactor studies and kinetic modelling of the H_2/O_2 reaction. Int. J. Chem. Kinet. **31**, 113–125 (1999)
2. J. Li, Z. Zhao, A. Kazakov, F.L. Dryer, An updated comprehensive kinetic model of hydrogen combustion. Int. J. Chem. Kinet. **36**(10), 566–575 (2004)
3. B.J. Lee, S.H. Chung, Stabilization of lifted tribrachial flames in laminar non-premixed jets. Combust. Flame **109**, 163–172 (1997)
4. O. Savas, S.R. Gollahalli, Stability of lifted laminar round gas-jet flame. J. Fluid Mech. **165**, 297–318 (1986)
5. W.M. Pitts, Assessment of theories for the behavior and blowout of lifted turbulent jet diffusion flames. Combust. Flame **22**(1), 809–816 (1989)
6. K.M. Lyons, Toward an understanding of the stabilization mechanisms of lifted turbulent jet flames: experiments. Prog. Energy Combust. Sci. **33**(2), 211–231 (2007)
7. L. Muniz, M.G. Mungal, Instantaneous flame-stabilization velocities in lifted-jet diffusion flames. Combust. Flame **111**, 16–31 (1997)
8. L. Vervisch, G.R. Ruetsch, A. Linan, Effects of heat release on triple flames. Phys. Fluids **7**, 1447–1454 (1995)
9. D. Veynante, L. Vervisch, T. Poinsot, A. Linan, G.R. Ruetsch, Triple flame structure and diffusion flame stabilization. in *Proceedings of Summer Program*, Center for Turbulence Research, NASA Ames/Stanford University (1994)
10. N. Peters, F.A. Williams, Lift-off characteristics of turbulent jet diffusion flames. AIAA J. **21**, 423–429 (1983)
11. J. Janicka, N. Peters, Prediction of turbulent jet diffusion flame lift-off using PDF transport equation. Combust. Flame **19**(1), 367–374 (1982)
12. N. Peters, Partially premixed diffusion flamelets in non-premixed turbulent combustion. Combust. Flame **20**(1), 353–360 (1985)
13. W.M. Pitts, Large-scale turbulent structures and the stabilisation of lifted turbulent jet diffusion flames. Combust. Flame **23**(1), 661–668 (1991)
14. J.E. Broadwell, J.A. Werner, M. Dahm, G. Mungal, Blowout of turbulent diffusion flames. Proc. Combust. Inst. **20**(1), 303–310 (1985)

15. R.W. Schefer, M. Namazian, J. Kelly, Stabilization of lifted turbulent-jet flames. Combust. Flame **99**, 75–86 (1994)
16. J.B. Kelman, A.R. Masri, Laser imaging in the stabilisation region of turbulent lifted flames. Combust. Sci. Technol. **135**, 117–134 (1998)
17. M.M. Tacke, D. Geyer, E.P. Hassel, J. Janicka, A detailed investigation of the stabilization point of lifted turbulent diffusion flames. Proc. Combust. Inst. **27**, 1157–1165 (1998)
18. R. Cabra, T. Myhrvold, Simultaneous laser Raman-Rayleigh-LIF measurements and numerical modeling results of a lifted turbulent H2/N2 jet flame in a vitiated coflow. Proc. Combust. Inst. **29**, 1881–1888 (2002)
19. T. Echekki, K.G. Gupta, Hydrogen autoignition in a turbulent jet with preheated co-flow air. Int. J. Hydrog. Energy **34**, 8352–8377 (2009)
20. M. O'Conaire, H.J. Curran, J.M. Simmie, W.J. Pitz, C.K. Westbrook, A comprehensive modeling study of hydrogen oxidation. Int. J. Chem. Kinet. **36**, 603–622 (2004)
21. C.T. Bowman, R.K. Hanson, D.F. Davidson, Gardiner Jr, V. Lissianski, G.P. Smith, D.M. Golden, M. Goldenberg, M. Frenklach, Gri-Mech 2.11. http://www.me.berkeley.edu/gri-mech/ 1999
22. W.P. Jones, S.N. Martinez, Large eddy simulation of autoignition with a subgrid probability density function method. Combust. Flame **150**, 170–187 (2007)
23. S.S. Patwardhan, S. De, K.N. Lakshmisha, B.N. Raghunandan, CMC simulations of lifted turbulent jet flame in a vitiated coflow. Proc. Combust. Inst. **32**, 1705–1712 (2009)
24. S. De, A. De, A. Jaiswal, A. Dash, Stabilization of lifted hydrogen jet diffusion flame in a vitiated co-flow: effects of jet and coflow velocities, coflow temperature and mixing. Int. J. Hydrog. Energy **41**, 15026–15042 (2016)
25. A.E. Sayed, R.A. Fraser, Consistent conditional moment closure modelling of a lifted turbulent jet flame using the presumed ß-PDF approach. J. Combust. (2014). https://doi.org/10.1155/2014/507459
26. S.N. Martinez, A. Kronenburg, Flame stabilization mechanisms in lifted flames. Flow Turbulent Combust. **87**, 377–406 (2001)
27. I. Stankovic, B. Merci, LES-CMC simulations of turbulent lifted hydrogen flame in a vitiated air co-flow. Therm. Sci. **17**, 763–772 (2013)
28. T. Myhrvold, I.S. Ertesvag, I.R. Gran, R. Cabra, A numerical investigation of a lifted H_2/N_2 turbulent jet flame in a vitiated coflow. Combust. Sci. Technol. **178**, 1001–1030 (2006)
29. A.R. Masri, R. Cao, S.B. Pope, G.M. Goldin, PDF calculations of turbulent lifted flames of H_2/N_2 fuel issuing into a vitiated coflow. Combust Theory Model **8**, 1–22 (2004)
30. R.R. Cao, S.B. Pope, A.R. Masri, Turbulent lifted flames in a vitiated coflow investigated using joint PDF calculations. Combust. Flame **142**, 438–453 (2005)
31. R.L. Gordon, A.R. Masri, S.B. Pope, G.M. Goldin, A numerical study of auto-ignition in turbulent lifted flames issuing into a vitiated co-flow. Combust Theory Model **11**, 351–376 (2007)
32. I. Stankovic, B. Merci, Analysis of auto-ignition of heated hydrogen–air mixtures with different detailed reaction mechanisms. Combust Theory Model **15**(3), 409–436 (2011)
33. C.S. Yoo, R. Sankaran, J.H. Chen, Three-dimensional direct numerical simulation of a turbulent lifted hydrogen jet flame in heated coflow: flame stabilization and structure. J. Fluid Mech. **640**, 453–481 (2009)
34. K. Luo, H. Wang, F. Yi, J. Fan, Direct numerical simulation study of an experimental lifted H_2/N2 Flame. Part 1: validation and flame structure. Energy Fuels **26**, 6118–6127 (2012)
35. H. Wang, K. Luo, F. Yi, J. Fan, Direct numerical simulation study of an experimental lifted H_2/N2 flame. Part 2: flame stabilization. Energy Fuels **26**, 4830–4839 (2012)
36. E. Knudsen, H. Pitsch, Capabilities and limitations of multi-regime flamelet combustion models. Combust. Flame **159**, 242–264 (2012)
37. Y. Wu, C. Cao, Y.E. Taohong, L.I.N. Qizhao, A new multi-dimensional flamelet generated manifolds approach for approximating partially premixed flame structure. J. Therm. Sci. Technol. JPN **10**(1), 1–15 (2015)

38. W. Xu, L. Yujian, L. Kun, J. Hanhui, F. Jianren, LES of pulverized coal combustion with a multi-regime flamelet model. Fuel **188**, 661–671 (2017)
39. M.U. Göktolga, J.A. van Oijen, L.P.H. de Goey, Modeling MILD combustion using a novel multistage FGM method. Proc. Combust. Inst. **36**(3), 4269–4277 (2016)
40. P.D. Nguyen, L. Vervisch, V. Subramanian, P. Domingo, Multidimensional flamelet-generated manifolds for partially premixed combustion. Combust. Flame **157**(1), 43–61 (2010)
41. B. Fiorina, O. Gicquel, L. Vervisch, S. Carpentier, N. Darabiha, Approximating the chemical structure of partially premixed and diffusion counterflow flames using FPI flamelet tabulation. Combust. Flame **140**(3), 147–160 (2005)
42. J.A. Van Oijen, R.J.M. Bastiaans, L.P.H. de Goey, Low-dimensional manifolds in direct numerical simulations of premixed turbulent flames. Proc. Combust. Inst. **31**, 1377–1384 (2007)
43. P. Domingo, L. Vervisch, D. Veynante, Large-eddy simulation of a lifted methane jet flame in a vitiated coflow. Combust. Flame **152**, 415–432 (2008)
44. M. Ihme, Y.C. See, Large-eddy simulation of a turbulent lifted flame in a vitiated co-flow. Flow Turbulent Combust **87**, 407–423 (2011)
45. J.A. Van Oijen, A. Donini, R.J.M. Bastiaans, J.H.M. ten Thije Boonkkamp, L.P.H. de Goey, State-of-the-art in premixed combustion modeling using flamelet generated manifolds. Prog. Energy Combust. Sci. **57**, 30–74 (2016)
46. ANSYS Fluent 16.0 User's guide, Canonsburg, PA, USA
47. R. Yadav, A. De, S. Jain, A hybrid flamelet generated manifold model for modeling partially premixed turbulent combustion flames. in *Proceedings of ASME TurboExpo* (2017). GT2017-65030
48. R.J. Kee, F.M. Rupley, J.A. Miller, Chemkin-II: a Fortran chemical kinetics package for the analysis of gas-phase chemical kinetics. No. SAND-89-8009. Sandia National Labs., Livermore, CA (1989)

Part IV
Two-Phase Modeling and Experiments

Mixing Dynamics in Interacting Vortices

Saptarshi Basu, Swetaprovo Chaudhuri, Baki M. Cetegen and Abhishek Saha

Abstract Mixing dynamics arising from the interaction of two convecting line vortices of different strengths and time delays have been investigated in this chapter. Experimentally obtained planar laser induced fluorescence images of acetone vortices mixing in air are thoroughly substantiated with computational modeling and analysis. Same and opposite direction of rotation, generating vortex pairs and couples were found to augment and dissipate species mixing respectively as indicated by global scalar statistics like mean and variance. Local mixing characteristics are analyzed using joint probability density function of vorticity and species concentration. This quantitatively showed how mixing is dissipated or augmented for individual vortices and at different vorticity magnitudes, even by the decaying and apparently indiscernible presence of a favorable or unfavorable direction of rotation of the preceding vortex.

1 Introduction

Flow induced mixing of different fluids (gas or liquid) is a fundamental process in many natural and engineered applications [1–10]. As an example, turbulent mixing

S. Basu (✉)
Department of Mechanical Engineering, Indian Institute of Science Bangalore, Bengaluru, India
e-mail: sbasu@iisc.ac.in

S. Chaudhuri
Department of Aerospace Engineering, Indian Institute of Science Bangalore, Bengaluru, India

B. M. Cetegen
Department of Mechanical Engineering, University of Connecticut Storrs, Connecticut, USA

A. Saha
Department of Mechanical and Aerospace Engineering, Princeton University, Princeton, NJ, USA

A. K. Runchal et al. (eds.), *Energy for Propulsion*, Green Energy and Technology,
https://doi.org/10.1007/978-981-10-7473-8_13

is ubiquitous in nature and is employed in many thermal systems such as fuel premixers, gas turbine combustion chambers, etc. Interacting vortices/eddies constitute units of turbulence and hence are abundantly seen in physical sciences like in atmospheric studies, thermal plumes [10], volcanoes, engineering applications like bluff body separated flows, microfluidics and biological processes like cardiac flows or propulsion mechanism of sea mammals. These vortices can be of different strengths and sizes. Organized, coherent vortices and their interactions assist in scalar transport and mixing augmentation in both reacting and non-reacting flows encountered in different applications [10].

One methodology commonly adopted in order to achieve an in-depth understanding of the complicated mixing dynamics in turbulent flows, has been to study the mixing characteristics in a flow field perturbed by vortical structures such as isolated vortices and interacting vortices such as vortex pairs, vortex couples, etc. Controlled interaction of vortices of a given strength and timing serves an important problem in understanding the fundamental steps to understand mixing processes in turbulent scalar transport.

Passive scalar mixing by vortices in non-reacting convective diffusive flows have been a subject of continued interest in the fluid dynamics community. Excellent reviews on the analysis of vortical flows can be found in the works of Renard et al. [10]. The complexity of mixing phenomenon introduced in the domain of a vortical flow makes this an active research area as of date as evidenced by many recent works on this subject. Villermaux and Duplat [11] showed mixing by aggregation in an experimental study of a stirred scalar reactor. Meunier and Villemaux [12] described the mixing phenomenon in a vortex by considering advective-diffusive behavior of a scalar blob in the deformation field of an axisymmetric vortex. Bajer et al. [13] analytically studied the mixing in a vortex with focus on the enhancement of scalar destruction at the vortex core. In microfluidic devices operating at a regime of very low Reynolds number, vortex interaction is a widely used method for passive mixing. Recently, Long et al. [14] discussed the usage of vortical flows in a 3D micro mixer. Sritharan et al. [15] also used vortex induced flows to enhance mixing in microchannels. They used acoustic pulses to generate streaming within the flow, which resulted in interacting vortices enhancing mixing.

Motivated by these computational, analytical and experimental studies, we previously reported a computational study [16] of vortical mixing in a gaseous flow field in the same experimental configuration of Cetegen [17]. However, the previous experimental and computation efforts involved scalar mixing analysis in a single isolated vortex. A more useful and fundamental problem is the fluid mechanics and mixing characteristics of interaction of two interacting vortices created in a laminar diffusion layer in the same experimental configuration of Cetegen [17]. This chapter concerns qualitative experimental results substantiated with the computational modeling and quantitative analysis. The modeling effort used the same experimental configuration for the ease of qualitative comparison. In the following sections, we first describe the experimental methods followed by a brief description of the computational model.

2 Problem Description: Experimental and Computational

An experimental and computational study of scalar mixing in the field of two successively generated counter rotating laminar vortices at the interface of two gas streams (one seeded with acetone and another unseeded) flowing parallel to each other in a rectangular flow channel is presented (Fig. 1). The isolated vortex evolutions and mixing in the field of a single vortex has been studied in detail by Cetegen [17]. In this study, a line vortex was initiated by rapidly increasing one of the stream velocities in relation to the other in otherwise equal velocity, co-flowing streams separated upstream by a splitter plate (Fig. 1a). Using planar laser induced fluorescence of acetone, the mixing field was characterized in terms of probability density distributions of the passive scalar and its moments. In the present study, a second vortex rotating in the opposite direction is created in a similar fashion in the other or the same gas stream with variable time lag with respect to the first vortex (Fig. 1a). Interacting vortices introduced in opposite streams are referred to as vortex couple (studied both experimentally and computationally) while the vortex generated in the same gas stream is called vortex pair (only studied computationally). The dynamics of the interacting vortices depends on several key variables, namely a.) the relative strengths (A1/A2 in Fig. 1a), the time delay between the introduction of the two vortices and c.) the side from which the second vortex is introduced (vortex pair or couple).

Planar laser induced fluorescence of acetone is utilized to visualize the scalar concentration distribution in this flow field. Similar flow field and scalar distribution was analyzed using a computational model to quantify the mixing dynamics in dual vortex configuration. A parametric study was conducted to determine the effects of time lag and convection time on scalar mixing characteristics in the interaction zone of the two vortices. Figure 1a top (bottom) row shows the configuration in which the second vortex is introduced on the opposite (same) side as the first vortex with a controlled phase lag. Commercial code fluent 6.3.16 was used for all the computational simulations. Using experiments and computations, we have attempted to visualize and analyze the mixing dynamics in a flow field of interacting vortices, respectively.

2.1 Experimental Setup

The experimental setup and methodology has been described by Cetegen [17] in details and is restated here for completeness. The same experimental setup [17] was modified to account for the generation of two vortices. The experimental setup [17] comprises of a two dimensional convergent nozzle divided into two equal zones by a splitter plate as schematically shown in Fig. 1b. The dimension of each nozzle segment is 2.0×6.0 cm at the exit plane. The optical test section which extends beyond the nozzle exit is enclosed by four quartz windows (10 cm long). Flow conditioning

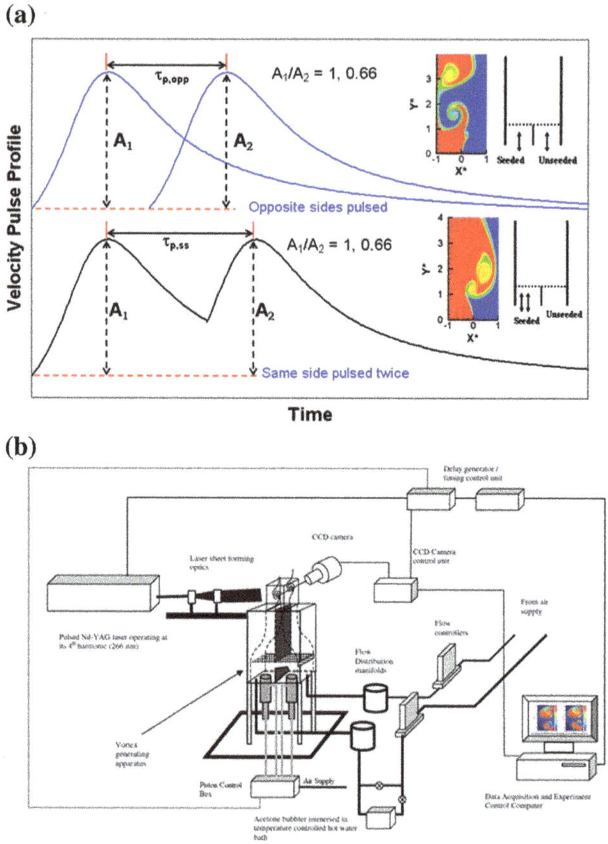

Fig. 1 **a** Pulse profiles (same side [vortex pair] and opposite sides [vortex couple]) for generating the interacting vortices, **b** Experimental setup for laser induced fluorescence of acetone (Adapted from 17)

is accomplished in the nozzle section using multiple layers of honeycombs and wire meshes. At the bottom of both of the chambers were 5 cm diameter cylindrical volumes into which two Teflon pistons (one for each chamber) were fitted. The Teflon pistons were actuated by electro pneumatic solenoids with a maximum stroke length of 26 mm. Different magnitudes of air pressure actuating the solenoid valves were utilized to initiate vortices with different magnitudes of circulation. Both the nozzle sections (for seeded and unseeded flow) were fed separately by metered compressed dry bottled air (Airgas industrial grade) using electronic mass flow controllers (Porter Instruments, Model 202). One of the air streams was partially bypassed and bubbled through an acetone filled container that was maintained at a constant temperature of 60 °C by immersing into a temperature controlled water bath (Fischer Scientific, Isotemp 105). This resulted in effective seeding of one of the air stream. Dynamic adjustment and control of the water bath temperature and flow bypass was carried out to ensure LIF signal with good signal to noise ratio. The vortex generation was initiated from both the acetone-seeded air stream and the unseeded air stream at different levels of phase lags.

2.1.1 Planar Laser-Induced Fluorescence of Acetone

Laser induced fluorescence from acetone vapor was used as the preferred experimental technique in this study [17]. Normally seeding the flow with particles (like smoke) and using Mie scattering for concentration measurement is not a suitable technique for the current study. Smoke does not exhibit molecular diffusion and hence is not an inappropriate marker of mixing dynamics. Acetone fluorescence was selected because it acts as a molecular tag with diffusivity similar to the self-diffusion coefficient of air. Laser induced acetone fluorescence has been widely used for concentration and temperature measurements in both reacting and non-reacting gas flows. In this implementation as described by Cetegen [17], the fourth harmonic beam ($\lambda =$ 266 nm) of a pulsed Nd-YAG laser (Continuum YG-681-10) was expanded to form a 100 μm thick, 30 mm high laser light sheet. The laser sheet illuminated the mid-span of the flow in the optically accessible section of the experimental setup (Fig. 1b). The laser pulse duration was about 10 ns while the pulse energy was about 17–20 mJ. The fluorescence was collected using f = 2.8 optics onto a 1024 × 1024 pixel CCD camera (PI-MAX) coupled to an external intensifier (GEN2, P43). The vortex generation and image acquisition tasks were synchronized using two electronic delay generators (Stanford Instruments Model DG 535). Piston motion and velocity were measured by a monochromatic high speed camera, Motion Scope (Redlake Imaging, Model 8000), to determine the linear motion of the marked piston shafts for different phase lags between the two pistons profiles to undertake a parametric study of the vortical interaction as a function of time lag between their generation. Sample PLIF images are shown in Fig. 2 with their computational counterparts (described later).

The timing of the vortex imaging has been shown in Fig. 3. The laser fire signal at 10 Hz was used as the master signal to coordinate the experiment's timing. The camera and the intensifier were phase locked to the laser pulse using one delay generator (DG1) capturing every fourth pulse due the readout time delay. Vortex generation in both the chambers were initiated by externally triggering the electro pneumatic piston solenoid controller slaved to the laser fire signal by using a second delay generator (DG2) pulsed by the first delay generator (DG1) controlling the camera. The time delay between the initiations of the two vortex pulses, i.e. delay between the respective motion starting times between two pistons, were also adjusted using the second delay generator (DG2) and this delay will be called phase lag on delay generator $\Delta\tau_{p1-p2}$ henceforth. Another time delay was introduced for the camera and intensifier gate to coincide with one of the laser pulses illuminating the vortex in the test section. The time delays between DG1 and DG2 were slowly varied to capture vortices at different convection distances in the test section. Detailed timing diagram for a particular case $\Delta\tau_{p1-p2} = 0.015$ s is shown in Fig. 3. The study is primarily restricted to the visualization of the acetone PLIF contours and more

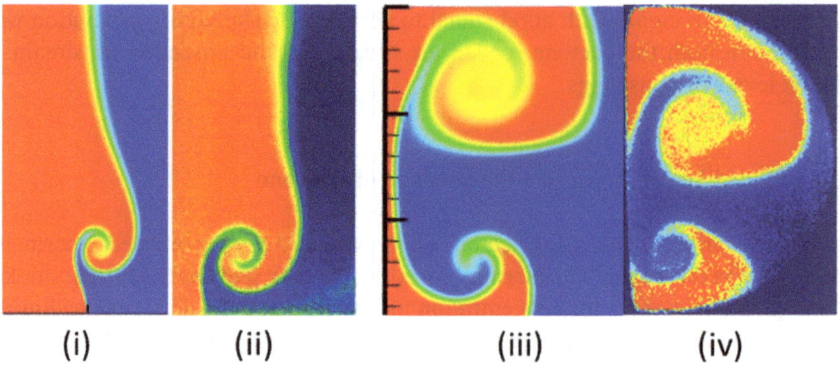

Fig. 2 **i** Species concentration (mass fraction) contour for single vortex (computational) [16], **ii** Species concentration (mass fraction) contour for single vortex (experimental) [17], **iii** Species concentration (mass fraction) contour for interacting vortices (computational), **iv** Species concentration (mass fraction) contour for interacting vortices (experimental)

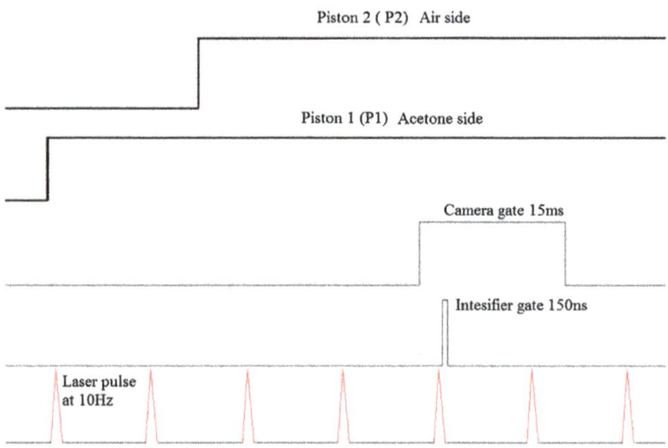

Fig. 3 Timing diagram for a sample vortex couple, $\tau_{P1-P2} = 0.015$ s

detailed mixing analysis based on behavior of concentration pdfs or joint pdfs is left for a future study. Figure 4 shows the P1 and P2 velocity at different $\Delta\tau_{p1-p2}$. It can be seen that the piston pulse profiles show reasonably good similarity with the theoretical pulse profiles (used in the computational study) shown in Fig. 1a. The peak values, growth and decay rates of the experimental pulse profile (Fig. 4) are also similar to that of the theoretical counterpart (Fig. 1a).

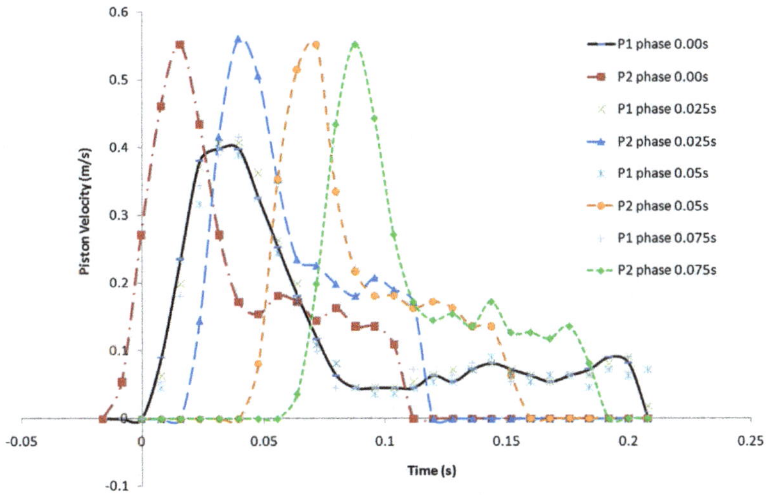

Fig. 4 Piston velocity time series

2.2 *Computational Model*

The computational model presented here and in our earlier work [16] simulates the laminar diffusion layer/wake formed by two gaseous flow streams (air in this case) having similar velocities being brought together at the trailing edge of a splitter plate as described in the previous section. The computational pulse profiles (Fig. 1a) mimics the volumetric displacement of a piston in the flow chamber used to create the line vortex in the experiments.

The extent of interaction of the vortices is a function of the time lag of the vortex generating pulses and their strengths. The mixing induced by the vortices is also dependent on the vortex configurations of vortex couple (opposite sign) or vortex pair (same sign) as indicated in Fig. 1a.

The governing equations for unsteady two-dimensional flow field in normalized form have been reported by Basu et al. [16] and are briefly restated here for clarity.

$$\text{Continuity:}\ \frac{\partial \tilde{\rho}}{\partial t*} + \nabla \cdot [\tilde{\rho}\tilde{V}] = 0 \tag{1}$$

$$\text{Momentum}:\ \tilde{\rho}\frac{\partial \tilde{V}}{\partial t*} + (\tilde{\rho}\tilde{V} \cdot \nabla)\tilde{V} = -Eu\nabla\tilde{P} + \frac{1}{\text{Re}}\nabla[\tilde{\mu}\nabla\tilde{V}] \tag{2}$$

$$\text{Species}:\ \frac{\partial \tilde{Y}}{\partial t*} + \tilde{V} \cdot \nabla\tilde{Y} = \frac{1}{\text{Re}Sc}\nabla \cdot (\tilde{D}\nabla\tilde{Y}) \tag{3}$$

$$\text{Energy}: \frac{\partial \tilde{T}}{\partial t*} + \tilde{V} \cdot \nabla \tilde{T} = \frac{1}{\text{Re Pr}} \nabla \cdot (\tilde{k} \nabla \tilde{T}) \tag{4}$$

where $\tilde{V} = [u/U_c, v/U_c]$ is the normalized velocity vector with u and v being the velocity components in the stream wise and cross stream directions. The convective velocity of the gas streams on either side of the splitter plate was denoted as U_c, which is the nominal velocity of the flow before the application of the pulse. $t* = tU_c/w$, $\tilde{\rho} = \rho/\rho_a(T_o)$, $\tilde{P} = P/P_o$, $\tilde{Y} = Y/Y_{s,i}$, and $\tilde{T} = T/T_o$ are the normalized time, density, pressure, mass fraction of the seed, and temperature, respectively, with P_0=1 atm and T_0=300 K [16]. Subscripts "a" and "s" refer to the unseeded air stream and acetone seeded air stream as in the experiments [16, 17]. "i" denotes the inlet values [16]. Transport coefficients $\tilde{\mu} = \mu/\mu_a(T_o)$, $\tilde{D} = D_{s,a}/[D_{s,a}(T_o)]$ and $\tilde{k} = k/k_a(T_o)$ are all normalized with their values at the reference temperature T_0=300 K. The non-dimensional quantities appearing in the governing equations are Euler number, $Eu = P_o/(\rho_a U_c^2)$, Reynolds number, $\text{Re} = (\rho_a U_c w)/\mu_a$, Schmidt number, $Sc = \upsilon_a/D_{s,a}$, and Prandtl number, $\text{Pr} = \upsilon_a/\alpha_a$ all evaluated at T_0. Convection velocity, U_c and channel width, w were utilized in the normalization [16]. Velocity, U_c at the two inlets (seeded or unseeded in Fig. 1a) was constant across the channel width before the initiation of the vortex generating velocity pulses. The transient velocity pulse profiles shown in Fig. 1a is provided as input boundary conditions to the above model. Further details of the numerical schemes used, boundary conditions, mesh size selection and convergence could be found in the work on single vortex mixing by Basu et al. [16]. The one gas stream is assumed to be seeded with acetone at a mass fraction of 0.25 at the exit plane.

The mixing enhancement as seen visually in Fig. 2 (evidenced in both experiments and computations) has been analyzed using two methods as will be described later. τ_p is defined as physical time delay normalized by convective time scale (w/U_c).

It is to be noted that the experimental and computational data show good agreement for the single vortex as shown in Fig. 2, parts i–ii. For the interacting vortices, the experimental and computational data show reasonably good agreement as well (Fig. 2, parts iii–iv).

3 Experimental Observations

In this section, a parametric experimental study of the double vortex interaction based on varying τ_{P1-P2} is presented. The time difference between the occurrence of peak velocities of P1 and P2 will be referred to as $t_{p,exp}$ henceforth. It is to be recognized that $t_{p,exp}$ (time difference between peak piston velocities) is not equal to the τ_{P1-P2} (time difference between input TTL signals, hence the different notations) in general due to slightly different inertia of the piston actuation systems. For most cases we found $\tau_{P1-P2} = t_{p,exp} + 0.025$ s.

Figure 5 shows the mixing layer formed between the acetone and air stream in the absence of vortex generating pulses. The rounding off at the top and bottom is

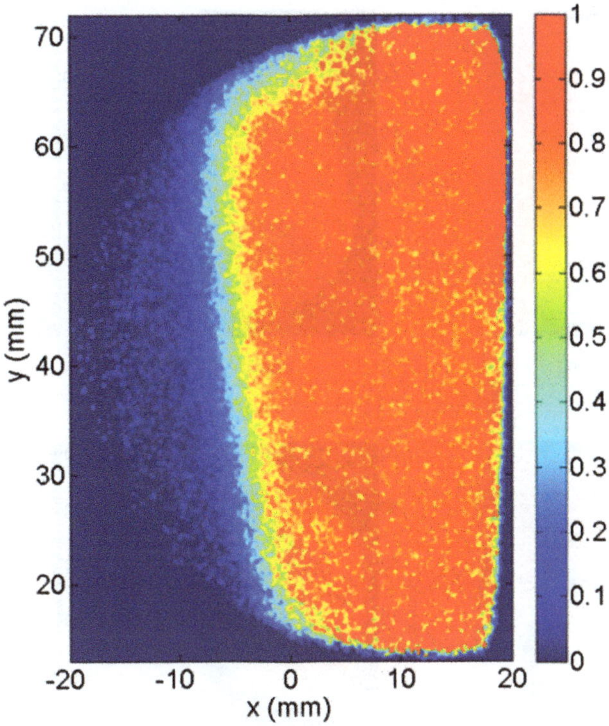

Fig. 5 Mixing layer between acetone and air streams

attributed to the clipping due to the circular intensifier field of view. Figure 6 shows the evolution of a single vortex rotating clockwise created by pulsing the acetone side. First proceeding along the top row, left to right: any image is obtained 2 ms after its immediate left neighbor showing the evolution and convection of the vortex along the other stream. The pulse strength is intentionally made stronger than that for a well formed vortex such that during its interaction with the counter-rotating vortex as studied subsequently, its vorticity is not dissipated to negligible values. Figure 7 shows the counterclockwise rotating vortex pulsed from the air side. These are very similar to those reported by Cetegen [17]. These two figures (Figs. 6 and 7) show the isolated vortex evolution, all the parameters of these are individually identical to their interacting counterparts described hence-forth. Figure 8 shows the evolution of two interacting vortices for $\tau_{P1-P2} = 0.075$ s. First the acetone side is pulsed and 0.075 s afterwards the air side is pulsed creating two counter-rotating vortices. It is observed that the time delay of $\tau_{P1-P2} = 0.075$ s, and $t_{p,exp} = 0.05$ s is too long to create a strong interaction between the two vortices, given the observation that coherent vortex structures of the individual vortices are preserved. However, it could be observed that the circulation strength which could be considered proportional to the growth rate of their diameters is reduced in comparison to their isolated counterparts as shown

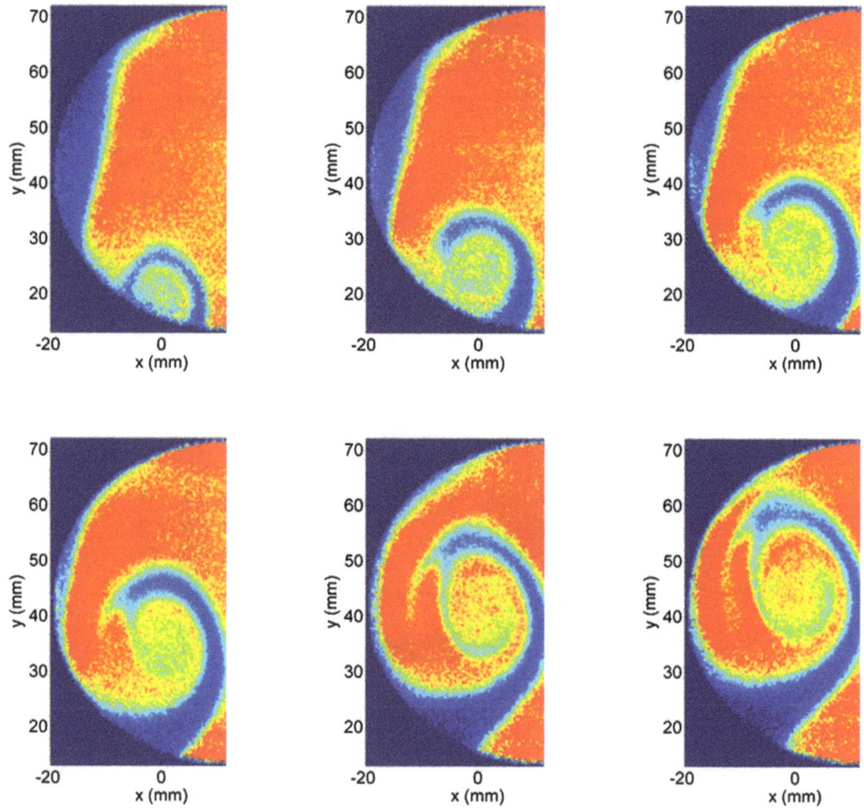

Fig. 6 PLIF profiles for acetone side pulsed vortex

in Figs. 4 and 5. Figure 9 shows remarkable repeatability of the interaction images obtained from different experimental realizations for a particular case of $\tau_{P1-P2} = 0.075$. However, for the very strongly interacting case shown later for e.g. $\tau_{P1-P2} = 0.025$ s, turbulence generation diminishes the repeatability to a large extent.

Figure 10 shows an interesting case of $\tau_{P1-P2} = 0.05$ s, and $t_{p,exp} = 0.03$ s. Here the time delay between the two vortices is smaller and a counterclockwise vortex is forced into an already existing clockwise circulation field. In the first few images of Fig. 10, apparently there is a Kelvin Helmholtz type of instability manifested by formation of multiple vortices with growing sizes. This phenomenon could be understood from the piston velocity time history for this case shown in Fig. 11. From, Fig. 12 it is clear that the velocity difference between the two fluid streams (air side and acetone seeded side) is growing which results in the formation of a Kelvin Helmholtz type instability with increasing amplitude. This increasing velocity difference causes a small timescale velocity perturbation which is superposed on the existing nominal velocity deficit across the mixing layer. This perturbation also leads

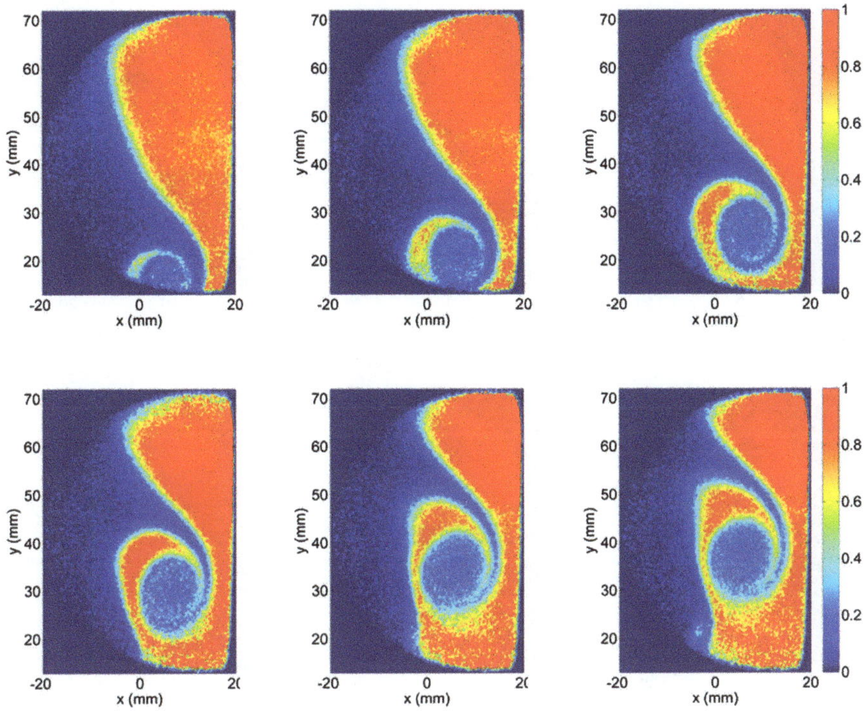

Fig. 7 PLIF profiles for air side pulsed vortex

to formation of small timescale eddy due to KH type instability. Finally when steady flow on the air side is reached, all the small vortices rotate as a part of a larger counterclockwise circulating field as shown in later images of Fig. 10. Figure 12 shows the simultaneous generation of two vortices τ_{P1-P2} = 0.025 s, or $t_{p,exp}$ = 0.00 s, which results in disorganized turbulent structures without any net circulation as might be expected from Biot Savart's law. Figure 13 shows the case where the air side pulse precedes the acetone side pulse with τ_{P1-P2} = 0.00 s, and $t_{p,exp}$ = − 0.025 s. The experimental results provide a good insight into the general mixing and flow structures of the interacting vortices. It is imperative to use the computational model for a few selected cases to show quantitatively the main features of mixing in this type of vortical flow field.

4 Computational Results

The computational data are reported primarily for the following cases, (a) vortex pair with A1/A2 (pulse strength ratios of the two vortices) = 1 and 0.66 with a phase lag of τ = 1.4 and (b) vortex couple with A1/A2 (pulse strength ratios of the two

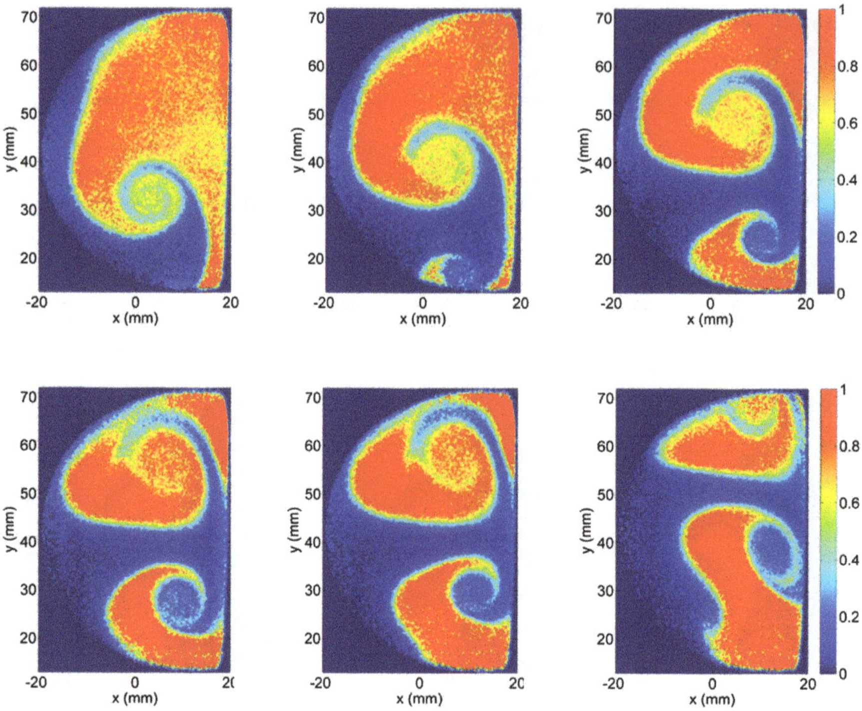

Fig. 8 $\Delta\tau_{P1-P2} = 0.075$ s

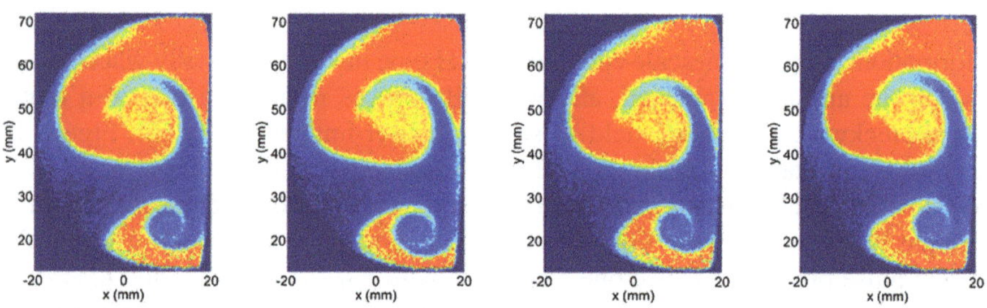

Fig. 9 Repeatability study of a particular mode of interaction for $\Delta\tau_{P1-P2} = 0.075$ s

vortices) = 1 and 0.66 with a phase lag of $\tau = 1.75$. The numerical conditions were chosen similar to the experimental conditions as possible in terms to phase delay. This represents a small subset of the experimental conditions. As described before the computational data shows reasonably good visual agreement with the experimental PLIF contours (Fig. 2). This gave us confidence to propose a mixing enhancement analysis based on probability density functions using the computational data to gain insight into the characteristics of species mixing in interacting vortices.

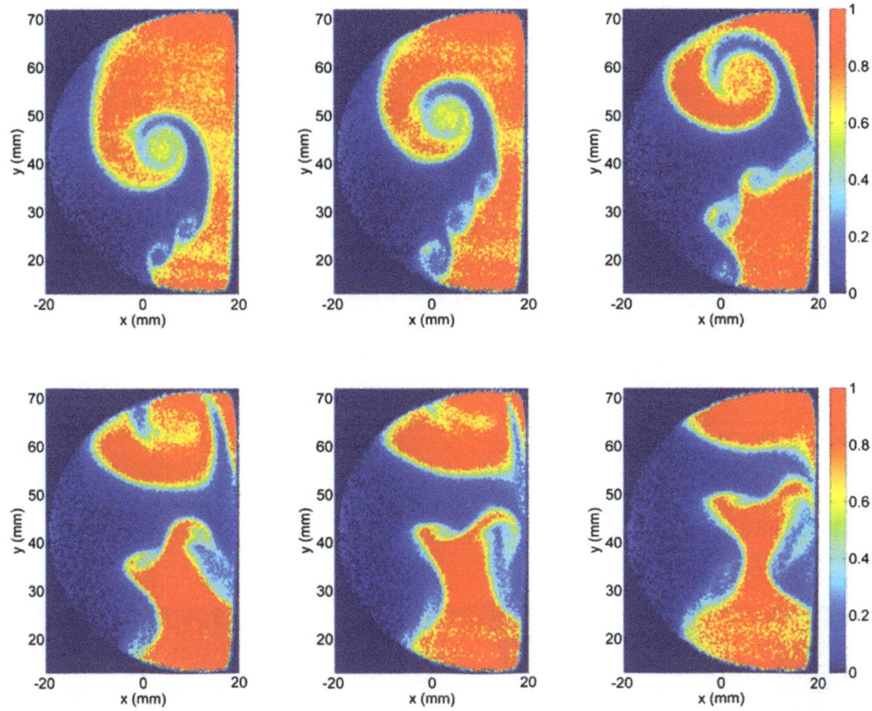

Fig. 10 $\Delta\tau_{P1-P2} = 0.050$ s

Figure 14a shows the temporal evolution of the species mass fraction profiles for vortices generated from opposite sides at a phase lag of $\tau_p = 1.4$. It can be seen that for the till $t^* = 8.4$ following the generation of the first pulse, the vortices show minimum interaction and predominantly show a single vortex. This is because even though the first vortex is strong at this point the second vortex is very weak due to the phase lag of $\tau_p = 1.4$. However after $t^* = 9$, both the vortices start showing strong interactions as exhibited in Fig. 14b. The first vortex is deformed by the rotation induced by the second vortex particularly in the vortical arms away from the vortex core. Significant thinning of the diffusion layer is also observed due to stretching induced by the counter rotating vortices. It is also observed that the walls obstruct the growth of the vortices leading to further thinning of the diffusion layer connecting the two vortices. Figure 14b shows the visual evidence of the growth of the vortices at different time instants. It is noticed that till $t^* = 9$, the vortex concentration profile matches closely the single vortex profile reported by Basu et al. [16]. However at later time instants, the second vortex closely interacts with the first and merges with one another. From $t^* = 16$ onwards, the cores of the two vortices become indistinguishable and represents a predominantly diffused core without any intricate windings of the vortical arms.

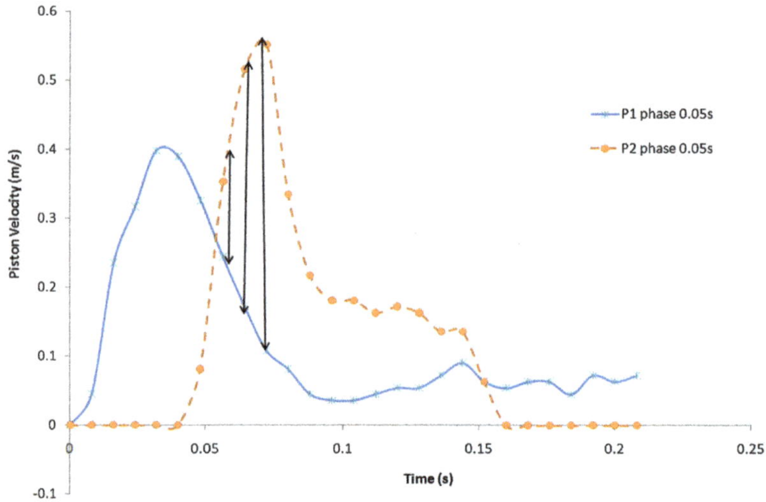

Fig. 11 Relative motions of P1 and P2 showing the increasing velocity difference between two pistons $\Delta\tau_{P1-P2} = 0.050$ s

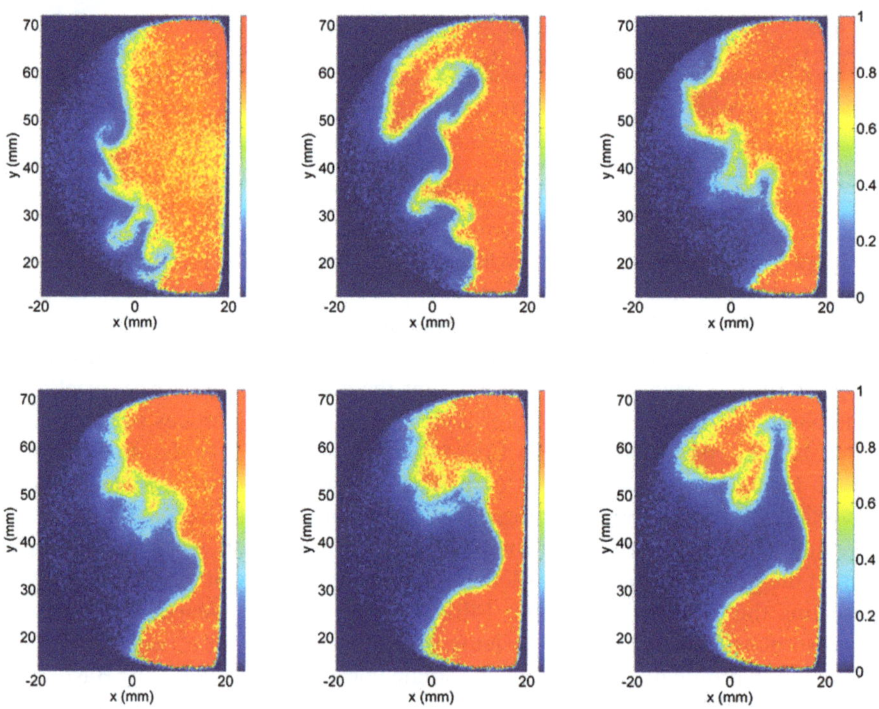

Fig. 12 $\Delta\tau_{P1-P2} = 0.025$ s

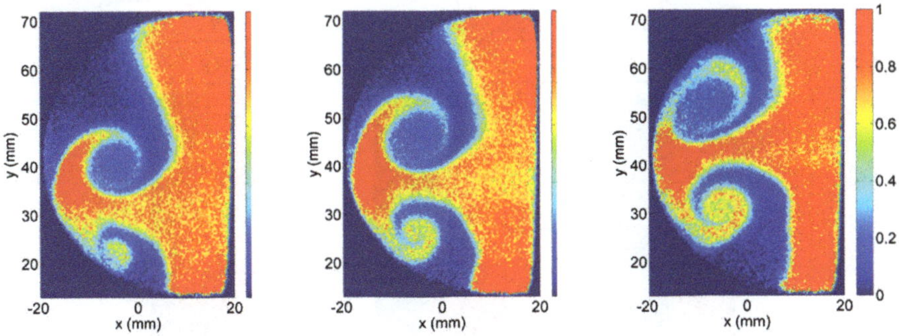

Fig. 13 $\Delta\tau_{P1-P2} = 0.000$ s

5 Mixing Enhancement Analysis

5.1 *Probability Density Function Representation*

The scalar mixing in the field of a vortex can be determined by evaluating the spatial probability density function (pdf) of the scalar concentration P(c) from spatial concentration distributions [17]. The pdfs were calculated over a spatial region (both lateral and streamwise extent) that is substantially larger than the total vortex span of the two interacting vortices [17]. These obtained pdfs are renormalized such that $\int_{c=0}^{c=1} P(c) = 1$ where c = 0 and c = 1 corresponding to the scalar concentrations in unseeded and seeded streams respectively. In Fig. 15 mean and variance of 'c' are compared for two different cases. These two parameters are defined as

$$\left.\begin{aligned} \bar{c} &= \frac{\int_{c=0}^{c=1} P(c) \cdot c \cdot dc}{\int_{c=0}^{c=1} P(c) \cdot dc} \\ c_{\text{var}} &= \frac{\int_{c=0}^{c=1} P(c) \cdot (c - \bar{c})^2 \cdot dc}{\int_{c=0}^{c=1} P(c) \cdot dc} \end{aligned}\right\} \quad (5)$$

Further details of the methodology for calculating these two parameters have been discussed by Cetegen [8, 17] for isolated vortices. For both conditions $\bar{c}$ shows a self-similar trend. It maintains a constant value before the first pulse induces a vortical mixing which enhances $\bar{c}$. However, after the second pulse, the trend for $\bar{c}$ becomes different for vortex couples and pairs. For vortex pair (Fig. 14a), second pulse induces mixing in same direction which further increases the value of $\bar{c}$. On the other hand the vortex couple (Fig. 14b) induces a second vortex rotating in the opposite direction of the first vortex. This generates a secondary peak of high probability at lower concentration resulting in reduction of $\bar{c}$. c_{var} is a parameter which measures the non-uniformity of the seed concentration within the flow field. c_{var} plots shown in Fig. 15

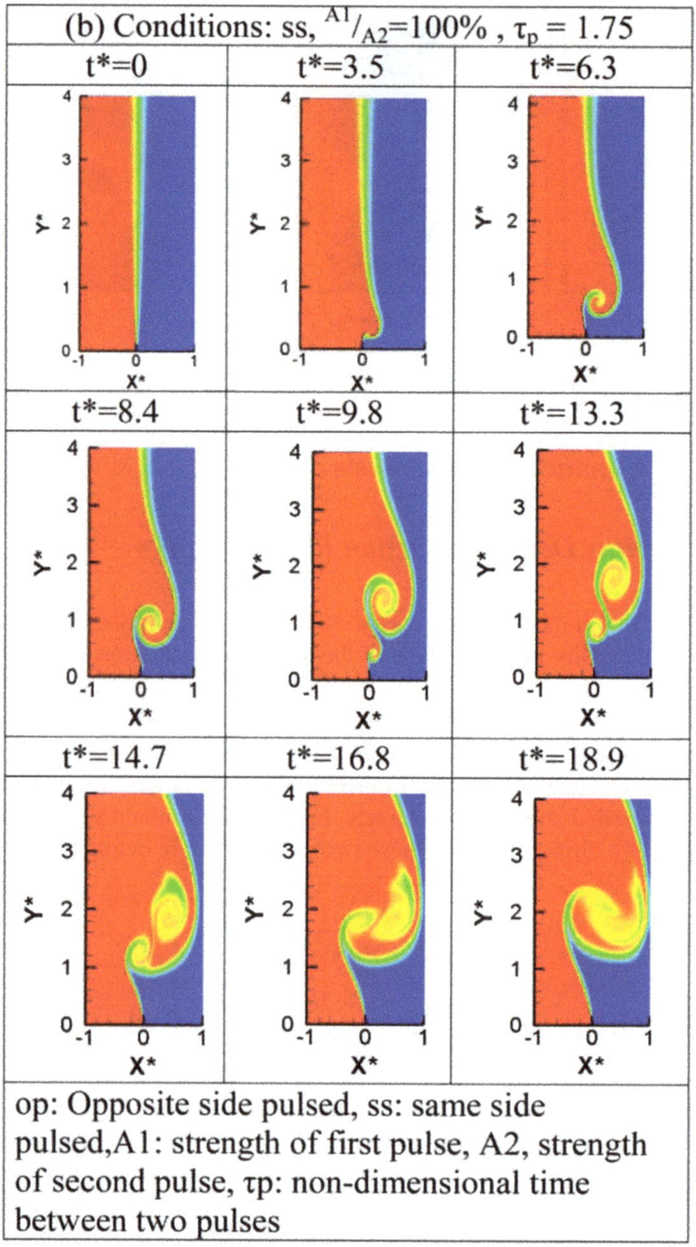

Fig. 14 Computational results: **a** A chronological visual representation of the interaction of two vortices (vortex couple) formed when seeded (acetone) and unseeded (air) sides were subsequently pulsed, **b** A chronological visual representation of the interaction of two vortices formed (vortex pair) when acetone (seeded side) was pulsed twice as the flow of air remained constant

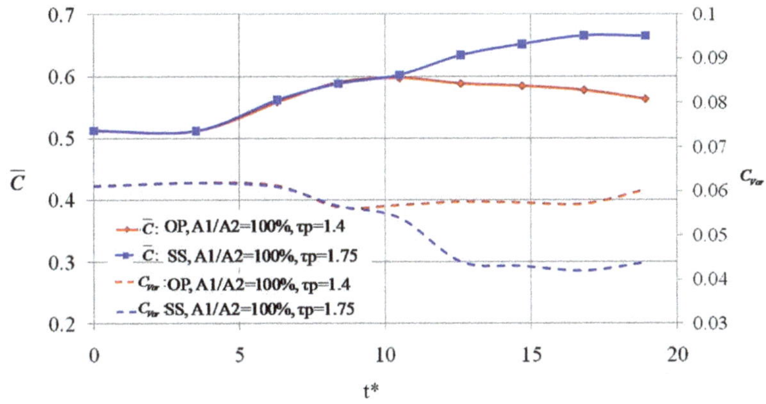

Fig. 15 Mean and variance of the scalar mass fraction c for different pulse strengths and time lags in the interacting vortices

suggest that c_{var} maintains almost a constant value for vortex couple, indicating that the mixing is very localized near the vortex cores. However, for vortex pair, a sharp decrease in the value of c_{var} is observed as the second vortex is generated. Although analysis with probability density function identifies global mixing patterns, it fails to correlate the effect of vorticity and corresponding mixing dynamics. This requires a different approach with conditional probability density function.

5.2 *Conditional Probability Density Function*

Statistical measures need to be introduced to quantify mixing in both molecular as well as advective sense for the interacting vortices while preserving their individual identities. This could be achieved, i.e. a high degree of mixing of a passive scalar in a vortical flow could be considered, if the following criteria are satisfied.

- The spatial gradients of the passive scalar have disappeared within the vortex i.e. a high degree of mixing can be considered to have occurred if species concentration gradients are low, indicating substantial diffusion has already taken place, a signature of molecular mixing. This can be observed in the scalar pdf when a single sharp peak is obtained at the intermediate levels of scalar concentration, suggesting homogeneity of the species concentration field. This is the conventional choice of reporting mixing behavior.
- Topologically, the species concentration and vorticity distribution functional forms are nearly similar for Schmidt numbers greater than unity. In other words, species concentration iso-lines which are spirals approach the shape of the circular vorticity iso-lines, indicating that sufficient advective transport of species have taken place. Motivated by the geometric similarity of the vorticity and species concentration

contour shapes a joint pdf of vorticity and species can be introduced to be a suitable marker of vortical mixing. This joint pdf can be suitably normalized by the pdf of vorticity to remove any ambiguity originating from considering the joint pdf which is dependent on many case specific parameters like area of interrogation, high vorticity regions not of interest in the current study such as boundary layers etc.

Thus for proper mixing both of the above mentioned criteria are satisfied if the probability density function $f(Y_i|\omega) \rightarrow 1$ in non-extremum species concentrations. For homogenous species and vorticity condition (say at very later stages of exponential decay of a Oseen vortex when $\omega \rightarrow 0$, and species field is nearly homogenous) the conditional probability function (cpdf) structures degenerates to a point in the typical contour plots as in Figs. 16 and 17 and attains a maximum attainable value of 1 indicating maximum degree of mixing. So, the cpdf is essentially a measure of mixing.

This is presented extensively in Figs. 16a, b and 18a, b for the different modes of interaction already mentioned to elucidate the evolution and nature of mixing in such interactions. Figure 16 presents as a matrix of individual images, where the first column shows species concentration iso-lines (black lines) superimposed with vorticity contours (jet color scale) in a rectangular region of interest selected within the computational domain at a particular instant of time. The time instants are reported on the left of the first columns for each row. The second column shows the corresponding cpdf of species concentration of acetone conditioned on vorticity at that time instant. Figure 16a shows the case of oppositely pulsed vortices of equal strength. At $t^* = 6.3$, the first row shows an unmixed state as evident from the low-magnitudes and non-coherent distribution of the cpdf. The subsequent time instants in Fig. 16a at $t^* = 10.5$, 14.7 and 18.9, progressively higher degree of mixing for the first vortex is observed in form of high levels of probability density in the negative vorticity regions of the cpdf. But in the last two images a second vortex with positive vorticity has entered the interrogation window. Even though visually its structure is similar to the first one at initial times, its vorticity is lower and is characterized by very weak mixing as observed from the time instants at $t^* = 14.7$ and 18.9. This is due the fact that the first vortex leaves a weakly rotational flow field, in which the second vortex is convected while rotating in opposite direction, i.e. vortex couple configuration. This substantially lowers its circulation strength and dissipates mixing even though the residual vorticity is weak and almost indiscernible in the contour images or the cpdf plots. This makes these mixing phenomena highly non-trivial and shows that it is strongly affected by prior history of vortex interaction. This mixing behavior is also unique as the cpdf has nearly same probability density over the entire range of species concentrations within the vortex: a feature which is not observed for the first vortex even at its initial stages. To substantiate this finding and to establish that this is due to effect of the first vortex and not a numerical artifact, a vortex with the exactly same parameters and pulsed at the same time instant and inlet velocity time histories but in the absence of a first vortex was simulated and its characteristics were compared with those of the second vortex. This is shown in Fig. 17 where we

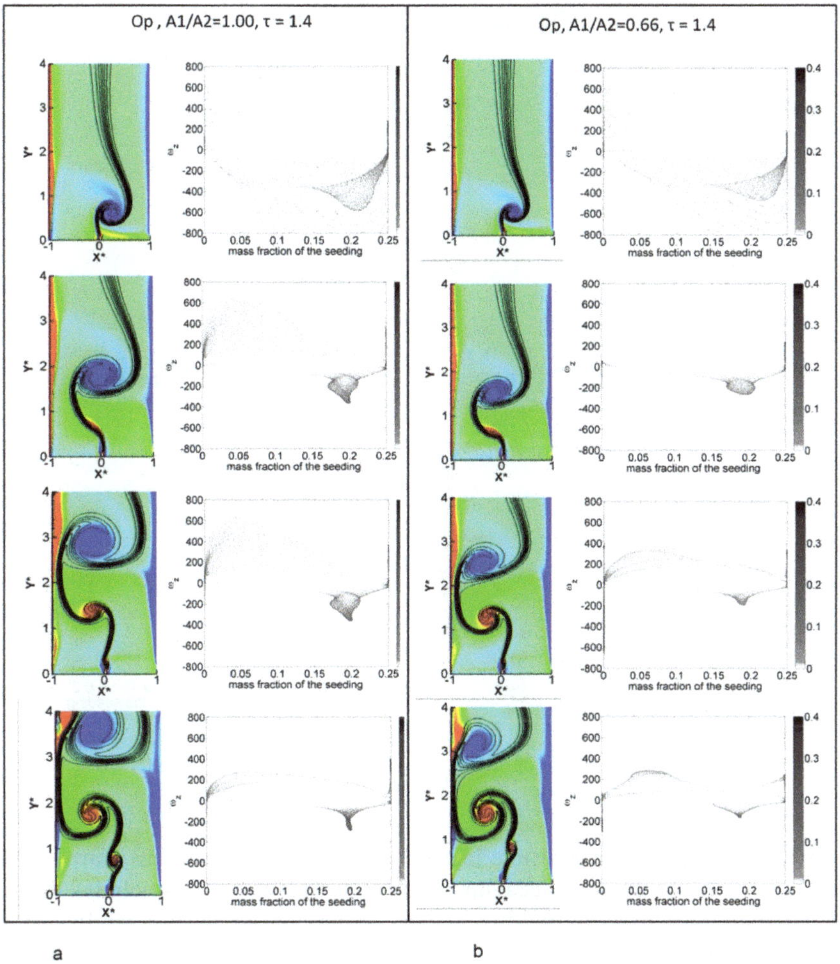

Fig. 16 Vortex couple: **a** Species concentration (mass fraction) isolines (black lines) superimposed with vorticity contours (jet color scale) in a rectangular region of interest selected within the computational domain at different time instants, **b** cpdf of concentration of seeding conditioned on vorticity at different time instants

observe a much higher degree of mixing for the individual vortex when it is not preceded by an oppositely rotating vortex. Figure 16b shows the contours and cpdfs of species concentration and vorticity for the oppositely pulsed but different pulse strength vortices. As observed from later stages of interaction at $t^*=14.7$ and 18.9 mixing is slightly enhanced for the second vortex in comparison to the one for equal strength case (Fig. 16a $t^*=14.7$ and 18.9) since the pulse strength for the preceding oppositely rotating vortex is reduced. If the counter-rotation was simultaneous it could be expected that vorticity would be damped by Biot Savart's law. However

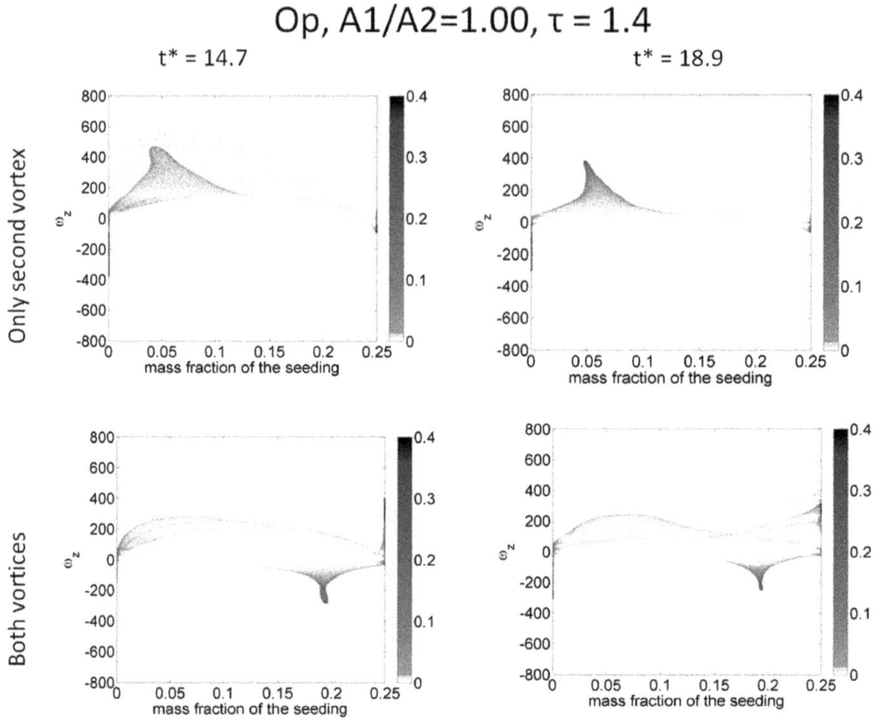

Fig. 17 Comparison of the degree of mixing for single isolated vortex and interacting vortices

all the cases reported have a time lag such that the first vortex has convected and second vortex evolves in its decaying influence. Thus it can be concluded that even the exponentially decaying field of a "convecting" vortex is sufficiently strong to dissipate mixing in a subsequent vortex if it is rotating in an opposite direction, i.e. vortex couple configuration.

Enhancement of mixing occurs when a second vortex spins in the same direction as that of its predecessor, i.e. vortex pair configuration. This is shown in Fig. 18a and Fig. 18b with equal and unequal strengths respectively. Figure 18a at $t^* = 10.5$, 14.7 and 18.9 show that two cpdf structures evolve corresponding to two vortices produced, with the structure on the right corresponding to the first vortex and the one on the left corresponding to the second. Counterintuitive to the general understanding of the development and decay of a vortex, the second vortex shows higher degree of mixing though the first one had a longer development time and intuitively would have better mixing characteristics. However, once again the second vortex evolves in an already strained flow and scalar field as left behind by the roll up of the first vortex. Due to velocity induction the two vortices rotate about each other resulting in additional species advection. Moreover, due the close proximity of the two vortices, the first vortex essentially compresses the second vortex, resulting in higher straining

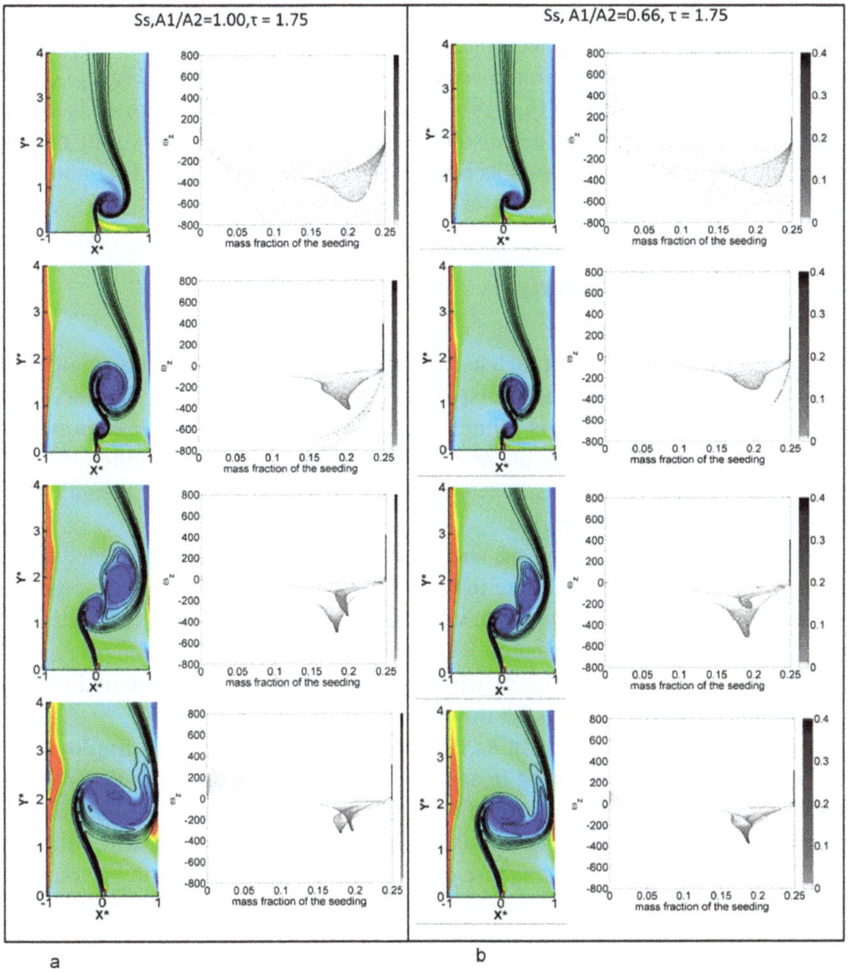

Fig. 18 Vortex pair: **a** Species concentration (mass fraction) isolines (black lines) superimposed with vorticity contours (jet color scale) in a rectangular region of interest selected within the computational domain at different time instants, **b** cpdf of concentration of seeding conditioned on vorticity at different time instants

of the second. This produces better mixing of the species in the second and hence higher probability densities in the cpdf. That the second vortex is better mixed than its isolated counterpart is evident by com-paring the two interaction time instants at $t^* = 14.7$ and 6.3. At around the same point of their evolution time, the strained second vortex at t^*= 14.7 is smaller in size with a more homogenous species field than the isolated one at t^*=6.3. This characteristic was not evident without the mixing parameter which quantifies the degree of mixing for each of the structures. At a later time step as shown at t^*= 18.9 the cpdf structures approach each other closely with similar mixing characteristics. Still, the second vortex shows a better mixed state as evident from the concentrated high value of the mixing parameter which suggests forced mixing within the second vortex by the first one. For the different pulse strength case shown in Fig. 18b, it is observed that the first vortex is distorted by the velocity induction effect of the second vortex hence the overall mixing is determined by that of the second as observed from the cpdfs at $t^{*}=14.7$ and 18.9.

5.3 Strain Rate Analysis

Strain analysis in another approach used by different authors in the context of vortical mixing. In a vortical field molecular diffusion takes place in the interface of two streams of flow. This mixing is primarily caused by stretching and transportation of the interface of two fluids. The phenomena of stretching can be quantified in terms of strain rate, which is defined as, $\varepsilon = \frac{1}{l}\frac{\Delta l}{\Delta t}$, where l is the length, Δl change in length in Δt time and ε is the strain rate.

Figures 5, 6, 7, 8, 9, 10, 11, 12 and 14 show that seeded and un-seeded sides do not have a clear line interface due to molecular diffusion. It can be observed in the figures that concentration of acetone varies from 0 (air-side) to 0.25 (acetone-seeded side) at the interface of two streams. Therefore strain analysis has been performed on an iso-concentration line of 0.125, which is the average of maximum and minimum acetone concentrations within the flow field.

To calculate instantaneous strains, a numerical technique has been employed using MATLAB. Data files containing instantaneous concentration distribution within the flow field has been extracted from Fluent results. Subsequently, a numerical code has been used to find out the successive grid points having acetone concentration of 0.125. Linear distances between these grid points eventually results in the total length of iso-concentration line. A typical iso-contour line for an arbitrary time instance (t^j) has been shown in Fig. 19. The line has been constructed by joining all the grid points containing acetone concentration of 0.125. Following this, the code calculated the distances between successive nodes. For example, in Fig. 19, on the iso concentration line, two nodes have been shown as node k and node k + 1. In the inset, the nodes are zoomed into show the process of length calculation. The horizontal and vertical projection of the distance between two nodes, Δx_k and Δy_k has been found out by subtracting coordinates of two nodes.

$$\left.\begin{aligned} \Delta x_k &= x_{k+1} - x_k \\ \Delta y_k &= y_{k+1} - y_k \end{aligned}\right\} \tag{6}$$

where x_k and y_k are co-ordinates of 'k'th node. The linear distance (Δl_k) between two nodes has been found out employing Pythagoras rule.

$$\Delta l_k = \sqrt{(\Delta x_k)^2 + (\Delta y_k)^2} \tag{7}$$

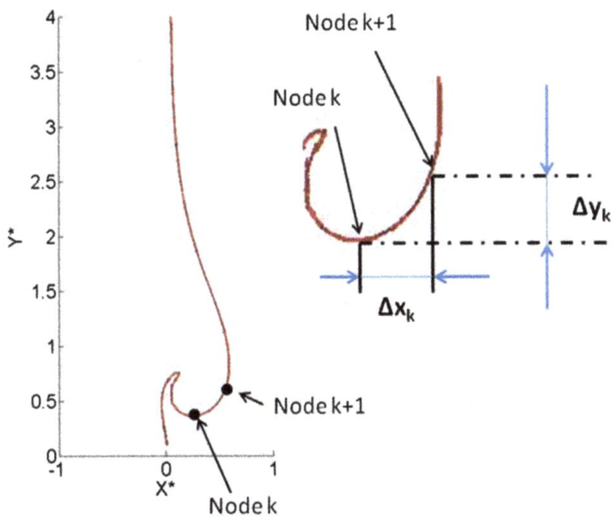

Fig. 19 Process of calculating the iso-concentration lines

This process has been repeated for all nodes present on the iso concentration line and finally the total length of the line has been calculated by summing the values.

$$\left.\begin{aligned} len_{xx} &= \sum_k |\Delta x_k| \\ len_{yy} &= \sum_k |\Delta y_k| \\ len &= \sum_k |\Delta l_k| \end{aligned}\right\} \tag{8}$$

where len_{xx} =horizontal (x-direction) projection of iso-concentration line, len_{yy} =vertical (y-direction) projection of iso-concentration line, len =length of iso-concentration line.

This method has been repeated for all time instances to calculate a progressive elongation of iso-concentration line. Figure 19 shows these iso-concentration lines at different time instances for condition of opposite side pulsed with same strength with time lag between pulses of $\tau_p = 1.4$.

After determination of instantaneous lengths, the strain has been calculated. It was necessary to calculate directional strain as well as total strain. They are calculated by the following formulae:

$$\left.\begin{aligned} \varepsilon_{xx}^{j+1} &= \frac{1}{len_{xx}^{j}} \frac{(len_{xx}^{j+1} - len_{xx}^{j})}{(t^{j+1} - t^{j})} \\ \varepsilon_{yy}^{j+1} &= \frac{1}{len_{yy}^{j}} \frac{(len_{yy}^{j+1} - len_{yy}^{j})}{(t^{j+1} - t^{j})} \\ \varepsilon^{j+1} &= \frac{1}{len^{j}} \frac{(len^{j+1} - len^{j})}{(t^{j+1} - t^{j})} \end{aligned}\right\} \tag{9}$$

where

len^j length of iso-concentration line at 'j'th time instant
len_{xx}^j horizontal (x-direction) projection of iso-concentration line at jth time instant
len_{xx}^j vertical (y-direction) projection of iso-concetration line at jth time instant
t^j absolute time at jth time instant
ε^j instantaneous total strain rate at jth time instant
ε_{xx}^j instantaneous horizontal (x-direction) strain rate at jth time instant
ε_{yy}^j instantaneous vertical (y-direction) strain rate at jth time instant.

5.3.1 Opposite Sides Pulsed, Same Pulse Strength, $\tau_p = 1.4$

Figure 20a depicts evolution of iso-concentration line for this condition. This figure clearly shows stretching of iso-concentration line in both horizontal (x) and vertical (y) direction. However, it can be observed that the stretching in horizontal direction is severe compared to other direction. Figure 20b shows temporal growth of iso-concentration line (*len*) and its two projections (len_{xx}, and len_{yy}). All the lines are observed to stretch with time. However, it can be noted that between $t^* = 5$ and 10 there is a rapid growth, which is artifact of initiation and development of the first vortex. Similarly, there is another phase of rapid growth at the later stage which corresponds to the second vortex. The two vortices are separated by a considerable amount of time, resulting in two distinct phase of higher growth rate. This has also been reflected in Fig. 19c, where the strain rates are compared. Both total strain (ε) and vertical strain (ε_{yy}) have two wide spread humps showing the effect of two vortices. Initially (at time = 0) the iso-concentration line was vertical resulting in zero horizontal projection (len_{xx}). This has been reflected in Fig. 19c with very high initial horizontal strain rate (ε_{xx}). However, with time this strain rate decays. This is because flow structure gets trapped between two side walls limiting the potential to expand in horizontal direction.

5.3.2 Opposite Sides Pulsed, Weak First Pulse, $\tau_p = 1.4$

Evolution of iso-concentration line in Fig. 21a shows less vertical stretching compared to the case with same pulse strength. Figure 21b shows pattern of iso-concentration line and its projection at different time instances. It is evident that length of iso-concentration line (*len*) and vertical projection (len_{yy}) shows a smaller

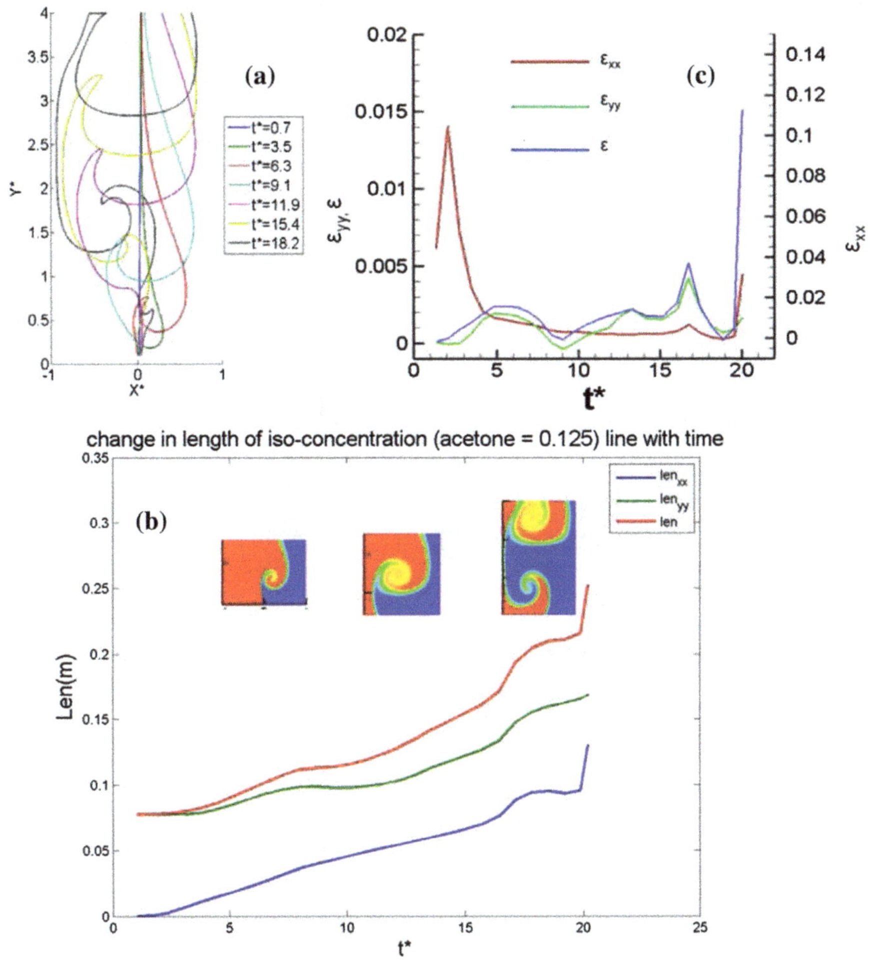

Fig. 20 **a** Instantaneous iso-concentration (c = 0.125) line at different time instances, **b** change in iso-concentration length over time period with instantaneous vortex structure, **c** change in strain rates for Op, A1/A2 = 100%, $\tau_p = 1.4$

increase in growth during generation of first vortex, while later on they exhibit a faster growth during second vortex. This is because of the fact that first vortex is weaker compared to the second one. This phenomenon has been observed in strain rate plots as shown in Fig. 21c. Both ε and ε_{yy} showed two separate zones of higher strain rates. Each of them corresponds to two separate vortices. However, it can be noted that the increase in strain rate due to stretching of iso-concentration line is weaker in case of first vortex compared to the second one. Other component of strain rate, ε_{xx} shows a similar trend like previous cases. This strain rate gets limited due to the wall effect.

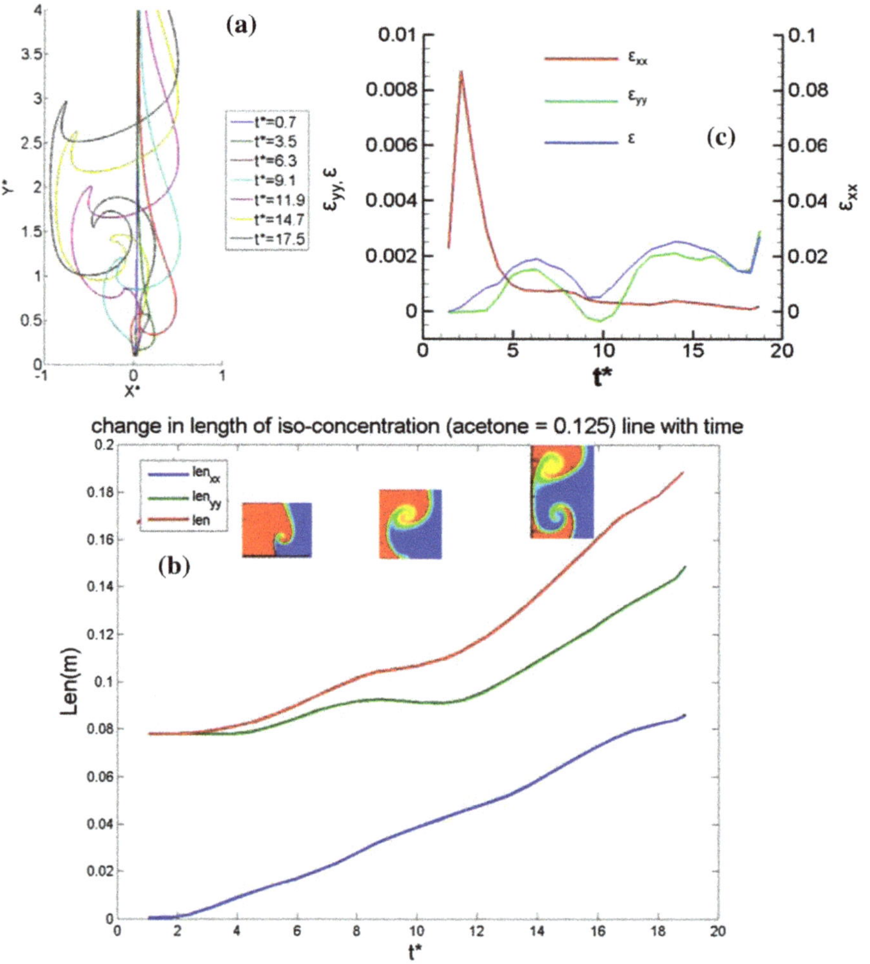

Fig. 21 **a** Instantaneous iso-concentration (c = 0.125) line at different time instances, **b** change in iso-concentration length over time period with instantaneous vortex structure, **c** change in strain rates for Op, A1/A2 = 66%, τp = 1.4

6 Summary

In this chapter we have investigated an interesting mixing phenomenon that arises from the interaction of two convecting line vortices generated at successive time instants. Same and opposite direction of rotation generating vortex pairs and couples augments and dissipates species mixing respectively as suggested by global scalar statistics like mean and variance. Local mixing due to vortical flow is investigated by introduction of a statistical parameter defined as cpdf of the scalar for a given vorticity.

This quantitatively shows how mixing is dissipated or augmented for individual vortices and at different vorticity magnitudes, even by the decaying and apparently indiscernible presence of a favorable or unfavorable direction of rotation of the preceding vortex. Since mixing is primarily caused by stretching and transportation of the interface of two fluids, detailed strain rate analysis of the iso-concentration lines shows phases of rapid growth corresponding to the initiation and development of the two vortices.

References

1. S.B. Pope, *Turbulent Flows* (Cambridge University Press, Cambridge, 2000)
2. F.E. Marble, Growth of a diffusion flame the field of a vortex, in *Recent Advances in Aerospace Sciences*, ed. by C. Casci, C. Bruno (Plenum, New York, 1985), pp. 395–413
3. A.R. Karagozian, F.E. Marble, Study of mixing and reaction in the field of a vortex. Combust. Sci. Technol. **45**, 65 (1986)
4. C.J. Rutland, J. Ferziger, Simulations of flame-vortex interactions. Combust. Flame **84**, 343 (1991)
5. J.C. Lee, C.E. Frouzakis, K. Boulouchos, Numerical study of opposed-jet H2/air diffusion flame—vortex interactions. Combust. Sci. Technol. **158**, 365 (2000)
6. A. Laverdant, S. Candel, A numerical analysis of a diffusion flame vortex interaction. Combust. Sci. Technol. **60**, 79 (1989)
7. A. Laverdant, S. Candel, Computation of diffusion and premixed flames rolled-up in vortex structures. J. Propul. Power **5**, 139 (1989)
8. B.M. Cetegen, W.A. Sirignano, Study of mixing and reaction in the field of a vortex. Combust. Sci. Technol. **72**, 157 (1990)
9. N. Peters, F.A. Williams, Premixed combustion in a vortex, in *Symposium (International) on Combustion*, vol. 22, pp. 495 (1988)
10. P.H. Renard, D. Thevenin, J.C. Rolon, S. Candel, Dynamics of flame/vortex interactions. Prog. Energy Combust. Sci. **26**, 225 (2000)
11. E. Villermaux, J. Duplat, Mixing as an aggregation process. Phys. Rev. Lett. **91**, 18 (2003)
12. P. Meunier, E. Villermaux, How vortices mix. J. Fluid Mech. **476**, 213–222 (2003)
13. K. Bajer, A.P. Bossom, A. Gilbert, Accelerated diffusion in the center of a vortex. J. Fluid Mech. **437**, 395–411 (2001)
14. M. Long, M.A. Sprague, A.A. Grimes, B.D. Rich, M. Khine, Appl. Phys. Letts **94**, 133501 (2009)
15. K. Sritharan, C.J. Strobl, M.F. Schneider, A. Wixforth, Z. Guttenberg, Appl. Phys. Lett. **88**, 054102 (2006)
16. S. Basu, T.J. Barber, B.M. Cetegen, Computational study of scalar mixing in the field of a gaseous laminar line vortex. Phys. Fluids **19**, 053601 (2007)
17. B.M. Cetegen, Scalar mixing in the field of a gaseous laminar line vortex. Exp. Fluids **40**, 967 (2006)

Numerical Modeling of Aerosol Transport and Dynamics

P. S. Rajagopal, Manish Joshi, Janki Shinde, S. Anand, Akshai K. Runchal, B. K. Sapra, Y. S. Mayya and Madhukar M. Rao

Abstract Flow and aerosol transport and dynamics in a reaction chamber, part of an aerosol generation system, is analyzed by coupling Computational Fluid Dynamics (CFD) and Aerosol Dynamic Equation. A predictable parametric aerosol output from reaction chamber is desirable for different contexts. The effect of residence time of the aerosol particles and mixing characteristics of the flow on the aerosol size distribution is analyzed using the ANSWER finite volume CFD code. The ANSWER uses the steady state turbulent flow field to solve the General Dynamics Equations (GDE) for the aerosol particles. The GDE includes mechanisms such as coagulation, gravitational settling and thermophoretic drift etc. A volume (and mass) preserving nodal method is implemented to model particle distribution changes due to coagulation. The modules modeling coagulation and gravitational settling were validated with respect to analytical solutions taken from the literature. The size distribution in reaction chamber design is seen to be robust for various flow scenarios at inlet number concentration of $1 \times 10^{12}/m^3$. This is explained by flow time scale being much smaller than coagulation time scale. At higher inlet concentration of $1 \times 10^{15}/m^3$ the average size distribution and outlet size distribution are significantly shifted from the inlet distribution, due to the much lower coagulation time scale. A noticeable difference between no or low swirl and high swirl flow is seen. A secondary ring inlet within the reaction chamber was seen to lead to identical aerosol distribution for different flow scenarios even at higher inlet concentration.

P. S. Rajagopal (✉) · J. Shinde · A. K. Runchal · M. M. Rao
CFDVRi, Mount View Complex, McLeod Ganj, Dharamshala, Himachal Pradesh 176219, India
e-mail: rajagopal@cfdvrinstitute.org

A. K. Runchal
e-mail: runchal@cfdvrinstitute.org

M. Joshi · S. Anand · B. K. Sapra
Bhabha Atomic Research Centre, Mumbai, India
e-mail: mjoshi@barc.gov.in

Y. S. Mayya
Indian Institute of Technology - Bombay, Mumbai, India

A. K. Runchal et al. (eds.), *Energy for Propulsion*, Green Energy and Technology,
https://doi.org/10.1007/978-981-10-7473-8_14

1 Introduction

Aerosols are two phase metastable systems consisting of solid or liquid particles suspended in a gaseous medium. The particles can be formed primarily (by direct emission or disintegration etc.) or by secondary processes such as gas to particle conversions. Their applicability and importance spans to several domains covering medical, industrial, cosmetic and atmospheric engineering and sciences. On account of their vital role in energy balance of earth (via interactions with solar radiation), potential to affect inhalation hazard and their growing importance in efficient drug delivery; understanding of their parameters/characteristics has become crucial in recent times. Atmospheric aerosol research has mostly focused on observations and interpretations of aerosol processes which are complex due to the involvement of multiple sources and dynamic interactions. On the other hand, industrial aerosol research aims at study of specific aspects collimating towards control of generation, transport and evolution characteristics. Apart from normal or desired processes, aerosol particles are also released in industrial accident scenarios such as in reactor containment vessels [30]. For the context of atmosphere as well as industrial releases, evolution of aerosol concentration is required to be quantified. Control of formation and evolution characteristics is also crucial for design of devices such as those used for medical applications. Parameters of the output of such devices (e.g. concentration, size distribution) can be controlled by optimizing the flow rate and geometry of the device. At the same time, accuracy of the measured/predicted characteristics of the aerosol particles could be important for decision making. For example, prediction of concentration, transport parameters and deposition of soot formed (and evolving) during fire helps in control measures. The soot concentration in the gas phase affects visibility and also impacts design decision related to safety [20]. For such cases where extrinsic contributors also play role in evolution of properties, extensive understanding and incorporation of dynamic effects in prediction models is crucial. Due to high temperature gradients and the turbulent nature of fire, thermophoretic deposition and turbulent diffusion effects are needed to be considered for accurate simulation of soot transport and deposition.

Evolution of aerosol size distribution takes place due to various processes which changes the properties of the distribution function. These processes include coagulation (agglomeration), nucleation, condensation/evaporation and deposition. Coagulation is characterized by a coagulation kernel, which is a function that gives the collision frequency of two particles of given sizes. Coagulation kernels are available for different colliding processes including Brownian diffusion and gravitational settling. These kernels can further be different for different aerosol sizes (relative to mean free path of the gas) such as continuum regime, free molecular regime and transition regime [27]. Analytic solutions for coagulation equation of aerosol particles (integro-differential equation) only exists for few restrictive cases [3]. Several numerical schemes have been adopted for solving this equation for describing coagulation effects for realistic aerosol scenarios. The method of moments and its variants [5, 19] track the time evolution of moments of an assumed distribution. The parameters characterizing the distribution (mean, standard deviation etc.) are then extracted from the moments. The accuracy of these methods depends on how well the assumed distribution resembles a real situation [30]. In the sectional method [11], the particle size range is divided into a finite number of bins. The number of particles that fall within consecutive bins is then tracked. The related nodal method [17, 24] uses zero width discrete nodes. A splitting factor distributes newly agglomerated particles between existing nodes in a volume (and mass) conserving fashion. Aerosol drift and deposition due to external forces such as gravity, temperature gradient etc. can be incorporated as drift fluxes in the aerosol dynamic equation [13, 20].

Computational fluid dynamics (CFD) has played an instrumental role in understanding flow effects and processes in a multitude of studies. The conclusions drawn from such studies have been used in designing several controlling parameters and preventing/mitigating failure conditions. In recent times, CFD usage has also been extended to research and application fields focused on simulant dispersion, transport and deposition. CFD has been used for improving designs of aerosol instruments such as impactors and cyclones [12, 14, 18, 29]. In another domain, CFD has been successfully implemented for predicting radon concentration distribution and related effects in chamber as well as room environment [2, 4]. Although several studies have used CFD for aerosol related research, only few are focused on coupling aerosol coagulation framework with flow equations.

This study discusses the development of coupling of aerosol dynamics equation (incorporated coagulation, gravitational settling, thermophoresis and diffusion) with a computational fluid dynamics code. Numerical scheme implemented for aerosol transport and dynamics combined particle transport with a nodal approach for aerosol dynamics and a drift flux model for gravitational settling. The coagulation and settling terms were treated explicitly in time. Validation of standalone numerical approaches for coagulation and settling has also been presented. The aerosol dynamics coupled CFD code was then applied for a predictive study for aerosol evolution and transport in a typical reaction chamber under flow conditions. The results obtained from these cases have been presented and interpreted by comparing time scales of the convection and coagulation processes.

2 Aerosol Transport and Dynamics Model

Evolution of number distribution function characterizing dynamics and transport of aerosol particles can be described mathematically in terms of General dynamic equation [9, 30] as

$$\frac{\partial n(d_p, r, t)}{\partial t} + \nabla \cdot [Un(d_p, r, t)] = \nabla \cdot \{D(r, t)\nabla_n(d_p, r, t)\}$$
$$+ \frac{1}{2}\int_0^{d_p} K(d_p^{'}, d_p - d_p^{'})n(d_p^{'}, r, t)n(d_p - d_p^{'}, r, t)d(d_p^{'})$$
$$- n(d_p, r, t)\int_0^{\infty} K(d_p, d_p^{'})n(d_p^{'}, r, t)d(d_p^{'})$$
$$+ S(d_p, r, t) - \lambda n(d_p, r, t)$$
$$- \nabla \cdot [U_{drift} n(d_p, r, t)] \quad (1)$$

where d_p and d'_p are particle diameters; n (d_p, r, t) is the spatially (r) and temporally (t) varying number concentration distribution function for particle diameter d_p; U is the gas phase velocity; D is the particle diffusivity; K is the collision frequency between particles of different sizes; S is the source term arising from nucleation and direct emission; λ is the decay rate of the species; U_{drift} is the total drift velocity of the aerosol particles due to various mechanisms like gravitational settling, thermophoresis, turbophoresis etc.

In the above equation, terms on the left-hand side include the temporal change in number distribution function and convective transport. The first term on the right side of the equation accounts for the diffusive transport of aerosol particles under the flow conditions. Coagulation terms appear as integrals in the GDE. The first integral in the GDE represents the rate of formation of new particles of a particular size due to collision of 2 particles of smaller sizes. The second integral represents the rate of depletion due to particles of a particular size combining with another particle.

The particle diffusivity used in the above equation is also a function of particle size and is given by

$$D = \frac{k_b T C_c}{3\pi \mu_g d_p} \quad (2)$$

where k_b is the Boltzmann constant, T is the temperature, μ_g is the gas viscosity, and C_c is the Cunningham slip factor [6, 8]

$$C_c = 1 + Kn * (2.514 + 0.8 * \exp(-0.55/Kn)) \quad (3)$$

where Kn, Knudsen number, is the ratio of the mean free path of the gas particles (λ) to the particle diameter (d_p) and is given as

$$Kn = \frac{2\lambda}{d_p} \quad (4)$$

2.1 Coagulation

Coagulation is the process of aerosol particles colliding with each other and forming a bigger particle. The frequency with which such collisions occur depends on the forces acting on the particles and on the size of the colliding particles. The function expressing this collision frequency is called the Coagulation Kernel. Examples of coagulation kernels are the Brownian Kernel and the Gravitational Kernel etc.

Various forms of the Brownian kernel exist for the continuum regime (Kn $\ll$ 1), free molecular regime (Kn $\gg$ 1) and the transition regime (Kn $\approx$ 1). In all studies reported in this work, the Fuchs kernel [10, 15] was employed. The Fuchs kernel Kij for collision frequency between 2 particles i and j is given by

$$K_{ij} = \frac{2\pi(D_i + D_j)(d_{pi} + d_{pj})}{\frac{d_{pi}+d_{pj}}{d_{pi}+d_{pj}+2\sqrt{\delta_i^2+\delta_j^2}} + \frac{8(D_i+D_j)}{(d_{pi}+d_{pj})\sqrt{\bar{v}_{pi}^2+\bar{v}_{pj}^2}}} \tag{5}$$

$$\bar{v}_{pi} = \sqrt{\frac{8k_bT}{m_i\pi}} \tag{6}$$

$$\delta_i = \frac{1}{3d_{pi}\lambda_i}\left[(d_{pi} + \lambda_i)^2 - (d_{pi}^2 + \lambda_i^2)^{\frac{3}{2}}\right] - d_{pi} \tag{7}$$

$$\lambda_i = \frac{8D_i}{\pi v_{pi}} \tag{8}$$

where d_{pi} and d_{pj} are particle diameters, D_i and D_j are particle diffusivities, v_{pi} and v_{pj} are thermal velocity of air molecules of same size as the aerosol particles, and δ_i and δ_j are distance from particle surface to continuum/free molecular regime dividing surface, λ_{pi} and λ_{pj} are particle mean free path, m_i is the mass of particle i.

The nodal approach [24] was used to parameterize integration terms of coagulation, with the range of aerosol particle sizes divided into a finite number of zero width nodes. The ANSWER [1] CFD code allows for aerosol particle size range to be divided into arbitrary number of equally or unequally spaced divisions. In this study, the particle size range was taken to be logarithmically equally spaced. Since the nodes are of zero-width, the size of new particles formed by collision may fall between existing nodes. This was resolved by using a splitting factor [24] that assigns the particles in a mass conserving fashion to the nodes adjacent to the new particle size.

2.2 Gravitational Settling

Aerosol particles experience the downward force of gravity under simultaneous counter drag. This force is proportional to the particle mass, so the bigger particles experience a larger force. The gravitational settling velocity v_g is given by [13, 20]

$$v_g = g\tau \tag{9}$$

where, τ is relaxation time,

$$\tau = \frac{C_c \rho_g d_p^2}{18\mu_g} \tag{10}$$

and ρ_g is gas density.

Only gravitational settling was used as a part of drift flux term in this study. Additionally, source term for the cases considered here are imposed lognormal distribution with defined properties.

3 Code Validation

The validation of the coagulation and gravitational settling models used in the ANSWER CFD code was carried out by comparison with published numerical and analytical solutions.

3.1 Coagulation

To validate the coagulation module, we consider two cases of particle coagulation in a well mixed chamber. It is assumed that due to well mixed nature of the flow, the aerosol concentration is homogeneous. Except coagulation, all other transport and aerosol mechanism were turned off. In the GDE (Eq. 1), all terms drop out except the time derivative of the number concentration, the coagulation integral terms, and the source term. The collision frequency determining the particle coagulation was taken as Fuchs kernel (Eq. 5).

In case A, the chamber is assumed to contain an initially homogeneously mixed lognormal distribution [16, 21] of particles (Eq. 11) undergoing coagulation. Without any injection or removal of particles (by deposition or ventilation), coagulation will increase the concentration of larger sized particles and decrease the size of smaller sized particles.

$$n(v, t) = \frac{N_0}{3\ln(\sigma)\sqrt{2\pi}} \exp\left\{\frac{-\ln^2(v/v_m)}{18\ln^2(\sigma)}\right\}\frac{1}{v} \tag{11}$$

In above equation, n(v, t) is the number concentration of a particle of volume v (per m^3), N_0 is the total number concentration of particles at the inlet (per m^3), v_m is

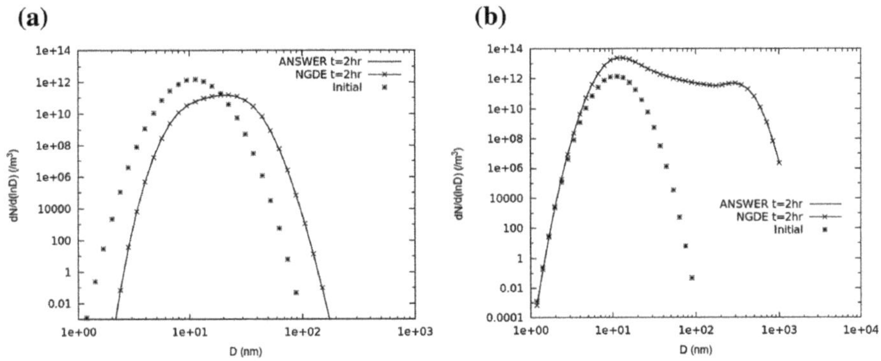

Fig. 1 **a** ANSWER versus NGDE comparison for initial log-normal distribution **b** ANSWER versus NGDE comparison for continuous injection of aerosol

the mean particle volume of the distribution and σ is the geometric standard deviation (GSD) of the distribution.

For case a, the initial number concentration of particles was taken to be $10^{12}/m^3$, with mean diameter of the lognormal distribution as 10 nm and GSD of 1.3. For case b, a continuous injection of particles into the chamber at the rate of $10^{12}/m^3/s$, with a mean diameter of 10 nm and GSD of 1.3 was taken. For both case a and case b, aerosol particles in the size range 1 nm to 1 μm were divided into 41 nodes, with the particles sizes distributed in a geometric ratio.

The results for case a and b from the ANSWER coagulation module were compared to the results from the published NGDE code [24] for identical conditions. Figure 1a shows the initial condition and predictions for case a after 2 h of physical time. It is seen that the peak of the distribution has moved to the right indicating an increase in particle size due to coagulation. Since no aerosol is removed from the chamber, conservation of aerosol particle mass leads to an overall decrease in the peak of the size distribution.

Figure 1b shows the initial condition (and continuous source distribution) and size distribution after 2 h for case b. It is seen that the peak concentration has increased due to a continuous injection of aerosol. The distribution also becomes bi-modal indicating the formation of coagulation peak.

It is seen that the ANSWER solution matches results from the NGDE code well for both case a and case b.

3.2 *Gravitational Settling*

The gravitational settling module in ANSWER was validated by comparison with the analytical solution for deposition efficiency for 2-D laminar flow in a duct [23]. In a 2-D rectangular duct of height h and Length L, the deposition efficiency E (ratio of particle mass deposited by gravitational settling to total mass coming into the duct) for particles of a particular size for flow with mean velocity of U is given as (Eq. 12) [23]

$$E = \frac{1}{2}\frac{V_s L}{Uh} \tag{12}$$

where V_S is the gravitational settling velocity (Eq. 9) of a particle.

For the ANSWER simulation, the velocity profile in the duct was set to be parabolic with a mean flow speed U. Particles of different sizes were assumed to come into the duct through the entire height of the inlet at constant concentration.

Figure 2 shows a comparison of deposition efficiency for the ANSWER simulation with the analytical solution (Eq. 12). As can be seen, code predictions matched closely with the analytical results.

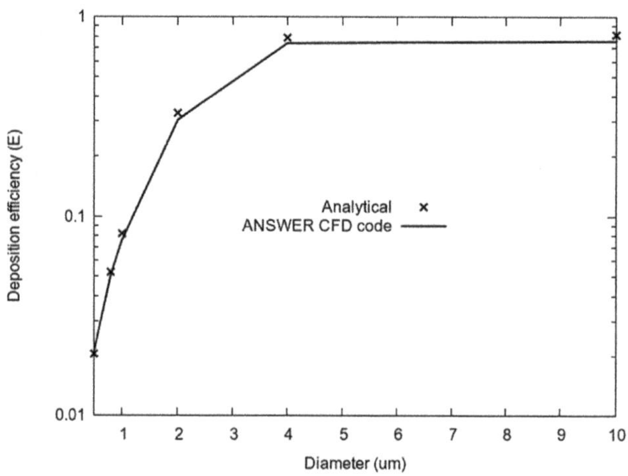

Fig. 2 Deposition efficiency of gravitational settling of particles for laminar flow in a 2D duct

4 Methodology and Problem Definition

The objective was to compute the concentration of aerosol particles of different sizes as a function of time and space in the domain of interest. The transport and dynamics of aerosol are governed by the GDE (Eq. 1). The solution of the GDE requires the flow field.

A steady state flow field is obtained using the ANSWER [1] CFD software. ANSWER is a finite volume based CFD code [22, 26, 28] that solves the full 3-D Navier-Stokes equations and heat and mass transfer, for incompressible and compressible flows with arbitrary polyhedral grids. ANSWER also has various turbulence models including k-ε and LES.

Once the flow field was obtained, ANSWER was used to compute the aerosol transport and dynamics. A separate transport equation was solved for each aerosol particle size class k at every time step (the particle size range being divided into $k = 1$ to m classes). ANSWER solves for N_k's, which are the particle number concentrations in the size range $(k - ½)$ to $(k + ½)$.

The GDE is an integro-differential equation due to the presence of the coagulation terms. The integral terms were treated explicitly in time and appear as source terms on the right hand side.

Settling terms such as gravitational settling, thermophoresis etc. were incorporated into the GDE as drift flux terms. The appropriate drift velocity was calculated at each cell face for every active settling. The settling terms were also handled explicitly in time in the code.

Since coagulation kernels are functions of temperature, the kernels will be recomputed at every cell and at each time step for cases where the temperature is a spatially and temporally varying field.

5 Case Studies

5.1 Reaction Chamber

Reaction chambers are used for aerosol experiments either for providing controlled conditions or for generating desired outlet parameters. Particles are introduced into reaction chamber to obtain a spatially uniform concentration. The size distribution of the aerosol at the outlet is also important for coupling to other sections. In this case study, a cylindrical chamber of diameter 0.4 m and length 1.4 m was considered. The inlet and outlet were both of diameter 100 mm. The carrier gas was assumed to carry the aerosol particles at the inlet. The inlet size distribution of the aerosol was assumed to be log-normal. The carrier gas properties were taken to be density $\rho = 1\ kg/m^3$ and viscosity $\mu = 2 \times 10^{-5}$ kg/m/s.

Both 2-D Axi-symmetric and 3-D simulations under three different flow scenarios were considered. The first scenario had only axial flow at the inlet. In order to enhance mixing and increase the residence time of the particles inside the domain, two further scenarios were considered with an added swirl (tangential) component to the inlet flow. The swirl strengths for the two additional scenarios were 20% (low swirl) and 80% (high swirl) of the axial flow speed respectively. The design of the chamber was modified for 3-D simulations to further enhance mixing under all flow conditions used in 2-D case.

5.1.1 Case 1: 2-D Axi-Symmetric Simulation

The inlet carrier gas flow rate is taken to be 10 kg/h for this simulation. The Reynolds number (Re) based on the inlet dimension came out to be 7911. The flow is in the turbulent regime and the standard k-ε model was used. The inlet turbulence boundary conditions were taken to be

Turbulent kinetic energy: $k = 3/2(U_0 I)^2$

Turbulent length scale: $l = 10\%$ of inlet diameter

Where U_0 is the inlet flow speed,

I (turbulent intensity) $= 0.16 \times Re^{-1/8}$ (CFD Online [7].

The finite volume grid contained 195 and 28 cells in the X (axial) and Y (radial) directions (5460 in 2-D domain). ANSWER CFD code was used to obtain a converged flow solution, with momentum residuals $< 1 \times 10^{-8}$ at convergence. Figure 3 shows the computational domain.

For the aerosol transport, the inlet aerosol total number concentration was taken to be $1 \times 10^{12}/m^3$, with a mean particle diameter of 100 nm and GSD = 1.5. The minimum and maximum aerosol particle sizes were 1 nm and 1 μm (1000 nm) respectively, with 25 size classes distributed logarithmically equally between them. Gravitational settling is significant beyond particle sizes of 1 μm. Below this size, settling can be ignored and the simulation can be carried out in the 2-D axi-symmetric mode. The transport simulations were carried out in transient mode for 2 h to ensure the particle concentrations had reached a steady state within the domain.

Figure 4 shows the total number concentration of particles in the domain as a function of time. The total number concentration of particles reaches an equilibrium value within 5 min for the high swirl scenario, but takes 20 min for the no swirl and low swirl scenarios. It is also seen from Fig. 4 that the total number of particles is roughly the same for the cases with no swirl and low swirl, while the 80% swirl scenario shows a slightly higher equilibrium total number of particles. This indicates the high swirl scenario convects the particles to the outlet much faster, reducing the time the particles have to undergo coagulation.

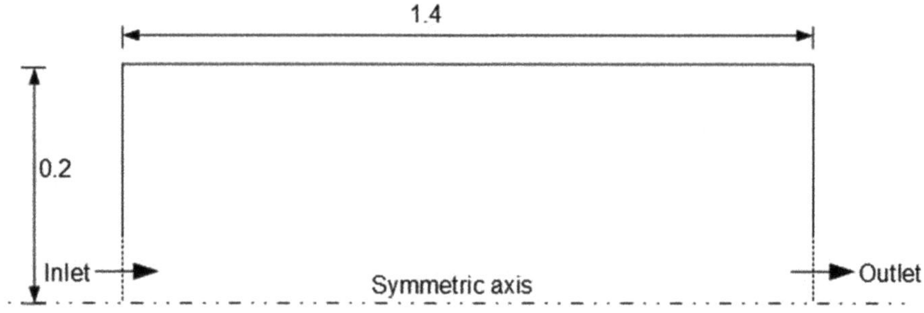

Fig. 3 Computational Domain for case 1

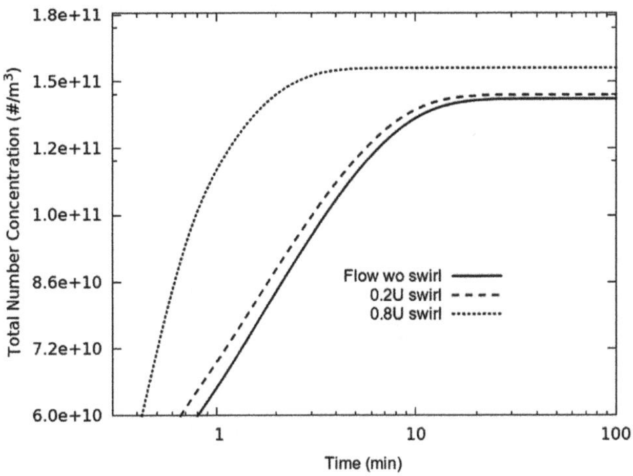

Fig. 4 Total number concentration of particles for $N_0 = 1 \times 10^{12}/m^3$

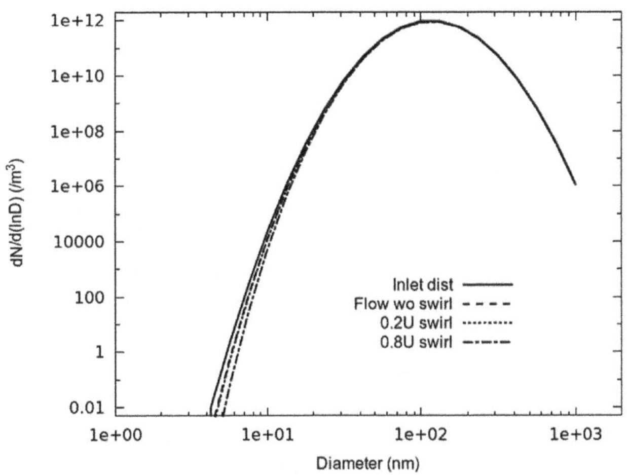

Fig. 5 Particle size distribution at outlet for $N_0 = 1 \times 10^{12}/m^3$

Figure 5 shows the particle size distribution at the outlet and Fig. 6 shows the average size distribution in the domain after 2 h for the same case.

It is seen in Figs. 5 and 6 that the flow conditions do not affect the particle size distributions. There is a minor change in concentrations at lower particle sizes but the peak and the width of the distribution are unaffected.

This can be understood by examining the time scales of various processes involved in the dynamical evolution. The convective time scale, t_{conv}, is roughly L/U_0, where L is the domain length. For the case under consideration, t_{conv} is of the order of 100 s.

The coagulation time scale is given by [5]

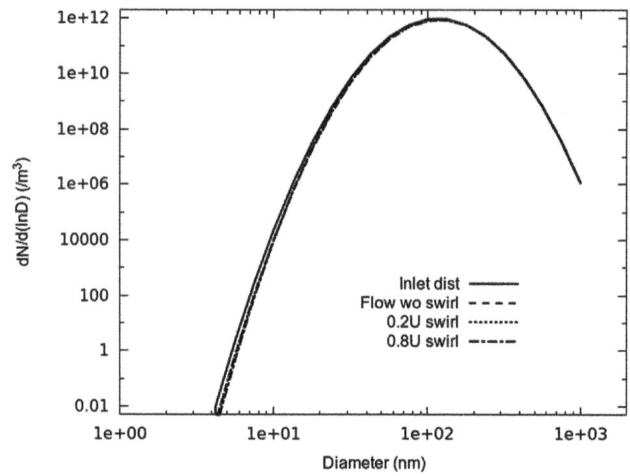

Fig. 6 Average particle size distribution in the domain for $N_0 = 1 \times 10^{12}/m^3$ after 2 h

$$t_{coag} = \frac{2}{K_{\min} N} \tag{13}$$

where K_{min} is the minimum value of the coagulation kernel, and N is the number of particles in the domain. For the case of $N_0 = 1 \times 10^{12}/m^3$, the coagulation time scale is of the order of 10^5 s. Therefore the flow conditions have little effect on the size distribution.

Further simulations were carried out with a higher inlet total number concentration of $N_0 = 1 \times 10^{15}/m^3$. In this scenario, the coagulation time scale is roughly the same as the convection time scale.

Figure 7 shows the time evolution of the total number of particles within the domain.

The effect of the flow now be clearly seen in Fig. 8 showing the outlet size distribution. The peak of the distribution has moved from 100 nm at the inlet to 200 nm at outlet for no swirl and 20% swirl scenarios, and closer to 300 nm at outlet for the 80% swirl scenario.

The swirl is limited to the inlet diameter and dies down quickly in the radial direction. The flow in the domain away from the axis is mostly unaffected by the swirl and this is seen in Fig. 9 where the average size distribution was unaffected by the flow conditions.

5.1.2 Case 2: Full 3-D Simulation

A full 3-D simulation was also carried out for a reaction chamber design. The chamber diameter is 0.4 m and length is 1.4 m. To enhance mixing, a secondary ring inlet with a number of periodically located holes is placed 400 mm downstream of the

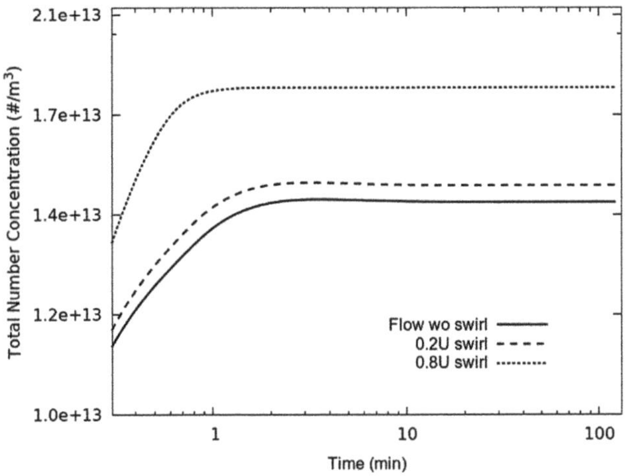

Fig. 7 Total number concentration of particles for $N_0 = 1 \times 10^{15}/m^3$

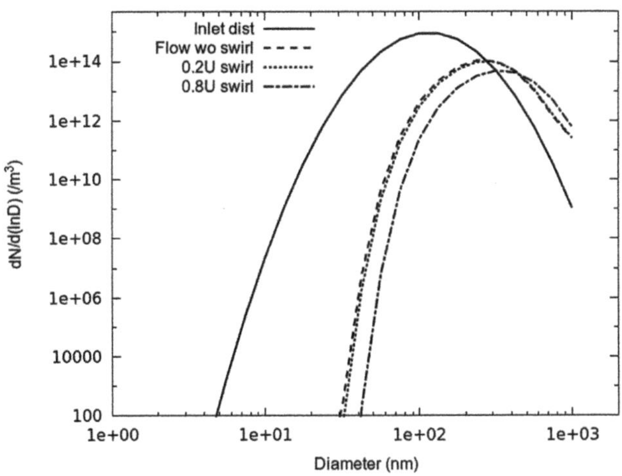

Fig. 8 Particle size distribution at outlet for $N_0 = 1 \times 10^{15}/m^3$

inlet. The inlet carrier gas flow rate is taken to be 10 kg/h. The flow rate through this secondary inlet is also 10 kg/h. The turbulence boundary conditions are applied as in Case 1.

The finite volume grid contained 77,153 cells in 3-D domain. ANSWER CFD code was used to obtain a converged flow solution, with momentum residuals $< 1 \times 10^{-8}$ at convergence.

Figure 10 shows the computational domain.

Figure 11 shows the inner details of the secondary inlet (Ring portion) in Fig. 10.

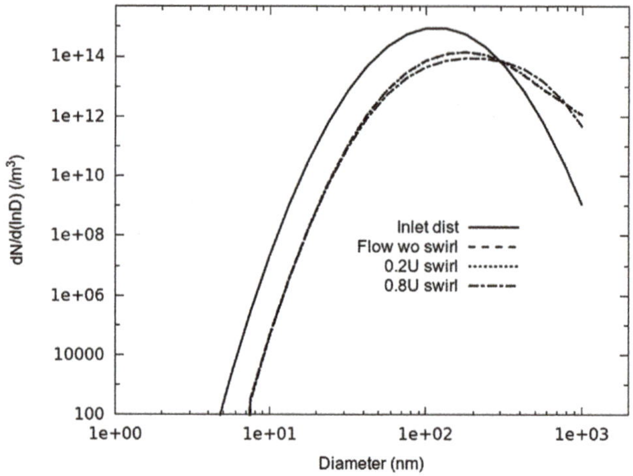

Fig. 9 Average particle size distribution in the domain for $N_0 = 1 \times 10^{15}/m^3$ after 2 h

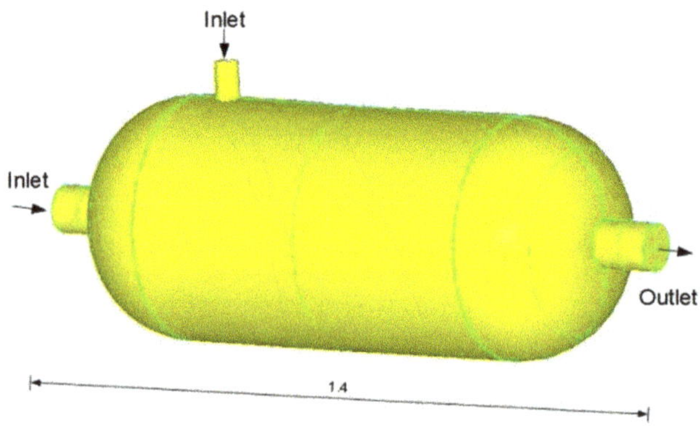

Fig. 10 Schematic of Reaction Chamber for Case 2 showing inlets and outlet

Figure 12 shows the total number concentration of the particles in the domain as a function of time for $N_0 = 1 \times 10^{12}/m^3$. It is now seen that, unlike in the 2-D axi-symmetric simulations, the equilibrium total number concentration of particles is slightly lesser for flow with higher swirl compared to no swirl and 20% swirl scenarios. This also shows that the high inlet swirl and secondary ring flow interact to increase residence time for particles, leading to greater coagulation.

Figures 13 and 14 show the average and outlet size distributions respectively. It is seen that the flow conditions have no effect on the distribution and that the peak has shifted very little from the inlet conditions. The shift is however more than what can be seen in the 2-D axi-symmetric simulations, indicating that effect of the flow in the secondary inlet was to slightly increase the residence time of particles inside the chamber.

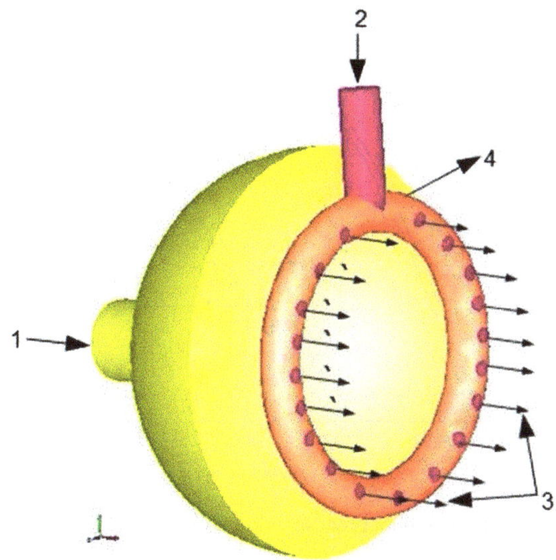

Fig. 11 Secondary inlet ring details in Reaction chamber: 1. Primary inlet 2. Secondary inlet 3. Circumferential hole on ring, and 4. Ring

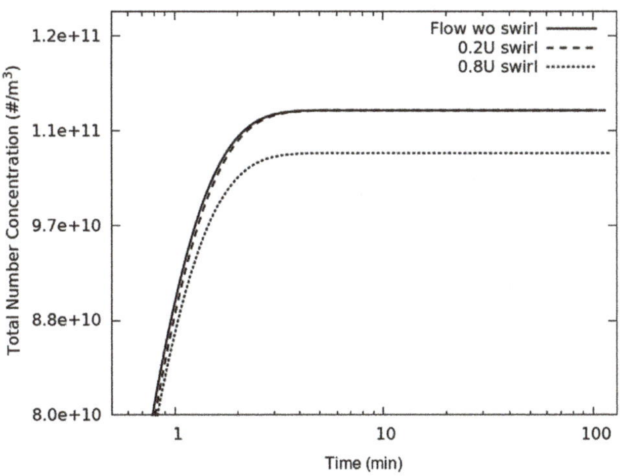

Fig. 12 Total number concentration of particles for $N_0 = 1 \times 10^{12}/m^3$

For higher inlet total concentration of $N_0 = 1 \times 10^{15}/m^3$, the total number of particles in the domain reaches an equilibrium value very quickly, as seen in Fig. 15. The higher swirl flow scenario is again seen to have a slightly lower equilibrium value.

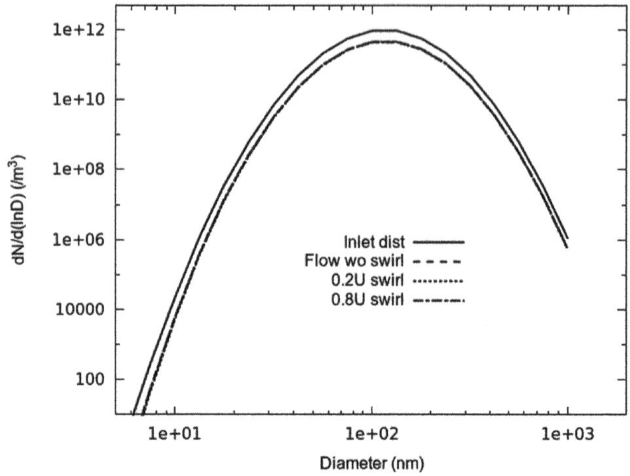

Fig. 13 Particle size distribution at outlet for $N_0 = 1 \times 10^{12}/m^3$

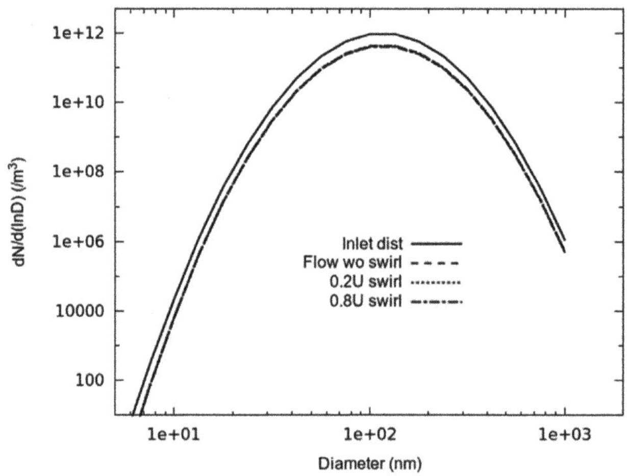

Fig. 14 Average particle size distribution in the domain for $N_0 = 1 \times 10^{12}/m^3$ after 2 h

Figures 16 and 17 show the average and outlet size distributions respectively. While the peak of the distribution has shifted to the right from the inlet conditions, all the flow scenarios lead to same location of the peak. The outlet size distribution shows a slight shift towards the right for smaller particle sizes under the high inlet swirl condition.

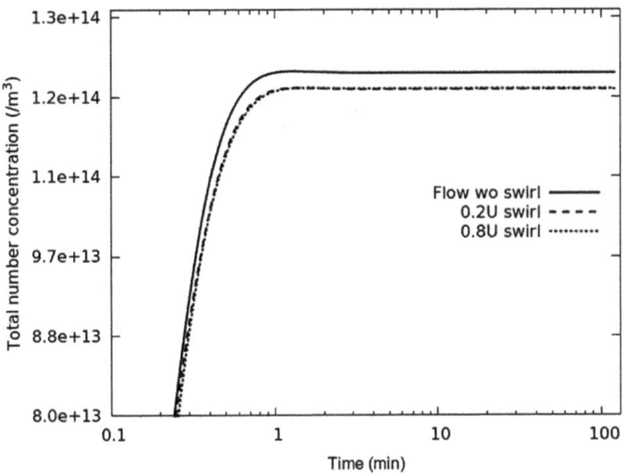

Fig. 15 Total number concentration of particles for $N_0 = 1 \times 10^{15}/m^3$

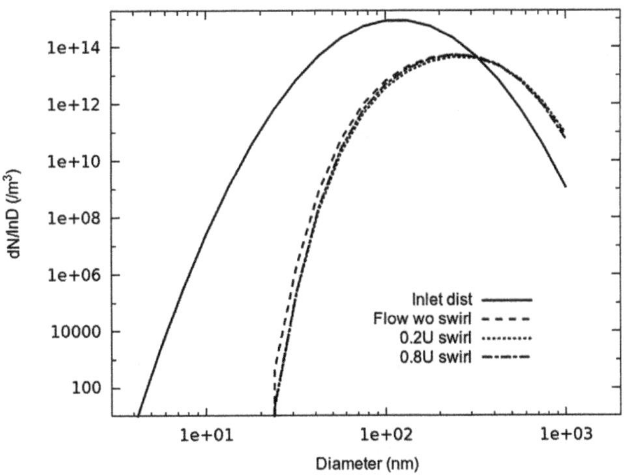

Fig. 16 Particle size distribution at outlet for $N_0 = 1 \times 10^{15}/m^3$

The difference in results between the axi-symmetric and full 3-D simulations indicate that the effect of the secondary inlet was to enhance mixing so that the size distribution is very similar for all flow scenarios considered.

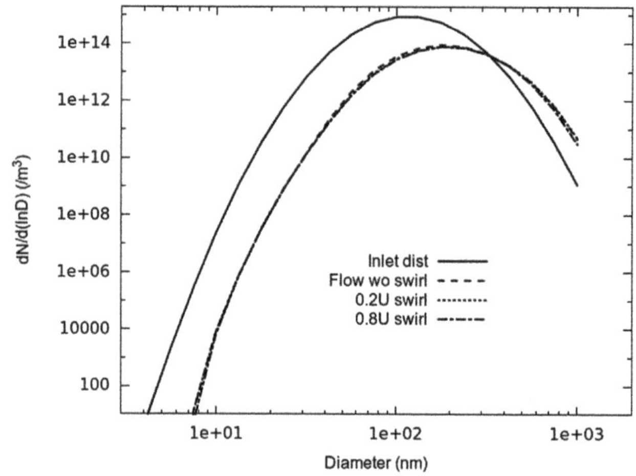

Fig. 17 Average particle size distribution in the domain for $N_0 = 1 \times 10^{15}/m^3$ after 2 h

6 Conclusions

ANSWER CFD code was coupled with an aerosol module incorporating aerosol dynamic equation in a CFD platform. The modified code uses coagulation, gravitational settling, aerosol diffusion and thermophoresis as part of GDE. The standalone modules were validated using published analytical solutions. Aerosol transport and dynamics was simulated using the modified ANSWER CFD code. These simulations were carried out for a typical reaction chamber design to understand the effect of different flow conditions and aerosol total number concentrations on the size distribution within the domain and at the outlet. The simulations showed that:

1. The reaction chamber design is robust with respect to the flow conditions for aerosol inlet total number concentrations of $1 \times 10^{12}/m^3$ and $1 \times 10^{15}/m^3$. This can be explained in terms of the time scale of convection being much less than coagulation time scale.
2. For higher aerosol inlet concentrations, significant shift was seen in the average and outlet distributions from inlet distributions.
3. The presence of the secondary ring inlet was seen to lead to identical aerosol distribution for different flow scenarios. Its effect might be to enhance mixing within the domain. Further study under different configurations of the secondary ring inlet is required to understand its impact.

References

1. ACRi, ANSWER® and CFDStudio® "KEYWORD COMMANDS", VERSION 6.43.0 (2016)

2. T.K. Agarwal, B.K. Sahoo, J.J. Gaware, M. Joshi, B.K. Sapra, CFD based simulation of thoron (220 Rn) concentration in a delay chamber for mitigation application. J. Environ. Radioact. **136**, 16–21 (2014)
3. S. Anand, Y.S. Mayya, Coagulation in a diffusing Gaussian aerosol puff: Comparison of analytical approximations with numerical solutions. J. Aerosol Sci. **40**(4), 348–361 (2009)
4. N. Chauhan, R.P. Chauhan, M. Joshi, T.K. Agarwal, P. Aggarwal, B.K. Sahoo, Study of indoor radon distribution using measurements and CFD modeling. J. Environ. Radioact. **136**, 105–111 (2014)
5. E.R. Cohen, E.U. Vaughan, Approximate solution of the equations for aerosol agglomeration. J. Colloid Interface Sci. **35**(4), 612–623 (1971)
6. Cunningham, E., On the velocity of steady fall of spherical particles through fluid medium, in *Proceedings of the Royal Society of London*. Series A, Containing Papers of a Mathematical and Physical Character, vol. 83, no. 563, pp. 357–365 (1910)
7. CFD Online, Turbulence Intensity (2016), https://www.cfd-online.com/Wiki/Turbulence_intensity. Accessed 1 Oct 2016
8. C.N. Davies (ed.), *Aerosol science (Vol. 1102)* (Academic Press, London, 1966)
9. Friedlander, S.K.1., *Smoke, Dust, and Haze: Fundamentals of Aerosol Dynamics*, 2nd edn. (Oxford University Press, New York, 2000)
10. N.A. Fuchs, *The Mechanics of Aerosols 1964* (Pagamon, New York, 1964)
11. F. Gelbard, Y. Tambour, J.H. Seinfeld, Sectional representations for simulating aerosol dynamics. J. Colloid Interface Sci. **76**(2), 541–556 (1980)
12. Gimbun, J., Chuah, T. G., Fakhru'l-Razi, A., and Choong, T.S., The influence of temperature and inlet velocity on cyclone pressure drop: a CFD study. Chem. Eng. Process. Process Intensif. **44**(1), 7–12 (2005)
13. A. Guha, Transport and deposition of particles in turbulent and laminar flow. Annu. Rev. Fluid Mech. **40**, 311–341 (2008)
14. W.D. Griffiths, F. Boysan, Computational fluid dynamics (CFD) and empirical modelling of the performance of a number of cyclone samplers. J. Aerosol Sci. **27**(2), 281–304 (1996)
15. Jacobson, M.Z., *Fundamentals of Atmospheric Modeling* (Cambridge university press, 2005)
16. K.W. Lee, J. Chen, J.A. Gieseke, Log-normally preserving size distribution for Brownian coagulation in the free-molecule regime. Aerosol Sci. Technol. **3**(1), 53–62 (1984)
17. K.E. Lehtinen, M.R. Zachariah, Self-preserving theory for the volume distribution of particles undergoing Brownian coagulation. J. Colloid Interface Sci. **242**(2), 314–318 (2001)
18. P.W. Longest, M. Hindle, S.D. Choudhuri, P.R. Byron, Numerical simulations of capillary aerosol generation: CFD model development and comparisons with experimental data. Aerosol Sci. Technol. **41**(10), 952–973 (2007)
19. R. McGraw, Description of aerosol dynamics by the quadrature method of moments. Aerosol Sci. Technol. **27**(2), 255–265 (1997)
20. K.J. Overholt, J.E. Floyd, O.A. Ezekoye, Computational modeling and validation of aerosol deposition in ventilation ducts. Fire Technol. **52**(1), 149–166 (2016)
21. E. Otto, H. Fissan, S.H. Park, K.W. Lee, The log-normal size distribution theory of Brownian aerosol coagulation for the entire particle size range: part II—analytical solution using Dahneke's coagulation kernel. J. Aerosol Sci. **30**(1), 17–34 (1999)
22. Patankar, S., *Numerical Heat Transfer and Fluid Flow* (CRC press, 1980)
23. J. Pich, Theory of gravitational deposition of particles from laminar flows in channels. J. Aerosol Sci. **3**(5), 351–361 (1972)
24. A. Prakash, A.P. Bapat, M.R. Zachariah, A simple numerical algorithm and software for solution of nucleation, surface growth, and coagulation problems. Aerosol Sci. Technol. **37**(11), 892–898 (2003)
25. J. Pyykönen, J. Jokiniemi, Computational fluid dynamics based sectional aerosol modelling schemes. J. Aerosol Sci. **31**, 531–550 (2000)
26. A.K. Runchal, CONDIF: A modified central-difference scheme for convective flows. Int. J. Numer. Meth. Eng. **24**(8), 1593–1608 (1987)

27. Seinfeld, J.H., Pandis, S.N., *Atmospheric Chemistry and Physics: From Air Pollution to Climate Change* (Wiley, 2016)
28. Versteeg, H.K., Malalasekera, W., *An Introduction to Computational Fluid Dynamics: The Finite Volume Method* (Pearson Education, 2007)
29. S. Vinchurkar, P.W. Longest, J. Peart, CFD simulations of the Andersen cascade impactor: Model development and effects of aerosol charge. J. Aerosol Sci. **40**(9), 807–822 (2009)
30. M.M.R. Williams, Nuclear aerosol behavior during reactor accidents. Prog. Nucl. Energy **23**(2), 101–108 (1990)

Fire Testing and Study of Liquid Pool Fire in Multiple Compartments

Pavan Kumar Sharma, Anoop Kumar Raut and Abhijit Kushari

Abstract Many experimental/numerical studies on single compartment or single room fires have been reported in the literature but few studies are available for multiple compartments. There is a need for fire studies in multiple compartments as it may help in predicting fire impact in buildup environments with multiple compartments. In the present work, fire experiments were conducted in a real life two-storey compartmentalized concrete building. Aviation Turbine Fuel (ATF) was used as the liquid fuel to generate pool fire. The work, focused on the thermal hydraulic aspects like heat distribution to the corresponding walls, temperature near walls, vertical distribution of temperature close to fire and comparison of heat-flux and temperature at specified locations in the interconnected rooms. The study elucidates the impact of ventilation on various thermal and hydraulic aspects of fire.

Keywords Multi-compartment · Pool fire · Thermal-hydraulics · Wall heat flux

1 Introduction

The loss of lives and property damage because of fire accidents will always remain a subject of global concern and scientific research, whether it is a fire accident in a building, in an office or in a nuclear power plant. The degree of concern increases with the importance of that building: for example, fire in a nuclear plant is more of a concern than fire in a commercial building because the former causes loss of life and property as well as long-term environmental impact on its surroundings. This motivation for fire research has varied vastly from investigating fire scenarios to

P. K. Sharma
Reactor Safety Division, Bhabha Atomic Research Centre, Trombay, Mumbai 400085, India

A. K. Raut · A. Kushari (✉)
Department of Aerospace Engineering, Indian Institute of Technology Kanpur, Kanpur 208016, India
e-mail: akushari@iitk.ac.in

A. K. Runchal et al. (eds.), *Energy for Propulsion*, Green Energy and Technology,
https://doi.org/10.1007/978-981-10-7473-8_15

understanding the production of toxic species and behavior of fire and heat spreading during building fires. Studying compartment fires is a part of that motivation where researchers are more interested to study the fire behavior and its effect in a confined enclosure, which replicates an actual fire scenario in a building. The behavior of fires in building structures with interconnected multiple compartments often differs in fundamental ways from fires in a single confined space. Such fires have been poorly explored so far. The work by Luo and Beck is one of those few previous studies which were performed in a multi-room environment [1]. The current work adds to this literature.

Room fire or compartment fire is treated as enclosure fire. As defined by Karlsson [2], enclosure fire is the fire in a confined space, which is controlled by confined space for air availability and thermal environment; and this affects the burning rate, growth and life period of fire. So, while designing the fire safety models; fire growth, temperature, heat and duration required for fire control are some of the aspects that always need to be considered [2]. Materials used in building construction and plastic materials used in building interiors are some other factors that should not be overlooked [3].

Life Period of fires in compartment start with growth phase followed by fully developed fire during development phase and then burns out with decay phase [4]. Fire ignited in development phase grows in size from a small to large fire; if it is not suppressed it grows to its maximum size which depends on the amount of fuel supply and amount of oxygen available [5–11]. The development of fire depends on: (a) size and shape of compartment [10], (b) amount of fuel supply [10], (c) amount and availability of other combustible material around fire (which feeds the fire and causes it to grow), (d) provision and amount of ventilation in the compartment that regulates the air flow which can influence the thermal energy distribution [12, 13] and (e) type of building construction materials used [2]. Fire development phase goes through flashover transition, which requires major attention because it is the phase where fire can be suppressed before it affects the safety of human life and structure [5]. After flashover, when fire becomes fully developed, heat energy released in the enclosure is at its greatest and sometimes it is limited because of unavailability of sufficient oxygen. In case of insufficient oxygen, the fire becomes oxygen starved and starts to decay.

In compartment fires with growth of fire, heat transfer occurs by the process of radiation, convection and conduction but it is usually dominated by radiation [2]. The soot is the carrier of heat because the soot particles produced by combustion radiate the heat energy in all direction. Ventilation plays an important role in fire sustenance as the lack of oxygen can lead to burnout; on the other hand, it generates the path for smoke and soot to travel from one compartment to another. In buildings, doors and windows are generally used to provide natural ventilation and they become the passage for air entrainment to enclosures. The limited supply of oxygen in a closed enclosure creates chances of generation of the carbon mono-oxide (CO) gas if any oxygen hungry equipment is running nearby [8].

Heat flow, convection, combustion, radiation, soot generation and burnout phenomenon are also the part of prediction for evaluation and generation of a fire model. During comparison between computational results and available experimental results, thermal radiation and soot radiation are two of the significant factors that play important role in accurate prediction of fire scenarios [14]. Smoke flow and temperature distribution during ventilation-controlled fire in compartments are dependent on ambient wind flow direction and speed. During fire incidents in tall buildings, ambient wind blow/flow affects smoke dispersion and fire dynamics. Ambient wind entering the compartment has two different effects: it supplies fresh air for entrainment, which provides more oxygen to promote fire growth, and because of entrainment of air, it promotes cooling process through convective heat transfer [15]. Many works available in literature are proof of how experimental and simulated results were used for confident predictions of firefighting by creating real life fire conditions. The real scale fire studies help fire fighters to understand the phenomenon of fire growth and smoke filling in multiple compartments [16, 17]. There are many experimental studies; in which full-scale fires in real scale rooms were studied, which replicates real life fire scenarios like in an office fire [18] or a storehouse fire [19]. These studies tell us that these types of full-scale experimental studies also help to modify fire safety policies and regulations. Full-scale experimental studies on single enclosures fires are also conducted in different weather conditions as per the geographical location of the building [17]. Work by Byström et al. [17] is an example of experimental studies where the fire studies were done at very low ambient temperature in a real scale building where horizontal and vertical compartmentalization were present.

With the goal to improve our understanding of fire and plume dynamics in multiple compartments, a research program was initiated for the study of associated thermal hydraulic aspects in multiple compartments, for a controlled fire burning in one compartment. A two-storey building with four full-scale rooms (interconnected multiple compartments) was used for a range of experiments. The heat and smoke transport to multiple compartments was studied with detailed measurements of heat-flux on boundaries, temperatures at different points, measurement of velocity and pressure of fire plume entraining through an opening at the ceiling, soot concentration and detailed measurement of increment in level of carbonaceous gases with depletion of Oxygen.

2 Description of the Facility and Instrumentation

At IIT Kanpur, an extensive fire testing facility, the first of its kind in India, has been developed to study single or multi-compartment fire scenarios and understand various aspects of fire dynamics, propagation and control. The state of the art instrumentation available in the facility is being used for fundamental and applied studies involving fire, combustion and fire safety issues. We have used two liquid fuels to generate two different types of pool fires, Aviation Turbine Fuel (ATF) and n-heptane. However, only the results pertaining to oil (ATF) fire is presented in this paper. The details of other experiments can be found in [20]. Figure 1 represents the state-of-the-art fire facility being used for the study of compartment fires, which contains four real

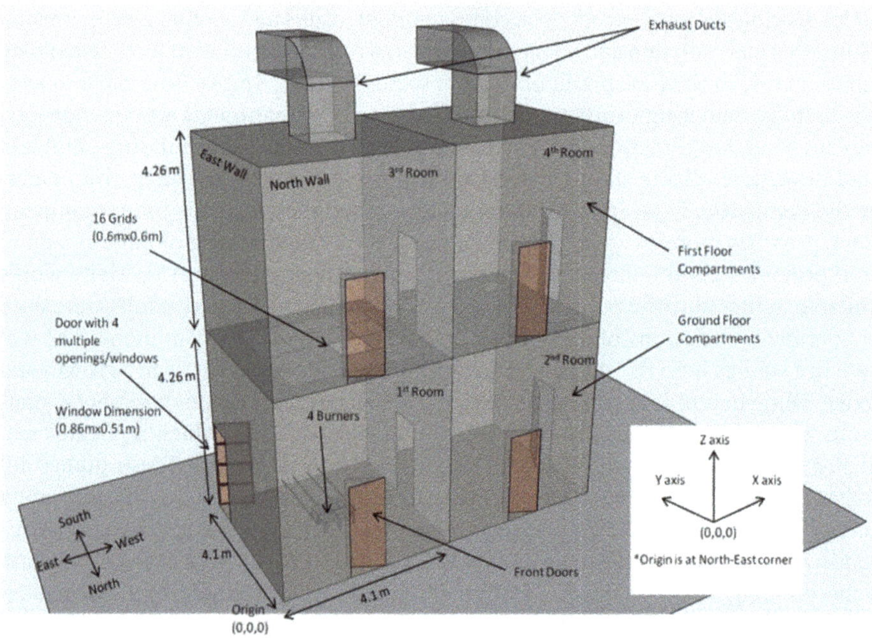

Fig. 1 Schematic of multi-compartment of fire testing facility at IITK

scale rooms in a two-story building. The figure shows all of the four compartments (/rooms), in which first and second rooms are on ground floor while third and fourth are above them. All the rooms have same dimensions. All rooms have an inspection door opening (0.86×2.06 m) on their North sidewall. The first compartment has multiple windows on its East-wall at its Southeast corner, which is built by dividing a door into four equal parts. Dimension of each window is 0.84×0.51 m ($L \times H$). The adjacent compartments are interconnected through a middle door between them. The ceilings of the two top compartments are connected with an exhaust duct. Many slots are present at different locations in all of the rooms for positioning the sensors.

In the test reported here, free ventilation was maintained in the compartments. In this phase, all the 4 rooms in the facility were used for study; the first room where fires were ignited is the most important room for this study as this room has most of the sensors. The other rooms also have the sensors which were used to compare the data collected from fire room with other inter-connected rooms as the adjacent compartments are affected and influenced by the thermal-hydraulic affects due to room fire.

Measurement of the heat flux on the walls was one of our main goal. Therefore, we have installed different types of heat flux sensors on the walls of the room at multiple locations. There are 5 types of heat flux sensors mainly used for this study. The HFP01 is a wall mounting round plate type sensor that measures the conductive heat flux over the wall. The SBG01 is Schmidt-Boelter heat flux meter to study the

Table 1 Location and Coordinates of heat flux sensors

Heat flux sensor				
No.	Sensor name	Room[a]	Location	Coordinates[b] (x, y, z)
1	SBG01	Fire room	Ceiling-center	1.98, 2.05, 4.2
2	SBG01		Floor	1.2, 1.54, 0.38
3	HF03	Fire room	South wall	2.06, 4.09, 1.34
4	HF03		North wall	2.63, 0, 1.35
5	RC01	Fire room	East wall	0, 2.03, 2.72
6	RC01	Side room (/Top room)	West wall	R-2 (4.09+4.01, 2.54, 2.71) R-3 (0, 2.54, 4.26+2.71)
7	HFP01	Fire room	East wall, E8	0, 3.78, 2.46
8	HFP01	Fire room	East wall-center	0, 2.03, 2.1
9	HFP01		South wall-center	2.06, 4.09, 2.25
10	HFP01		North wall-center	1.93, 0, 2.22
11	HFP01		West wall-center	4, 2.04, 2.34
12	HFP01	Side room	East wall-center	4.23, 1.96, 2.4
13	HFP01	Side room	West wall-center	4.09+4.01, 2.0, 2.32
14	HFP01	Top room	East wall-center	0, 2.11, 4.2+2.29

[a]For convenience room-1 is named as fire-room. Room-2 is named as side-room and room-3 is named as top-room
[b]Origin of the coordinate system is the North-East corner of the fire room

fire and fire resistance and HF03 is a portable high range heat flux sensor. The RC01 is a square plate radiation-convection heat flux sensor. Details of the locations of the heat flux sensors are provided in Table 1.

In two-zone fire models, generally a room is divided into two different zones, the upper zone and the lower zone. The upper zone is the hotter region affected by hot air floating beneath the ceiling while the lower zone is colder and heavier. Upon heating, the air from the lower zone moves towards the upper zone establishing a convection current. We have divided the room in two zones to study the distinctive behavior of each. The wall sensors (HFP01) are mounted at the center of the walls to measure the heat flux at the center point of zone boundaries. A set of HF03 sensors was mounted in the lower zone of north and south walls of the fire-room. Two SBG01 were mounted near the pool for fire monitoring, one was mounted on a frame kept near the pool facing upward and another one hanging from the ceiling center facing downward. Two RC01 or rad-con sensors were mounted at the upper zone of the east wall of fire-room and west wall of side-room to measure heat flux by radiation and convection on compartment boundaries.

Table 2 Location and coordinates of thermocouples

	Room	Location of thermocouple	Coordinates (x, y, z)
1	Fire room	East-wall	0, 2.03, 2.72
2	Side room/top room	West-wall East-wall	R-2 (4.09+4.01, 2.54, 2.71) R-3 (0, 2.54, 4.26+2.71)
3	Fire room	Pole, P4 (4.13 m above)	2.0, 2.70, 4.14
4	Fire room	Pole, P3 (3.13 m above)	2.0, 2.70, 3.14
5	Fire room	Pole, P2 (2.13 m above)	2.0, 2.70, 2.14
6	Fire room	Pole, P1 (1.13 m above)	2.0, 2.70, 1.14
7	Top room	center of room (2.13 m above floor)	2.0, 2.10, 4.26+2.14
8	Side room	center of room (2.13 m above floor)	4.09+2.0, 2.1, 2.14
9	Fire room	East-South corner @E8, above windows	0, 3.82, 2.46
10	Outside the fire room	S1 (2.13 m above floor)	Behind the south wall at South-East corner
11	Fire room	South-West corner @S4	4.08, 3.50, 1.40
12	Fire room	West-wall, within wall above door	4.09, 2.0, 2.13

Table 2 shows the list of the thermocouples with their location and coordinates. K type thermocouples of 3 mm thickness were used at different places to record the temperature variation at respective points with respect to time and with respect to the fire growth. A vertical pole was installed at the center of the fire-room (near the pool) into which 4 thermocouples were mounted to measure the temperature variation of the fire plume rising upward. The 1st thermocouple was 1.13 m above the ground, the 2nd, 3rd and 4th thermocouple were at 2.13, 3.13 and 4.13 m above the ground. Two thermocouples were hung from the ceiling at a height of 2.13 m above floor at the center of side-room and top-room. This arrangement allowed us to compare the temperature variation in all 3 rooms at the same point.

An NDIR based gas analyser that can measure NO, CO, CO_2, CH_4 and O_2 gases was installed outside of fire-room. The emission gas probe was kept at the north-east corner of the fire room 0.5 m above the ground. Output results from analyser were displayed on analyser's display panel and output current in mA was acquired via the data acquisition module. A 5-hole probe was used to measure the pressure and velocity profiles at the ceiling opening of 0.5×0.5 m between room 1 (ground floor)

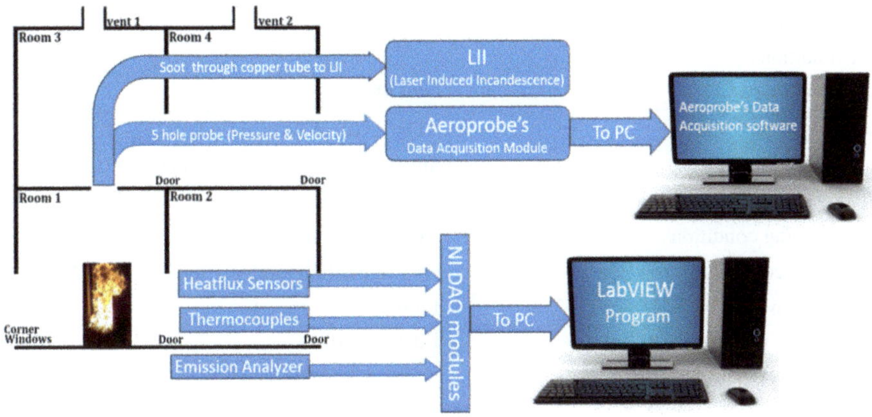

Fig. 2 Schematic of experimental setup

and 3 (first floor) at the centre of the ceiling at different points on a grid. The pressure data for a probe was acquired using a 36-port pressure scanner.

Various data acquisition modules were used to acquire analog outputs from these sensors, which were in physical units, using LabVIEW software for further processing. Further details of the measurements and instrumentation can be found in [20]. A schematic of the experimental setup is presented and an image of the pool fire is shown in Fig. 2.

3 Results and Discussion

In the series of liquid fires, the first test involved the study of small oil pool fire. For this test ATF (aviation turbine fuel) was used as the combustible liquid fuel. We name this as oil fire because ATF is used as aviation oil and is similar to kerosene oil used in industries and households. A small cylindrical vessel of 0.32 m diameter was kept at a height of 0.35 m above the ground on a stand. Initially, the pool was partially filled with water, and then ATF was poured to fill the vessel. All the windows and front door were kept open to support free ventilation during fire. The fire was ignited using a small cardboard paper dipped in ATF and placed over the surface of the liquid; as the paper cardboard ignited it heated up the rest of fuel over the surface and the fire spread over the surface. The estimated heat release rate [4] from this fire was 200 kW. Test conditions are tabulated in Table 3.

Different measurements, e.g., heat flux, temperature, etc. were taken at different locations and the results are discussed in this section. First, we will discuss the heat flux variations at different locations. Figure 3a shows the heat flux measured at the ceiling and the floor level. This plot shows that the maximum heat flux measured at the ceiling was 1.06 kW/m^2 at 420 s while that at the floor was 0.6 kW/m^2 at 430 s.

Table 3 Details of the test

Test conditions	
Fire type	Pool fire
Fuel type	ATF (Aviation turbine fuel)
Fuel burner type	Small cylindrical vessel of diameter 0.32 m
Windows/opening condition	All windows open
Door opening condition	All doors open
Ventilation condition	Free ventilation (ambient temperature 23 °C)
Test duration	1200 s
Data rate (of processed results)	1 data per 10 s

The fact that the heat flux measured at the ceiling is almost double that at the floor suggests the well-known dominance of buoyant convection in free burning pool fires. After reaching the maximum, the heat flux started to decline at both the locations which suggests that fire stabilized and then started to decay at a very slow rate after 420 s. Due to the excess amount of fuel, the fire did not extinguish till 1200 s which was the specified time for the data acquisition. The opening in the ceiling ensures that there is no accumulation of heat close to the ceiling. The measured heat flux close to the ceiling exhibits a low frequency fluctuation due to the "Puffing" of burned products out of the ceiling opening. Inherent buoyancy driven fluctuation of the flame also contribute to this fluctuation in the heat flux.

Figure 3b shows the heat flux measured on the two opposite walls (north and south wall) of fire room, which lies in the lower zone of the fire plume. At 460 s, the maximum heat flux measured on north was 0.4 kW/m^2 while that on south wall was 0.42 kW/m^2. Due to the symmetry of the room, both the sensors received almost the same amount of heat flux. Figure 3c shows the heat flux at the center of the east, south, north and west walls of fire room. The maximum heat flux measured on the north, south and east walls at 520 s were 221.3, 229 and 151 W/m^2, respectively. Whereas that at the West wall was 209 W/m^2 at 580 s. The north wall, having the open door and hence source of combustion air, has received higher heat flux compared to others. On the other hand, the east wall did not have any ventilation opening, resulting in lower heat flux due to the lower concentration of combustion oxygen in that side of the room. The similarities in values are mainly because of the geometric symmetry and differences are due to the presence of openings, which affects the hot air distribution in the compartment. The fire seeks a source of oxygen to sustain itself and hence tilts towards the door or window, resulting in higher heat fluxes closer to the openings and lower fluxes away from the openings. Figure 3d shows the heat flux measured at the center of walls of 3 rooms; i.e., the east wall of fire room, west and east wall of side room and east wall of top room. The maximum heat flux was measured on the fire room's east wall (163 W/m^2 at 530 s). On the side room's east wall, the heat flux was 19.5 W/m^2 at 760 s, on top room's east wall the heat flux was 98.6 W/m^2 at 610 s and on side room's west wall was 18 W/m^2. Thus, it can be

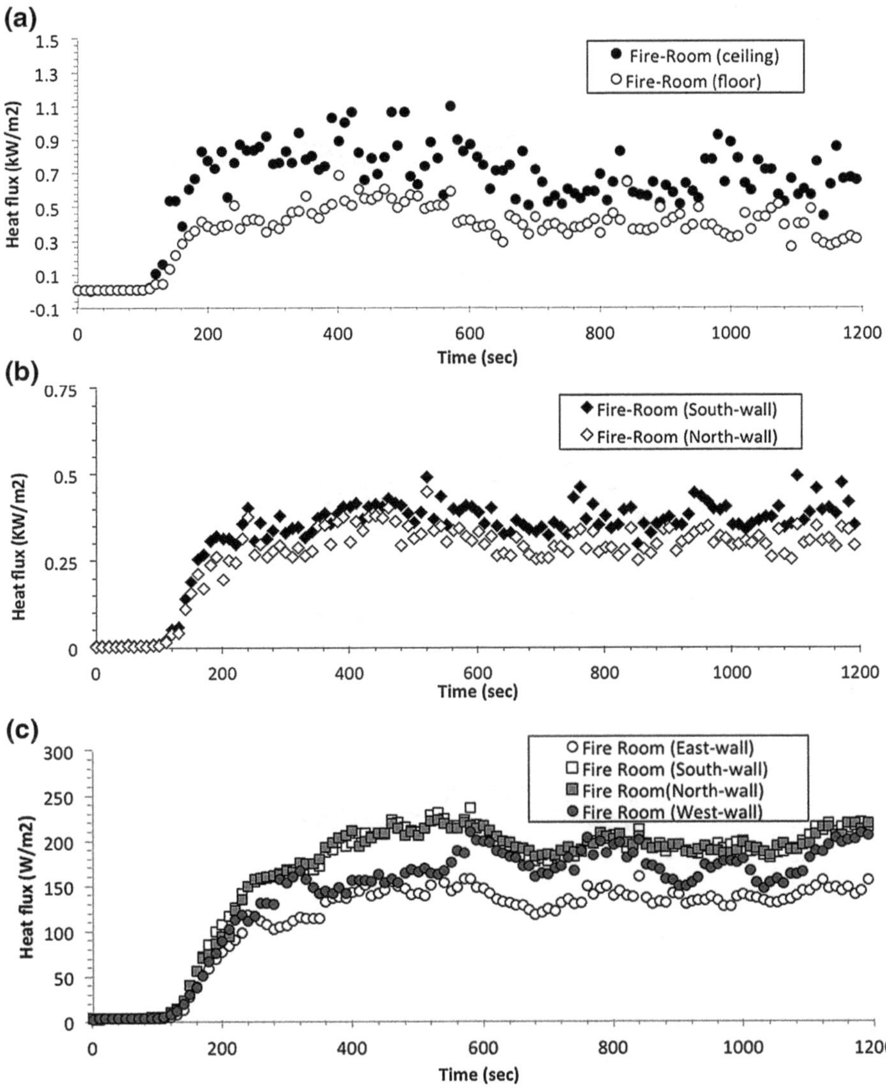

Fig. 3 **a** Total heat-flux on floor (close to fire) and at ceiling (above the fire) in fire-room. **b** Total heat-flux on lower zone of two opposite side of boundary walls (North wall and South wall) of fire room. **c** Comparison between heat fluxes at center point of all 4 walls (North, South, East and West wall) in fire room. **d** Comparison between heat flux at centre point on boundary walls of three interconnected rooms (fire room, side room and top room). **e** Convective heat-flux at upper zone of end boundaries of fire-room and side-room. **f** Total heat-flux at upper zone of end boundaries of fire-room and side-room. **g** Convective heat transfer coefficient "h" at upper zone of boundary wall of fire-room

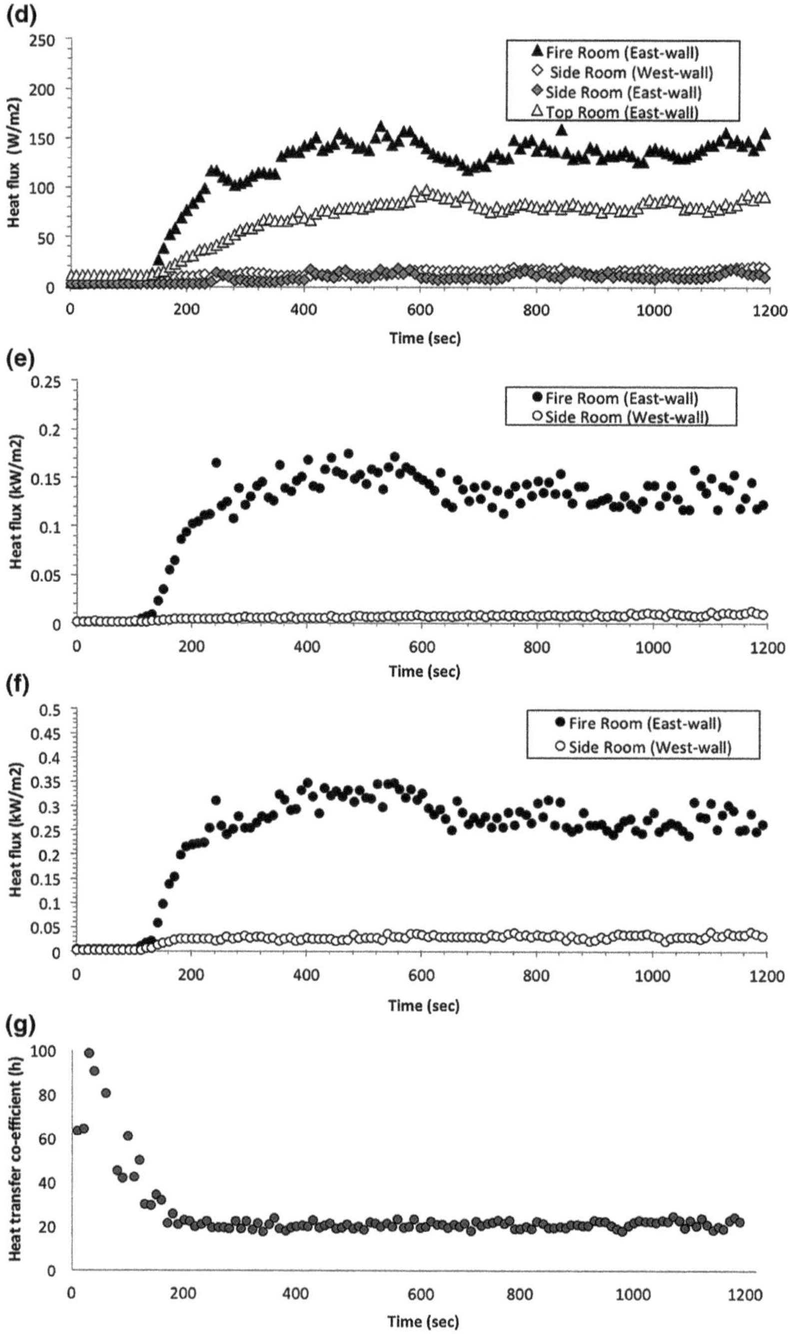

Fig. 3 (continued)

concluded that the fire room walls received the highest amount of heat flux followed by the top room and then the side room suggesting once again the dominance of buoyancy driven upward convection of heat. The west wall of the side room or the other end boundary of the lower compartment has received the least heat flux among all, which is justified because of its location. Figure 3e shows the convective heat flux measured on the east wall of the fire room and west wall of the side room. The maximum heat flux received by east wall of fire room was 0.175 kW/m^2 at 470 s while on the west wall of the side room, the heat flux was only 0.0137 kW/m^2. This is much lower than the value on the same wall in the side of the fire room. The heat flux from the fire room to the side room is primarily conductive and hence the convective heat flux in the side room is very low even though there was an opening between the two rooms. Figure 3f shows the total heat flux measured at the same location as mentioned above in Fig. 3e. The maximum heat flux received by east wall of the fire room was 0.34 kW/m^2 at 400 s and on the west wall of the side room, the heat flux was 0.04 kW/m^2 at 760 s. The side room is not expected to receive any radiative heat flux. Therefore, since the maximum convective heat flux was 0.0137 kW/m^2, the conductive heat flux through the wall is 0.0263 kW/m^2, i.e. more than double the convective heat flux. Figure 3g shows the convective heat transfer coefficient or "h" in the upper zone of the boundary wall (east wall) of fire room. The coefficient starts with a relatively high value at the beginning but decreases rapidly during the fire growth stage stabilizing thereafter to a constant value of about 20 W/m^2 K.

Figure 4a shows the temperature variation of fire plume rising upward in vertical direction plotted using data from four thermocouples, which were mounted on a vertical pole at every 1-m interval starting from 1.13 m. The maximum temperature recorded was 41.3 °C at 1.13 m, 56.4 °C at 2.13 m, 56.6 °C at 3.13 and 60.4 °C at 4.13 m. This shows that with increase in height the temperature of the plume is increasing, as is common in buoyancy driven fires. Figure 4b shows the comparison between 3 thermocouples at 2.13 m above floor level placed in 3 different rooms. The highest temperature was measured in the fire room at 41.3 °C, the next highest temperature was measured in the top room at 37.6 and the temperature measured in side room was 24.18 °C, reached after a time delay of 1180 s due to the low rates of heat transfer or heat flux as discussed in Fig. 3. Fire plume affects fire room the most whereas after striking the ceiling it entrains into the top compartment (or room) and very small amount of hot air enters the side room from the upper zone and from the lower zone (doors and windows) cold air entrains into the fire room to sustain the fire. Figure 4c shows the temperature variation of air close to the walls. The temperature comparison between the upper zone (above the doorway on East wall) and the lower zone (on South-West corner) of the fire-room shows that the maximum temperature recorded in the upper zone of the east wall was 49.7 °C while that measured on the south wall was only 28.7 °C. The temperature increase in the upper zone of the east wall is higher than the growth in lower zone of south wall. Figure 4d shows the temperature variation between the end boundaries of the two rooms (east wall of fire room and top room). The highest temperature measured in fire room was 45.4 °C and in the side room was 25.8 °C, which is also reached at a much later time. The difference in temperature and the time is quite obvious, as the temperature growth on the boundary of the fire room will be faster compared to side room, due to larger heat fluxes in the fire-room. The hot air travelling from one end of the fire room to other end of the side room will take some time and meanwhile it will keep losing the heat to the surrounding. Thus, the temperature at the end boundary of side room is not affected much by presence of fire.

The measured data shows a rapid increase in wall heat flux followed by a gradual drift during the fire development and sustenance stage as seen in Fig. 3. However, the fire affects only the fire room and the top rooms, with almost no impact in the side room suggesting the dominance of buoyancy in fire propagation.

Figures 5a–c show the plots obtained from emission measurements. The average value of CO during peak period of fire was 0, CO_2 was 0.2 vol.% and O_2 decreased by 0.3 vol.%. The case being with free ventilation is expected to have minimal variation in emission parameters with almost complete combustion of the fuel vapour resulting in the absence of CO. Figures 6a–c show the velocity and pressure of fire plume entraining the top compartment. The maximum velocity of plume measured was 2.76 m/s. The maximum dynamic pressure measured was 13 Pa at 210 s. The higher values of axial velocity (Fig. 6a) as compared to other velocity components confirms the fact that the fire is primarily buoyancy driven and hence the side rooms are not affected by the fire plume as has been discussed earlier. The fluctuation of the velocity data with time suggests "puffing" of the burned products out of the fire room at a low frequency of about 0.05 Hz.

The data presented so far establishes the primacy of buoyancy in the thermal-hydraulic aspects of fire in a room. The side rooms are much safer as compared to the top rooms vis-à-vis the fire room. The emission parameters are well within the acceptable limits in case of well-ventilated fires and the wall heat fluxes are also not very large. Therefore, proper thermal stress management of the ceiling is expected to make a building fire safe. However, in order to study the impact of fire on walls, corner fires and the fires in the vicinity of the walls must also be studied. The present chapter provides an extensive data base of the designers to validate the fire safety codes.

4 Conclusions

Detailed measurement of wall heat flux, temperature, pressure, velocity and emissions were carried out for a small pool fire of ATF in a multi-compartment fire facility under well-ventilated conditions. It was observed that the primary impact of the fire was in the fire room and in the room just above the burning fire suggesting the dominance of bouncy on fire plume dynamics and growth. The fire had very minimal effect in the side rooms. The growth of the fire plume is accompanied by a rapid increase in wall heat flux and temperature in the fire room followed by a steady state. The data presented in this paper can be a good source for the validation of fire simulation codes.

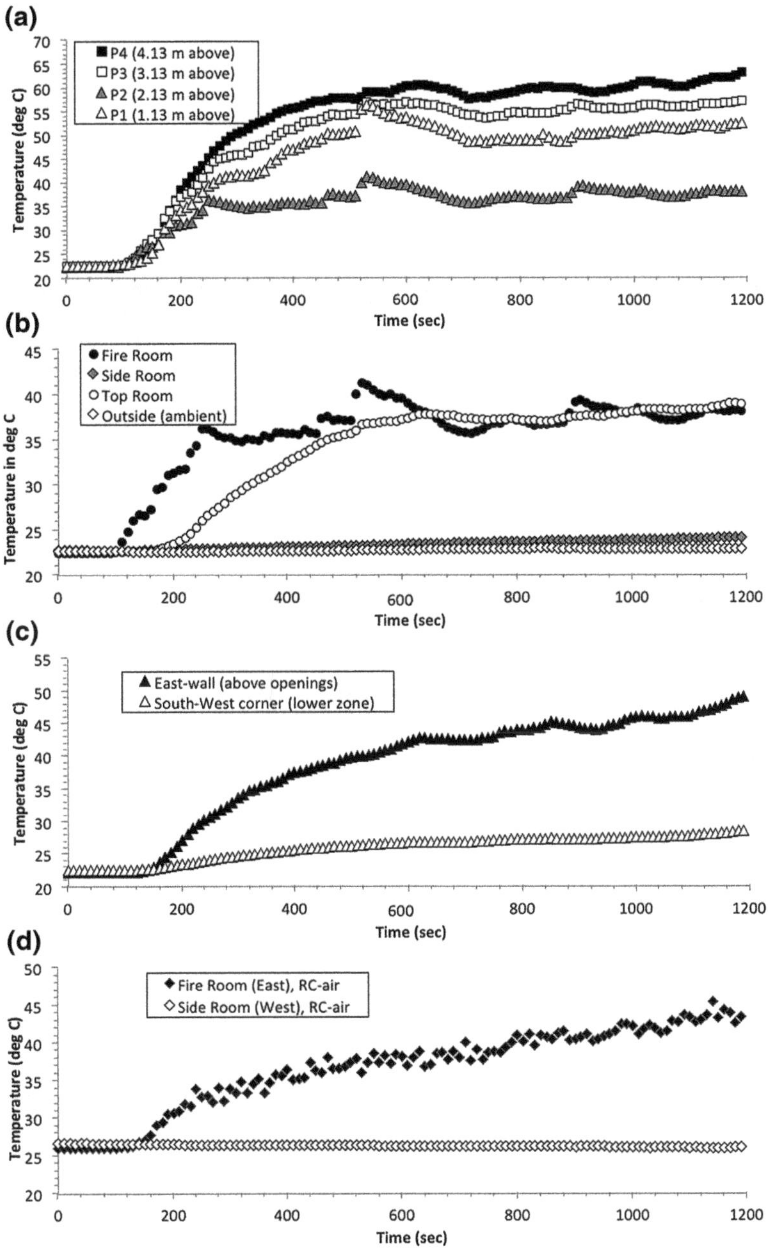

Fig. 4 **a** Temperature variation of fire plume rising upward in vertical direction above floor. **b** Comparison between temperatures at centre point of all three interconnected rooms. **c** Temperature variation closed to wall between upper zone (above doorway on East wall) and lower zone (on South-West corner) of fire-room. **d** Temperature variation between end boundaries of ground floor rooms (fire-room and side-room)

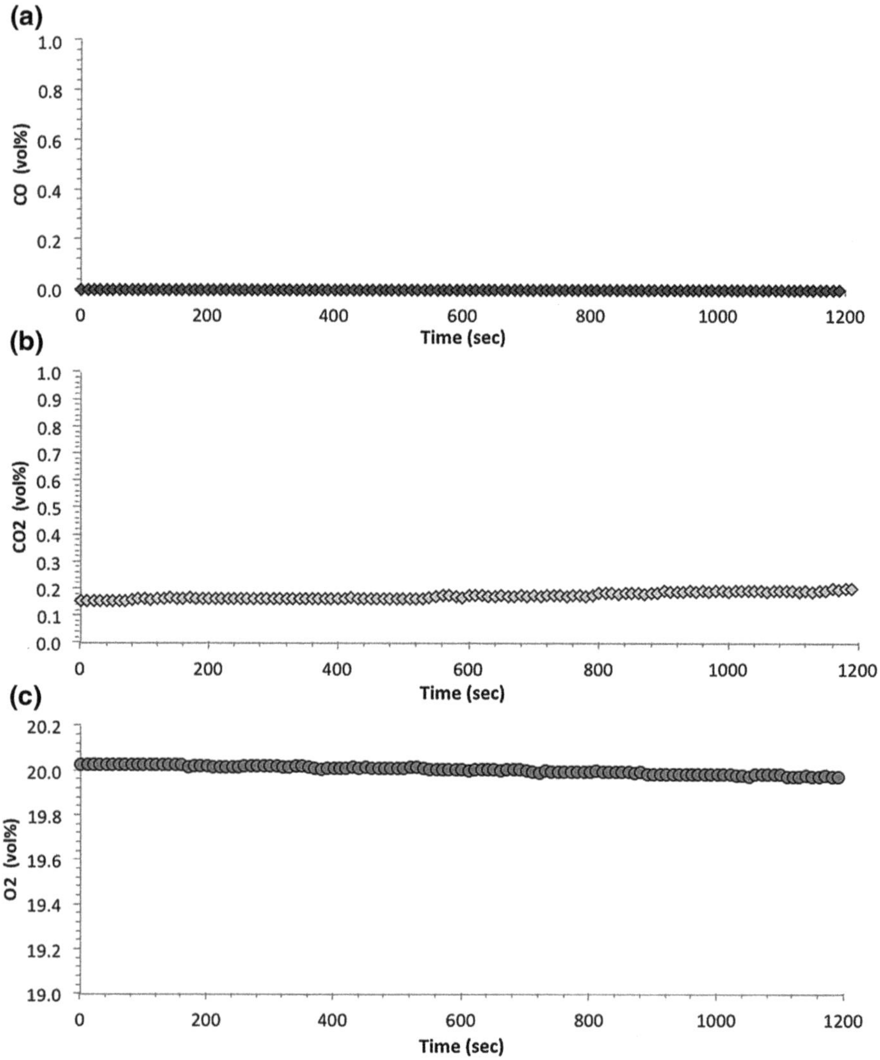

Fig. 5 **a** CO concentration in vol.%. **b** CO_2 concentration in vol.%. **c** O_2 concentration in vol.%

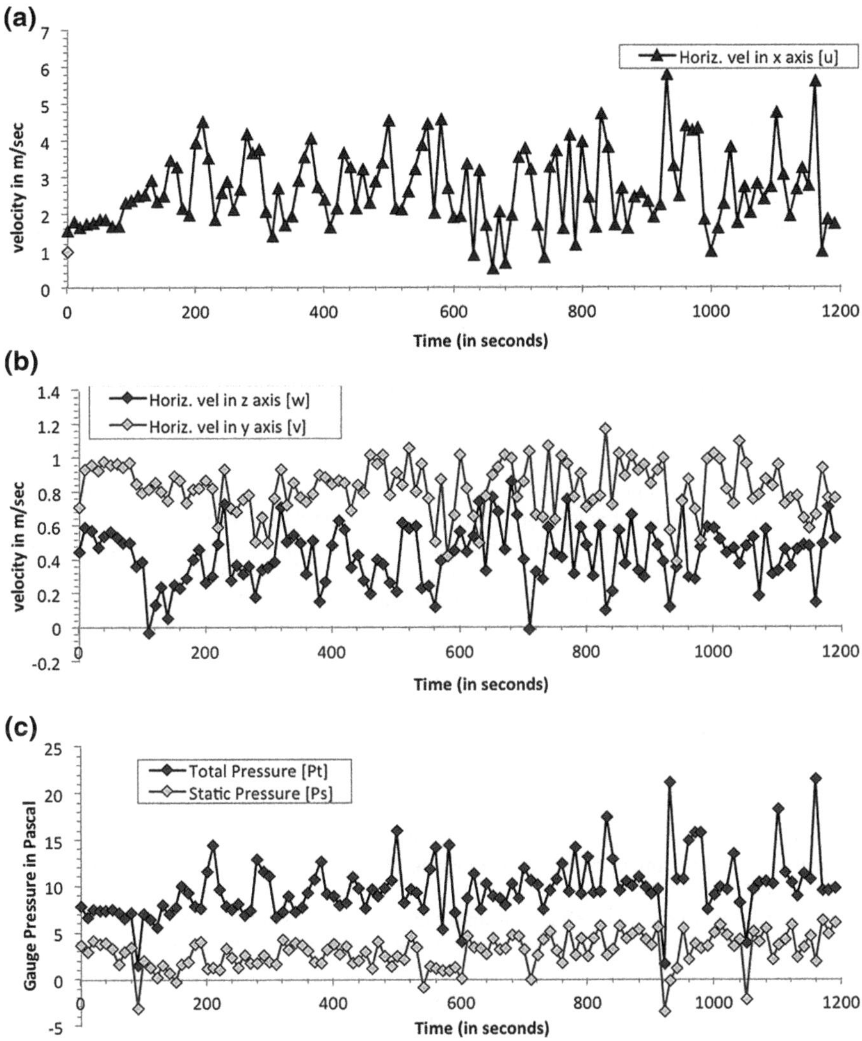

Fig. 6 **a** Axial velocity of the fire plume at the centre of the ceiling opening. **b** Planer velocities (along y and z) of the fire plume at the centre of the ceiling opening. **c** Gauge pressure measured at the centre of the ceiling opening

References

M. Luo, V. Beck, The fire environment in a multi-room building: comparison of predicted and experimental results. Fire Saf. J. **23**, 413–438 (1994). https://doi.org/10.1016/0379-7112(94)90006-x

B. Karlsson, J.G. Quintiere, *Enclosure Fire Dynamics* (CRC Press LLC, 1999), http://dx.doi.org/10.1201/9781420050219

W.K. Chow, C.W. Leung, Necessity of testing fire behavior of plastic materials under flashover. Polym. Testing **25**, 853–858 (2006). https://doi.org/10.1016/j.polymertesting.2006.03.012

D.D. Drysdale, *An Introduction to Fire Dynamics* (Wiley, 1985), http://dx.doi.org/10.1002/9781119975465

V. Babrauksa, Free burning fires. Fire Saf. J. **11**, 33–51 (1986). https://doi.org/10.1016/-0379-7112(86)90051-2

V. Novozhilov, Flashover control under fire suppression conditions. Fire Saf. J. **36**, 641–660 (2001). https://doi.org/10.1016/s0379-7112(01)00019-4

F.W. Mowrer, Enclosure smoke filling revisited. Fire Saf. J. **33**, 93–114 (1999). https://doi.org/10.1016/s0379-7112(99)00023-5

W.K. Chow, W.Y. Hung, On the fire safety for internal voids in highrise buildings. Build. Environ. **38**, 1317–1325 (2003). https://doi.org/10.1016/s0360-1323(03)00114-8

N.P. Bryner, G.W. Mulholland, Smoke emission and burning rates for urban structures. Atmos. Environ. **25**, 2553–2562 (1991). https://doi.org/10.1016/0960-1686(91)90172-4

E.H. Yii, A.H. Buchanan, C.M. Fleischmann, Simulating the effects of fuel type and geometry on post-flashover fire temperatures. Fire Saf. J. **41**, 62–75 (2006). https://doi.org/10.1016/j.firesaf.2005.09.001

V. Babrauskas, Heat of combustion and potential heat, in *Chapter 8, Heat Release in Fires, Elsevier Applied Science*, ed. by S. J. eds. (Grayson, NY, 1992), 207–223 pp

K. Klobut, K. Sirien, Air flows measured in large openings in a horizontal partition. Build. Environ. **29**(3), 325–335 (1994). https://doi.org/10.1016/0360-1323(94)90030-2

P. Tuomaala, J. Rahola, Combined air flow and thermal simulation of buildings. Build. Environ. **30**(2), 255–265 (1995). https://doi.org/10.1016/0360-1323(94)00044-s

G.H. Yeoh, R.K.K. Yuen, S.C.P. Chueng, W.K. Kwok, On modeling combustion, radiation and soot processes in compartment fires. Build. Environ. **38**, 771–785 (2003). https://doi.org/10.1016/s0360-1323(03)00022-2

H. Chen, N. Liu, W. Chow, Wind effects on smoke motion and temperature of ventilation-controlled fire in a two-vent compartment. Build. Environ. **44**, 2521–2526 (2009). https://doi.org/10.1016/j.buildenv.2009.04.008

D. Mackay, T. Barber, G.H. Yeoh, Experimental and computational studies of compartment fire behavior training scenarios. Build. Environ. **45**, 2620–2628 (2010). https://doi.org/10.1016/j.buildenv.2010.05.021

A. Byström, X. Cheng, U. Wickström, M. Veljkovic, Full-scale experimental and numerical studies on compartment fire under low ambient temperature. Build. Environ. **51**, 255–262 (2012). https://doi.org/10.1016/j.buildenv.2011.11.010

C.J. Chen, W.D. Hsieh, W.C. Hu, C.M. Lai, T.H. Lin, Experimental investigation and numerical simulation of a furnished office fire. Build. Environ. **45**, 2735–2742 (2010). https://doi.org/10.1016/j.buildenv.2010.06.003

P. Yang, X. Tan, W. Xin, Experimental study and numerical simulation for a storehouse fire accident. Build. Environ. **46**, 1445–1459 (2011). https://doi.org/10.1016/j.buildenv.2011.01.012

P.K. Sharma, A. Raut, A. Kushari, *Study of Small Scale Room Fires, DAE/AE/20090034/2* (Department of Aerospace Engineering, IIT Kanpur, 2015)

Study of Liquid Breakup Mechanism for Application of Solid Rocket Propulsion

Ryoichi Amano and Yi-Hsin Yen

Abstract Aluminum-based solid fuel is widely used in solid rocket propulsion system. During combustion, the solid fuel transforms into liquid status oxidant and agglomerate into droplets which impinge to the inner wall of solid rocket motor (SRM) nozzle and result in erosion problem. When a droplet breaks up into smaller size droplet results in less inertia which has a higher chance to following the exhaust gas stream instead of impinging to the inner wall of the nozzle. In this study, the liquid breakup is achieved by changing the fluid property of surface tension. The result of a liquid breakup is obtained using computational fluid dynamics (CFD) simulation of large eddy simulations (LES) and compared with experiment. The result presents the reduction of droplet size by changing the liquid surface tension. The mechanism of droplet breakup is discussed which is found to be due to the lower Laplace pressure or droplet bounding pressure lead to the lower surface tension of the liquid.

Keywords A liquid breakup · Two-phase flow · LES · Solid rocket motor

1 Introduction

The aluminum-based propellant solid rocket propulsion system is widely used in aerospace industry due to its low operational cost and readiness to manufacture, In propellant combustion stage, aluminum-based propellant oxidize into aluminum oxidant or alumina which is liquid status under the high-temperature environment (2,700–3,000 K) in combustion chamber [1]. The liquid status alumina droplet produced after combustion reaction can be agglomerated into a larger size of a droplet which accelerates by the exhaust gas and melts propellant flow through the combus-

R. Amano (✉) · Y.-H. Yen
Mechanical Engineering Department, University of Wisconsin-Milwaukee, Milwaukee, WI 53211, USA
e-mail: amano@uwm.edu

Y.-H. Yen
e-mail: yeny@uwm.edu

A. K. Runchal et al. (eds.), *Energy for Propulsion*, Green Energy and Technology,
https://doi.org/10.1007/978-981-10-7473-8_16

tion chamber to the exit nozzle. This process results in nozzle erosion problem by alumina droplet impingement [2]. The erosion changes nozzle throat geometry and leads to throat area increase. In this study, liquid breakup behavior is investigated by varying the liquid surface tension. The result is compared to experiment and LES simulation. The result of the experiments and simulations were compared with image superimpose technique.

As aluminum is used metallic solid fuel with characters of high performance, economical and less toxic to the organism and less environmental impact. The combustion of aluminum-based propellant with ammonium perchlorate (AP) produce alumina (Al2O3), which is in liquid status with size ranging from 1 to 500 μm in the combustion chamber. AP tends to agglomerate and from alumina slag, during those slag and alumina particles impinge on the nozzle throat lead to erosion and remove material on the inner wall of the nozzle. The erosion process changes the geometry of nozzle especially the nozzle throat and the nozzle performance decreases due to increasing nozzle throat diameter or reduce the exit/throat area ratio. Once throat/exit area ratio reduction, the optimized nozzle geometry start to deform, and the exhaust gas velocity dropped which affect rocket performance [3, 4].

Smaller-sized particles are more accessible to follow flow stream of carrier gas due to higher projection area to mass ratio. Thus, the smaller particle will have more chance to avoid impinging on the nozzle wall during rapid direction changing the curvature of flow stream near the nozzle entrance. Therefore, the reduction of the average alumina particle size will be a potential approach to reduce erosion problem of SRM and further improve the performance and burn time [5]. Combustion chamber pressure highly dominates the agglomerate particle size inside the combustion chamber; it is confirmed that higher the pressure reduces the agglomerate particle size. However, the means of increasing the chamber pressure to reduce agglomerate particle size will be limited due fuel burn rate, thrust control. The rocket structure will need to take into concern at the same time for target operation chamber pressure modification. The study of liquid breakup mechanism could provide a better understanding of reducing agglomerate particle size which is the purpose of this study.

2 Experimental and CFD Setting-up

2.1 Experimental and CFD Geometry

The experimental set-up is shown in Fig. 1.

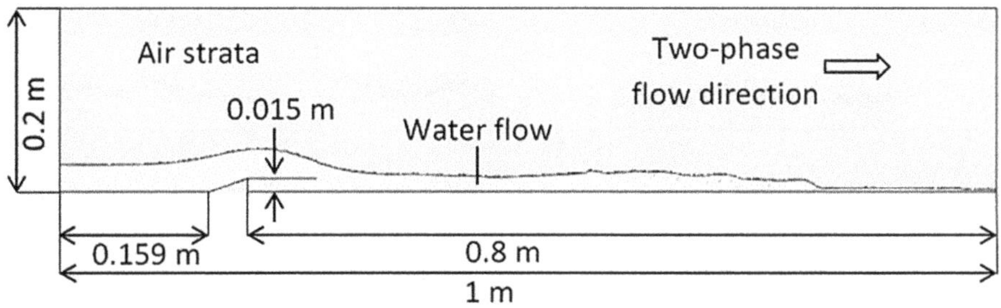

Fig. 1 Test section

2.2 Structure of Numerical Tool

The Computational Fluid Dynamic (CFD) approach we use is Large Eddy Simulation (LES) to simulation gas-liquid interacting flow.

2.3 Large Eddy Simulation

The turbulence structure is shown in Fig. 2. We could observe that there are the multiple structures of turbulence. The large eddy simulation (LES) technique solves turbulence behavior accurately. Therefore LES could better predict the turbulent flow behavior. However, compared to Reynolds-Averaged Navier-Stokes (RANS) solver LES required finer mesh to address useful eddy sizes, and this makes LES becomes more computationally expensive to RANS solver. Due to the computer technology rapidly growth in last two decades.

The set of the equations of Navier-Stokes equations are given by

$$\frac{\partial u_i}{\partial x_i} = 0 \tag{1}$$

$$\frac{\partial u_i}{\partial t} = -\frac{\partial u_i u_j}{\partial x_j} + X_i - \frac{1}{\rho}\frac{\partial p}{\partial x_i} + \nu\frac{\partial^2 u_i}{\partial x_j^2} \tag{2}$$

For the large-eddy simulation, the equations need to be spatially filtered into the resolved-scale component and subgrid-scale (SGS) component, the resolved-scale component is defined as

$$\tilde{v}_i(x_i) = \iiint v_i(x_i^{'})G(x_i - x_i^{'})dx_i^{'} \tag{3}$$

and

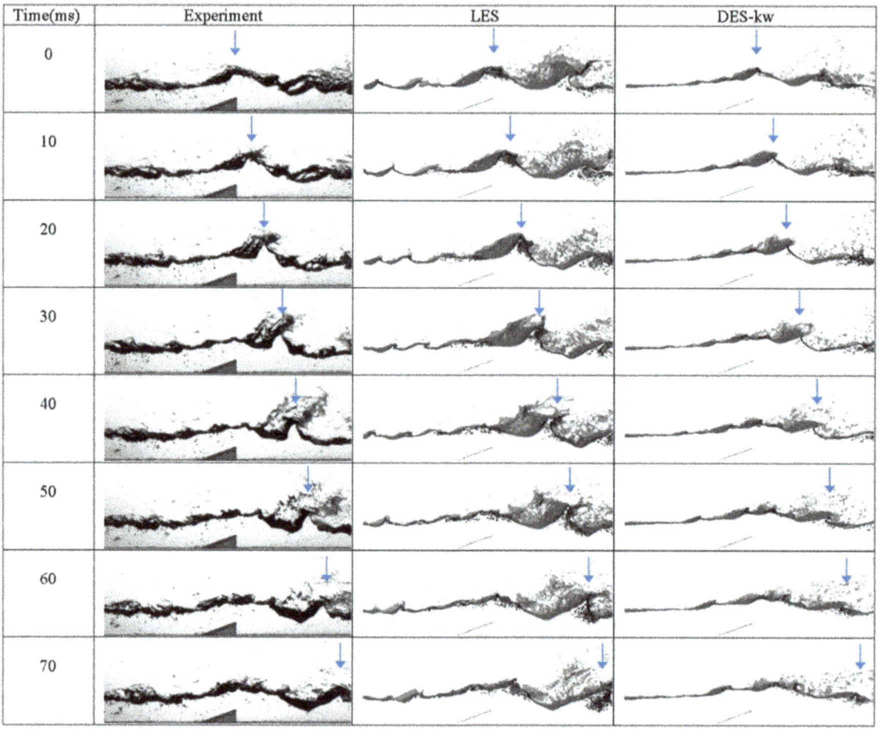

Fig. 2 Liquid breakup comparison of experiment and CFD method of LES and DES [1]

$$\frac{\partial \tilde{u}_i}{\partial t} = -\frac{\partial \tilde{u}_i \tilde{u}_j}{\partial x_j} - \frac{\partial \tau_{ij}}{\partial x_j} + X_i - \frac{1}{\rho}\frac{\partial \tilde{p}}{\partial x_i} + \nu \frac{\partial^2 \tilde{u}_i}{\partial x_j^2} \tag{4}$$

Note: $\tau_{ij} = \widetilde{u_i u_j} - \tilde{u}_i \tilde{u}_j$.

Smagorinsky Subgrid Scale model [6] is the earliest LES Subgrid Scale model which applies mixing length hypothesis to model the subgrid-scale stresses. Over several decades of development, the subgrid-scale model evolves better in predicting flow field. The subgrid model [6] uses the turbulent viscosity as:

$$\mu_t = \rho \Delta^2 S_w \tag{5}$$

3 Breakup Mechanism

As shown in Fig. 1a, water surface tension (yellow arrow) that holds the liquid in a spherical shape to maintain least surface potential, the additional pressure inside droplet caused by surface tension is described as Laplace pressure. Figure 1b is when

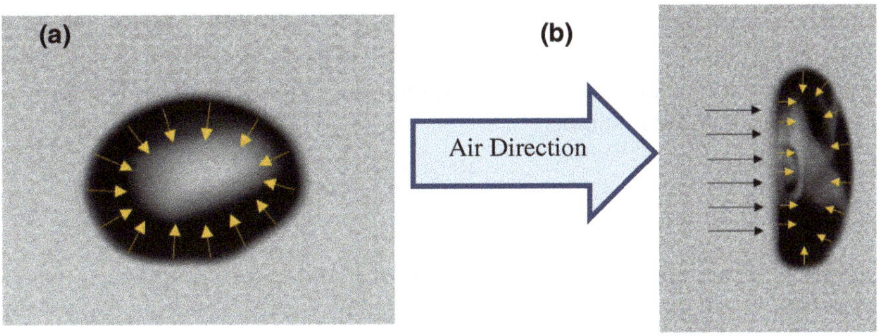

Fig. 3 Laplace pressure and stagnation pressure (Photo by UW-Milwaukee Wind Tunnel Lab)

air interacts with droplet with a certain speed, the stagnation pressure (black arrow) act on the surface of the droplet and force it to deform Fig. 3. It depends on the balance of those two principal force balance whether the droplet will break up or not.

The difference in pressure between in and outside of a liquid droplet caused by surface tension is expressed as Laplace pressure which could be given using Young-Laplace equation shown in the following.

$$\Delta \mathrm{P} = -\sigma \nabla \cdot n \tag{6}$$

where σ is the surface tension. The Laplace pressure is the product of surface tension and divergence of the normal vector of droplet's surface. By expanding Young-Laplace equation in spherical coordinate, we could express the Laplace pressure in Eq. (7), and where r is radius of droplet

$$\Delta \mathrm{P} = \frac{2\sigma}{r} \tag{7}$$

The stagnation pressure of moving air flow that is responsible in deforming the droplet is described by Eq. (8)

$$P_{Stag.} = \frac{1}{2}\rho_{air}v^2 \tag{8}$$

where ρ_{air} is density of air and v is the relative air speed to the droplet. By rearranging Eqs. (7) and (8), we could find the breakup pressure ratio (BPR) of stagnation and Laplace pressure shown as following:

$$\mathrm{BPR} = \frac{P_{\mathrm{Stag.}}}{P_{\mathrm{Lap.}}} = \frac{\rho_{\mathrm{air}}v^2 r}{4\sigma} \tag{9}$$

In the above expression, the ratio of the breakup pressure is proportional to the air density, square of two-phase velocity difference, the radius of droplet and inverse proportional to the surface tension of the liquid. The expression of breakup pressure ratio is similar to Weber Number.

$$We = \frac{\rho_{air} v^2 r}{\sigma} \tag{10}$$

Weber number represents the tendency of a liquid to break up due to balance between gas-phase inertia to liquid surface tension force [7]. When the Weber number becomes high, the force due to surface tension is predominant over the inertial forces deformation. This force, then, leads to the liquid breakup. Weber number of 5 or higher usually marks the threshold of a breakup. Weber number represents the tendency of a liquid to break up due to balance between gas-phase inertia to liquid surface tension force [7]. When the Weber number becomes large, the surface tension is dominant over the deforming inertial forces of air, which causes the liquid breakup process. Weber number of 5 or higher usually marks the threshold of a breakup.

4 Flow Channel Experiment Result and Discussion

4.1 Liquid Breakup Analysis

The overall overlapping images from experiment and simulation are 6,540 and 2,000 frames, respectively. A featured superimpose image illustration is shown in Fig. 4. A bubble recirculation region points out in part (a) in Fig. 5, we can see when air velocity increases more bubble will be engulfed into the correction part after the ramp as demonstrated. The smooth transition region in part (b) of Fig. 5 could be considered as the droplet distribution area, the darker the color means, the higher chance of liquid droplet could be observed. The flow streamlines shown in Fig. 4 is demonstrated in Fig. 5. In this figure, it is seen that the outer boundary of contour lines and exam the difference between cases. The recirculation region could also be observed if there is one in the case.

The superimposed image of reduced surface tension water and contour map are demonstrated in Figs. 6 and 7. For the event of air velocity = 20 m/s cases, both experiment and simulation show the similar result in superimposing an image, no bubble recirculation was observed, and both cases show same liquid droplet distribution region, and so does the outline of the contour map. In the event of air velocity = 30 m/s, no bubble recirculation was observed in both experiment and simulation as well. However, the droplet distribution region is larger from test case than simulation; it is also evident that in an experimental test case the droplet distribution has a smoother transition from the water surface to air and has wider distribution range after the ramp. These results also come from the contour map that the outline

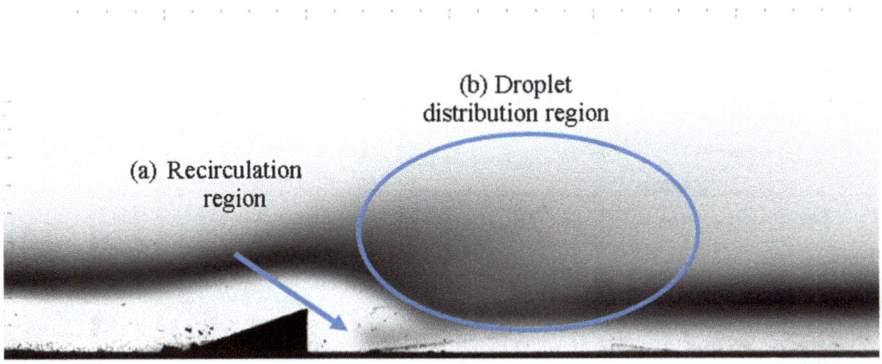

Fig. 4 Featured location of superimposing an image

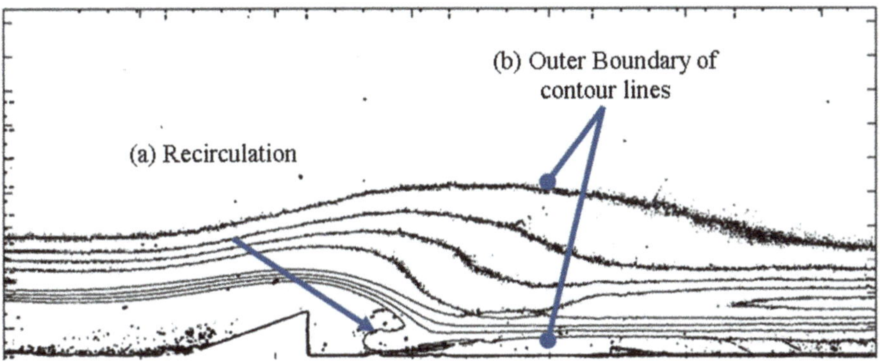

Fig. 5 Contour map of the superimposed image of Fig. 4

of the experiment is larger than simulation. Same as the regular water group case that air carries the droplet and move further after the ramp compared to result from the simulation. For the event of air velocity = 40 m/s, bubble recirculation region could be observed in both experiment and simulation. However, the result from simulation shows overprediction of bubble recirculation compares to the experimental tests. The droplet distribution area in the experimental test shows the more smooth transition from water to air and also occupy wider range compared to the result of the simulation. It is also observed in contour map that the droplet distribution region in the experiment has a full spectrum from an outline. From contour map, the over predict recirculation region is also being observed.

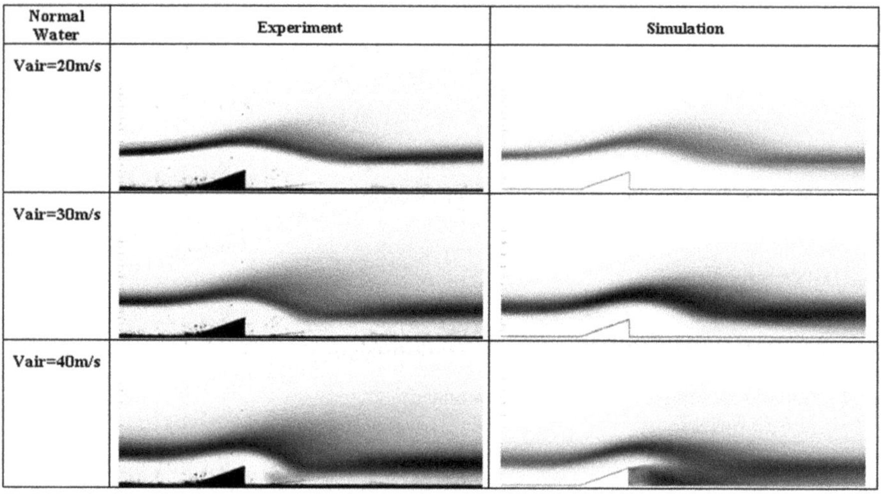

Fig. 6 Superimpose images of normal water ($\sigma = 0.074$ N/m) cases for experiment (left) and simulation (right) for air velocity Vair = 20 m/s, Vair = 30 m/s and Vair = 40 m/s [2]

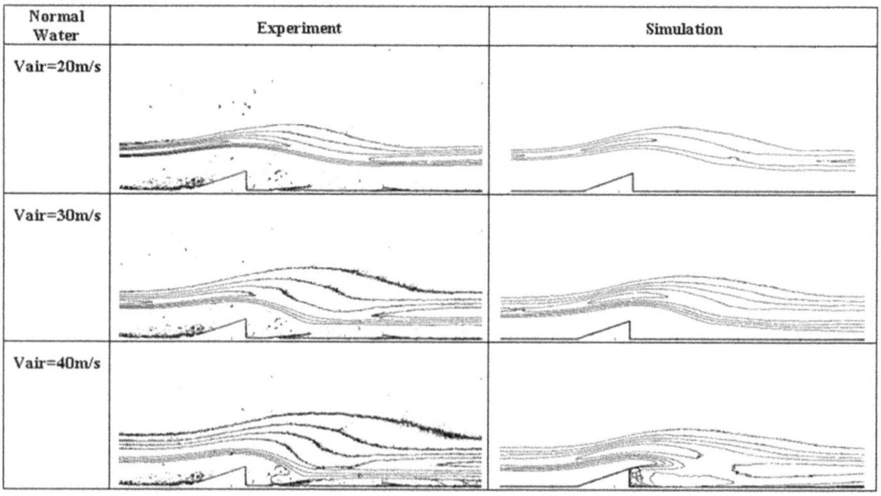

Fig. 7 Contour map of regular water ($\sigma = 0.074$ N/m) cases from Fig. 6

4.2 Welch Analysis

The frequency data extraction location is 20 mm behind the ramp where wave crest could reach maximum average height, and the wave position fluctuates as a function of time (see Fig. 8).

The flow channel simulation wave location data are shown in Fig. 9 for cases of Vair = 20 m/s. Data of regular water and reduced surface tension water are arranged

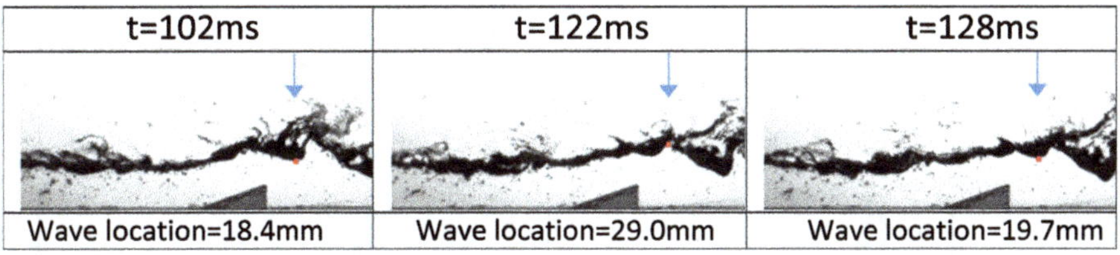

Fig. 8 Frequency history data extraction location from the experiment

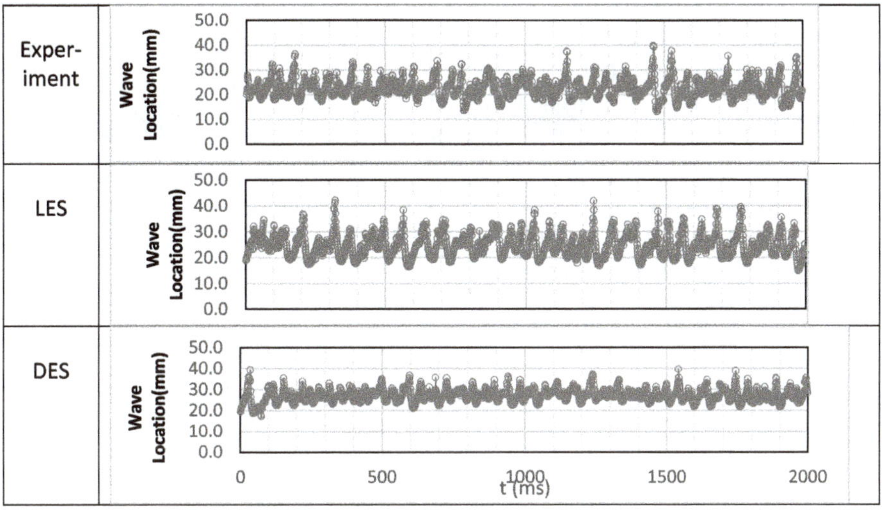

Fig. 9 Wave location history of experiment, LES and DES [1]

in the same figure from top to bottom. It could be observed that higher air velocity will result in lower the wave location and the reason result to this phenomenon is due to higher air velocity increase the breakup activity, more water break up into droplet from the main water body and resulting in decreasing water level. It could also be observed that for the case of air velocity = 20 m/s.

Since Detached eddy simulation (DES) is a modification of Reynolds-Averaged Navier-Stock (RANS) model with judgment criteria to enable Large Eddy Simulation (LES). The behavior of DES will range between LES and URANS, and this phenomenon could be observed.

Figure 10 shows experimentally obtained results and the data processed from the simulation results. Both sets correlate well each other.

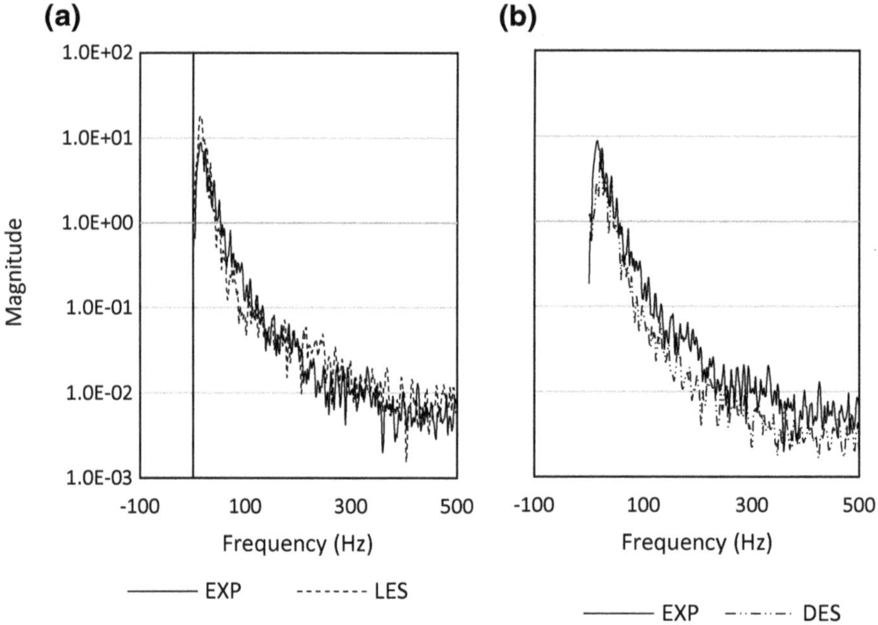

Fig. 10 Welch frequency analysis of experiment to LES and experiment to DES

4.3 Flow Channel Experiment Droplet Volume to Size Distribution

The flow channel experiment droplet volume to size distribution comparison between regular water and reduced surface tension water were tested. High-speed camera resolution could recognize the minimum size of the droplet is 0.24 mm per pixel. Regarding data accuracy, for any droplet equivalent diameter smaller than 0.6 mm are filter out. In this figure, volume to size distribution per frame is the averaged values from 6,540 frames for each case. The volume unit is marked in mm3. It could be observed that cases from different velocity setting are clear to distinguish, as shown in the figure, group (a) of Vair = 40 m/s setup has the most volume to size distribution curve throughout all range of droplet diameter. Group (b) of Vair = 30 m/s data distribution set in the middle and group (c) of Vair = 20 m/s has the least volume of size distribution for all range of droplet sizes. High air velocity causes more breakup behavior. Therefore more water is carried by droplet beside carried by main water body.

It could be observed that for cases with different surface tension, the cases with reduced surface tension water has more volume to size distribution than the cases of ordinary water cases. Less surface tension could also explain the reason reduces the surface energy. The surface energy of liquid provides bounding force for liquid stay together. When a surface tension decreases, the less bounding force acting on liquid

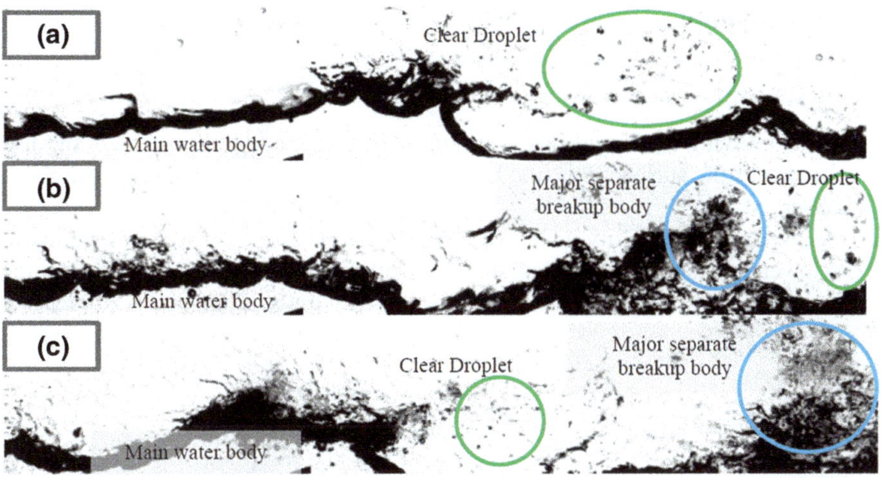

Fig. 11 Droplet distinguish demonstration for cases form **a** Vair = 20 m/s, **b** Vair = 30 m/s and **c** Vair = 40 m/s [2]

result more breakup when external disturbance like moving working on the surface of the liquid. The volume to size distribution difference between regular water and surface tension is evident when air velocity is lower (such as 20 m/s), however it becomes less significant when air velocity increases to 30 m/s and when air velocity reaches to 40 m/s, the difference becomes not so clear to observe expect in small droplet diameter region of 0.6–0.8 mm. This phenomenon could be explained by experiment quality as shown in Fig. 11. When air velocity = 20 m/s, the droplet breakup activity is gentle, and droplet from a liquid breakup is clear to distinguish. When air velocity reaches to 30 m/s, the liquid breakup action behaves differently, the liquid breakup begins from wave crest reaches to the top of the ramp where the water starts to deform and break up into primary separate breakup body from the main water body. The primary separate breakup body keeps breakup into a smaller liquid droplet. The situation get worse when air velocity reaches to 40 m/s, when air velocity increases, more volume of water is belong to the significant separate breakup body and the fuzzy region make the difference between regular water and reduced surface tension water becomes not so bright.

Although it is not easy to distinguish the difference between regular water and reduced surface tension water in the cases of air velocity = 40 m/s, in the event of air velocity of 20 m/s and 30 m/s, it could confirm that liquid with reduced surface tension will increase liquid breakup activity. To further confirm the correlation between the liquid breakup and liquid surface tension, a precise, controlled single droplet breakup experiment was conducted and documented in Chap. 8 which provide more aspect to liquid breakup study.

5 Conclusions

Through this study, the following conclusions emerged.

(1) By employing the liquid break-up model, it was found that the lower the frequency of the surface flow, the higher the magnitude of the droplet power was observed.
(2) As the gas flow velocity increases, the wake of the droplets coalesces, while for a lower flow speed of the gas, the size of the droplet can be kept small in the range of 1000 micros with the gas flow velocity in the order of 20–40 m/s.

References

1. Y.-H. Yena, R. Amano, Investigation of alumina flow breakup process in solid rocket propulsion chamber. in *54th AIAA Aerospace Sciences Meeting*, 2016, p. 0498
2. P. Thakre, V. Yang, Chemical erosion of refractory-metal nozzle inserts in solid-propellant rocket motors. J. Propul. Power **25**, 40–50 (2009)
3. Y. Xiao, R. Amano, T. Cai, J. Li, New method to determine the velocities of particles on a solid propellant surface in a solid rocket motor. J. Heat Transf. **127**, 1057–1061 (2005)
4. Y. Xiao, R. Amano, T. Cai, J. Li, G. He, Particle velocity on solid-propellant surface using X-ray real-time radiography. AIAA J. **41**, 1763–1770 (2003)
5. A. Bandera, F. Maggi, L. DeLuca, Agglomeration of aluminized solid rocket propellants. AIAA Pap. **5439**, 2009 (2009)
6. M. Germano, U. Piomelli, P. Moin, W.H. Cabot, A dynamic subgrid-scale eddy viscosity model. Phys. Fluids **3**, 1760 (1991). https://doi.org/10.1063/1.857955
7. D.D. Joseph, Breakup of a liquid drop suddenly exposed to a high-speed airstream. Minneapolis, MN (1999)

Part V
Renewable Energy and Fuels; Production and Utilization

Thermochemical Reforming of Wastes to Renewable Fuels

K. G. Burra and Ashwani K. Gupta

Abstract Growing population and consequential rise in energy demand are contributors to overdependence of carbon based fossil fuels combustion for various applications which continues to increase the atmospheric CO_2 to unrecyclable levels leading to anthropogenic global warming from insufficient CO_2 capture and sequestration, and thus incomplete carbon cycle. Depleting fossil fuel reserves and concerns of CO_2 emissions from fossil fuel combustion, along with the concerns of improper waste disposal poses great global challenges that need to be addressed for energy and environment sustainability. Many techniques for energy and fuel production from biomass and solid wastes have been examined in the recent past of which thermochemical reformation of wastes are dominant, compared to biochemical processes such as anaerobic digestion, as they provide high reaction rates from their high operational temperatures. This chapter serves the purpose of providing detailed scenario of thermochemical processes starting with classification to include pyrolysis, gasification, and hydrothermal conversion techniques. Gasification techniques offer efficient and effective transformation of solid biomass and wastes into gas/liquid fuels and value added materials. This technique offers clean energy production at high efficiency compared to other transformation techniques via syngas which can be used for combined heat and power generation, production of fuels for transportation using Fischer Tropsch synthesis, and production of value added chemicals. Challenges of gasification, which include tar residuals and low-grade feedstocks are explained in detail in the chapter including catalytic and sorption based tar removal techniques. Low grade or high moisture content feedstocks may need drying before gasification which can significantly lower the economic value and efficiency of gasification that depends on net energy density of the feedstock. Hydrothermal processing is beneficial for the conversion of high moisture content feedstock such as wet grass, algal biomass, municipal waste, and sludge to bio-oils, which can further be refined to produce liquid biofuels that helps to reduce fossil fuel requirement of gasoline, diesel, and other fuels used for transportation, energy,

K. G. Burra · A. K. Gupta (✉)
The Combustion Laboratory, Department of Mechanical Engineering,
University of Maryland, College Park, MD 20742, USA
e-mail: akgupta@umd.edu

A. K. Runchal et al. (eds.), *Energy for Propulsion*, Green Energy and Technology,
https://doi.org/10.1007/978-981-10-7473-8_17

power purposes. Other thermochemical methods such as fast pyrolysis have also been examined during the past couple of decades for the production of bio-oils for biofuel synthesis. Catalytic conversion techniques for refining of the bio-crude and bio-oils produced from liquefaction and fast pyrolysis are also discussed with focus on hetero-atom removal such as hydrodeoxygenation and the challenges associated with it. This chapter provides a review on the various thermochemical reformation techniques, their advantages and drawbacks. It emphasizes on informing various advancements in terms of the reactors used, the operational parameters that control the reactions and the proposed reaction pathways for these techniques. A focus in this chapter on state of the art global scenario to develop these processes includes catalytic reforming of their products to achieve enhanced quality products and their corresponding challenges to produce clean and sustainable energy, fuels and value added products.

Keywords Pyrolysis · Gasification · Hydrothermal conversion · Catalytic reforming · Hydrodeoxygenation · Wastes and biomass · Syngas · Biofuels

Nomenclature

MSW	Municipal Solid Wastes
HD (LD)	PE High Density (Low Density) Polyethylene
PAH	Poly Aromatic Hydrocarbon
FCC	Fluid Catalytic Cracking
FT	Fischer Tropsch
HMF	5-Hydroxymethylfurfural
HDO	Hydro De-oxygenation
HDS	Hydro De-sulfurization
HTL	Hydrothermal Liquefaction
DFT	Density Functional Theory
FTIR	Fourier Transform Infra-red
SEM	Scanning Electron Microscope
DMF	Dimethyl Furan

1 Introduction

The advent of modern medicine and improved nutritional knowledge has helped to increase the average human age and global population that has resulted in increased use of available global resources. Coinciding with the industrial revolution, the global population increased from 1 billion in the 1800s [1] to over 7 billion by 2017. The global population is estimated to increase by additional 1 billion by 2030. The pursuit of economic growth for the betterment of living standards has led to increased energy consumption. Fossil fuel mining producing coal, natural gas, and petroleum based fuels have been the backbone for realization of the industrial revolution. This led up to the present day infrastructure which is heavily reliant on these reserves to serve the energy and transportation fuel demands. The recent statistics regarding global energy supply sources reveals the dominance of fossil fuel contribution [2], see Fig. 1. When these fossil fuels are used for energy production, the emissions in the form of NO_x, CO, unburnt hydrocarbons and CO_2 are inevitable from all combustion systems. While the process of creating fossil fuel reserves spans over millions of years, the rate of its consumption is exceedingly high in the order of decades. The only major CO_2 sequestering process in nature corresponds to photosynthesis whose CO_2 consumption rate cannot currently match the CO_2 production rate. This incompleteness in carbon cycle has led to gradual rise in the atmospheric CO_2 content [3]. This additional CO_2 present in the atmosphere has led to added absorption of solar radiation, called greenhouse effect, causing a rise in global temperature, also called global warming. The significant contribution of fossil fuels to global CO_2 emissions from the combustion of fuels can be seen from Fig. 2 [2]. The correlation between rising fossil fuel consumption and anthropogenic global warming has led to rising interest in renewable fuel resources. With global warming concerns clouding the fossil fuel usage due to their inherent inability to serve as sustainable source calls for alternate clean and sustainable fuel and energy production. While major growth in hydroelectric power plants took place to meet the need for renewable energy production, its availability on global terms is rather limited. Its scalability to offset the fossil fuel usage without major replacements in the available infrastructure, especially in transportation and industrial sector, is limited. A radical surge in the development of clean, sustainable and scalable renewable energy production is necessary to offset the fossil fuel contribution which currently aggregates to near 90% of the world energy consumption. Biofuels and waste correspond to only about 10% and hydro energy correspond to some 2.4%.

Parallel to the issue of fossil fuel usage for energy and fuel production, the solid waste disposal issue is also on the rise with the continuous increase in world population. The continuously increasing amounts of waste generation are not being balanced by the similar efforts on rate of recycling or recovery. Municipal solid waste is one of the leading forms of waste. Typical major components of MSW and their contribution are given in Fig. 3 [4]. While waste components such as those of metal and glass can be recycled, these components in 2014 corresponded to only 13% of total MSW in the USA. Most of the remaining materials are either reformed to energy via combustion or discarded to landfills. On an aggregate, more than 50% of the solid wastes are discarded into landfills in the USA, see Fig. 4 [4]. The carbonaceous wastes such as plastic, leather, wood, rubber, yard trimmings, food, and paper wastes can serve as potential feedstocks for energy generation using either combustion or converted to fuel via chemical synthesis. These wastes supplemented with other feedstocks, such as, woody biomass, algae and sea weed biomass, and human and animal wastes, require much shorter time span for their energy generation compared to the existing fossil fuel reserves. Biomass and other bio-wastes feedstocks are nearly carbon neu-

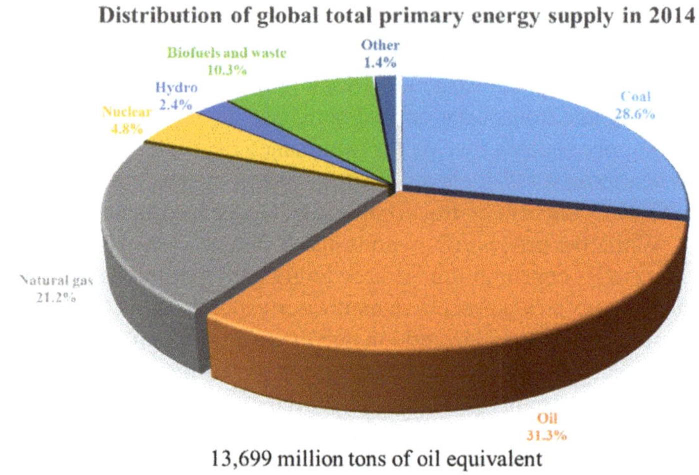

Fig. 1 World primary energy supply in 2014 [2]

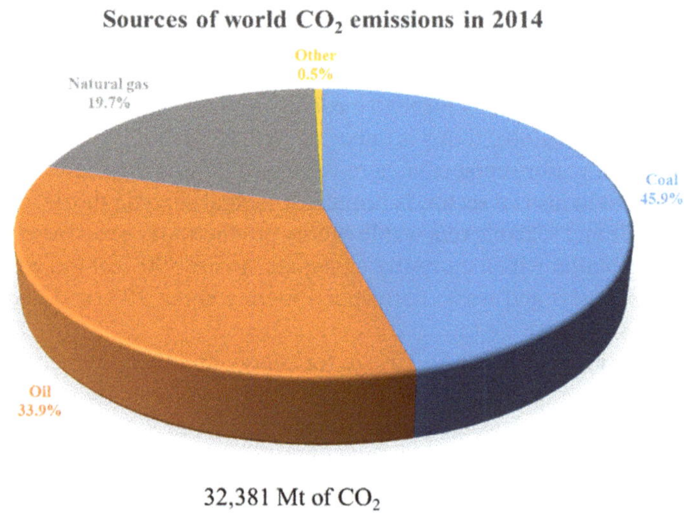

Fig. 2 World greenhouse gas emissions and sources of contribution [2]

tral. However, they are less carbon positive in the case of products from fossil fuel reserves such as plastic, textile and synthetic rubber and other polymer wastes.

The biomass and carbonaceous solid or sludge wastes given above can be broadly classified based on the source of feedstock generation, see Fig. 5. While this classification helps in realizing the sources, a more detailed classification is necessary that includes more detailed parameters of the wastes, such as composition and energy content. Woody biomass, paper and agricultural residue and other biomass and biowastes can be characterized as lignocellulosic biomass due to their composition

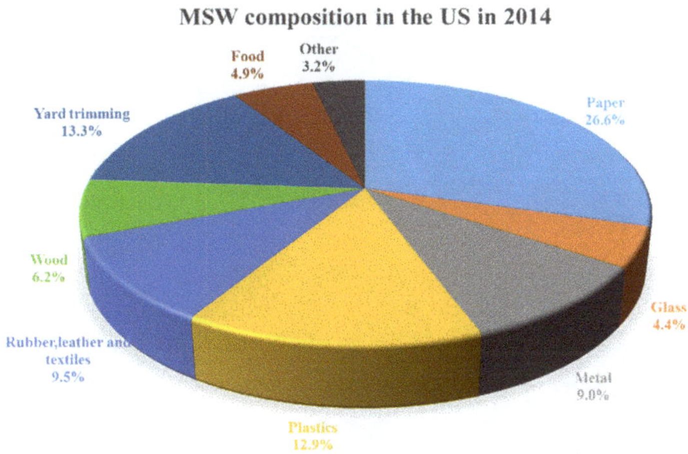

Fig. 3 Typical composition of municipal solid wastes in the USA [4]

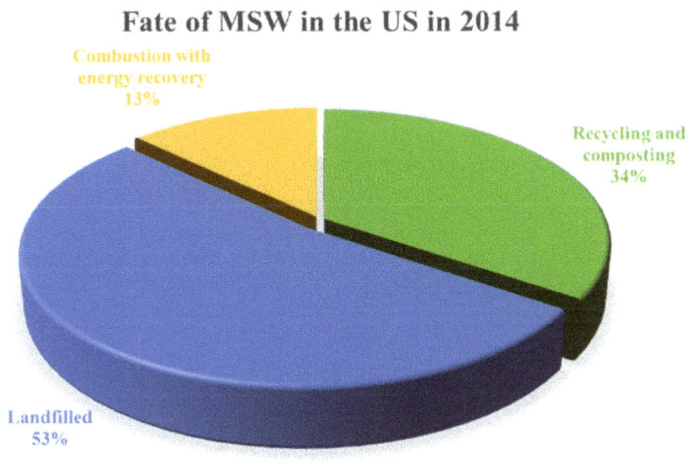

Fig. 4 Fate of municipal solid wastes in the USA [4]

constituting predominantly as cellulose, hemicellulose, and lignin. Details about these components and other extractives will be discussed later. Typical biomass and waste types and their lignocellulosic content is given in Fig. 6 [5]. Although these biomasses are similar in their generic lignocellulosic content, they may vary significantly in the pre-processed moisture and ash content, and thus energy content which significantly dictates the processing technique suitable for their reformation to energy, fuel or value-added chemical products. Plastic and textile wastes are typically classified based on their monomeric composition, energy content, and usage, see Fig. 7. Low moisture content in these plastic wastes helps to justify ignoring its moisture content as an important parameter unlike in biomass wastes wherein widely

Fig. 5 Classification of feedstock based on the source

different amounts of moisture can be found, up to 80% by weight. Plastics have very high energy density compared to biomass.

2 Biomass and Solid Waste Feedstocks

We limit our focus here to solid carbonaceous biomass and solid wastes acquired from municipal solid wastes, agricultural and industrial wastes. Lignocellulosic biomass constitution is dominated by cellulose, hemicellulose, and lignin along with other extractives and inorganic materials. All these saccharides based biomass are bio polymers with varying degree of polymerization and monomeric composition. Typical characteristics of cellulose, hemicellulose, and lignin are summarized in Fig. 8 [8].

2.1 Cellulose

Cellulosic composition in the biomass corresponds to the most abundant natural saccharide polymer $(C_6H_{10}O_5)_n$ with degree of polymerization of around 10,000 and

Lignocellulosic material	Cellulose (dry wt. %)	Hemicellulose (dry wt. %)	Lignin (dry wt. %)
Hardwood stems	40-55	24-40	18-25
Softwood stems	45-50	25-35	25-35
Nut shells	25-30	25-30	30-40
Corn cobs	45	35	15
Grasses	25-40	35-50	10-30
Paper	85-99	0	0-15
Wheat straw	30	50	15
Sorted refuse	60	20	20
Leaves	15-20	80-85	0
Cotton seed hairs	80-95	5-20	0
Newspaper	40-55	25-40	18-30
Waste papers from chemical pulps	60-70	10-20	5-10
Primary wastewater solids	8-15	NA	24-29
Swine waste	6	28	NA
Solid cattle manure	1.6-4.7	1.4-3.3	2.7-5.7
Coastal Bermuda grass	25	35.7	6.4
Switch grass	45	31.4	12

Fig. 6 Typical lignocellulosic content in the various feedstocks [5]

	HDPE and LDPE	Poly styrene	Poly carbonate	Poly propylene	Polyethylene terephthalate (PET)	Poly hexamethylen eadipamide or Nylon	Poly methyl methacrylat e (PMMA)	Polyvinyl chloride (PVC)
Estimated LHV (kJ/g)	44-47	41.96	30.32	42.66	23.22	28.76	25.1	20.5
Structure	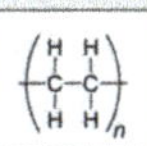	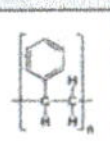	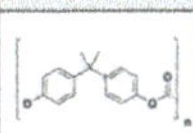	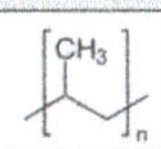	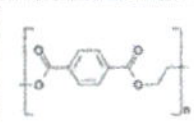	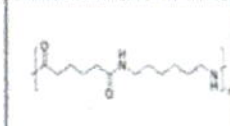	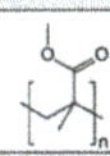	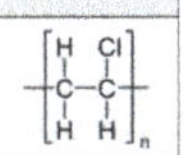
Uses	Containers, bags, film wraps	Packing peanuts, bottles	Windows, electronics, bottles	Piping, insulation	Textiles, containers	Textiles,	Glass alternative	Piping, cable insulation, construction

Fig. 7 Energy content and monomers of different types of plastics [6, 7]

molecular weight around 500,000 g/mol. It is the carbohydrate portion of biomass that forms to provide structural strength. It consists of D-glucopyranose units with β-1, 4 glycoside linkage between the monomers. Intermolecular H-bonding in cellulose makes it insoluble in water at room temperature and the solubility increases with increase in temperature making it completely soluble at 603 K in subcritical water [9].

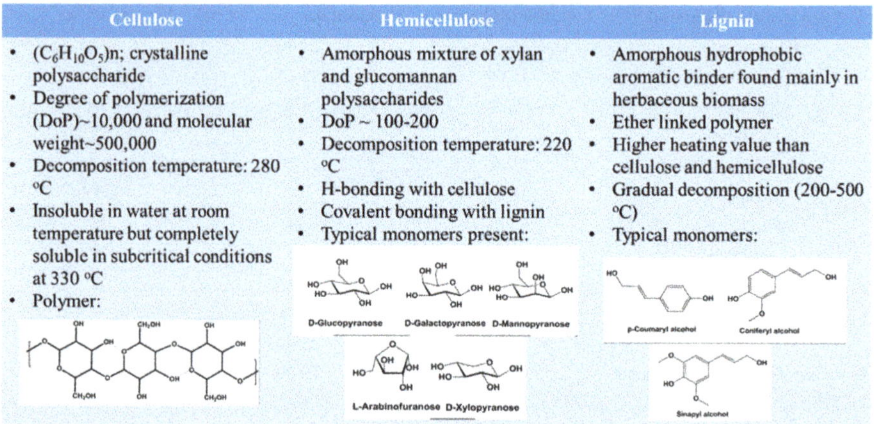

Cellulose	Hemicellulose	Lignin
• $(C_6H_{10}O_5)n$; crystalline polysaccharide • Degree of polymerization (DoP)~10,000 and molecular weight~500,000 • Decomposition temperature: 280 °C • Insoluble in water at room temperature but completely soluble in subcritical conditions at 330 °C • Polymer:	• Amorphous mixture of xylan and glucomannan polysaccharides • DoP ~ 100-200 • Decomposition temperature: 220 °C • H-bonding with cellulose • Covalent bonding with lignin • Typical monomers present:	• Amorphous hydrophobic aromatic binder found mainly in herbaceous biomass • Ether linked polymer • Higher heating value than cellulose and hemicellulose • Gradual decomposition (200-500 °C) • Typical monomers:

Fig. 8 Typical characteristics of lignocellulosic components [8]

2.2 *Hemicellulose*

Hemicellulose is the other dominant carbohydrate portion of biomass which is found predominantly in woods than in herbs to foster structural strength. This, unlike cellulose, is a heteropolysaccharide with a low degree of polymerization of around 200 or less. Usual constituent monomers are D-glucopyranose, D-galactopyranose, D-mannopyranose, L-arabinofuranose, D-xylopyranose monomers, see Fig. 8 [9]. It also forms H-bonding with cellulose and covalent bonds with lignin, but the low molecular weight makes it soluble in water. Thermally it is more unstable than cellulose.

2.3 *Lignin*

While cellulose and hemicellulose are cycloalkane based polymers, lignin is the aromatic phenol derivative binder compound that provides stability along with hydrophobicity to a plant and thus possess low solubility in water. It is the most stable among the lignocellulosic components and is characteristic of gradual thermal decomposition. The aromaticity makes it denser in energy content compared to cellulose and hemicellulose. Monomers of lignin are given in Fig. 8 [9]. Ethyl and methyl phenol derivatives are the usual decomposition products from lignin.

2.4 Biomass Moisture Content

Biomass and bio-wastes although are lignocellulosic derivatives; they vary significantly in terms of their moisture content in the form of absorbed moisture and trapped water content based on their origination. The energy density is significantly affected by this moisture content making it an important decision-making parameter in choosing the right thermochemical conversion pathway to minimize the energy requirement for drying and without any loss of throughput to the system. Effect of moisture content on different conversion techniques are discussed later in this chapter.

2.5 Inorganic and Other Extractive Content

Biomass can come from vast variety of sources extending from forest wood to chicken manure. The inorganic and other content can vary significantly although the total variation with respect to lignocellulosic content is relatively low. Typical ash and N content in different biomass sources are summarized in Fig. 9 [10]. The total inorganic content is high especially in manure wastes due to their source of origin and the mineral intake of the animal under consideration. Organic extractives other than lignocellulose content include fats, waxes, protein and amino acids, phenolics, simple sugars, pectin, mucilage, gums, resins, terpenes, starches, glycosides, saponins and essential oils [9]. Their very low content makes them of less interest for thermal processing, but for consideration of pollutants such as NH_3, H_2S or HCN emissions during thermochemical conversion they are essential in order to selectively minimize the unwanted products.

Feedstock (dry basis)	Lignocellulose	Macro algae	Microalgae	Manures	Sewage sludge
Ash%	3-8	15-25	7-26	10-20	20-50
H/C	1.2	1.2	1.6	1.5	1.6
O%	35-45	25-40	25-30	35-45	50
N%	0.5-3	3-7	5-9	3-6	3-8
HHV MJ/kg	12-20	10-20	25-30	10-20	14

Fig. 9 Characteristic ultimate analysis of biomass and wastes [10]

2.6 Plastic Constituents

Characteristic monomers of plastics, commonly listed based on their usage, are given in Fig. 7. Almost all of the plastic and textile polymers in their high dense state are insoluble in water and contain very low inorganic matter. This makes plastic wastes maintain high energy density and thus high heating value compared to any other wastes, making them valuable candidate for thermochemical conversion although they are not carbon neutral. Secondary usage of the waste plastics for energy, fuel, and even value added chemicals products makes them an ideal supplement to biomass based energy production.

2.7 Biofuel Based Conversion Pathways

Majority of the techniques practiced for both thermochemical and biochemical processing are summarized in Fig. 10 that are based on the nature of the feedstock used and the products produced. This also includes even the first generation of feedstocks which are not recommended as they adversely affect the food stock production in pursuit of meeting energy demands which is a devastating pathway that can be avoided by choosing second generation of feedstock and wastes which are in dire need of disposal. While biodegradation and synthesis processes are promising that have achieved significant place in large scale production for example, bio-ethanol in Brazil, relatively low throughput due to long reaction times, compared to thermochemical processes, makes them less attractive for our quest to replace the fossil fuels. Thermochemical processes such as those given in Fig. 10 (modified from [11]) are all relatively rapid and provide high throughput in waste disposal and product synthesis although product selectivity in thermochemical processes inherently is not superior as compared to enzymatic processing. Thermochemical conversion is essential and a preferred pathway for waste disposal compared to the existing methods of direct combustion via incineration for energy recovery or direct landfill disposal. Incineration produces high levels of unwanted pollutants that must be captured and treated prior to the discharge of gases to the environment and disposal of the remaining residue to landfills. While landfills are not preferred as they lack sustainability due to land requirement, odors and leaching ability, direct combustion of solid wastes for energy recovery is not preferred as combustion of solids inherently limits reaction rates and combustion times to impact flame temperature that also limits the energy efficiency. Thermochemical conversion to intermediates such as syngas, pyrolytic oils or bio-oils is preferred because of higher heating value of the processed products in the form of syngas and other liquid fuels. However, some energy is required during such conversion. Chosen the right pathway, the thermochemical conversion energy input can be minimized to only about 10–15% of the energy content in the waste. Along with increased product yield, and flexibility, it can provide higher aggregate energy recovery at higher efficiency than direct solid waste combustion based on

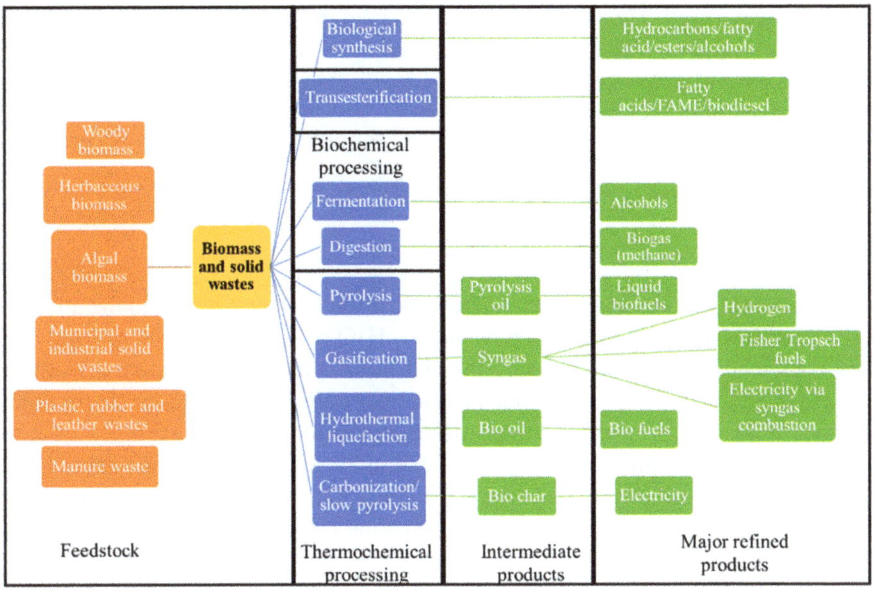

Fig. 10 Biomass and solid waste conversion pathways for energy and fuel production

energy recovery. The product yield consisting of liquid fuels and syngas also offers energy production pathways for cleaner combustion and energy transformation with minimal change to existing infrastructure. Such flexibility in both product selectivity along with feedstock selection makes thermochemical processing pathways attractive for the conversion of biomass and solid wastes that have wide variation in their chemical and physical composition.

3 Thermochemical Processing

Classification of thermochemical processing carbonaceous feedstock can be classified in terms of combination of operational temperature, heating rate and pressure as these results determine the selectivity of phase in the products. Solid product obtained from biomass conversion is bio-char, a graphite like product rich in carbon content although the net yield and quality of the product is dictated by the operational conditions. Gaseous products from thermochemical conversion of biomass and wastes are predominantly H_2, CO, CO_2, CH_4, C_2 and C_3 hydrocarbons whose composition and yield depends on operational conditions. While the solid and gaseous products are predictable, the liquid products and their compositional selectivity varies sensitively corresponding to both the operational conditions as well as the type of feedstock used. More about the selectivity will be discussed later in this section with classification on the pathways used. The three major thermochemical techniques under discussion are gasification, pyrolysis and hydrothermal conversion. Many derivatives of these techniques have been investigated which will be discussed along with the advantages and challenges with such modifications.

3.1 Pyrolysis

Process of pyrolysis includes thermal decomposition of the feedstock in the absence of any kind of oxidizing agent such as air/O_2, H_2O or CO_2. Pyrolysis of solid carbonaceous feedstock results in char, gas and liquid yields with relative selectivity during conversion that critically depends on temperature, heating rate, pressure, vapor residence time, catalyst used and feedstock composition. While the temperatures in pyrolysis are significantly lower than gasification, the main classifying parameter in pyrolysis is the heating rate, and vapor residence time in the thermal zone. Heating rate refers to rise in temperature per unit time (second) while vapor residence time refers to conditions wherein the oxygenate vapors formed during pyrolysis are instantly removed or allowed to stay during the controlled residence time. Fast pyrolysis may need maintaining residence times as low as 2 s at moderately low temperatures of 773 K. Low temperature and long residence times favor the formation of solids in the form of char. High temperatures and long residence times favor the formation of gases from the high degree of cracking of the feedstock. But, moderate temperatures and short residence time results in liquid yield formation usually called bio-oil. Typical residence times in different pyrolysis processes favoring different extremes in the products are given in Fig. 11 [12]. The short residence times also consider the contact time on char formed which should be reduced to avoid cracking of vapors into low molecular weight hydrocarbon gases. To avoid secondary vapor phase reactions, the collected vapors undergo rapid cooling to obtain enhanced yield of bio-oil. Vapor quenching is efficiently performed by collection of vapors in an immiscible hydrocarbon solvent [12]. We limit our discussion in this chapter to fast pyrolysis since the focus is on synthesis to produce energy and fuels for transportation. The desire for coping with very high heating rate limits the feedstocks to moisture content less than 10% and particle sizes around 2 mm or less as heating rate is the rate limiting step in fast pyrolysis of biomass. It also helps to reduce water content in the bio-oil. The pyrolytic mechanisms of individual contents of lignocellulosic material are summarized in Fig. 12 [13]. Of the biomass, stability due to aromaticity content in lignin is the major source of tar and char produced. At low heating rates, cellulose pyrolysis leads to dehydration and depolymerization leading to levoglucosan (a substance formed by heating carbohydrates that has an unusual heterocyclic bicycle-octane structure). At high heating rates they can be minimized favoring liquid products. Lignin pyrolysis leads to phenolic derivatives such as guaiacol.

The effect of temperature on relative yields of organics (oil), char, gas and water is shown in Fig. 13 [14]. Optimal temperature for high bio-oil yield is seen to be

Process	Temperature	Vapor residence time	Liquid yield (% wt.)	Solid (% wt.)	Gas (% wt.)
Fast pyrolysis	~ 773 K	~1 s	75%	12% char	13%
Intermediate pyrolysis	~773 K	10-30 s	50% in 2 phases	25% char	25%
Carbonization (slow)	~673 K	~hours-days	30%	35% char	35%
Gasification	> 973 K	Same as char	5%	10% char	85%
Torrefaction (slow)	~563 K	Solid res. Time ~ 10-60 min	0%; 5% if condensed	80% solid	20%

Fig. 11 Typical residence times associated with different modes of pyrolysis and gasification

Cellulose	Hemicellulose	Lignin
• Starts with water removal and depolymerization • Anhydride to levoglucosan formation at high heating rates • Low heating rates-> dehydration reactions and depolymerization causing charring at ~575 K • Higher heating rates reduce char formation by increasing tar via hindering dehydration and increasing conversion to levoglucosan and increasing depolymerization of unstable species	• Decomposes readily compared to cellulose • More volatiles and less tar and char compared to cellulose • Deacetylation in hemicellulose -> major contributor to acetic acid production and dehydration producing water and furfural • Xylans especially present in hardwoods degrade to water, methanol, formic, acetic and propionic acids, hydroxyl-1-propanone, hydroxyl-1-butanone, 2-furfural • More aqueous products compared to the rest of constituents	• Harder to dehydrate compared to cellulose • Chief products are phenolic derivatives such as guaiacol formed from cracking of phenyl propane units • Major contributor to tar and charcoal production among the three • Between 423 K and 573 K, cleavage of aryl-alkyl-ether linkages occurs • Ether bonds dominant in lignin compared to acetyls in cellulose and hemicellulose dehydrate to form phenols

Fig. 12 Characteristic features of cellulose, hemicellulose and lignin content during pyrolysis

around 500 °C (773 K). Temperature below this value leads to repolymerization to enhance char formation, but higher than 773 K lead to enhanced cracking to produce more gas yields. The liquid products formed are much influenced by the temperature at which reaction occurs. At approximately 673 K mixed oxygenates are favored; around 773 K phenolic esters are favored; 873 K leads to alkyl phenolics; from 973 K starts the formation of heterocyclic ethers; above 1073 K polyaromatic hydrocarbons (PAH) and larger compounds are formed leading to further aromatization and dehydrogenation to further enhance carbonization [8, 13]. High temperatures leads to insignificant amounts of liquid yields and high yields of char and gases from the carbonization.

Different types of reactors have been investigated to realize fast pyrolysis. However, the most common is fluidized bed pyrolysis. This is established to provide bio-oil yields as high as 70–75% (by wt.). But fluidization needs large amounts of inert gas causing vapor dilution and difficulty to capture vapors. This in turn requires larger reactor sizes. As the residence times of char and vapors in a contemporary

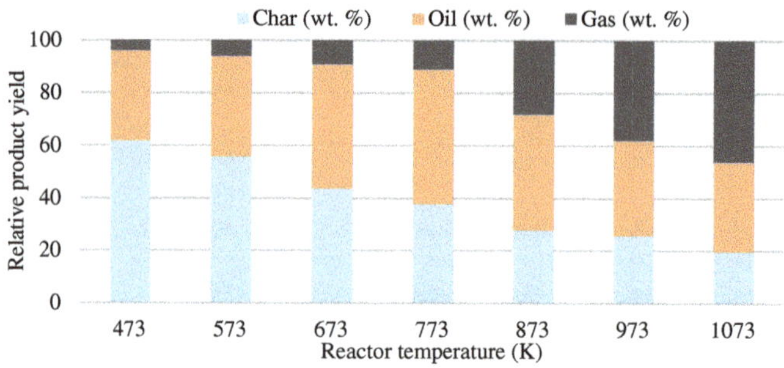

Fig. 13 Effect of temperature on relative yields in fast pyrolysis [14]

fluidized bed are similar, enhanced cracking of the vapors occurs causing increased gas yields. This can be mitigated by removing char particles using either cyclones or hot gas filtration, although sticky phase of this mixture clogs the filter used in hot gas filtration. Heat required for fluidized bed pyrolysis is supplied either as electrical heating or from the recirculation of hot and inert heated bed material from the combustion of char and portion of bio-oil. Other complex reactor designs, such as rotating cone pyrolysis, have also been investigated to reduce the requirement of inert gas for fluidization by rotating the reactor and thus lowering the reactor size. Heat provided in such reactors is similar to fluidized bed that utilizes recycling of heat from the hot bed obtained from char combustion zone. The restriction on particle size to achieve high heating rate can be discounted by using ablative pyrolysis. Although its reactor design and complexity are questionable in terms of scalability due to its direct dependence on the reactor surface area, high relative centrifugal velocity is achieved to mechanically pressurize the feedstock against the hot reactor walls. This pathway avoids the need for inert gas. Another way the above restrictions of inert gas and the requirement of feed flexibility can be achieved is from vacuum pyrolysis. As the name suggests, it involves maintaining vacuum in the reactor that makes it easy to remove vapors and also yields less char which directly depends on pressure. Much of the reactor design research is a topic that is still under current investigations and away from industrial scalability due to the restrictive need for high grade feedstocks in terms of moisture and ash content along with high heating value. But, for the high grade feedstocks available, such as wood pellets, or even plastic wastes, this process is a worthy pathway to obtain liquid fuels and chemicals. Detailed reviews of pyrolysis reactors and yields are available in the literature including their origins and outputs [8, 12].

Vapors formed during pyrolysis are rapidly condensed to obtain non-thermodynamic equilibrium products in liquid phase. This is referred to as bio-oil which also contains water. It is usually dark brown in color due to the presence of micro char particles (that can be removed by hot vapor filtration). It is a micro-emulsion of aqueous products from cellulose and hemicellulose and discontinuous lignin derived monomers and oligomers stabilized together by forces such as H-bonding. The stability of bio-oil is one of the critical issues in fast pyrolysis which dictates the ageing of bio-oil causing increase in viscosity and phase separation with time. Depending on storage temperature, bio-oil ageing includes tertiary reactions such as esterification, or transesterification, phenol or aldol condensation, etherification. High oxygenate content along with inseparable water content results in many undesired properties to bio-oil such as low energy density that can be only 50% of that of petroleum based liquid fuels (wt. basis), high melting and boiling points, and high viscosity along with high acidity (pH ~2.5) due to the presence of carboxylic acid. Bio-oil contains large variety of oxygenate functional groups such as aldehydes and ketones, alcohols, phenols, furans, aliphatic and aromatic hydrocarbons, carboxylic acids and esters and sugars. These cannot be separated by distillation or any other heating processes as attempt to vaporize bio-oil leads to solid residue formation. Other issues with these undesired properties and their causes are summarized in Fig. 14 [12]. Addition of stable solvents such as methanol and acetone increased the stability of bio-oil and lowered the viscosity. Influence of many factors on the yield and selectivity of liquid products formed is not well understood. One of such factors is the presence of ions. The presence of alkali ions such as K, is seen to favor char formation and removing such ions has led to increased liquid yields. Addition of zeolites during pyrolysis also favors cracking, decarboxylation, dehydration, and decarbonylation and thus reducing liquid yields. Increasing basicity by addition of NaOH reportedly increased oil yields [8]. Due to such complications in the stability of bio-oil during its formation, the refining of bio-oil into bio-fuels or extraction of chemical requires additional downstream processing that is separate from the pyrolysis reactor so that one can easily select a suitable catalyst and control the operating conditions. Downstream refining reactions will be discussed after the introduction to hydrothermal processing since the generic problem with bio-oil is similar, although more complicated as compared to the liquid products from hydrothermal processing.

4 Gasification

Gasification comprises of thermal decomposition of solid carbonaceous material including wastes in the presence of oxygen containing gasifying agents such as steam, CO_2 or O_2/air at temperatures greater than 1000 K (approximately) to obtain high yield of synthesis gas (called syngas) which is a mixture of H_2, CO, CO_2, H_2O, CH_4, C_2, C_3 along with minimum yields of tar or char (liquid and solid yields). Typical reaction pathway observed in gasification are given in Fig. 15 [15]. The solid feedstock initially is converted into oxygenated vapors including CO_2, H_2O and CO along with char whose relative yield depends on heating rate and operational temperature. These reactions basically correspond to thermal decomposition via depolymerization and cracking reactions. This primary heterogeneous pyrolysis is followed by thermal cracking to reform to lower molecular weight hydrocarbons along with CO, CO_2, and H_2 as well as side repolymerization reactions that to form heavy molecular weight wax, such as polyaromatic hydrocarbons, also called tar, depending on the reactor operational temperature. Low temperatures favor the tar

Characteristic	Cause	Effects
Acidity (Low pH)	Organic acids formation	Vessel and pipe corrode
Aging	Continued secondary reactions including repolymerization	Increase in viscosity with time due to secondary reactions such as condensation; potential phase separation
Alkali metals	Collected in char, High ash feed can lead to incomplete solids separation	Catalyst poisoning; Solid deposition in combustion (damage to turbines and combustor); Erosion and corrosion; Slag formation
Char	Incomplete char separation	Aging of oil; Sedimentation; blockage of filters, catalyst, fuel injector; Alkali metal poisoning
Chlorine	Contaminant content in biomass	Catalyst poisoning in upgrading process
Feedstock Contamination and solids (sand)	Poor preprocessing and material separation	Can increase particulate carryover; Sedimentation; corrosion and erosion; blockage
Poor distillability	Reactive mixtures from pyrolysis	The reactive mixtures start to react at 353 K and decompose above 373 K
High viscosity	Aging and product composition dependent	High pressure drop which means high reactor and pumping costs; temperature sensitivity
Low H/C	Low H/C in the source (typically high lignin)	More difficulty in upgrading due to higher stability of typically aromatics
Materials incompatibility	Phenols and aromatics	Seals and gaskets damaged
Low hydrocarbon miscibility	High O/C ratio due to the oxygenates	Difficulty to use the existing refinery infrastructure
Nitrogen	Biomass contaminants or proteins	Catalyst poisoning and NO_x during combustion
High O/C	Depending on the feedstock	Poor stability and low hydrocarbon miscibility
Phase separation	High feed water/ash; poor char separation	Layering; poor mixing
Structure	Emulsion based due to rapid de-polymerization and quenching	Susceptible to aging and thus increasing viscosity and phase separation
Toxicity	Biopolymer degradation products	Small human toxicity
Sulfur	Source contaminants; rubber	Catalyst poisoning
Temperature sensitivity	Incomplete reactions due to rapid quenching	Decomposition and phase separation above 373 K; Irreversible aging causing rise in viscosity and phase separation above 333 K;

Fig. 14 Key characteristics of bio-oil, their cause and effects [12]

formation while high temperatures greater than 1300 K have reported no residual tar yields. This combination of low molecular weight hydrocarbons, along with the tar and char react with the gasifying agents to result in gaseous composition dictated by thermodynamic equilibrium. The major and simplified set of reactions governing final composition of gases formed is summarized in Fig. 16. For simplicity, the reaction pathway is drawn linear. But in reality, the intermediates such as the oxygenate vapors can also react with either the gasifying agents, char or light hydrocarbons in complicated pathways of intermediate reactions.

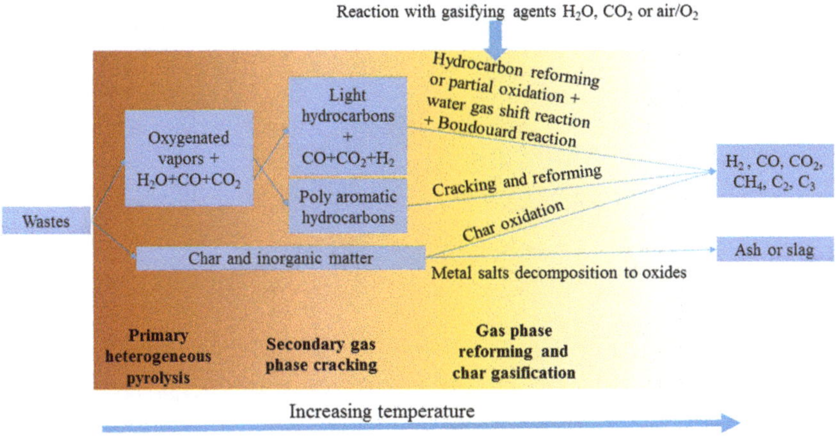

Fig. 15 Simplified reaction pathways in gasification

Combustion:

$$C+0.5O_2 \rightarrow CO;\ \Delta H^o_R=-110.5\ kJ/mol$$

$$CO+O_2 \rightarrow CO_2;\ \Delta H^o_R= -283\ kJ/mol$$

$$H_2+0.5O_2 \rightarrow H_2O;\ \Delta H^o_R=-241.8\ kJ/mol$$

$$CH_4+2O_2 \rightarrow CO_2+2H_2O;\ \Delta H^o_R=-802.23\ kJ/mol$$

Reforming:

Water gas shift reaction: $CO+H_2O \leftrightarrow CO_2+H_2;\ \Delta H^o_R=-41.2\ kJ/mol$

Steam reforming: $CH_4+H_2O \leftrightarrow CO+3H_2;\ \Delta H^o_R= 206.17\ kJ/mol$

Dry reforming: $CH_4+CO_2 \leftrightarrow 2CO+2H_2;\ \Delta H^o_R= 247.37\ kJ/mol$

Char reactions:

Boudouard reaction: $CO_2+C \leftrightarrow 2CO;\ \Delta H^o_R= 172.5\ kJ/mol$

Char-steam gasification: $C+H_2O \leftrightarrow CO+H_2;\ \Delta H^o_R= 131.3\ kJ/mol$

Methanation: $C+2H_2 \leftrightarrow CH_4;\ \Delta H^o_R= 74.9\ kJ/mol$

Thermal cracking: $C_nH_m \rightarrow C_{n-a}H_{m-x}+C_aH_b+0.5(x-b)H_2;\ \Delta H^o_R>0$

Fig. 16 Major equilibrium reactions considered in gasification

The dependence of gasifying agent on the product yield and selectivity along with the advantages and disadvantages are summarized in Fig. 17. Comparison of gasification rate of food waste using steam and carbon dioxide measured in a semi-batch reactor is shown in Fig. 18. In general, the gasifying agent is chosen depending on the downstream application or on the availability of reactants. Steam gasification can be used to gain syngas with high H_2 content while dry/CO_2 gasification enhances CO content in the syngas. Steam or CO_2 gasification reactions are endothermic which can be maintained either by external heat or by using air/O_2 at very fuel-rich conditions. The dependence of gasifying agent and temperature on the syngas composition can be found in refs. [18, 19]. Effect of temperature on gasifying rates of food waste examined in semi-batch reactor is shown in Fig. 19. While operational temperatures can vary from 1000 to 1500 K for different gasifying reactors used, the industrial gasifiers often operate at low to atmospheric pressures. The reason

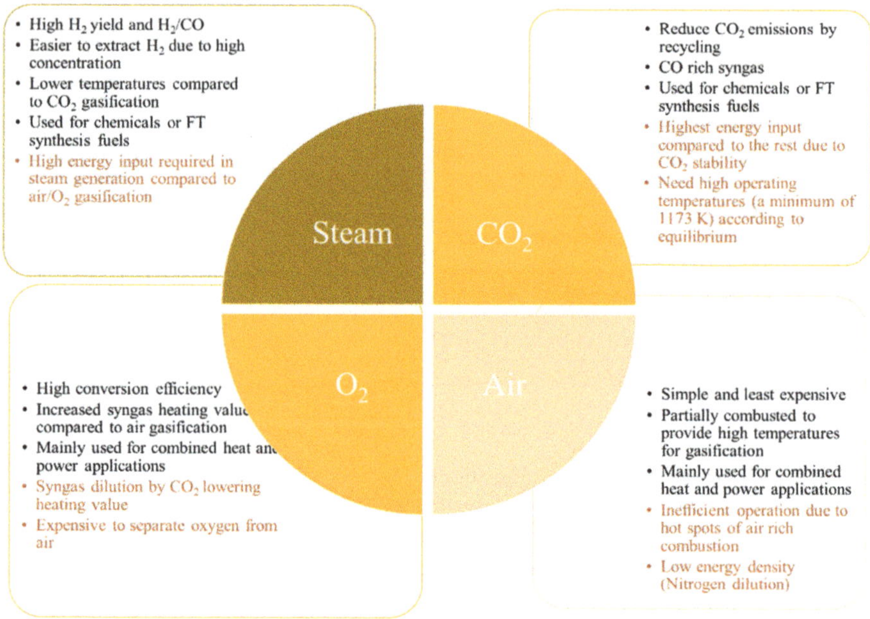

Fig. 17 Effect of gasifying agent on the product yield with advantages and issues

for choosing low to atmospheric pressures is that, although high pressures enhance char gasification and lowers tar yields, it produces methane rich syngas which is less desired compared to H_2 and CO. High pressures also pose significant restrictions due to the material handling and high pressure feeding of the feedstock making it less attractive for industrial operational conditions. The experimental investigation of temperature dependency on gases yields have matched well with the equilibrium results due to the high temperatures at which gasifiers are operated resulting in equilibrium driven reactions.

Different types of reactors have been investigated and also scaled-up for gasification which include fixed and fluidized bed reactors depending on the desired downstream applications. Typical schematics comparing the major gasifier designs can is shown in Fig. 20 for fixed bed reactors, and Fig. 21 for fluidized bed and entrained flow gasifiers. Figure 22 provides a summary of the widely used reactors and their characteristics. Typical industrial gasifier contains combustion zone wherein oxygen/air content is high that allows for the combustion of some of the reactant feedstock or their products, such as char. This zone is followed by the reforming or gasification zone wherein the feedstock drying, pyrolysis and reformation occurs with gasifying agent. The order of combustion zone and gasification zone depends on the design configuration of the gasifier. In countercurrent flows as in updraft gasifiers, combustion occurs near the air feed while syngas is collected at the feedstock feed. However, in co-current or downdraft gasifiers, the air or some other

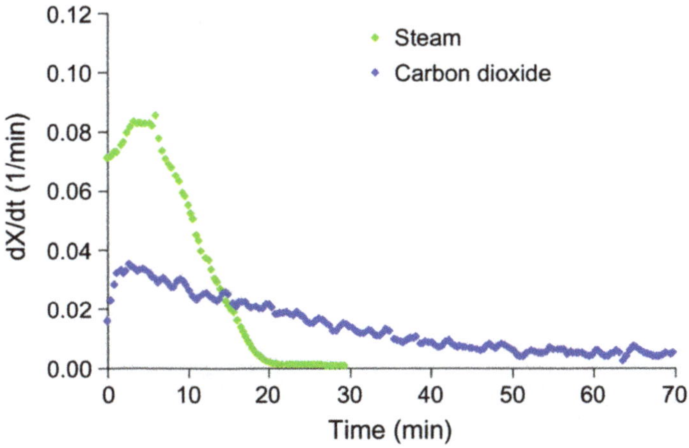

Fig. 18 Evolution of conversion rate in gasification of food wastes using steam and CO_2 as gasifying agent (X = conversion and T = 1173 K) [16]

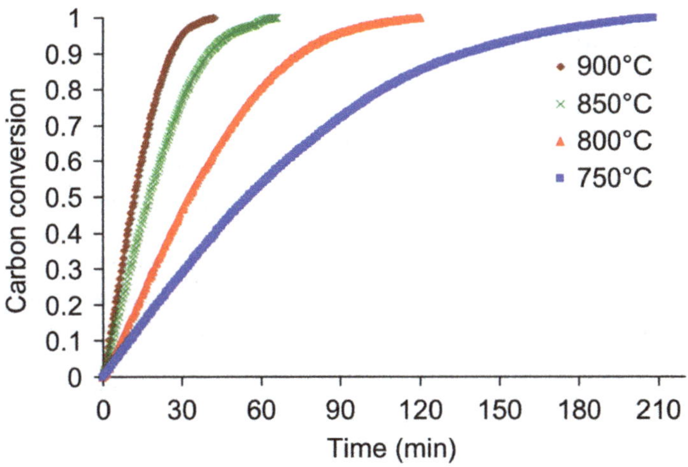

Fig. 19 Carbon conversion rate of steam gasification of food waste at different temperatures [17]

gasifying agents are added in same direction as the feedstock but after the drying and close to pyrolysis zone, both along with the gasification zones, are supported by the interior combustion zone close to the air intake. Different derivatives of these reactors can be found in the literature [20], but the genesis of choosing such derivatives lies in solving specific application issues inherent in conventional form of such reactor, for example, derivatives of cooling techniques used to quench the reaction. Some derivatives use water to cool the reaction using a heat exchanger which provides high energy efficiency in the form of heat recovery but relatively low syngas quality. In contrast, others have used water spraying to condense the products providing high

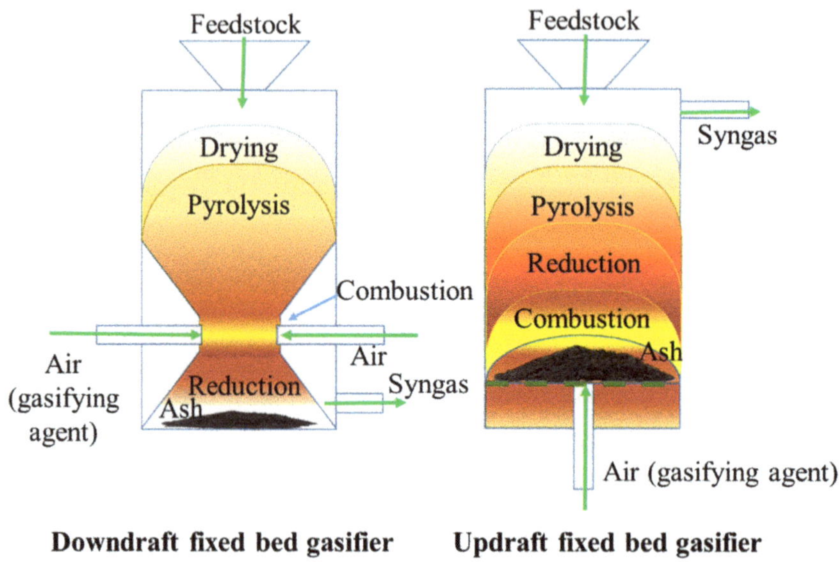

Fig. 20 Simplified diagram of updraft and downdraft fixed bed gasifiers

quality syngas by avoiding backward reactions. However, its energy efficiency is lower due to lack of heat recovery in such water cooling systems. While fixed bed reactors were needed to be operated at high temperatures, fluidized bed reactors due to the inherent uniformity allows operation at relatively low temperatures to provide higher energy efficiency along with feed flexibility. In contrast, entrained flow reactors are made for high conversion at high temperature and pressure which provides none to limited tar and char yields but makes the feedstock particle size, energy density, and mode of heat recovery crucial. Both fixed bed and entrained flow reactors can operate in slagging mode which means that the ash formed after gasification, due to high temperatures, melt in the reactor that can be collected as molten slag instead of solid particulates. This is useful to avoid syngas filtration and also provide easier means of ash handling. Low temperature operation helps to maintain higher energy efficiency and reduce the criticality of feedstock size and form, but tar and char formation tendency is higher in such condition which then calls for downstream gas cleaning or recycling. Char formed during gasification even though can react with the gasifying agents via water gas reaction or Boudouard reaction, industry prefers to recycle the char particles into the combustion zone of the gasifier to burn and provide energy for the endothermic steam or dry gasification processes downstream in the gasifier.

The choice of feedstock also poses some limitations. Investigations into different types of feedstocks for gasification can be found in literature which include wood chips, pellets, paper and cardboard feeds, waste water sludge, chicken manure and others [21–24]. Figure 23 summarizes hydrogen gas yield per unit feedstock mass

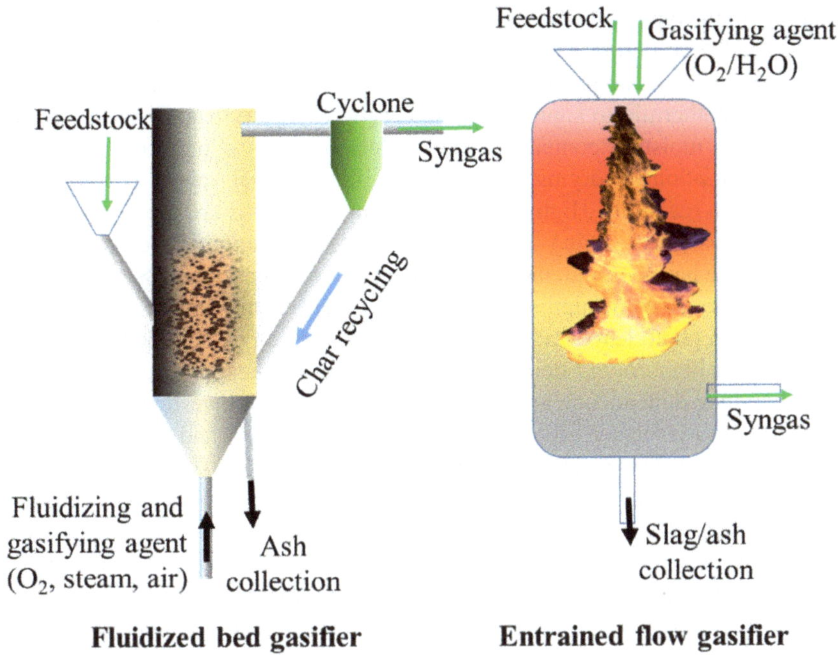

Fig. 21 Simplified diagram of fluidized bed and entrained flow gasifiers

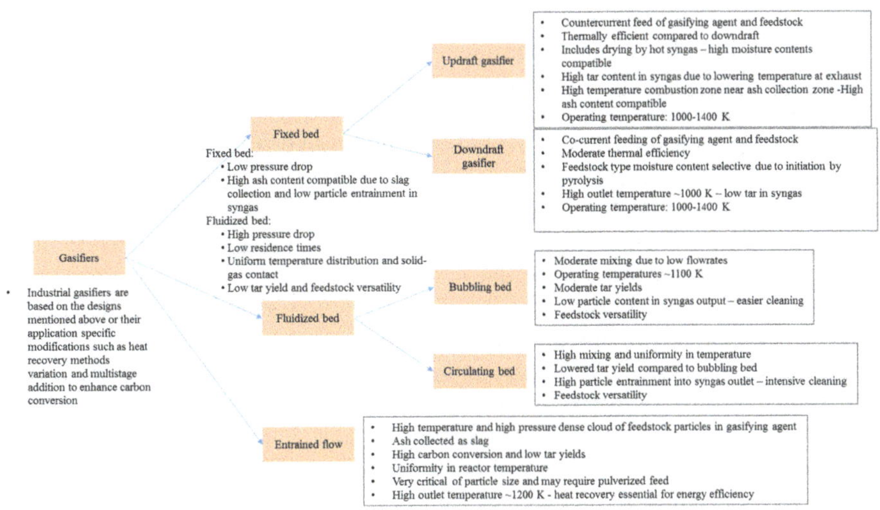

Fig. 22 Classification of major industrial gasifiers and their characteristics

from steam gasification of different types of feedstocks in lab scale semi-batch reactor. The cumulative yield differed in both yield and the dependence of temperature depending on the type of feedstock used. Moderately high energy density is required from the feedstock before it reaches the gasifier. High moisture content feedstock such as municipal waste sludge, yard trimming, or food waste, makes the process extremely endothermic to significantly lower the energy efficiency of the process. Relatively low moisture content of 10% or less by mass is preferred to limit the energy consumption for drying within the gasifier to lower values as compared to the gasification reactions. High moisture content feedstocks are pre-dried either by sun drying, waste heat drying or electric drying. If the supplied feedstock straight from the source has very high moisture content, energy analysis is essential to determine if the energy consumption for drying of such drying can be maintained by the gasification plant. Low grade feedstocks as above can be complemented by co-gasifying with high energy density wastes such as plastics, and rubber. Not only does this reduce dependence on the supply of feedstocks, investigations into such gasification reported synergistic effects increasing the syngas and hydrogen yield by mutual interaction. This can be seen especially between biomass such as wood chips and plastics such as polyethylene in [25]. In the case of moisture diluted feedstocks, although such feedstock can be gasified, there are other thermochemical pathways such as hydrothermal processing which is inherently more useful and energy efficient to reform high moisture biomass. Other diluent that lowers the energy density of feedstock is inorganics that can be determined from its ash content. Ash content as high as 20% (based on dry mass basis) can be seen in feedstocks such as chicken or other animal manure, food waste and others [19]. This dilution does not have as significant effect on the energy efficiency as moisture dilution, but very high ash content complicates the gasifier design. The ash collection route now becomes critical in these conditions. Fluidized bed gasifiers are amongst the contemporary choice for high ash content feedstock as the temperature is low and ash can easily be collected. Metal oxide components of the ash was reported to provide catalytic enhancement in biomass gasification. This can be exploited in fluidized bed gasification of high ash containing feedstock in which the bed material can be the ash formed that enhances the syngas yield and quality. The other reason for the choice of fluidized bed for low-grade feedstock is their high efficiency at low temperatures compared to other gasifiers. The low-grade feedstock, due to insufficient energy density provides challenges to achieve high temperatures, as in entrained flow gasifiers. This then makes fluidized bed as good prominent choice.

One of the main issues of low temperature operation in gasification is the formation of tar, which effectively results in low carbon conversion to gaseous product yields. Tar content along with gaseous hydrocarbon content in syngas are not desirable products. Characteristic composition of tar is shown in Fig. 24 [27]. Tars must be avoided to limit clogging of pipes and loss of potential fuel energy yield along with avoidance of carcinogenetic emissions from the combustion of tar particles and improve the quality of syngas for fuel production, efficient combustion and for use in chemical plants. Different techniques for tar elimination have been explored [27–34]. Some of them include, catalytic cracking or cracking on char bed, or thermal cracking by high temperature operation. These techniques were examined both in situ over the catalytic fluidized bed in a gasifier or as part of downstream syngas cleaning.

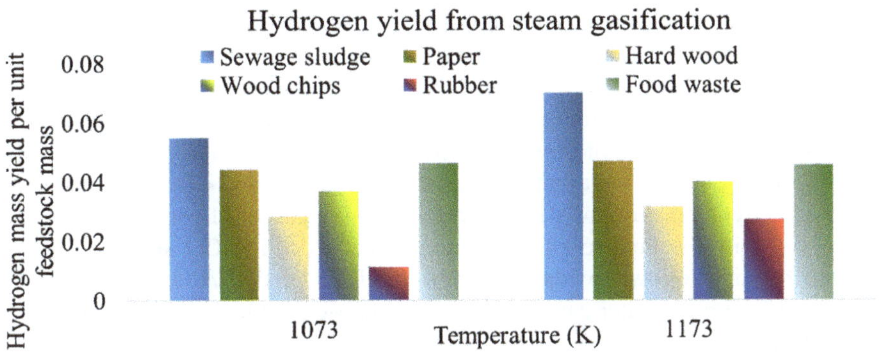

Fig. 23 Hydrogen gas yield from steam gasification of different feedstocks at different temperatures [21, 24, 26]

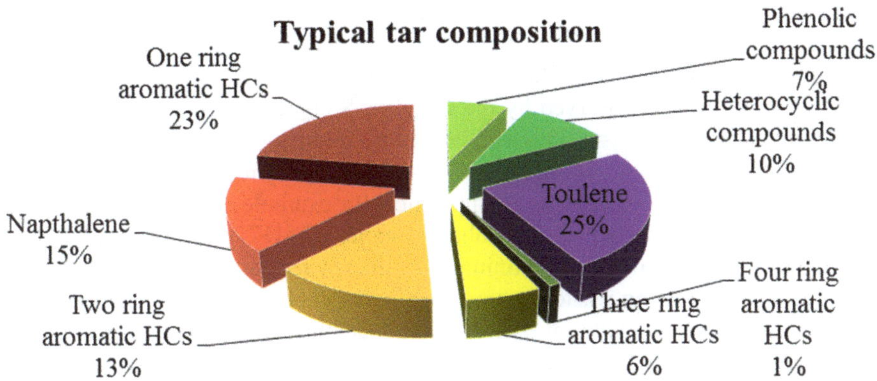

Fig. 24 Typical composition of tar yield during gasification at relatively lower temperature [27]

Downstream catalysts used for tar cracking are hydrocarbon active metals which are usually transition metals; most common being Ni, Pt, Pd, Ru, Rh, Fe supported on acidic and highly porous metal oxides such as alumino-silicates or zeolites, or zirconia and others. Alkali carbonates such as K_2CO_3, Na_2CO_3, and $CaCO_3$ were also reported to be effective in tar cracking. Petroleum research has also helped in this aspect. Examples include, fluid catalytic cracking (FCC) catalysts such as ion exchanged zeolites (H-ZSM-5 and others) that have been found to be very effective with high activity towards tar cracking. In situ tar cracking in the gasifier can be performed efficiently on mineral based catalysts such as dolomite, magnesite, calcite, olivine, clay minerals as they although are lower in activity compared to synthetic catalysts, their availability, structural strength and commercial feasibility makes them a much preferred option.

Syngas is the product value of interest from gasification process. Syngas is a gaseous fuel with medium to high heating value that can be used for thermal energy production from the combustion of syngas. Syngas is an intermediate product that can be converted downstream to multiple high demand value added chemicals or liquid fuels that can be used for transportation fuels, see Fig. 25 [35]. Although heat or power generation does not need any complicated downstream processing other than particulate and char filtration, situation is much different for fuels or chemical synthesis. This varied use of syngas is dependent critically on its chemical composition especially H_2/CO molar ratio. Downstream cleaning and reforming of syngas includes catalytic cracking or steam reforming of the gaseous hydrocarbons (CH_4, C_2 and C_3) with the final result being a combination of H_2, CO and CO_2. Water gas shift reaction follows to adjust the ratio of H_2/CO and H_2/CO_2. For hydrogen production, complete water gas shift reaction is carried out leaving a mixture of H_2 with CO_2. CO_2 in this mixture can be removed by physisorption basis using pressure swing adsorption or chemisorption basis using for example CaO for carbon capture as $CaCO_3$ as shown in Fig. 26 [29]. H_2 can also be separated from CO_2 using novel membrane separation techniques such as Pd based, holey-graphene based and other novel membranes. This is an area of active research in the field of hydrogen production and H based fuel cells. Other impurities such as HCN, NH_3, H_2S or HCl are also part of the raw syngas produced from extractive content such as amino acids and other impurities as observed by the Cl, N or S content from ultimate analysis of different biomass or waste samples [29]. These impurities can be removed either by dry scrubbing with conversion of $Ca(OH)_2$ to $CaCl_2$ or amine solvent extraction. Wet scrubbing can also remove these impurities; for example, HCl is highly soluble in water and can addition of Na_2CO_3 enhances capture of HCl during wet scrubbing. Research is in progress on how to minimize the energy requirement and design complications while lowering waste streams as in wet scrubbing for syngas cleaning to remove these impurities with lower energy and reactant consumption/solvent recovery. Novel biochemical techniques are also under development such as, removal of H_2S by reacting with $Fe_2(SO_4)_3$ which can be regenerated by biological oxidation with the help of Thiobacillus ferro-oxidants [29] as shown in Fig. 21. Final products from syngas can be chosen depending on the market requirements, see Fig. 25. Fisher-Tropsch synthesis is useful as the product offers a replacement for all kinds of transportation fuel with almost no additional energy required for the synthesis of the syngas [36]. The emissions from burning such fuels are essentially from sequestered by the flora which completes the carbon cycle in a shorter time scale making the process sustainable. Hydrogen production from syngas for fuel cells can also provide a clean energy option for mobility, but considering the existing infrastructure and its ability to accommodate FT fuels make it an ideal pathway provided one can suffice with the other energy parameters required and feedstock availability.

5 Hydrothermal Processing

Hydrothermal processing is quite different from the above discussed thermochemical techniques. The process includes conversion of biomass or carbonaceous feedstock into oil, gas and char in the presence of sub-critical or super-critical water. The idea

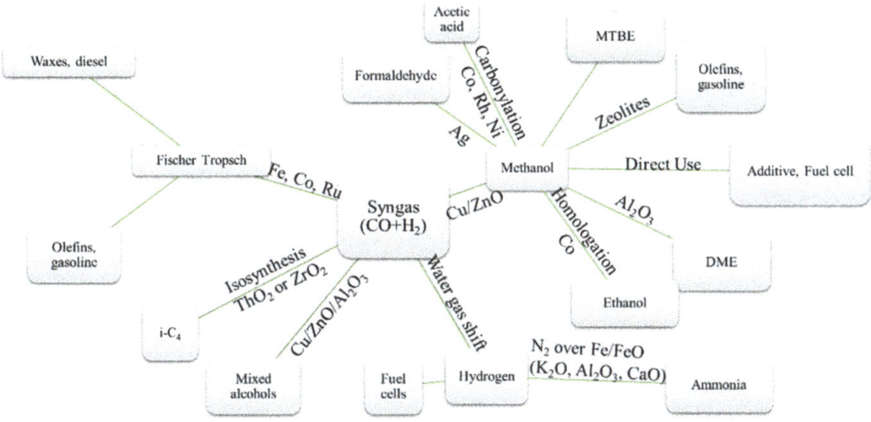

Fig. 25 Syngas and its downstream derivative products

CO_2 capture:	$CO_2 + CaO \leftrightarrow CaCO_3$
HCl capture:	$2HCl + Ca(OH)_2 \leftrightarrow CaCl_2 + 2H_2O$
H_2S capture:	$H_2S + Fe_2(SO_4)_3 \rightarrow 2FeSO_4 + H_2SO_4 + S\downarrow$
Biological recovery:	$2FeSO_4 + H_2SO_4 + 0.5O_2 \rightarrow Fe_2(SO_4)_3 + H_2O$

Fig. 26 Chemical transformation during syngas cleaning

for such technique stems from the properties of water near the critical point (647 K and 22.1 MPa). Near critical point, water has low dielectric point and increased ionic product which increases its solvent capabilities with respect to non-polar substances. Details of changes in properties can be seen in Fig. 27 [37]. The high diffusion rates and low viscosity can be achieved at the same time along with the ability of water for H-donation [9]. The inherent basis of this process being the water phase reaction that makes it superior to reform high moisture content and low ranked feedstocks such as food waste, municipal sludge and algae as compared to gasification or pyrolysis. Majorly investigated derivative of this technique include hydrothermal liquefaction. Investigations on supercritical gasification and carbonization are given [38].

Sub-critical conditions (373–674 K at sufficient pressure for water to remain in liquid phase) are used to produce high yields of bio-oil in hydrothermal liquefaction (HTL) while supercritical phase is used to further breakdown the biomass to produce gaseous yields in supercritical gasification which is usually used for power generation applications or for the disposal of toxic wastes. Low temperatures or pressures cause production of bio-oils with high viscosity but high temperatures and pressures results in gasification of the materials. High pressures and temperatures need to be balanced to achieve high oil yields with sufficient reaction rates. Unlike pyrolysis, heating rate does not account for any measurable significance in hydrothermal processing, but

H_2O	Ambient liquid	Sub-critical phase	Super-critical phase	
Temperature (K)	298	523	673	673
Pressure (MPa)	0.1	5	25	50
Density (g/cm^3)	0.997	0.80	0.17	0.58
Dielectric constant	78.5	27.1	5.9	10.5
pK_{su}	14.0	11.2	19.4	11.9
Heat capacity (kJ/(kg.K)	4.22	4.86	13	6.8
Dynamic viscosity (mPa.s)	0.89	0.11	0.03	0.07
Thermal conductivity (mW/(m.K)	608	620	160	438

Fig. 27 Variation in the properties of water under normal, sub-critical and super-critical conditions

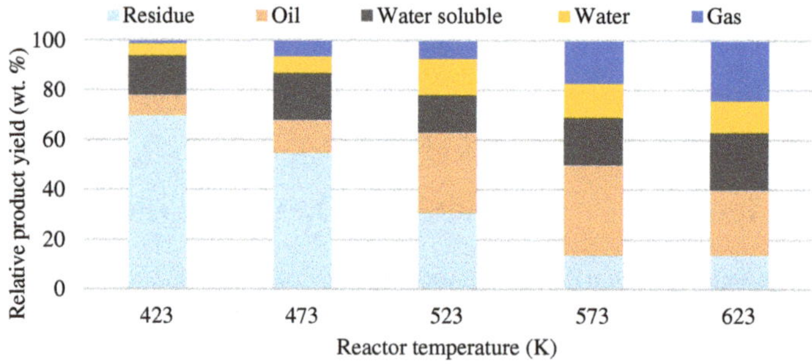

Fig. 28 Effect of temperature on relative product yield in hydrothermal liquefaction

residence time control is critical to maintain a balance between heavy oil and light oil yields to avoid hydrocarbon gas formation. The effect of temperature on the relative product yield is shown in Fig. 28 [39, 40].

The reaction pathway for conversion of lignocellulosic feedstock in subcritical water starts with dispersion of biomass in the water followed by hydrolysis led depolymerization to oligomers to monomers. These monomers undergo decarboxylation, dehydration, and decarbonylation along with other side reactions. Cellulose converts to glucose monomers which degrades to form acetic acid, acetaldehyde, and other aldehyde and furfural derivatives. Hemicellulose forms aldehyde and acid derivatives such as glyceraldehyde, formic acid, lactic acid. High thermal stability of lignin causes hydrolysis a difficult task leading to the formation of phenolic compounds. Decarboxylation and dehydration reactions do occur in hydrothermal liquefaction unlike in pyrolysis which is dominated by cracking reactions. This achieves low oxygen content in bio-crude from liquefaction compared to bio-oil from fast pyrolysis, to offer better stability, low viscosity, and high heating value. These properties offer fuel like properties compared to pyrolysis oil. Comparison of bio-oil from

	Pyrolysis	Liquefaction
Drying	Needed	Not needed
Pressure (MPa)	0.1-0.5	5-20
Temperature (K)	643-800	273-673
Catalyst	Not needed	Sometimes needed
Heating value	Low (~17 MJ/kg)	High (~30 MJ/kg)
O/C	High	Low
Aging	Susceptible	No
Water content	High	Low
Viscosity	Low	High
Upgrade	Difficult	Relatively easier

Fig. 29 Comparison of fast pyrolysis and hydrothermal liquefaction processes and their product characteristics

fast pyrolysis and bio-crude from hydrothermal liquefaction is shown in Fig. 29 [41]. Even though it is better than pyrolysis oil, oxygen content is not insignificant and needs downstream refining before liquid fuel applications. The product liquid has both oil content and aqueous content which can be separated out using solvent extraction technique. Comparative reviews of reactors used and upgrading of liquid yields from pyrolysis and liquefaction are available [39, 42–44]. The reactor design in hydrothermal liquefaction is relatively simple compared to fast pyrolysis and it is reported to be operable in both batch and continuous modes [10].

Catalytic enhancement can be achieved in liquefaction by addition of Arrhenius acid for dehydration via acid hydrolysis [45]. Quality of products achieved can be enhanced by addition of solvents especially H-donor solvents such as tetralin to water which can enhance the alkane yield by lowering oxygen content. The oil produced from hydrothermal liquefaction has comparable energy density to that of petroleum crude and can replace it provided the refining portion of the bio-crude can become industrially established similar to that of petroleum industry. Catalysts were also found to enhance the bio-crude yields although proper mechanisms for such enhancement is complicated to simulate or fully understand the conversion. Investigations on different types of feedstocks and catalysts can be found in the literature. Fe salts such as $FeSO_4$, and FeS; Na_2CO_3 and K_2CO_3; or supported transition metal catalysts such as Ru/C have shown catalytic effect to enhance the bio-oil yields [41]. Further details of the influence of catalysts selection on individual reaction types is explained later. Fundamental study to determine the effect of catalysts on individual component yield of biomass liquefaction should be obtained. Such research of using model compounds or biomass components to simplify the process and observe the reaction pathway, both with and without catalysts are currently underway. Gasoline type hydrocarbon yield was achieved by cellulose liquefaction with in situ reduction in H_2 using catalysts such as Pt/Al_2O_3 [42, 46–49]. Separate upgrading such as bio-crude upgrading of the bio-algal oil in supercritical water with Pt/Al_2O_3 catalyst has also progressed.

Within the biomass processing techniques focus on production of HMF (5-Hydroxymethylfurfural) is high due to its capability to act as multifunctional precursor and replacement intermediate in many petrochemical and petroleum applications. Typical reaction pathway involving biomass processing for furfural based derivatives, avoiding additional side reactions, can be shown in Fig. 30 [50]. Model reactant representatives were chosen to focus on the characteristics of important products. Presence of high pressure solvent dissolves and depolymerizes the biopolymers such as polysaccharides, here pyranose form, into aldohexoses (glucose) and ketohexoses (fructose) with addition of water molecules. Thus formed hexoses in the presence of acid or base catalysts dehydrate to form furfural derivatives especially HMF. After aldol condensation, HMF can be refined in the presence of heterogeneous metal catalysts and hydrogen to undergo hydrogenolysis and hydrodeoxygenation to form alkane based fuels similar to diesel and other existing liquid fossil fuels. Investigations have revealed that the presence of aprotic solvents such as DMSO (Dimethyl sulfoxide) enhances the yield of HMF compared to due to the selective conversion of polymers to fructose form compared to glucose form as in water. Presence of acid catalysts enhances the dehydration reactions and avoids side reactions compared to base catalysts. Even though conversion of sugars such as fructose to liquid fuels is investigated and different strategies were proposed, an efficient pathway for conversion of a variable combination of all the monomeric hexoses and pentoses to liquid fuels is yet to be developed.

Conversion of biomass into liquid fuels can also be processed via lighter alkane intermediates such as hexane via alkane metathesis reactions as shown in Fig. 31, in the presence of organometallic catalysts [51]. But, the selective production of lighter alkanes from oxygenated hydrocarbons such as glucose derivatives (for example sorbitol) includes enhancement in selectivity in C-O bond cleavage compared to C-C bond breaking. Generic reaction strategy proposed for light alkane production is shown in Fig. 26 involving series of dehydration and hydrogenation reactions. Sorbitol that is produced as in Fig. 31 is first dehydrated in acidic catalysts especially alumina based catalysts such as zeolites to form enolic species which involves conversion of alcoholic group to C=O or C=C group. This enolic group in the presence of metallic catalytic species such as Pt, at high H_2 pressures and temperature ~500 K, converts to hexane involving multiple repetitions of this process of individual OH group in sorbitol. An important complication in such technique is the bond cleavage selectivity of the operational conditions and the catalytic species towards C-O compared to C-C. This selectivity helps to avoid C-C cleavage to form gaseous products as in gasification. This recursively coupled reactions was carried out in aqueous phase with catalysts usually metallic active phase such as Pt, Pd, Ni, Ru, Co supported on acidic metal oxide support such as Al_2O_3, zeolite, silica-alumina. Studies conducted showed that relative selectivity towards C-O bond breaking over C-C bond was higher with Ni and Ru compared to Pt and Pd, although detailed investigations are necessary to understand further details [50].

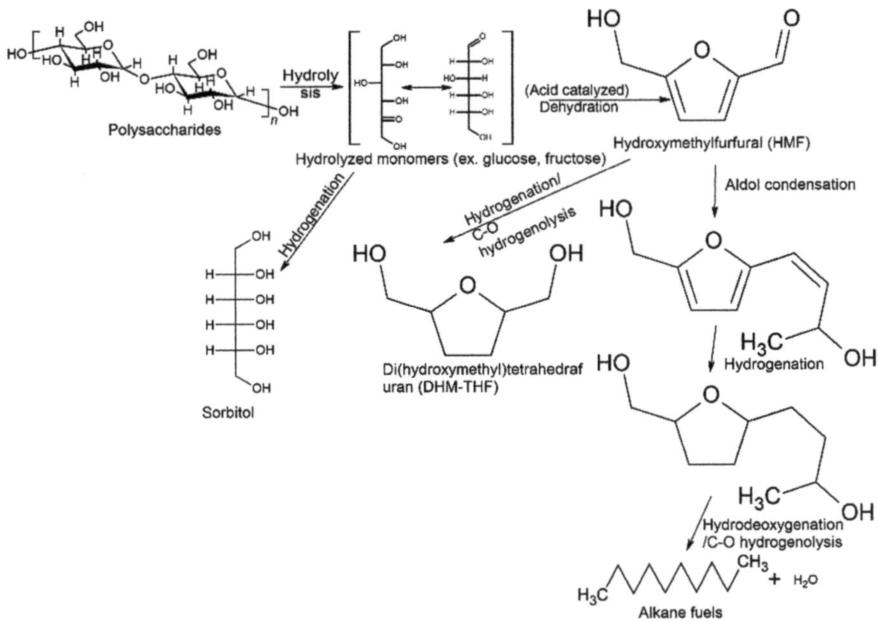

Fig. 30 Major modeled carbohydrate processing pathway for liquid alkane fuel production [50]

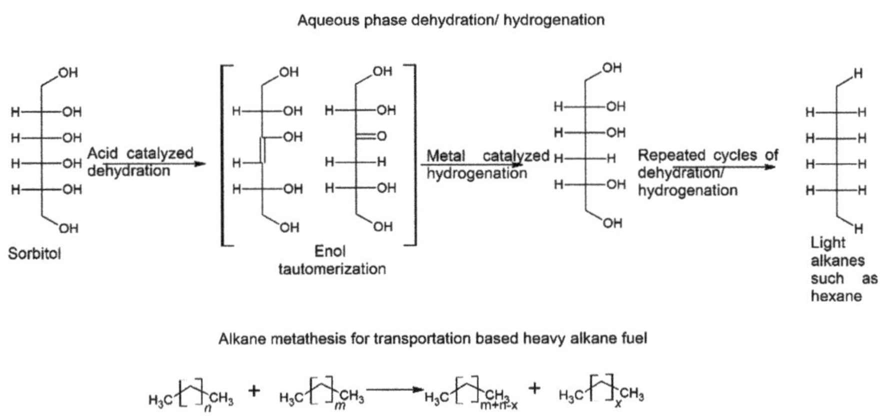

Fig. 31 Model hexose conversion for liquid fuel production via recursively coupled dehydration-hydrogenation reaction for light alkanes followed by alkane metathesis reaction [50, 51]

From the above reforming processes, we can observe the common importance on removal of oxygen by breaking C-O bond selectively without C-C bond breaking to avoid gaseous product formation. The oxygen content in bio-crude from hydrothermal liquefaction and bio-oil from fast pyrolysis needs to be removed to achieve high quality liquid hydrocarbons with low O/C and energy content which can then be catalytically upgraded with the current knowledge from petroleum industry that produces fuels to meet the demand in the existing infrastructure. Although existing understanding from petroleum industry in deoxygenation of these liquids is useful, it is observed not to be replicable to biofuels due to significantly higher oxygen content in bio-oil compared to petroleum crude. Hydro-treating of bio-crude in a similar technique as petroleum crude resulted in heavy residue and catalyst deactivation by coke. So, the subtle differences in catalyst design and operating conditions make a significant difference in deoxygenation of bio-crude compared to petroleum crude. In addition, oxygen content reduction need to be carried out, using for example, hydrodeoxygenation (HDO), a process designed on the similar lines as hydrodesulfurization (HDS) in crude oil. Oxygen removal using hydrodeoxygenation involves catalytic conversion of oxygen to water and with deoxygenated hydrocarbons in high pressure H_2, as seen above in the aqueous phase dehydration/hydrogenation pathway. Conventional catalysts imported from hydrodesulfurization are Co-Mo or Ni-Mo oxides or sulfides supported on alumina. Catalyst deactivation and investigations to improve selectivity and yield are still being pursued. Zeolite usage for cracking was also investigated to crack the bio-oil which can also reduce the oxygen content under certain conditions. Simplified investigation by deoxygenation of model compounds such as guaiacol, naphthol, tri-methyl phenol and others as phenolic representative is the current trend [44, 49, 52, 53]. Experimental investigations into these upgrading techniques involved both gas-phase and liquid phase conversion in batch and continuous modes and products analyzed using gas chromatography. In the pursuit of understanding the effect of catalyst lattice for optimal HDO, theoretical computations such as density functional theory (DFT) can also be found [54–60]. Catalytic studies advanced significantly over the recent decades in both experimentally which includes in situ analysis by FTIR, environmental SEM studies and other in situ analysis. Computational studies include the development of DFT for understanding the energetics of intermediates involved in heterogeneous catalysis. Research using the above techniques for understanding both the mechanisms of different components of bio-oil formation and upgrading is in progress and interferences between mixtures of multiple oxygenates still remains unknown even when the model compound analysis provided some insight into deoxygenation of single model compounds. Review of catalytic upgrading is given [9, 12, 42, 44, 53, 60–65].

Other applications of the products of hydrothermal liquefaction include the aqueous fraction of liquid yield, which can be steam reformed for syngas and hydrogen production. Aqueous components in bio-crude and bio-oil such as acetic acid were demonstrated to be effective in steam reforming to produce hydrogen. Sustainable pathway for value added chemical production can be established from HTL for resins, fertilizers, flavoring agents although it needs investigations to improve the selectivity of desired products. Sustained optimized production of upgraded alkane or DMF (dimethyl furan) based biofuels is essential for the goal of offsetting fossil fuels, especially in the transportation sector [9].

6 Conclusions

Strong global need exists to develop various techniques described here to meet the fuel and value added products demand at large industrial scales. The inherent advantage of discounting the need for major replacements in the energy demand infrastructure can be provided by thermochemical processed fuels as discussed and presented here. The thermochemical fuels produced along with high throughput make them highly scalable unlike biochemical techniques. Even though burning these will produce greenhouse gases, shorter sequestration time by biomass production and waste generation accommodates such emissions to makes it sustainable. It also provides self-sustainability in local or regional areas to support national economy, avoid unhealthy emissions that helps to mitigate foreign oil imports dependency and associated political influences. The need for pursuing thermochemical conversion of biomass and solid waste is essential for every nation on both local and global scales in our pursuit of clean and sustainable lifestyle. Fundamental and applied research on waste and biomass reforming to cleaner fuels can help assist our goal of clean and sustainable energy availability. The oxygen content in biofuels are higher than fossil fuels so that these fuels must be deoxygenated for use in current engines.

References

1. Population Reference Bureau. 2016 World Population Data Sheet. 2015 World Population Data Sheet 23 (2016), https://doi.org/10.2307/1972177
2. IEA. Key World Energy Statistics 2016. Statistics (Ber) 80 (2016), https://doi.org/10.1787/9789264039537-en
3. B. Metz, O. Davidson, H. de Coninck, M. Loos, L. Meyer, *IPCC Special Report on Carbon Dioxide Capture and Storage*, vol. 2 (Cambridge University Press, New York, NY, United States, 2011)
4. USEPA. Advancing sustainable materials management: 2014 fact sheet. United States Environmental Protection Agency, Office of Emergency Managemaent, Washington, DC 20460, 22 (2016)
5. Y. Sun, J. Cheng, Hydrolysis of lignocellulosic materials for ethanol production: a review (n.d)
6. M.-H. Cho, Y.-K. Choi, J.-S. Kim, Air gasification of PVC (polyvinyl chloride)-containing plastic waste in a two-stage gasifier using Ca-based additives and Ni-loaded activated carbon for the production of clean and hydrogen-rich producer gas. Energy **87**, 586–593 (2015). https://doi.org/10.1016/j.energy.2015.05.026
7. R.N. Walters, S.M. Hackett, R.E. Lyon, Heats of combustion of high temperature polymers. Fire Mater. **24**, 245–252 (2000). https://doi.org/10.1002/1099-1018(200009/10)24:5<245:AID-FAM744>3.0.CO;2-7
8. D. Mohan, C.U. Pittman, P.H. Steele, Pyrolysis of wood/ Biomass for bio-oil: a critical review. Energy Fuesl **20**, 848–889 (2006). https://doi.org/10.1021/ef0502397
9. K. Tekin, S. Karagöz, S. Bektaş, A review of hydrothermal biomass processing. Renew. Sustain. Energy Rev. **40**, 673–687 (2014). https://doi.org/10.1016/j.rser.2014.07.216
10. D.C. Elliott, P. Biller, A.B. Ross, A.J. Schmidt, S.B. Jones, Hydrothermal liquefaction of biomass: developments from batch to continuous process. Bioresour. Technol. **178**, 147–156 (2015). https://doi.org/10.1016/j.biortech.2014.09.132

11. D. Graff, S.C. Albers, A.M. Berklund, G.D. Graff, The rise and fall of innovation in biofuels (n.d)
12. A.V. Bridgwater, Review of fast pyrolysis of biomass and product upgrading. Biomass Bioenerg. **38**, 68–94 (2012). https://doi.org/10.1016/j.biombioe.2011.01.048
13. M. Balat, Energy sources, part A: recovery, utilization, and environmental effects mechanisms of thermochemical biomass conversion processes. Part 1: reactions of pyrolysis mechanisms of thermochemical biomass conversion processes. Part 1: reactions of pyrolysis (2017), https://doi.org/10.1080/15567030600817258
14. A.O. Akinola, Effect of temperature on product yield of pyrolysis of seven selected wood species in south west Nigeria **9359**, 176–181 (2016)
15. A.J. Toft, A comparison of integrated biomass to electricity systems. PhD thesis, Aston University, UK (1996)
16. I.I. Ahmed, A.K. Gupta, Kinetics of woodchips char gasification with steam and carbon dioxide. Appl. Energy **88**, 1613–1619 (2011). https://doi.org/10.1016/j.apenergy.2010.11.007
17. I.I. Ahmed, A.K. Gupta, Pyrolysis and gasification of food waste: syngas characteristics and char gasification kinetics. Appl. Energy **87**(1), 101–108 (2010)
18. M.S. Hussein, K.G. Burra, R.S. Amano, A.K. Gupta, Temperature and gasifying media effects on chicken manure pyrolysis and gasification. Fuel **202**, 36–45 (2017). https://doi.org/10.1016/j.fuel.2017.04.017
19. K.G. Burra, M.S. Hussein, R.S. Amano, A.K. Gupta, Syngas evolutionary behavior during chicken manure pyrolysis and air gasification. Appl. Energy **181** (2016). https://doi.org/10.1016/j.apenergy.2016.08.095
20. M.-K. Bahng, C. Mukarakate, D.J. Robichaud, M.R. Nimlos, Current technologies for analysis of biomass thermochemical processing: a review. Anal. Chim. Acta **651**, 117–138 (2009). https://doi.org/10.1016/j.aca.2009.08.016
21. I. Ahmed, A.K. Gupta, Syngas yield during pyrolysis and steam gasification of paper. Appl. Energy **86**, 1813–1821 (2009). https://doi.org/10.1016/j.apenergy.2009.01.025
22. I.I. Ahmed, A.K. Gupta, Kinetics of woodchips char gasification with steam and carbon dioxide. Appl. Energy **88**, 1613–1619 (2011). https://doi.org/10.1016/j.apenergy.2010.11.007
23. I. Ahmed, A.K. Gupta, Characteristics of cardboard and paper gasification with CO2. Appl. Energy **86**, 2626–2634 (2009). https://doi.org/10.1016/j.apenergy.2009.04.002
24. N. Nipattummakul, I. Ahmed, S. Kerdsuwan, A.K. Gupta, High temperature steam gasification of wastewater sludge. Appl. Energy **87**, 3729–3734 (2010). https://doi.org/10.1016/j.apenergy.2010.07.001
25. I.I. Ahmed, N. Nipattummakul, A.K. Gupta, Characteristics of syngas from co-gasification of polyethylene and woodchips. Appl. Energy **88**, 165–174 (2011). https://doi.org/10.1016/j.apenergy.2010.07.007
26. I.I. Ahmed, A.K. Gupta, Pyrolysis and gasification of food waste: syngas characteristics and char gasification kinetics. Appl. Energy **87**, 101–108 (2010). https://doi.org/10.1016/j.apenergy.2009.08.032
27. Z. Abu El-Rub, E.A. Bramer, G. Brem, Review of catalysts for tar elimination in biomass gasification processes. Ind. Eng. Chem. Res. **43**, 6911–6919 (2004). https://doi.org/10.1021/ie0498403
28. L. Devi, K.J. Ptasinski, F.J.J. Janssen, A review of the primary measures for tar elimination in biomass gasification processes. Biomass Bioenerg. **24**, 125–140 (2003). https://doi.org/10.1016/S0961-9534(02)00102-2
29. P.J. Woolcock, R.C. Brown, A review of cleaning technologies for biomass-derived syngas. Biomass Bioenerg. **52**, 54–84 (2013). https://doi.org/10.1016/j.biombioe.2013.02.036
30. D. Sutton, B. Kelleher, J.R.H. Ross, Review of literature on catalysts for biomass gasification. Fuel Process. Technol. **73**, 155–173 (2001). https://doi.org/10.1016/S0378-3820(01)00208-9
31. D.A. Bulushev, J.R.H. Ross, Catalysis for conversion of biomass to fuels via pyrolysis and gasification: a review. Catal. Today **171**, 1–13 (2011). https://doi.org/10.1016/j.cattod.2011.02.005

32. Palma C. Font, Modelling of tar formation and evolution for biomass gasification: a review. Appl. Energy **111**, 129–141 (2013). https://doi.org/10.1016/j.apenergy.2013.04.082
33. T.A. Milne, R.J. Evans, Biomass gasifier "tars": their nature, formation, and conversion. Golden, CO (1998). https://doi.org/10.2172/3726
34. J. Søren, U. Birk, Formation, decomposition and cracking of biomass tars in gasification. KgsLyngby Tech Univ Denmark Dep Mech Eng (2005)
35. P.L. Spath, D.C. Dayton, Preliminary screening—technical and economic assessment of synthesis gas to fuels and chemicals with emphasis on the potential for biomass-derived syngas. Natl. Renew. Energy Lab. 1–160 (2003). https://doi.org/10.2172/15006100
36. S.S. Ail, S. Dasappa, Biomass to liquid transportation fuel via Fischer Tropsch synthesis—technology review and current scenario. Renew. Sustain. Energy Rev. **58**, 267–286 (2016). https://doi.org/10.1016/j.rser.2015.12.143
37. S.S. Toor, L. Rosendahl, A. Rudolf, Hydrothermal liquefaction of biomass: a review of subcritical water technologies. Energy **36**, 2328–2342 (2011). https://doi.org/10.1016/j.energy.2011.03.013
38. J.A. Libra, K.S. Ro, C. Kammann, A. Funke, N.D. Berge, Y. Neubauer et al., Hydrothermal carbonization of biomass residuals: a comparative review of the chemistry, processes and applications of wet and dry pyrolysis. Biofuels **2**, 71–106 (2011). https://doi.org/10.4155/bfs.10.81
39. J. Akhtar, N. Aishah, S. Amin, A review on process conditions for optimum bio-oil yield in hydrothermal liquefaction of biomass. Renew. Sustain. Energy Rev. **15**, 1615–1624 (2010). https://doi.org/10.1016/j.rser.2010.11.054
40. M. Sugano, H. Takagi, K. Hirano, K. Mashimo, Hydrothermal liquefaction of plantation biomass with two kinds of wastewater from paper industry. J. Mater. Sci. **43**, 2476–2486 (2008). https://doi.org/10.1007/s10853-007-2106-8
41. A. Dimitriadis, S. Bezergianni, Hydrothermal liquefaction of various biomass and waste feedstocks for biocrude production: a state of the art review (2016). https://doi.org/10.1016/j.rser.2016.09.120
42. D.C. Elliott, Historical developments in hydroprocessing bio-oils (n.d.). https://doi.org/10.1021/ef070044u
43. L. Qiang, L. Wen-Zhi, Z. Xi-Feng, Overview of fuel properties of biomass fast pyrolysis oils. Energy Convers. Manag. **50**, 1376–1383 (2009). https://doi.org/10.1016/j.enconman.2009.01.001
44. M.M. Ahmad, M. Fitrir, R. Nordin, Azizan M. Tazli, Upgrading of bio-oil into high-value hydrocarbons via hydrodeoxygenation. Am. J. Appl. Sci. **7**, 746–755 (2010)
45. A.A. Peterson, F. Vogel, R.P. Lachance, M. Fröling, M.J. Antal, J.W. Tester, Thermochemical biofuel production in hydrothermal media: a review of sub-and supercritical water technologies (2008). https://doi.org/10.1039/b810100k
46. T.A. Wierzbicki, I.C. Lee, A.K. Gupta, Recent advances in catalytic oxidation and reformation of jet fuels. Appl. Energy **165**, 904–918 (2016). https://doi.org/10.1016/j.apenergy.2015.12.057
47. Q. Xia, Z. Chen, Y. Shao, X. Gong, H. Wang, X. Liu et al., Direct hydrodeoxygenation of raw woody biomass into liquid alkanes. Nat. Commun. **7**, 11162 (2016). https://doi.org/10.1038/ncomms11162
48. P.P. Peralta-Yahya, M. Ouellet, R. Chan, A. Mukhopadhyay, J.D. Keasling, T.S. Lee, Identification and microbial production of a terpene-based advanced biofuel. Nat. Commun. **2**, 483 (2011). https://doi.org/10.1038/ncomms1494
49. P.M. Mortensen, J.-D. Grunwaldt, P.A. Jensen, K.G. Knudsen, A.D. Jensen, A review of catalytic upgrading of bio-oil to engine fuels. Appl. Catal. A Gen. **407**, 1–19 (2011). https://doi.org/10.1016/j.apcata.2011.08.046
50. J.N. Chheda, G.W. Huber, J.A. Dumesic, Liquid-phase catalytic processing of biomass-derived oxygenated hydrocarbons to fuels and chemicals. Angew Chemie Int Ed **46**, 7164–7183 (2007). https://doi.org/10.1002/anie.200604274

51. A.S. Goldman, Catalytic alkane metathesis by tandem alkane dehydrogenation-olefin metathesis. Science **312**, 257–261 (2006). https://doi.org/10.1126/science.1123787
52. M.J. Roy, Hydrodeoxygenation of lignin model compounds via thermal catalytic reactions (2012)
53. Q. Bu, H. Lei, A.H. Zacher, L. Wang, S. Ren, J. Liang et al., A review of catalytic hydrodeoxygenation of lignin-derived phenols from biomass pyrolysis. Bioresour. Technol. **124**, 470–477 (2012). https://doi.org/10.1016/j.biortech.2012.08.089
54. Y. Yoon, R. Rousseau, R.S. Weber, D. Mei, J.A. Lercher, First-principles study of phenol hydrogenation on Pt and Ni catalysts in aqueous phase (n.d.). https://doi.org/10.1021/ja501592y
55. J.K. Nørskov, T. Bligaard, J. Rossmeisl, C.H. Christensen, Towards the computational design of solid catalysts. Nat. Chem. **1**, 37–46 (2009). https://doi.org/10.1038/nchem.121
56. C.A. Teles, R.C. Rabelo-Neto, J.R. de Lima, L.V. Mattos, D.E. Resasco, F.B. Noronha, The effect of metal type on hydrodeoxygenation of phenol over silica supported catalysts. Catal. Lett. (2016). https://doi.org/10.1007/s10562-016-1815-5
57. P.M. De Souza, R.C. Rabelo-Neto, L.E.P. Borges, G. Jacobs, B.H. Davis, D.E. Resasco et al., Hydrodeoxygenation of Phenol over Pd catalysts. Effect of support on reaction mechanism and catalyst deactivation (n.d.). https://doi.org/10.1021/acscatal.6b02022
58. D. Garcia-Pintos, J. Voss, A.D. Jensen, F. Studt, Hydrodeoxygenation of Phenol to Benzene and Cyclohexane on Rh(111) and Rh(211) surfaces: insights from density functional theory (n.d.). https://doi.org/10.1021/acs.jpcc.6b02970
59. A.M. Verma, N. Kishore, Molecular simulation DFT study on gas-phase hydrodeoxygenation of guaiacol by various reaction schemes DFT study on gas-phase hydrodeoxygenation of guaiacol by various reaction schemes. Mol. Simul. **43**, 141–153 (2017). https://doi.org/10.1080/08927022.2016.1239825
60. W. Yu, J.G. Chen, Reaction pathways of model compounds of biomass-derived oxygenates on Fe/Ni bimetallic surfaces (2015). https://doi.org/10.1016/j.susc.2015.01.009
61. B. Kunwar, H.N. Cheng, S.R. Chandrashekaran, B.K. Sharma, Plastics to fuel: a review. Renew. Sustain. Energy Rev. **54**, 421–428 (2016). https://doi.org/10.1016/j.rser.2015.10.015
62. H. Lee, H. Kim, M.J. Yu, C.H. Ko, J.-K. Jeon, J. Jae et al., Catalytic hydrodeoxygenation of bio-oil model compounds over Pt/HY catalyst. Sci. Rep. **6**, 28765 (2016). https://doi.org/10.1038/srep28765
63. E. Furimsky, Catalytic hydrodeoxygenation. Appl. Catal. A Gen. **199**, 147–190 (2000)
64. Q. Zhang, J. Chang, T. Wang, Y. Xu, Review of biomass pyrolysis oil properties and upgrading research. Energy Convers. Manag. **48**, 87–92 (2007). https://doi.org/10.1016/j.enconman.2006.05.010
65. Y. Zhu, M.J. Biddy, S.B. Jones, D.C. Elliott, A.J. Schmidt, Techno-economic analysis of liquid fuel production from woody biomass via hydrothermal liquefaction (HTL) and upgrading. Appl. Energy **129**, 384–394 (2014). https://doi.org/10.1016/j.apenergy.2014.03.053

General Concepts for Catalytic Synthesis of Biodiesel from Waste Cooking Oil

Yun Hin Taufiq-Yap and Nasar Mansir

Abstract The current global energy crisis is as a result of human population growth and technological advancements in developing countries. Beside the potential depletion of conventional fossil fuel reserves, fossil fuel is related with some environmental problems, i.e. global warming, high toxicity and non biodegradability, hence it is considered as non sustainable source of energy. Biofuels, particularly biodiesel is considered as one of the most promising and long-term energy source that can replace the fossil fuel. This is due to its biodegradability, non toxicity, safe and recyclable nature. Nevertheless, biodiesel production through the utilization of homogeneous catalytic system from food grade vegetable oil is no longer justifiable by industries in the near future, principally as a result of food versus fuel rivalry and other environmental issues associated to the conventional homogeneous catalytic system. Employing the waste cooking oil as feedstock to produce biodiesel using heterogeneous solid catalyst would help in making biodiesel fuel affordable and sustainable. This chapter review the general concepts of catalytic synthesis of biodiesel from waste cooking oil using heterogeneous catalysts and the problems associated with conventional catalysts usage for biodiesel production. The chapter also discussed the biodiesel quality assessment parameters.

Keywords Fossil fuel · Biodiesel · Waste cooking oil · Homogeneous catalyst Heterogeneous catalyst · Biodiesel properties

Y. H. Taufiq-Yap (✉) · N. Mansir
Faculty of Science, Catalysis Science and Technology Research Centre, Universiti Putra Malaysia, UPM Serdang, 43400 Seri Kembangan, Selangor, Malaysia
e-mail: taufiq@upm.edu.my

Y. H. Taufiq-Yap · N. Mansir
Faculty of Science, Department of Chemistry, Universiti Putra Malaysia, 43400 UPM Serdang, Seri Kembangan, Selangor, Malaysia

N. Mansir
Faculty of Science, Department of Chemistry, Federal University Dutse, Dutse 7156, Jigawa State, Nigeria

A. K. Runchal et al. (eds.), *Energy for Propulsion*, Green Energy and Technology,
https://doi.org/10.1007/978-981-10-7473-8_18

1 Introduction

Energy is generally considered as the backbone of modern scientific and technological advancements and therefore need to be harnessed in a sustainable way. Even though conventional fossil fuel will maintain to remain relevant as a major source of energy in the near future, but the oil reserves are imminently declining globally [1–5]. Besides, conventional fossil fuel is associated to some environmental and health problems such as Green house gases (GHGs), global warming, non biodegradability and poisonous exhaust from automobile [6]. The aforementioned problems triggered the search for renewable source of energy alternative to traditional fossil fuel. The recent discovery of biofuels would provide lasting solution to the problems of conventional fossil fuels. Biofuels are considered as alternative replacement to fossil fuels due to their less toxic emissions of carbon dioxide (CO_2), free sulphur content, biodegradability and feedstock availability [7, 8].

Biodiesel was found to be promising alternative considering its similarities in chemical properties with petroleum based diesel such as high flash point, pour point and improved cetane number [9, 10]. Study showed that Biodiesel demand is estimated to doubled or tripled by the year 2020 and beyond due to increase in population growth in developing countries and technological advancements [11, 12].

Commercialization of biodiesel production to outmatch the popular conventional fuel is the only way to solve the current and future energy challenges. Nevertheless, biodiesel production at industrial scale remains the major issue considering the nature of the feedstock and catalytic system used [1]. Employing low grade feedstock like palm fatty acid distillate (PFAD), *Jatropha curcas* oil and waste cooking oil for commercial production of biodiesel would help in reducing the high production cost of biodiesel due to its easy accessibility and low cost [13–15]. Nevertheless, the conventional catalytic system (homogeneous catalyst) could not work on these feedstock as they contain some impurities such as moisture and high FFA content [2, 16, 17]. Homogeneous liquid catalysts like Potassium hydroxide (KOH), Sodium hydroxide (NaOH), alkoxide and Tetraoxosulphate (VI) acid (H_2SO_4), can only be employed for biodiesel from high grade vegetable oils like palm oil, soybean oil and sunflower oil [18]. Nevertheless, the high cost of fresh vegetable oils, which account for more than 80% of total production cost of biodiesel, therefore, will make biodiesel production very expensive [11, 19, 20].

Separation and purification steps of biodiesel from the post reaction mixture would also make biodiesel very difficult as it requires considerable amount of energy and large amount of water during washing, hence polluting the environment by the generated waste water [21–23].

To overcome such issues, heterogeneous catalytic system is now considered as effective route for biodiesel production from low grade feedstock with least marked difficulty [24–26]. Moreover, heterogeneous catalysts can be recovered and reused several times before complete loss of activity, hence demonstrating more economical viability for commercial production of biodiesel [27, 28].

2 Feedstock for Biodiesel Production

Commercialization of biodiesel production is solely depends on the feedstock. The feedstock of biodiesel production can be categorised into three different categories. These include first generation, second generation and third generation feedstock.

2.1 The First Generation Feedstocks for Biodiesel

The first generation feedstocks are mainly consumable oils that are used as a source of food. This category contains mainly vegetable oils of high purity containing less than 1% free fatty acid. Example of these oils include palm oil, sunflower oil and soybean oil, were initially the most used feedstock for biodiesel synthesis [29]. The palm oil is commonly employed as feedstock for biodiesel production in Asia, particularly Malaysia and Indonesia. This is due to the abundant production of palm oil in the area. Malaysia is ranked as the second leading palm oil producer in the world [30]. Hence, in the advent of biodiesel, palm oil was used popularly as feedstock.

However, the continual usage of these feedstocks for biodiesel production will not be sustainable considering its high cost, and possible food/fuel competition which may pose great danger of hunger and starvation in the developing world. Moreover, the utilization of the available arable lands for the production of such feedstock may results to deforestation and damage to the ecosystem and consequently the ecological imbalance. Besides, the food grade oils are only fit to homogeneous liquid catalysts. However, these catalysts are difficult to separate after use and can pollute environment through water washing [2]. The aforementioned, are the major drawbacks for the utilization of edible oils for the commercialization of biodiesel [31–33].

2.2 Second Generation Feedstocks for Biodiesel

The second generation feedstocks are mainly the non-edible feedstocks [34]. These feedstocks are not suitable for human consumption as a result of toxic compositions present in the oils hence are considered as promising and sustainable [35]. Besides, the non-edible oils can successfully surmount the issue of potential food versus fuel competition. Moreover, non edible oil bearing crops can be cultivated on the unproductive/infertile land unfavourable for food crops cultivation. The non edible crops can grow along the roads, rail ways, follow lands and degraded forests [34]. Therefore, planting non edible crops can help in mitigating poverty in the rural areas at the same time contributing great impact for the sustainability of biodiesel production. The common examples of such feedstocks are *Jatropha curcas* oil, *Moringa oleifera* oil, *Mahua* oil, *Tobacco* oil, Karanja oil, Linseed oil etc. [29]. Waste cooking oil or used oil, PFAD, Yellow grease and animal fats are also considered as sustainable and promising second generation feedstocks for biodiesel production [36, 37].

2.3 *Microalgae Oil*

Microalgae oil is also non edible oil and considered as the third generation feedstock. Microalgae are the most promising, economically viable and inexhaustible feedstocks for the production of biodiesel when compared to second generation feedstock [38]. High oil content microalgae can potentially produce oil yield 25 times much higher when compared to that of conventional biodiesel bearing crops like palm oil, sunflower oil etc. Microalgae with an average oil production of at least 70% by weight of dry biomass needed only 0.1 m^2 year per kilogram biodiesel of land to produce 121,104 kg of biodiesel per year [39]. The above mentioned large production value of this plant made it to be recognized as excellent source of biodiesel feedstock [35]. Moreover, microalgae reproduce themselves and completing an entire life cycle within few days. Microalgae does not require fertilizer or even fresh water, as it can survive in salt or even contaminated sea or pond water. Herbicides and pesticides are not in the range of its requirement for cultivation. Beside the huge oil content, bio-compensation of CO_2 emissions is another added advantage of this plant.

Studies showed that, only 1 kg of dry algal biomass absorbs about 1.83 kg of CO_2 and therefore improving the air quality [39–44]. Nevertheless, despite the striking advantages of microalgae as a source of feedstock for biodiesel production, serious technological problems related to microalgae oil production should be addressed before they become credible contenders for biodiesel production in commercial scale [11].

3 Waste Cooking Oil as Viable and Sustainable Feedstock for Biodiesel Production

Generally, biofuels are meant to represent a potential source of renewable energy that could feasibly replace the conventional fossil fuels and provide eco-friendly fuel with less ecological problems (Global warming, Green house gas effects). Biodiesel is currently one of the most widely accepted alternative renewable sources of energy option to conventional fossil fuel. However, biodiesel production cost remains the major issue today. Current commercial production of biodiesel depends entirely on edible vegetable oil such as rapeseed (6.01 million ton), palm oil (6.34 million ton), Soybean oil (7.08 million ton) and some non-edible oil such as Castor oil and *Jatropha curcas* oil as major feedstocks worldwide [45, 46]. The use of conventional feedstocks (edible oils) for biodiesel accounts for more than 80% of total biodiesel

Table 1 Common fatty acid composition of plant based oil [55]

No.	Fatty acid	Formula	Structure	wt%
1.	Palmitic acid	$C_{16}H_{32}O_2$	16:0	7.3
2.	Stearic acid	$C_{18}H_{36}O_2$	18:0	4.0
3.	Oleic acid	$C_{18}H_{34}O_2$	18:1	26.9
4.	Linoleic acid	$C_{18}H_{32}O_2$	18:2	60.0
5.	Linolenic acid	$C_{18}H_{30}O_2$	18:3	0.5

production cost. Hence making biodiesel more expensive than the petroleum based diesel and therefore, not sustainable.

Moreover, continual usage of traditional edible oils as feedstock for biodiesel may result to future food versus fuel competition hence, violating one of the fundamental objective of biodiesel production [47]. *Jatropha curcas* oil is particularly remarkable feedstock for FAME production as it has about 40% oil content and can grow on infertile soil thereby, avoiding competition with arable land for food crops, however, *jatropha curcas* seeds are toxic in nature, hence the harvest requires severe labour [20]. Recent studies suggested that the production cost of biodiesel could be reduced to halve through the utilization of waste cooking oil in comparison to the high grade vegetable oils [20].

Currently, about 15 million tons of waste fried oil is annually been disposed into water bodies or land across the globe. Such large amount of low cost feedstock noteworthy special consideration, as it could bridge significant gap in current biodiesel demands [45]. Waste cooking oil when properly utilized can produce biodiesel with less CO_2 emission at relatively low cost and thereby making it economically and socially viable as a renewable fuel. Besides that, converting waste cooking oil into biodiesel could help significantly in reducing the environmental pollution caused by its disposal into water bodies and land [48].

Nevertheless, some chemical changes occur to waste cooking oil as a result of heating at high temperature during cooking/frying processes, which increases the level of FFA and some impurities like water content. Basically fresh vegetable oil contained triglyceride diglyceride, monoglyceride and FFA usually (<1%) [49]. High FFA and water contents in the feedstock hampers biodiesel production due to subsequent soap formation when basic catalyst is used [8, 50, 51]. These challenges must be taken into consideration before using it as a feedstock [52].

Different feedstocks for biodiesel synthesis have different fatty acid content and in turn can influence the biodiesel yield. Other fuel properties like acid value, cloud point, cetane number, pour point, flash point, oxidation stability, cold filter and plugging point could also be influence by FFA composition of the feedstock [53, 54]. Basically, the vegetable based oils comprised of five major fatty acids components as presented in Table 1, namely; Palmitic (16:0), Stearic (18:0), Oleic (18:1), Linoleic (18:2) and Linolenic (18:3).

Waste cooking oil has so far been regarded as one of the cheapest feedstock for biodiesel synthesis, with dual advantage [50]. Utilizing it as feedstock for biodiesel is at the same time making the environment cleaner for not being disposed into the water bodies that could pollute the water. Hence waste cooking oil is considered as economically viable and environmentally benign feedstock for biodiesel production on a commercial scale.

4 General Overview and Role of Catalyst in Biodiesel Synthesis

A catalyst is defined as a substance or material, that when added in the balanced chemical reaction speeds up the chemical reaction by lowering its activation energy to reach the chemical equilibrium between the reactants and products without changing the thermodynamic equilibrium of the system [56]. It is usually added in a minute amount in comparison to the quantity of the reactants. Under normal circumstance, the catalysts could not be consumed during the reaction and therefore unchanged after the reaction process.

However, in some cases the catalyst inhibits the reaction by being consumed and regenerated, while in other cases it seems not to include in the process and functions by high calibre of surface characteristics [57]. In biodiesel synthesis reactions, catalysts are classified into homogeneous and heterogeneous.

A homogeneous catalyst is having similar phase with the reactants while the heterogeneous catalyst system has different phase with the reaction medium.

Heterogeneous solid catalysts are now considered as the most feasible with regard to biodiesel production reaction process. This is due to their low cost, reusability and easy separation from the post reaction mixture and disposal after the reaction. It is very important to characterize these materials and classify their function carefully in terms of selectivity and reactivity in chemical reactions. Activity and selectivity of the catalysts are important characteristics that could be measured by performing the chemical reaction [58].

4.1 Catalysis

Catalysis can be defined as a technique representing the vital technology for accelerating the essential chemical conversion, which is a key to recognize environmentally friendly and commercially feasible reactions for transforming energy carriers to direct usable energy. Catalysis could also be defined as a technique used to perform chemical conversions in the wisest use of materials and energy leading to the acceleration of chemical reactions by the addition of a catalyst. Moreover, the use of heterogeneous catalysts for chemical conversions is not only decreasing the total energy input needed for the production processes, but also improves two considerable catalyst aspects, i.e. selectivity and thermal stability, thus leading to ecologically benign green technology.

4.2 *Industrial Application of a Catalyst*

The catalyst was first employed for industrial application by J. Roebuck in 1746 in the process of manufacturing lead chamber sulfuric acid. The applications of catalyst in large proportion have been adopted by the chemical industry since then. Initially, only pure components were used as catalyst, however, from the year 1900 to date multi component catalysts were studied and now employed in different chemical industries [59, 60].

Moreover, in the last 40 years, more than three hundred solid acid and base catalyst have been developed for various functions in the food and chemical industries. The structures and surface properties of the newly developed catalysts are being elucidated by newly developed methods and highly sophisticated techniques. The characterized solid base and acid catalysts are usually applied in different chemical reactions. The roles of acid and base properties are being measured for catalytic activities during a chemical reaction.

The heterogeneous acid and base catalysts are now considered largely as ecologically and economically viable in the field of catalysis. This is due to their advantages over homogeneous acid and base catalysts. The advantages include easy separation after use, less corrosive, ecologically benign, less disposal problem and reusability for more than one reaction cycle. Additionally, solid catalysts could be designed to possessed high activity, selectivity and stability. Hence, the emergence of heterogeneous solid acid and base catalysts is considered as a breakthrough in chemical industries, as it has solved most of the problems associated to homogeneous liquid catalysts.

Industrially, the catalyst plays a fundamental role in the improvement of sustainable reactions, which have primarily allowed the present and future universal production and exploitation of energy and chemicals, at the same time mitigating the harmful consequences to the surroundings and has the significance to key technology. Presently, larger industrial operations employed the use of modern catalytic system for the production of energy and chemicals. These operations include transesterification/esterification reaction for biodiesel, catalytic cracking of biofuels and fossil fuels, synthesis of ammonia and methyl alcohol. Other industrial application of catalyst for energy and chemical synthesis are; alkylation, isomerisation, dehydration and condensation, amination, cracking, etherification aromatization, hydration, hydrocracking, oligomerization and Polymerization [61]. Catalyst application at industrial scale, which account for about 85–90% of the output of the chemical manufactured through catalytic routes generated about $900 billion in products worldwide [62].

4.3 Homogeneous Catalysts for Biodiesel Production and Their Problems

Production of biodiesel in commercial scale from vegetable oil relies solely on homogeneous catalyst system considering their major advantage of reduced effect of mass transfer resistance, hence faster reaction efficiency due to high acid and basic sites distribution when compared to the heterogeneous catalysts. The homogeneous catalysts used in the synthesis of biodiesel could either be acidic or basic in nature. Basic homogeneous catalysts are the most frequently used catalysts to support the biodiesel production process at industrial scale. This is due to the faster reaction efficiency of base-catalysed transesterification reaction process when compared to acid catalysts, which generally react at slower rate; consequently, producing high biodiesel yield at mild reaction conditions [63]. However, the FFA and water content of the feedstock are the fundamental consideration on choosing the type of catalyst to be used in a particular biodiesel production reaction.

The frequently employed homogeneous catalysts for biodiesel synthesis are hydroxides and alkoxides of alkali metals such as potassium hydroxide (KOH), sodium hydroxide (NaOH), sodium methoxide (CH_3ONa) and potassium methoxide (CH_3OK) [64, 65]. In all of the listed homogeneous catalysts, KOH and NaOH are the most frequently employed catalysts for biodiesel production at commercial scale mainly due to their capacity to assist transesterification reaction at relatively mild reaction conditions and high reaction rate, which moves 4000 times faster when compared to acid catalyst reaction rate [66]. Excellent biodiesel yield is believed to be achieved at short reaction time, and most importantly, these catalysts are cheap and easy to prepare [63]. Several studies have been conducted on different feedstocks to produce biodiesel using homogeneous catalysts. Uzun et al. [65] reported the biodiesel yield of 96% at a reaction temperature of 50 °C, and 30 min reaction time when NaOH catalyst was used on soybean oil using methanol as reaction medium. Similar studies was conducted by Fadhil and Ali [67], where KOH catalyst was used to transesterify Heckel fish oil, and 97% of biodiesel yield was recorded at 32 °C reaction temperature, 60 min reaction time, 6:1 methanol to oil ratio and 5.5 wt% catalyst loading. Nevertheless, formation of soap and other unwanted emulsions as a result of FFA content (>1%) and other impurities from low grade feedstock is another problem that may hinder commercialization of biodiesel [68].

The prominent homogeneous acid catalysts for methanolysis are mineral acids such as Hydrochloric acid (HCl) and Sulphuric acid (H_2SO_4) [69]. Others are nitric acid (HNO_3), aluminium chloride ($AlCl_3$), organic sulfonic acid and phosphoric acid (H_3PO_4) [63]. Homogeneous acid catalysts generally have a slow rate of reaction and therefore, take long time during reaction process to esterify feedstock to biodiesel. Marchetti and Errazu [70] reported the biodiesel yield of 96% from model oil prepared using oleic acid and sunflower oil when H_2SO_4 was used as a catalyst at 55 °C reaction temperature, 6.126:1 methanol to oil molar ratio, 2.26 wt% catalyst loading and 240 min was the time taken to satisfactorily completes the reaction process. In a related study, Soriano et al. [23] reported that 98% yield of biodiesel was achieved when canola oil was subjected to methanolysis using $AlCl_3$ as a catalyst under the reaction temperature of 110 °C, catalyst loading of 5 wt%, methanol to oil molar ratio of 1:1 and 1080 min as reaction time. However, homogeneous catalysts made biodiesel production difficult and expensive due to the peculiar problems associated to these catalysts. The major drawbacks to homogeneous catalysts that severely affect the biodiesel production at industrial scale include high cost of production as these catalysts fit only high purity feedstock such as palm oil, sunflower oil, soybean oil, etc.

Besides, separation of catalysts after the reaction process involves regular water washing that requires a substantial amount of water and energy, which will make biodiesel production difficult and expensive. Consequently, the waste water generated during the washing of the catalyst would pollute the environment and subsequently distort the ecosystem. Hence, homogeneous catalysts are not environmentally and economically benign for the sustainable biodiesel production at industrial scale.

4.4 Heterogeneous Catalysts for Biodiesel Synthesis

Heterogeneous catalysts are mainly solid catalysts that are in different phase with the reacting medium. Solid catalysts can easily be recovered and reused more than once in a chemical reaction. These catalysts could be acidic, basic or bifunctional.

4.4.1 Solid Acid Catalysts for Biodiesel Synthesis

Recently, heterogeneous solid acid catalysts are receiving greater attention from scientific and industrial point of view for biodiesel production. This is due to their several advantages over their homogeneous counterparts. Heterogeneous solid acid catalysts when properly utilized are expected to yield outstanding grade biodiesel from vegetable based oils. Moreover, unlike homogeneous acid catalysts, heterogeneous acid catalysts are not associated with corrosion of biodiesel production plant/equipment [71]. Solid acid catalysts promote esterification process of oleic acid or PFAD to synthesize methyl ester and water as the major products under low temperature [72].

More importantly, heterogeneous acid catalysts could simultaneously accelerate both transesterification of triglycerides (TGs) and esterification reactions of FFA. The two reactions; transesterification and esterification are catalysed by Bronsted acid species and Lewis acid species respectively [63, 73, 74]. It is generally assumed that catalytic biodiesel synthesis reaction process occur in the same way for both homogeneous and heterogeneous acid catalysts except that, the heterogeneous acid catalysed reaction is influenced by the catalysts surface, which depends greatly on the interconnected system of large pores, surface acid sites and hydrophobicity [63]. Heterogeneous acid catalysts can esterify waste oil like palm fatty acid distillate which has over 85% FFA content and other low grade feedstocks like waste cooking oil to biodiesel [8].

Nevertheless, heterogeneous acid catalysts can only perform transesterification reaction under severe reaction conditions such as high reaction temperature, low reaction rate and long reaction period [2]. These drawbacks may probably make biodiesel production difficult and expensive. Several studies were reported on biodiesel production from low grade feedstocks using heterogeneous solid acid catalysts under severe reaction conditions (Table 2). Hence, the new route of catalytic system that can perform biodiesel production process on low grade feedstock under moderate reaction conditions is needed.

4.4.2 Solid Base Catalyst for Biodiesel Synthesis

Solid base catalysts constitute of alkaline metals and alkaline earth metals which made the group one and two of the periodic table. These catalysts have been employed as catalysts for biodiesel production process. Heterogeneous base catalysts could be an alkaline earth metal in a single form, alkali metal in a supportive form, other mixed oxides of alkaline earth metal or hydrotalcites and anionic ion-exchange resins.

Alkaline earth metal oxides catalysts if properly modified could have the capacity to assist transesterification reaction of low grade feedstock such as waste fried oil in a single reaction stage route, due to the possession of excellent basic potency [84]. The basicity potency of group II metal oxides and other hydroxides is advanced with the increase in the atomic number ranging from magnesium (Mg) in period II to Barium (Ba) in period VI, which considerably affects their catalytic potential for transesterification of triglyceride [85].

The source of basic centres in alkaline earth metal oxides might be linked to the existence of $M^{2+}$and O^{2-} ion pair as well as surface hydroxyl group in a different coordination environment. The surface oxygen of these oxides may possibly be affected by the surface composition of the solid catalysts and calcination temperature. Moreover, the specific surface area, level of leaching of the active site of the catalyst in the reaction intermediate, basicity and selectivity during the biodiesel reaction process are deemed to be considered as important features in choosing the suitable catalyst for biodiesel production [63].

The most commonly used basic heterogeneous catalyst for transesterification process is calcium oxide (CaO) [86]. This is due to the possession of higher basicity, low solubility, easy to handle and most importantly availability of CaO at relatively low cost. Additionally, CaO derives its popularity for catalysis as a result of its abundance in different natural resources in the form of calcium carbonate ($CaCO_3$). Apart from lime stone which is natural but non-renewable source of $CaCO_3$, other renewable natural sources are egg shells [87, 88], crab shells [87], capiz shells [89], snail shells [90], mussel shells [91] and oyster shells [92] among others. Proper employment of these shells for the synthesis of CaO will considerably reduce the over dependency on non-renewable lime stone as a source of CaO. Waste egg shell is believed to have a high content of $CaCO_3$ having a composition of about 97.1%. This shell when thermally treated at 900 °C for 2 h would result in the synthesis of CaO catalyst [63].

Table 2 Literature reports of solid acid catalyzed biodiesel production at high reaction conditions

Catalyst	Preparation method	Reaction conditions				Feedstocks	Yield (%)	Refs.
		Temperature (°C)	Reaction time (h)	Catalyst loading (wt%)	Methanol to oil molar ratio			
$Al_2O_3/TiO_2/ZnO$	Co-mixing	200	6	8	1:1	Colza oil	93.7	[5]
Ar-SBA-15	Not reported	260	2.5	1	20:1	Crude palm oil	90	[75]
$Al_2O_3/ZrO_2/WO_3$	Co-precipitation	175	1	4 g	40:1	Soybean	100	[76]
$Al_2O_3/TiO_2/ZnO$	Sol-gel	200	8	5	Not specified	Colza oil	93.7	[77]
$Al_2O_3/ZrO_2/WO_3$	Wet impregnation	200	20	4 g	4.5:1	Soybean oil	>90	[78]
Ar-SBA-15	Direct mixing	140	6	2	20:1	Crude palm oil	90	[75]
CsHPW	Precipitation	200	10	3	20:1	Oleic acid-Soybean mixture	90.4	[79]
$Cs_{2.5}H_{0.5}PW_{12}O_{40}$	Precipitation	260	1	3	40:1	Sesame oil	92	[80]
Carbonized and Sulfated asphalt	Carbonization	260	1	2.5	18.2:1	Waste cooking oil	89.9	[1]
$Mn_{3.5x}Zr_{0.5y}Al_xO_3$	Co-precipitation	150	5	2.5	14:1	WCO	>93	[81]
SO_4^{2-}/TiO_2-SiO_2	Impregnation	200	3	4	9:1	Used cooking oil	90	[6]
$TiO2/SO_4^{2-}$	Precipitation	230	2	8	12:1	Coconut oil Cotton oil	86.3 90	[75]
VOPO4.2H20	Not reported	150	1	2	1:1	Used cooking oil	80%	[82]
ZrO_2/SO_4^{2-}	Sulfonation	200	1	1	6:1	Palm kernel	90.3	[83]

Besides the cost-effective benefits, the capacity of CaO as a solid catalyst for biodiesel synthesis can match the conventional homogeneous catalysts having achieved outstanding conversion yield over some selected feedstocks. Kawashima et al. [93] reported 90% biodiesel yield when transesterification reaction was performed on rapeseed oil using CaO as catalyst and methanol as reaction medium, other reaction conditions used are 60 °C as reaction temperature, 0.1 g catalyst weight and 3 h reaction time. Similarly, Lee et al. [94] reported biodiesel yield of 86.75% using palm oil and obtuse horn derived CaO as feedstock and catalyst under the optimum reaction parameters of 5 wt% amount of catalyst, 6 h reaction time and 12:1 methanol to oil ratio. CaO material irrespective of its source, is unstable, hence, prone to poisoning effects when exposed to atmospheric medium i.e. carbon dioxide (CO_2) and moisture (H_2O), which would results in the formation of calcium carbonate ($CaCO_3$) and calcium hydroxide $Ca(OH)_2$ layers, respectively on the catalytic surface [95]. These effects distort and block the catalytic active sites hence lowering the activity of the catalysts and biodiesel yield [63].

Despite the encouraging results obtained using these catalysts, CaO catalyst alone could not transesterify impure oil such as waste fried oil and *Jatropha oil*. This is due to the existence of high FFA (>3%) composition and other impurities such as water in a single run process. In order to transesterify low grade feedstock like waste cooking oil using CaO catalyst, two stage process must be involved to reduce the FFA content to (<1%) [96]. These processes are considered as time and resource consuming, making biodiesel production difficult and expensive. Hence, for the CaO catalyst to work on low grade feedstock like waste cooking oil and Jatropha oil containing high FFA (>3%), there is need for further modification that will provide some acid centres so that to esterify the high FFA present in the feedstock for efficient and sustainable biodiesel production. The modification of basic heterogeneous catalysts like CaO for high activity on low grade oil could be doping with other metal oxides from transition group. Some transition metal oxides like MoO_3, WO_3, TiO_2 or ZrO_3 possessed some acidic properties that would enhance the catalytic activity during transesterification of low grade feedstock like waste cooking oil.

4.4.3 Bifunctional Solid Catalysts

Generally, separation and waste water disposal problems, corrosion of biodiesel refining equipments associated to homogeneous catalysts, increases concerned over the adoption of heterogeneous catalysts for biodiesel synthesis [2, 27]. Utilization of solid catalysts for biodiesel synthesis can overcome the separation problems. Solid base catalysts in which basic component comes from waste material like egg shell is considered as environmentally benign due to its biodegradable property hence can greatly reduce the environmental problems. Over the last few years, there were greater concerns over the heterogeneous catalysts whereby intensive studies were conducted for transesterification reactions over these catalysts, [49, 94, 97–99].

The best heterogeneous catalysts should have strong basic sites, large surface area, large pores, less toxicity, high stability, less corrosive effect, ease separation, high reusability, less leaching effect and low cost of preparation [8].

Mixed oxides catalysts are originated from metal oxides particularly alkali earth metal oxides, lanthanide metal oxides or other transition metal oxides. Mixed oxides catalysts could be basic, acidic or bi-functional (having both acidic and basic properties). Bi-functional heterogeneous catalyst if utilized properly could perform transesterification and esterification of TGs and FFA simultaneously under moderate reaction conditions, respectively. This is because; the bi-functional catalysts are having both basic and acidic sites and therefore can perform simultaneously esterification and transesterification reaction during biodiesel synthesis. Additionally, bi-functional catalysts can be modified to bring in the needed physicochemical properties that could easily overcome the problem of high FFA and moisture content present in the impure feedstock like waste cooking oil during the transesterification reaction [25]. Hence, bifunctional heterogeneous catalysts could become excellent option catalysts to overcome the associated problems with other catalysts. Over the recent years, several researches were undertaken using modified mixed oxide catalysts with bifunctional activity to produce biodiesel from high FFA feedstock under mild reaction conditions as presented in Table 3.

5 Biodiesel Synthesis and the Reaction Mechanisms

Fatty acid methyl ester (FAME) or conventionally biodiesel is generally synthesized by the transesterification reaction of triglycerides (TGs) or esterification of FFA. During transesterification reaction, 1 mol of triglycerides react with 3 mol of methanol, catalyst and temperature over a period of time to generate 3 mol of methyl ester (biodiesel) and glycerol as presented in Fig. 1. Chan and Fang [102, 103] reported the formation of transesterification reaction in the 3 step processes as depicted in Fig. 2. Transesterification reaction is reversible reaction process and therefore needs high amount of methanol in order to move the reaction forward [63]. The excess methanol and glycerol produced together with FAME at the end of the reaction could be separated using centrifuge [1]. Catalytic conversion of triglycerides using solid base catalyst is much faster than esterification reaction using solid acid catalyst. This is due to the slower catalytic activity of solid acid catalyst resulting from lesser level of catalyst contact with reaction mixture [102].

In esterification reaction, 1 mol of FFA reacts with 1 mol of methanol in the presence of an acid catalyst to form 1 mol of FAME and water as shown in Fig. 3. Chen and Fang [102] reported the lengthy reaction rate of the esterification reaction process as a result of lesser catalytic activity when compared to base catalysed transesterification reaction.

Table 3 Literature report on biodiesel production from low grade feedstock using bifunctional catalysts

Catalyst	Preparation method	Reaction conditions				Feedstocks	Yield (%)	Refs.
		Temperature (°C)	Reaction time (h)	Catalyst loading (wt%)	Methanol to oil molar ratio			
Bi_2O_3–La_2O_3	Impregnation	150	4	2	15:1	Jatropha curas oil	93	[7]
CaO/La_2O_3	Co-precipitation	65	6	4	24:1	Jatropha curas oil	86.51	[100]
CaO-ZrO_2	Impregnation	65	2	10	30:1	Waste cooking oil	92.1	[82]
Cs-Zr/Al_2O_3	Impregnation	65	3	3	20:1	Waste cooking oil	90	[101]
Mo-Mn/γ-Al_2O_3-15wt% MgO	Wet impregnation method	100	4	4 g	27:1	Waste cooking oil	91.4	[25]
SO_4^{2-}/SnO_2-SiO_2	Impregnation	150	3	4	15:1	Waste cooking oil	>92.3	[21]

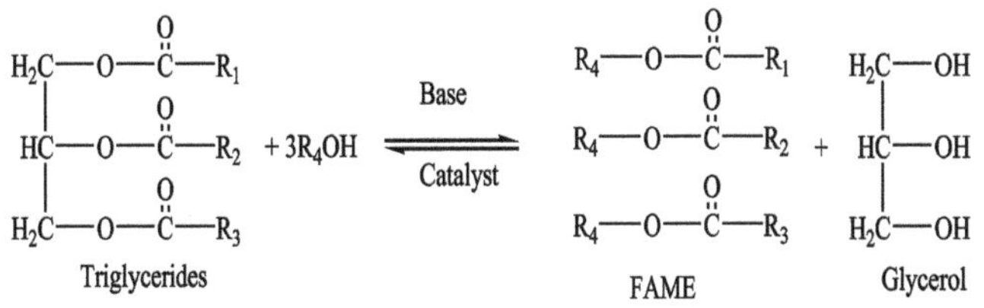

Fig. 1 General equation for transesterification reaction of biodiesel synthesis from triglycerides [63]

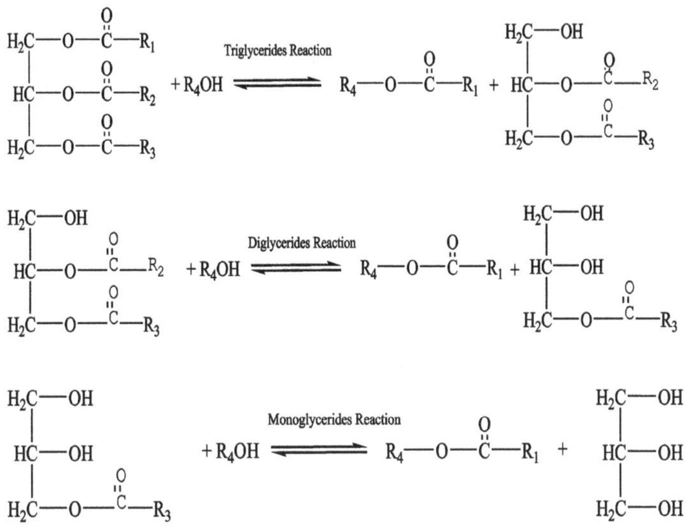

Fig. 2 Three step equations for transesterification reaction process of biodiesel production from triglycerides [63]

Fig. 3 Esterification reaction of biodiesel production from fatty acid

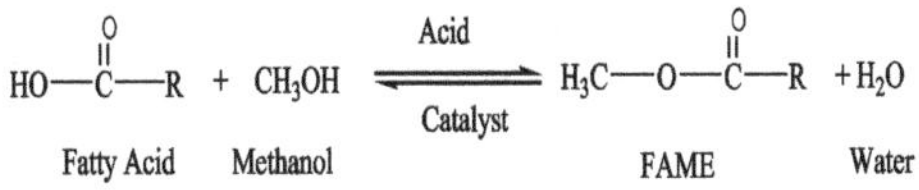

5.1 Mechanisms for the Transesterification Reaction Using Base Catalyst

The mechanisms for the transesterification and esterification reaction processes of heterogeneous catalysts and homogeneous catalysts are similar. The transesterification reaction of triglycerides (TG) to FAME and glycerol using base catalysts involves the formation of methoxide ion by dissociation of metal hydroxide (KOH) or metal oxide (CaO) into the methanol and encapsulated into TG- Methoxide micro-emulsion, then followed by nucleophillic attack of CH_3O^- on the carbonyl group on TG, which leads to the formation of tetrahedral intermediates. The second step is the

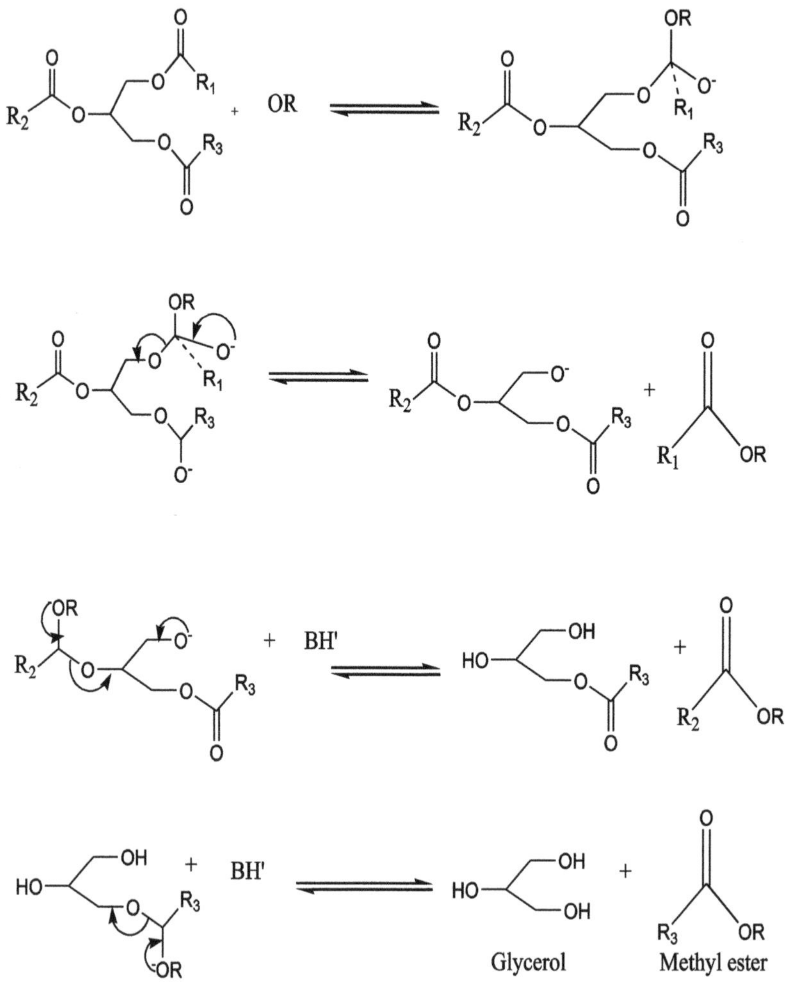

R_1, R_2, R_3 : Carbon chain of the fatty acids. R: alkyl group of methanol; OR: CH_3O^-; B: KOH: BH^+: $[K(OK)H]^+$

Fig. 4 Mechanism of transesterification for FAME production

intermediate break down to form methyl ester and diglycerides. The CH_3O^- active species are regenerated in the third step. The steps are repeated twice to produce monoglycerides, methyl ester and glycerol methyl ester, respectively as shown in Fig. 4.

5.2 Mechanism of Esterification Reactions Using Acid Catalyst

The mechanism for the esterification reaction of fatty acid using acid catalysts is depicted using ethanoic acid as a feedstock and sulfuric acid as catalyst in five steps reaction as presented in Fig. 5. In order to make the reaction mechanism simple and less confusing, all the mechanisms steps are shown in one way reaction.

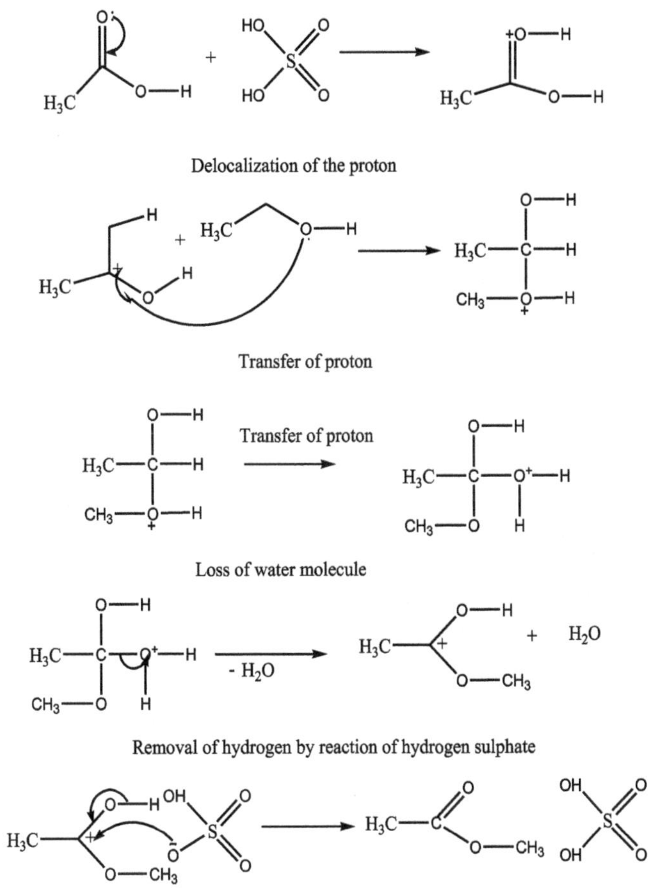

Fig. 5 Mechanism of esterification reaction for FAME production

6 Influence of Reaction Conditions for Biodiesel Synthesis

FAME is synthesized by either transesterification reaction process of triglycerides (major component of vegetable oil) or esterification reaction of lipid using basic or acid catalyst under certain reaction conditions. Biodiesel yield is mainly depended on the catalytic performance of the catalyst employed for the reaction. Moreover, the catalytic performance for the transesterification of triglycerides to FAME is influenced by five major factors, namely; temperature, amount of catalyst used, methanol to oil molar ratio, temperature of the reaction and time.

6.1 Methanol to Oil Molar Ratio

Methanol to oil molar ratio is one of the four main reaction conditions for the biodiesel reaction synthesis. This parameter plays an important role both in transesterification and esterification reaction process. During esterification of high FFA oil to biodiesel, which is also reversible, excess methanol is necessary in order to boost the reaction rate and biodiesel yield [4]. Conversions of triglycerides (TGs) to FAME occur when the reaction occur between the TGs molecules and the catalyst active site. At a molecular stage, TGs undergoes protonation at the carbonyl group to form the carbonation ion, which can also undergo a reaction to results in the formation of methyl esters [8]. Nevertheless, the catalyst active sites are submerged with excess methanol, which hampers the protonated TGs. Yan Li et al. (2010) reported that 90% of biodiesel yield was attained at the maximum of 18:1 methanol to oil ratio. No major increase in biodiesel yield was observed from 15:1 to 18:1 methanol to oil ratio hence 15:1 was adopted in view of economics and methanol recovery after the reaction [77]. The high amount of methanol during methanolysis was reported to be problematic during the separation process. This is due to the formation of emulsion from methyl ester and glycerol layers formed. The polar nature of the methanol made it to behave as emulsifier, which improves the formation of emulsion [4]. However, it is generally concluded that excess methanol increases the production yield and also enhances the interaction between TGs molecules and catalyst [14, 104].

6.2 Catalyst Loading

Catalysts are predominantly used during the catalytic conversion of triglycerides to biodiesel through the lowering of the activation energy and therefore, make the reaction faster. During the methanolysis of TGs or lipids, the amount of catalyst used in the reaction determines the overall yield of the produced biodiesel. The amount of catalyst used for a particular reaction is optimized to find the best loading for better biodiesel yield. Low amount of catalyst is employed in early reaction stages and the FAME yield is measured. The amount of catalyst would be increased in the subsequent reactions keeping other reaction parameters constant until the best FAME yield is recorded. Increase in amount of catalyst to 3 wt% during the transesterification reaction at 3 h reaction time was reported to enhance the overall FAME yield recording 90%. Subsequent increase of the loading to 4 wt% resulted to decrease in FAME yield to 83% [101]. Solubility of methanol was reported to be observed in high FFA content and other impure feedstocks like waste fried oil when low catalyst loading is used during methanolysis. Using high dosage of catalyst beyond the optimum loading during biodiesel synthesis results in declining of FAME yield through product absorption [105].

6.3 Reaction of Temperature

The transesterification reaction of triglycerides to FAME involves the use of reaction temperature, which provides the needed activation energy for the reaction to occur. During the transesterification reaction for biodiesel synthesis from vegetable based oil, increase in the reaction temperature helps the reaction to reach thermodynamic equilibrium and subsequently the rate of reaction, which facilitate the conversion of triglycerides to methyl ester. Moreover, insufficient reaction temperature during the transesterification reaction leads to incomplete conversion of triglycerides to biodiesel. Some studies reported that temperatures beyond optimal reaction temperatures lead to the formation of emulsions and therefore, diminishes the overall FAME yield conversion [7].

6.4 Reaction Time

In the catalytic conversion of triglycerides to biodiesel, time is considered as one of the fundamental parameters necessary for the reaction to occur, depending on the activity of the catalyst. Effect of reaction time is usually examined under the time range of 1–5 h, keeping other reaction conditions constant. Increase in reaction time enhances the conversion to triglycerides to FAME. Amani et al. [101] reported the considerable increase in FAME yield when reaction time was prolonged. According to their studies, the methyl ester yield was increased from 74% at 1 h to 90% at 3 h and 20:1 methanol to oil molar ratio. This phenomenon is attributed to slow mass transfer exhibited by heterogeneous catalyst reaction. Transesterification is reversible reaction and prolonging it beyond optimal reaction time may lead to moving the direction of the equilibrium backward and hence decrease the yield conversion.

Table 4 Diesel and biodiesel standard properties values in accordance with American Society for Testing and Materials (ASTM) [106]

Property	Diesel	Biodiesel
Standard	ASTMD975	ASTMD6751
Kinematic viscosity (mm^2/s)	1.9–4.1	1.9–6.0
Fuel composition	Hydrocarbon (C10–C21)	FAME (C12–C22)
Cetane number	40–55	48–60
Pour point (°C)	−35 to −15	−15 to 16
Cloud point (°C)	−15 to 5	−15 to 16
Flash point (°C)	60–80	100–170
Acid value (%)		
Water content (vol. %)	0.05	0.05
Density (g/cm^3)	0.85	0.88
Carbon (wt%)	87	77
Hydrogen content (wt%)	13	12
Oxygen content (wt%)	0	11

7 Biodiesel Quality Evaluation

The synthesized biodiesel can only be ascertained when it fulfils the standard fuel properties according to the accepted biodiesel fuel standard of American Society for Testing Materials (ASTM D751) and European standard (EN 14214). Table 4 presented the comparative summary of standard fuel properties between biodiesel and petroleum based diesel.

7.1 Kinematic Viscosity

Kinematic viscosity is one of the biodiesel fuel properties that are evaluated in the laboratory and ensure it falls within the accepted limit of biodiesel property standard before used in an internal combustion engines. Lower viscosity defined excellent kinematic viscosity of any synthesized biodiesel as it provides prevention against engine deposition. The kinematic viscosity of biodiesel is evaluated at a temperature of 40 °C based on ASTM and EN international standards. The accepted range of kinematic viscosity of biodiesel is 1.9–6.0 $mm^2\ s^{-1}$ as specified by ASTM D6751 and EN14214 biodiesel standards.

7.2 *Pour Point*

Pour point is considered to be one of the key cold flow properties of biodiesel fuel. It is defined as the lowest temperature at which the fluid/liquid stops flowing at lower temperature. Pour point is evaluated in the laboratory using cloud and pour point apparatus. The biodiesel sample is subjected to heating at a temperature of 45 °C in water bath and left to cool until the biodiesel sample reach down to ambient temperature (25 °C). Thereafter, the sample is introduced into the cloud and pour point equipment set at 0 °C. The analysis is performed by the subsequent decrease of the apparatus temperature to sub zero degrees and periodic check of biodiesel flow until the sample is no longer flows. The temperature at which the sample stops flow is then recorded as the pour point of the biodiesel sample. The standard pour point of biodiesel is specified to be in the range of −15 to 9 according to ASTM D6751.

7.3 *Flash Point*

Flammability is fundamental feature of any fuel particularly biodiesel. The flammability of biodiesel is referred as flash point (FP), which is define as the temperature required for a fuel to vaporises and sufficiently ignite and forms a flammable mixture with oxygen in controlled condition. The FP of biodiesel is evaluated based on ASTMD6751 and EN14214 standards, which is specified at the range of 100–170 °C.

7.4 *Other Fuel Properties*

Besides pour point, flash point and viscosity, acid value (mg/KOH), moisture content and (wt%) density (kg/m^3) are also fundamental in evaluating fuel properties of the synthesized biodiesel. The standard values of these properties must be in accordance to ASTM D6751 and EN14214 biodiesel standard specifications as depicted in Table 4.

8 Setbacks and Future Stance for Biodiesel Production

Taking into consideration of the current global energy challenges, there is need to urgently diversify into other possible sources of energy. The over dependent fossil fuel is no longer sustainable due to its un-renewable nature and environmental problems such as global warming. Renewable sources of energy like biofuels are considered as the best option to conventional fuel due to its biodegradability, less toxic and similar performance to fossil fuel.

Commercialization of FAME production to meet the global energy demand is the only solution to the current energy challenges. However, current FAME production is not sustainable considering the high production cost when high grade oil is employed as a feedstock and the conventional catalytic system, which is associated with separation and waste water generation problems. These problems could only be solved through the utilization of heterogeneous catalysts and low cost feedstock such as jatropha curas oil, PFAD and waste fried oil.

The conventional feedstock for FAME production is high grade oil with low FFA (<1%) content which can easily be converted to biodiesel at low reaction conditions. However, about 80% of total production cost of FAME will go to feedstock when high grade oil is used as a feedstock. Using Waste cooking oil as feedstock for FAME production will reduce the production cost to halve when compared to high grade vegetable oil as feedstock.

The heterogeneous solid catalyst could serve as the best option to homogeneous catalyst for the commercial FAME production considering the easy separation of the catalyst after the reaction process; more environmentally benign and more economical as it could be used more than once before it become inactive.

Acknowledgements The Authors would like to sincerely appreciate the University Putra Malaysia (UPM), Malaysia for the funding through the research group project GP-IPD/2016/9490400.

References

1. Q. Shu, Z. Nawaz, J. Gao, Y. Liao, Q. Zhang, D. Wang, J. Wang, Synthesis of biodiesel from a model waste oil feedstock using a carbon-based solid acid catalyst: reaction and separation. Bioresour. Technol. **101**, 5374–5384 (2010)
2. N. Mansir, Y.H. Taufiq-Yap, U. Rashid, I.M. Lokman, Investigation of heterogeneous solid acid catalyst performance on low grade feedstocks for biodiesel production: a review. Energy Convers. Manag. **141**, 171–182 (2017)
3. L. Bournay, D. Casanave, B. Delfort, G. Hillion, J.A. Chodorge, New heterogeneous process for biodiesel production: a way to improve the quality and the value of the crude glycerin produced by biodiesel plants. Catal. Today **106**, 190–192 (2010)
4. D.Y.C. Leung, Y. Guo, Transesterification of neat and used frying oil: optimization for biodiesel production. Fuel Process. Technol. **87**, 883–890 (2006)
5. B. Delfort, D. Le Pennec, C. Lendresse, Process for transesterification of vegetable oils or animal oils by means of heterogeneous catalysts based on zinc or bismuth, titanium and aluminum. United State patent 7,151,187 B2, 2006
6. B.-X. Peng, Q. Shu, J.-F. Wang, G.-R. Wang, D.-Z. Wang, M.-H. Han, Biodiesel production from waste oil feedstocks by solid acid catalysis. Process Saf. Environ. Prot. **86**, 441–447 (2008)
7. M.F. Rabiah Nizah, Y.H. Taufiq-Yap, U. Rashid, S.H. Teo, Z.A. Shajaratun Nur, A. Islam, Production of biodiesel from non-edible *Jatropha curcas* oil via transesterification using Bi2O3-La2O3 catalyst. Energy Convers. Manag. **88**, 1257–1262 (2014)
8. I.M. Lokman, U. Rashid, R. Yunus, Y.H. Taufiq-, Carbohydrate-derived solid acid catalysts for biodiesel production from low-cost feedstocks: a review. Catal. Rev. Sci. Eng. **56**, 187–219 (2014)

9. H.T. Siow, G. Motonobu, Y.H. Taufiq-yap, Biodiesel production from *Jatropha curcas* L. oil with Ca and La mixed oxide catalyst in near supercritical methanol conditions. J. Supercrit. Fluids **104**, 243–250 (2015)
10. A. Islam, Y.H. Taufiq-yap, E. Chan, M. Moniruzzaman, S. Islam, Advances in solid-catalytic and non-catalytic technologies for biodiesel production. Energy Convers. Manag. **88**, 1200–1218 (2014)
11. A. Galadima, O. Muraza, Biodiesel production from algae by using heterogeneous catalysts: a critical review. Energy **78**, 0–11 (2014)
12. F. Murphy, G. Devlin, R. Deverell, K. Mcdonnell, Potential to increase indigenous biodiesel production to help meet 2020 targets—an EU perspective with a focus on Ireland. Renew. Sustain. Energy Rev. **35**, 154–170 (2014)
13. J.H. Cho, H.S. Kim, S.W. Hong, Y. Yeo, A single step non-catalytic esterification of palm fatty acid distillate (PFAD) for biodiesel production. Fuel **93**, 373–380 (2012)
14. D. Yujaroen, M. Goto, M. Sasaki, A. Shotipruk, Esterification of palm fatty reaction, acid distillate (PFAD) in supercritical methanol: effect of hydrolysis on reactivity. Fuel **88**, 2011–2016 (2009)
15. S. Jain, M.P. Sharma, Kinetics of acid base catalyzed transesterification of *Jatropha curcas* oil. Bioresour. Technol. **101**, 7701–7706 (2010)
16. S. Yan, S.O. Salley, K.Y.S. Ng, Simultaneous transesterification and esterification of unrefined or waste oils over ZnO-La2O3 catalysts. Appl. Catal. A Gen. **353**, 203–212 (2009)
17. K. Jacobson, R. Gopinath, L.C. Meher, A.K. Dalai, Solid acid catalyzed biodiesel production from waste cooking oil. Appl. Catal. B Environ. **85**, 86–91 (2008)
18. Z. Wen, X. Yu, S. Tu, J. Yan, E. Dahlquist, Synthesis of biodiesel from vegetable oil with methanol catalyzed by Li-doped magnesium oxide catalysts. Appl. Energy **87**, 743–748 (2010)
19. M.M.K. Bhuiya, M.G. Rasul, M.M.K. Khan, N. Ashwath, A.K. Azad, M.A. Hazrat, Second generation biodiesel: potential alternative to-edible oil-derived biodiesel. Energy Procedia. **61**, 1969–1972 (2014)
20. J.C. Escobar, E.S. Lora, O.J. Venturini, E.E. Yáñez, E.F. Castillo, O. Almazan, Biofuels: environment, technology and food security. Renew. Sustain. Energy Rev. **13**, 1275–1287 (2009)
21. M.K. Lam, K.T. Lee, A.R. Mohamed, Sulfated tin oxide as solid superacid catalyst for transesterification of waste cooking oil: an optimization study. Appl. Catal. B Environ. **93**, 134–139 (2009)
22. M. Di Serio, M. Ledda, M. Cozzolino, G. Minutillo, R. Tesser, Transesterification of soybean oil to biodiesel by using heterogeneous basic catalysts. Ind. Eng. Chem. Res. **45**, 3009–3014 (2006)
23. N.U. Soriano, R. Venditti, D.S. Argyropoulos, Biodiesel synthesis via homogeneous Lewis acid-catalyzed transesterification. Fuel **88**, 560–565 (2010)
24. M.K. Jothiramalingam, R. Wang, Review of recent developments in solid acid, base, and enzyme catalysts (heterogeneous) for biodiesel production via transesterification. Ind. Eng. Chem. Res. **48**, 6162–6172 (2009)
25. M. Farooq, A. Ramli, D. Subbarao, Biodiesel production from waste cooking oil using bifunctional heterogeneous solid catalysts. J. Clean. Prod. **59**, 131–140 (2013)
26. A.K. Endalew, Y. Kiros, R. Zanzi, Heterogeneous catalysis for biodiesel production from *Jatropha curcas* oil (JCO). Energy **36**, 2693–2700 (2011)
27. F.H. Alhassan, R. Yunus, U. Rashid, K. Sirat, A. Islam, H.V. Lee, H. Yup, Production of biodiesel from mixed waste vegetable oils using Ferric hydrogen sulphate as an effective reusable heterogeneous solid acid catalyst. Appl. Catal. A Gen. **456** 182–187 (2013)
28. S. Sirisomboonchai, M. Abuduwayiti, G. Guan, C. Samart, S. Abliz, X. Hao, K. Kusakabe, A. Abudula, Biodiesel production from waste cooking oil using calcined scallop shell as catalyst. Energy Convers. Manag. **95**, 242–247 (2015)
29. A.E. Atabani, A.S. Silitonga, H.C. Ong, T.M.I. Mahlia, H.H. Masjuki, I.A. Badruddin, H. Fayaz, Non-edible vegetable oils: a critical evaluation of oil extraction, fatty acid compositions, biodiesel production, characteristics, engine performance and emissions production. Renew. Sustain. Energy Rev. **18**, 211–245 (2013)

30. T. Mielke, Conference & exhibition price outlook (POC), price outlook palm lauric oils impacts from global vegetable oil mark. A fundamental approach, Palm Lauric Oils, Kuala Lumpur, Malaysia, 2010
31. Y.C. Sharma, B. Singh, S.N. Upadhyay, Advancements in development and characterization of biodiesel: a review. Fuel **87**, 2355–2373 (2008)
32. T. Thamsiiriroj, J.D. Murphy, Is it better to import palm oil from Thailand to produce biodiesel in Ireland than to produce biodiesel from the indigenious iris rape seed. Appl. Energy **86**, 595–604 (2009)
33. G. Hammond, S. Kallu, M. McManus, Development of biofuels for the UK automotive market. Appl. Energy **85**, 506–516 (2008)
34. I.B. Banković-Ilić, O.S. Stamenković, V.B. Veljković, Biodiesel production from non-edible plant oils. Renew. Sustain. Energy Rev. **16**, 3621–3647 (2012)
35. A.L. Ahmad, N. Yasin, C. Derek, J. Lim, Microalgae as a sustainable energy source for biodiesel production: a review. Renew. Sustain. Energy Rev. **15**, 584–593 (2011)
36. A. Demirbas, Progress and recent trends in biodiesel fuels. Energy Convers. Manag. **50**, 14–34 (2009)
37. S.P. Singh, D. Singh, Biodiesel production through the use of different sources and characterization of oils and their esters as the substitute of biodiesel: a review. Renew. Sustain. Energy Rev. **14**, 200–216 (2010)
38. T.M. Mata, A.A. Martins, N.S. Caetano, Microalgae for biodiesel production and other applications: a review. Renew. Sustain. Energy Rev. **14**, 217–232 (2010)
39. Z.S. Cabanelas, F.A. Chinalia, C.O. Souza, J.A. Perales, P.F. Almeida, J.I. Druzian, I.A. Nascimento, From waste to energy: microalgae production in waste water and glycerol. Appl. Energy **109**, 283–290 (2013)
40. R.A. Lee, J.M. Lavoie, From first to third generation biofuels: challenges of producing a commodity from a biomass of increasing complexity. Anim. Front. **3**, 6–11 (2013)
41. R. Dillascneider, C. Steinweg, R. Rosello-Sastre, C. Posten, Biofuels from microalgae: photoconversion efficiency during lipid accumulation. Bioresour. Technol. **26**, 647–654 (2013)
42. Y. Chisti, Biodiesel from microalgae beats bioethanol. Trends Biotechnol. **26**, 126–131 (2008)
43. M.K. Ji, R.A.I. Abou-Shanab, S.H. Kim, E.S. Salama, S.H. Lee, A.N. Kabra, Y.S. Lee, S. Hong, B.H. Jeon, Cultivation of microalgae specie in tertiary municipal waste water supplemented with CO2 for nutrient removal and biomass production. Ecol. Eng. **58**, 142–148 (2013)
44. I. Rawat, V. Bhola, R.R. Kumar, F. Bux, Improving the feasibility of producing biofuels from microalgae using waste water. Environ. Technol. **34**, 1765–1775 (2013)
45. A.F. Lee, J.A. Bennett, J.C. Manayil, K. Wilson, Heterogeneous catalysis for sustainable biodiesel production via esterification and transesterification. Chem. Soc. Rev. **43**, 7887–7916 (2014)
46. A.F. Lee, J.A. Bennett, J.C. Manayil, K. Wilson, Heterogeneous catalysis for sustainable biodiesel production via esterification and transesterification. R. Soc. Chem. 1–30 (2014)
47. R. Sarin, M. Sharma, S. Sinharay, R.K. Malhotra, Jatropha —palm biodiesel blends: an optimum mix for Asia. Fuel **86**, 1365–1371 (2007)
48. C. Daniel, M. De Araújo, C.C. De Andrade, E. De Souza, F.A. Dupas, Biodiesel production from used cooking oil: a review. Renew. Sustain. Energy Rev. **27**, 445–452 (2013)
49. M. Verziua, S.M. Comana, R. Richards, V.I. Parvulescu, Transesterification of vegetable oils over CaO catalysts. Catal. Today. **167**, 64–70 (2011)
50. T. Tuong, V. Tran, S. Kaiprommarat, S. Kongparakul, P. Reubroycharoen, G. Guan, M. Huan, C. Samart, Green biodiesel production from waste cooking oil using an environmentally benign acid catalyst. Waste Manag. **52**, 367–374 (2016)
51. I.M. Atadashi, M.K. Aroua, A.R. AbdulAziz, N.M.N. Sulaiman, The effects of water on biodiesel production and refining technologies: a review. Renew. Sustain. Energy Rev. **16**, 3456–3470 (2012)
52. U. Rashid, M. Ibrahim, S. Yasin, R. Yunus, Y.H. Taufiq-Yap, G. Knothe, Biodiesel from Citrus reticulate (Mandarin orange) seed oil, a potential non-food feedstock. Ind. Crop. Prod. **45**, 355–359 (2013)

53. F. Motasemi, F.N. Ani, A review on microwave-assisted production of biodiesel. Renew. Sustain. Energy Rev. **16**, 4719–4733 (2012)
54. K. Nakajima, M. Hara, Amorphous carbon with SO_3H groups as a solid brønsted acid catalyst. Am. Chem. Soc. Catal. **2**, 1296–1304 (2012)
55. T. Muppaneni, H.K. Reddy, S. Ponnusamy, P.D. Patil, Y.Q. Sun, P. Dailey, S.G. Deng, Optimization of biodiesel production from palm oil under supercritical ethanol conditions using hexane as co-solvent: a response surface methodology approach. Fuel **107**, 633–640 (2013)
56. A. Peter, D.P. Julio, *Physical Chemistry*, 9th edn (W.H Freeman and Company, New York, 2010)
57. G. Ertl, H. Knözinger, J. Weitkamp, *Preparation of Solid Catalysts* (Wiley, 2008)
58. J.R. Anderson, K.C. Pratt, *Introduction to Characterisation and Testing of Catalysts* (Academic Press, 1985)
59. I. Leach, E. Bruce, Industrial catalysis: chemistry applied to your life-style and environment. *Applied Industrial Catalysis* (Academic press, New York, 1983)
60. G. Jacobs, B.H. Davis, Low temperature water-gas shift catalysts. Catalysis R. Soc. Chem. **20** (2007)
61. K. Anabe, W.F. Ho, Industrial application of solid acid and base catalysts. Appl. Catal. A Gen. **181**, 399–434 (1999)
62. A.J.B. Robertson, *Catalysis of Gas Reactions by Metals*, Landon, Lagos, London (1970)
63. M.R. Avhad, J.M. Marchetti, Innovation in solid heterogeneous catalysis for the generation of economically viable and ecofriendly biodiesel: a review. Catal. Rev. **4940**, 1–52 (2016)
64. J.M. Dias, M.C.M. Alvim-Ferraz, M.F. Almeida, Comparison of the performance of different homogeneous alkali catalysts during transesterification of waste and virgin oils and evaluation of biodiesel quality. Fuel **87**, 3572–3578 (2008)
65. B.B. Uzun, M. Kılıç, N. Özbay, A.E. Pütün, E. Pütün, Biodiesel production from waste frying oils: optimization of reaction parameters and determination of fuel properties. Energy **44**, 347–351 (2012)
66. G. Baskar, R. Aiswarya, Trends in catalytic production of biodiesel from various feedstocks. Renew. Sustain. Energy Rev. **57**, 496–504 (2016)
67. A.B. Fadhil, L.H. Ali, Alkaline-catalyzed transesterification of *Silurus triostegus* Heckel fish oil: optimization of transesterification parameters. Renew. Energy. **60**, 481–488 (2013)
68. O. Nur Syazwani, U. Rashid, Y.H. Taufiq Yap, Low-cost solid catalyst derived from waste Cyrtopleura costata (Angel Wing Shell) for biodiesel production using microalgae oil. Energy Convers. Manag. **101**, 749–756 (2015)
69. U. Rashid, F. Anwar, Production of biodiesel through optimized alkaline catalyzed oil, transesterification of rapeseed oil. Fuel **87**, 265–273 (2008)
70. J.M. Marchetti, A.F. Errazu, Esterification of free fatty acids using sulfuric acid as catalyst in the presence of triglycerides. Biomass Bioenergy **32**, 892–895 (2008)
71. A.M. Dehkordi, M. Ghasemi, Transesterification of waste cooking oil to biodiesel using Ca and Zr mixed oxides as heterogeneous base catalysts. Fuel Process. Technol. **97**, 45–51 (2012)
72. I.M. Lokman, M. Goto, U. Rashid, Y.H. Taufiq-yap, Sub- and supercritical esterification of palm fatty acid distillate with carbohydrate-derived solid acid catalyst. Chem. Eng. J. **284**, 872–878 (2016)
73. Y. Muhammad, W. Mohd, A. Wan, A.R.A. Aziz, Activity of solid acid catalysts for biodiesel production: a critical review. Appl. Catal. A Gen. **470**, 140–161 (2014)
74. D. Rattanaphra, A. Harvey, Simultaneous conversion of triglyceride/free fatty acid mixtures into biodiesel using sulfated zirconia. Top. Catal. **53**, 773–782 (2010)
75. L. Pesaresi, D.R. Brown, A.F. Lee, J.M. Montero, H. Williams, K. Wilson, Cs-doped H4SiW12O40 catalysts for biodiesel applications. Appl. Catal. A Gen. **360**, 50–58 (2009)
76. S. Furuta, H. Matsuhashi, K. Arata, Biodiesel fuel production with solid amorphous-zirconia catalysis in fixed bed reactor. Biomass Bioenerg. **30**, 870–873 (2006)
77. Y. Li, X. Zhang, L. Sun, J. Zhang, H. Xu, Fatty acid methyl ester synthesis catalyzed by solid superacid catalyst. Appl. Energy **87**, 156–159 (2010)

78. S. Furuta, M. Hiromi, A. Kazushi, Biodiesel fuel production with solid superacid catalysis in fixed bed reactor under atmospheric pressure. Catal. Commun. **5**, 721–723 (2004)
79. R. Sheikh, M. Choi, J. Im, Y. Park, Study on the solid acid catalysts in biodiesel production from high acid value oil. J. Ind. Eng. Chem. **19**, 1413–1419 (2013)
80. H. Shin, S. An, R. Sheikh, Y.H. Park, S. Bae, Transesterification of used vegetable oils with a Cs-doped heteropolyacid catalyst in supercritical methanol. Fuel **96**, 572–578 (2012)
81. H. Amani, Z. Ahmad, M. Asif, B.H. Hameed, Transesterification of waste cooking palm oil by MnZr with supported alumina as a potential heterogeneous catalyst. J. Ind. Eng. Chem. **20**, 4437–4442 (2014)
82. S. Ardizzone, C.L. Bianchi, G. Cappelletti, F. Porta, Liquid-phase catalytic activity of sulfated zirconia from sol–gel precursors: the role of the surface features. J. Catal. **227**, 470–478 (2004)
83. J. Jitputti, B. Kitiyanan, P. Rangsunvigit, K. Bunyakiat, L. Attanatho, P. Jenvanitpanjakul, Transesterification of crude palm kernel oil and crude coconut oil by different solid catalysts. Chem. Eng. J. **116**, 61–66 (2006)
84. Z. Kesic, I. Lukic, M. Zdujic, H. Liu, D. Skala, Mechanochemically synthesized CaO ZnO catalyst for biodiesel production. Procedia Eng. **42**, 1169–1178 (2012)
85. B.M. Weckhuysen, G. Mestl, M.P. Rosynek, T.R. Krawietz, J.F. Haw, J.H. Lunsford, Destructive adsorption of carbon tetrachloride on alkaline earth metal oxides. Phys. Chem. B. **102**, 3773–3778 (1998)
86. M.E. Borges, L. Díaz, Recent developments on heterogeneous catalysts for biodiesel production by oil esterification and transesterification reactions: a review. Renew. Sustain. Energy Rev. **16**, 2839–2849 (2012)
87. L.M. Correia, R.M.A. Saboya, J.A. de Sousa Campelo, N. Cecilia, C.L. Rodríguez-Castellón, E. Cavalcante Jr, R.S. Vieira, Characterization of calcium oxide catalysts from natural sources and their application in the transesterification of sun-flower oil. Bioresour. Technol. **151**, 207–213 (2014)
88. A. Navajas, T. Issariyakul, G. Arzamendi, L.M. Gandía, A.K. Dalai, Development of eggshell derived catalyst for transesterification of used cooking oil for biodiesel production. Asia Pacific J. Chem. Eng. **8**, 742–748 (2013)
89. W. Suryaputra, I. Winata, N. Indraswati, S. Ismadji, Waste capiz (Amusium crista-tum) shell as a new heterogeneous catalyst for biodiesel production. Renew. Energy **50**, 795–799 (2013)
90. A. Birla, B. Singh, S.N. Upadhyay, Y.C. Sharma, Kinetics studies of synthesis of biodiesel from waste frying oil using a heterogeneous catalyst derived from snail shell. Bioresour. Technol. **106**, 95–100 (2012)
91. R. Rezaei, M. Mohadesi, G.R. Moradi, Optimization of biodiesel production using waste mussel shell catalyst. Fuel **109**, 534–541 (2013)
92. N. Nakatani, H. Takamori, K. Takeda, H. Sakugawa, Transesterification of soybean oil using combusted oyster shell waste as a catalyst. Bioresour. Technol. **100**, 1510–1513 (2009)
93. A. Kawashima, K. Matsubara, K. Honda, Acceleration of catalytic activity of calcium oxide for biodiesel production. Bioresour. Technol. **100**, 696–700 (2009)
94. S. Lee, Y. Wong, Y. Tan, S. Yew, Transesterification of palm oil to biodiesel by using waste obtuse horn shell-derived CaO. Energy Convers. Manag. **93**, 282–288 (2015)
95. D.M. Marinkovi, M.V. Stankovi, A.V. Veli, J.M. Avramovi, M.R. Miladinovi, O.O. Stamenkovi, V.B. Veljkovi, M. Jovanovi, Calcium oxide as a promising heterogeneous catalyst for biodiesel production: current state and perspectives. Renew. Sustain. Energy Rev. **56**, 1387–1408 (2016)
96. Y. Hua, M. Omar, C. Nolasco-hipolito, Y.H. Taufiq-yap, Waste ostrich- and chicken-eggshells as heterogeneous base catalyst for biodiesel production from used cooking oil: catalyst characterization and biodiesel yield performance. Appl. Energy **160**, 58–70 (2015)
97. H.V. Lee, J.C. Juan, Y.H. Tau, Preparation and application of binary acid–base CaO–La2O3 catalyst for biodiesel production. Renew. Energy **74**, 124–132 (2015)
98. S.E. Mahesh, A. Ramanathan, K.M.M.S. Begum, A. Narayanan, Biodiesel production from waste cooking oil using KBr impregnated CaO as catalyst. Energy Convers. Manag. **91**, 442–450 (2015)

99. Y.C. Wong, Y.P. Tan, I. Ramli, H.S. Tee, Biodiesel production via transesterification of palm oil by using CaO–CeO2 mixed oxide catalysts. Fuel **162**, 288–293 (2015)
100. Y.H. Taufiq-yap, S. Hwa, U. Rashid, A. Islam, M. Zobir, Transesterification of *Jatropha curcas* crude oil to biodiesel on calcium lanthanum mixed oxide catalyst: effect of stoichiometric composition. Energy Convers. Manag. **88**, 1290–1296 (2014)
101. H. Amani, Z. Ahmad, B.H. Hameed, Highly active alumina-supported Cs–Zr mixed oxide catalysts for low-temperature transesterification of waste cooking oil. Appl. Catal. A Gen. **487**, 16–25 (2014)
102. G. Chen, B. Fang, Preparation of solid acid catalyst from glucose-starch mixture for biodiesel production. Bioresour. Technol. **102**, 2635–2640 (2011)
103. B.V. Vlada, M.J. Avramovi, S.O. Stamenkovi, Biodiesel production by ultrasound-assisted transesterification: state of the art and the perspectives. Renew. Sustain. Energy Rev. **16**, 1193–1209 (2012)
104. P. Nakpong, S. Wootthikanokkhan, High free fatty acid coconut oil as a potential energy, feedstock for biodiesel production in Thailand. Renew. Energy **35**, 1682–1687 (2010)
105. F. Ezebor, M. Khairuddean, A. Zuhairi, P. Lim, Oil palm trunk and sugarcane bagasse derived solid acid catalysts for rapid esterification of fatty acids and moisture-assisted transesterification of oils under pseudo-infinite methanol. Bioresour. Technol. **157**, 254–262 (2014)
106. E. Lotero, Y. Liu, D.E. Lopez, K. Suwannakarn, D.A. Bruce, J.G. Goodwin, Synthesis of biodiesel via acid catalysis. Ind. Eng. Chem. Res. **44**, 5353–5363 (2005)

Biofuels in Air and Ground Transportation

Hema Ramsurn and Ram B. Gupta

Abstract Use of biofuels in aviation and ground transportation is increasing. The advantages of biofuels include their broad availability, carbon neutrality, environmental friendliness with potential economic and social benefits for the local communities. Bioethanol and biodiesel are by far the most prominent because of their ease of substitution of the conventional fossil fuels. Due to stringent quality standards in the aviation industry, rigorous testing of the biofuels is crucial. A number of feedstocks and technological routes are being explored for biofuel production including hydrotreated vegetable oils, Fischer-Tropsch fuels, synthesized iso-paraffinic fuels and alcohol-to-jet fuels. Some commercial flights have started using biofuels. Even though bioethanol and biodiesel are being blended with conventional fuel, there are still technological and economical challenges that prevent this fuel type to completely replace fossil fuels. The concept of bio-refinery which will utilize all parts of the biomass and transform all the co-products to value-added chemicals will give the biofuel industry a more competitive economic edge. Biofuels are promising alternatives and their use will continue to increase in the near future.

1 Introduction

According to the Intergovernmental Panel on Climate Change (IPCC) [1], oil requirements for transportation will steadily increase by 1.3% each year till year 2030. In order to cater for this growing demand, several alternatives have been contemplated including energy efficient vehicle designs, enhancement of public transportation, and use of alternative (non-fossil) fuels and energy technologies. Furthermore, the combustion of fossil fuels has raised environmental concerns as the transportation sector alone accounts for 25% of the global CO_2 emissions. The use of biofuels will reduce

H. Ramsurn (✉)
The Russell School of Chemical Engineering, University of Tulsa, Tulsa, OK 74104, USA
e-mail: hema-ramsurn@utulsa.edu

R. B. Gupta
School of Engineering, Virginia Commonwealth University, Richmond, VA 23284, USA

A. K. Runchal et al. (eds.), *Energy for Propulsion*, Green Energy and Technology,
https://doi.org/10.1007/978-981-10-7473-8_19

greenhouse gas (GHG) emissions and dependency on imports which is a national security issue. Biofuels can be of added benefit to the developing countries where marginal lands can be used to cultivate biofuel crops boosting the rural economy [2]. Depending on the use, biofuels will however need to match the physico-chemical properties of the fossil fuel being replaced. In other words, biofuels should be able to blend with the existing petroleum-based fuels, combust in existing internal combustion engines, and be transported and distributed through the existing infrastructure. In this article, the use of biofuels for aviation and land transportation is discussed along with future prospects of complementing and/or replacing fossil fuels.

2 Biofuels in Aviation

According to Air Transport Action Group (ATAG) [3], 3.6 billion passengers travelled by air in year 2015 with 34.8 million scheduled commercial flights worldwide using 247 million tons of jet fuel emitting 781 million tons of CO_2, which is 2% of the global CO_2 emissions from all sources. Aviation is one of the fastest growing economic sectors, leading to unavoidable increase in fuel demand and environmental impact especially due to CO_2 and NO_x emissions. The modern aircrafts are already highly fuel-efficient and it is predicted that advances in air traffic management and engine efficiency will only reduce emission by 0.8% per year which will be insufficient to offset the increase in the emission due to increase in the passengers [4]. According to the International Civil Aviation Organization (ICAO) [5], fuel consumption could reach up to 852 million tonnes by 2050. Petroleum derived fuel such as kerosene, Jet A, Jet A-1 with a typical carbon chain-length of C_{11}-C_{13}, are currently being used for which alternatives are needed with similar characteristics. Biofuels need to deliver a high energy content per mass and volume (to reduce aircraft weight and drag), be thermally stable at low and high temperatures and have similar viscosity, surface tension, and ignition properties as current jet fuel. In the short-to medium-period, drop-in fuels (as opposed to new fuel types such as liquefied hydrogen) will need to be used so that the existing infrastructure could still be used. Figure 1 illustrates the reduction targets proposed by the aviation industry by the use of different mechanisms including the use of biofuels. It would require 426 million tonnes of biofuel to meet the GHG reduction goals; currently, less than 0.1% of global consumption is renewable in this sector [5].

Various efforts have been devoted to make progress in this area. In 2011, the U.S. Departments of Agriculture and Navy, through a joint venture "Farm-to-Fleet", invested $510 million in the private sector to produce drop-in aviation and marine biofuels. In 2013, the U.S. Department of Defense awarded $20.1 million contracts to Emerald Biofuels, Natures BioReserve, Fulcrum Biofuels, and Red Rock Biofuels for preliminary plant design as part of Phase I of the Defense Production Act Title III Advanced Drop-in Biofuels Production Project. Later in 2014, $210 million was awarded to Emerald Biofuels, Fulcrum BioEnergy, and Red Rock Biofuels for Phase II. These biorefineries are expected to produce a combined capacity of 84,000 tons

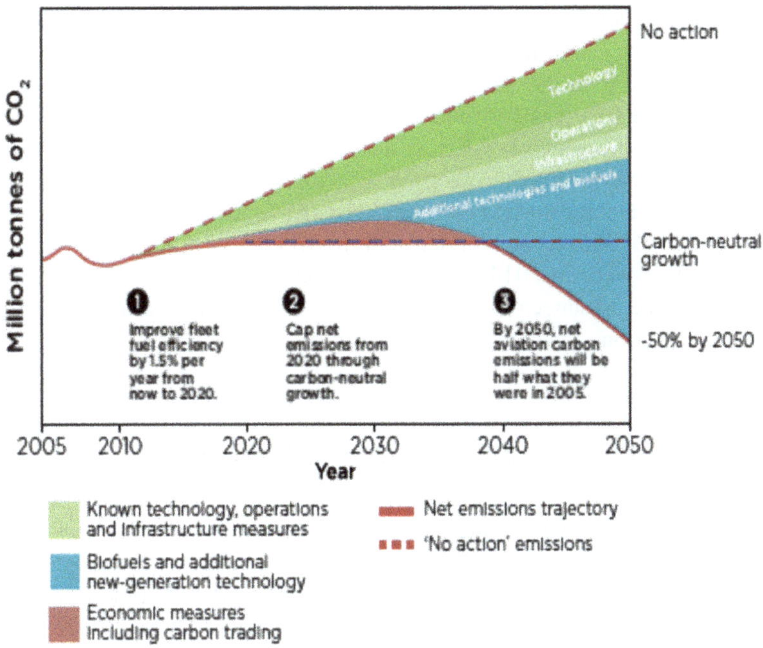

Fig. 1 Aviation industry's view on long-term emissions and their reduction through different pathways (Adapted from ATAG 2010 report) [6]

of jet fuels and marine diesel, at a competitive rate (less than \$4 per gallon) with 50% reduction of emissions. Another initiative was the "Farm-to-Fly" where USDA, Airlines for America and Boeing partnered to promote the production of 1 billion gallons (or about 3.3 million tons) of drop-in biofuels per year for use in the existing engines without modifications. There were other regional and national groups across the U.S. working on bringing stakeholders together with the aim of using certified biofuels in the aviation industry [7]. In the white paper "Roadmap to a Single European Transport Area" [8], the goals are set for about 40% sustainable low carbon fuels use in aviation by 2050. The initiative "European Advanced Biofuels Flight Path" aims at achieving a production of 2 million tons of biofuel per year for aviation by the year 2020 [9].

In June 2011, KLM Royal Dutch Airlines obtained special permission to fly the first passenger airplane between Amsterdam and Paris using a bio jet fuel derived from cooking oil. In 2015, more than 2000 flights have used alternative fuels. United Airlines announced an agreement to make a \$30 million equity investment in Fulcrum Bioenergy which converts municipal solid waste (MSW) into renewable jet fuel [10]. Since then in 2016, KLM has recently signed a 3 year contract to purchase biofuel for all its flights to Los Angeles. This fuel is produced by a local biorefinery AltAir Fuels and supplied by SkyNRG. After Norway, Los Angeles is the second airport that has included biofuel into its regular refueling process [11]. The different routes used to produce bio-jet fuels are characterized according to the feedstock, the conversion process or technology and the composition of the resulting fuel which dictates the amount of blending with the conventional jet fuel. As of now, there are four main types of biofuel alternatives for aviation applications: (1) hydrotreated vegetable oils (HVOs) or Hydro-processed Esters and Fatty Acids (HEFA) or Hydrotreated Renewable Jet (HRJ), (2) Fischer-Tropsch (FT) fuels, (3) Synthesized Iso-Paraffinic (SIP) fuel, and (4) alcohol-to-jet (ATJ) based on isobutanol. The first two have been certified (ASTM D7566) to be blended with fossil jet fuel at up to 50 vol.% while SIP has been approved (in 2014) to be blended at up to 10 vol.% [12]. The blend limit guarantees the presence of needed aromatics content in the biofuel for effective engine seals operation.

2.1 *Vegetable Oil-Based Fuels (HEFA, HVO, HRJ)*

These fuels are produced in refineries by hydrogen treatment of esters and fatty acids (i.e., triglycerides). Feedstock include edible plant oils (soybean, canola), non-edible plant oils (camelina, jatropha), algal oil, used/rendered cooking oil (yellow grease) and rendered animal fat (tallow). The vast majority of bio-jet currently available is mostly derived from these feedstocks. HEFA are generally paraffinic liquids, C_nH_{2n+2}, produced by hydrodeoxygenation of the above-mentioned oils and greases with propane and water as side products [13]. After the oils are cleaned to remove impurities, they are converted to shorter chain diesel-range paraffins through oxygen removal by reacting with hydrogen to convert any olefins to paraffins. The conversion increases the energy content and thermal and oxidative stability of the fuel. The diesel range paraffins are then cracked and isomerized to form paraffins with carbon numbers (C_{11}-C_{13}) in the jet fuel range. Because of their high-energy and high cetane number, the HEFA/HVO fuels can be used without blending and they also emit less GHG, NO_x and particulates and have better cold flow properties as compared to fossil fuel. They are also stable for storage and are resistant to microbial growth. If paraffin content is increased, the cold flow properties are further affected and depending on the feedstock used, there is a variability in the cetane number which in turn affects the fuel ignition. This is the reason why blending with conventional fuel is still the preferred option [14].

It should be noted that the chemical process that produces HEFA, can also produce bio-diesel which has a larger market with higher sale prices. Because of the regulations in California and agreement with airlines, only Honeywell UOP/Altair Fuels exclusively produces bio-jet fuels (currently at 40 million gallons or 134,000 tons per year level) while other HEPA facilities mainly produce diesel (Fig. 2). For example, Neste (Rotterdam: 800,000 tons/year, Singapore: 800,000 tons/year and Finland: 380,000 tons/year), Honeywell UOP & ENI (Italy: 300,000 tons/year), and Sinopec (China: 18,000 tons/year) produce bio-diesel but have an interest in producing bio-jet in the future. There are also a few start-ups which will be operational in 1–3 years: Pertamina (Indonesia: <23,000 tons/year), Honeywell UOP & SG Preston

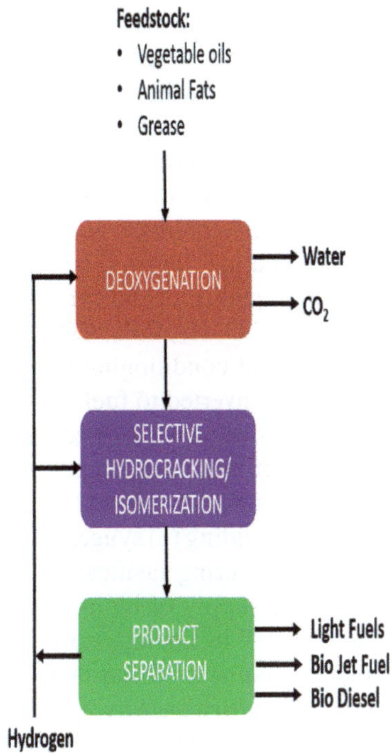

Fig. 2 Renewable jet fuel process

(Ohio: 400,000 tons/year, Indiana: 400,000 tons/year, Canada: 400,000 tons/year) and, Honeywell UOP & Petrixo (UAE: 500,000 tons/year) [15]. Out of various producers, Honeywell UOP technology seems to be the most developed whereby selective cracking with isomerization and hydrotreatment is used to maximize yield of jet fuel. This technology has been designed in such a way that existing refineries can be repurposed to target renewable jet fuel (RJF) [13b, 16]. Neste's technology, on the other hand, is geared towards optimum diesel production and will require adjustments for producing RJF.

2.2 Fischer-Tropsch (FT) Fuels

In the early 1990s, Sasol (South Africa) pioneered the first FT process to convert coal to liquid fuel for aviation. Based on the same principle, any biomass (e.g., forest and agricultural wastes, municipal solid wastes or MSW) can be gasified at very high temperature to form synthetic gas (a mixture of H_2 and CO) which is then catalytically converted through FT synthesis to produce liquid hydrocarbons. Here the main advantage is that irrespective of the feedstock, the composition of the gas would be predictable. FT fuel is distinct as it is non-toxic, has very low NO_x emissions, high cetane number, and low particulate matter, sulphur, and aromatics [17]. For producing synthesis gas, biomass is gasified at very high temperatures (typically, 1200 °C) using entrained flow reactors (similar to coal gasification technology) which have high efficiencies and low-tar production. Fibrous biomass feeding in such reactors is still an issue, however. To address this challenge, biomass is pretreated and then fed to the reactor. The second way to gasify biomass would be at a lower temperature (up to 900°) using bubbling or circulating fluidized bed reactors. The processes are thus simpler, easier to feed but the quality of the syngas decreases [18]. Biomass decomposes to syngas (a mixture of CO, H_2, CH_4, CO_2), tars and solid char. Here, the syngas requires more cleaning and conditioning to crack the tar, methane and light hydrocarbons so that it can be converted to fuel.

FT jet fuel derived from coal has been used since 1999 and blended up to 50 vol.% with fossil jet fuel while FT jet fuel from natural gas from Shell Technology has been used since 2012 and blended up to 25 vol.% [19]. However, using biomass as a feedstock has specific challenges including (a) syngas clean-up as biomass will also produce CO_2, methane, tar, inorganics during gasification, (b) catalyst contamination and deactivation, and (c) distributed supply of biomass which is inconsistent with the economies of scale needed for FT technology to be cost effective. A number of pilot and demonstration plants were started and then closed namely CHOREN, NSE Biofuels, Forest BtL, and Solena/British Airways Greensky project [19a, 20]. Other upcoming projects are BioTfueL [21] (200,000 t/year of RTF and diesel by co-gasification of biomass and coal, France 2020), Fulcrum Bioenergy (10 million gallon/year from MSW, Nevada 2019) [22], and Red Rock (16 million gal/year from forest and sawmill residues, Colorado 2017).

2.3 *Synthesized Iso-paraffinic (SIP) Bio-jet Fuel*

This bio-jet fuel is produced biochemically through sugar fermentation by microorganisms to produce farnesene, which is hydrogenated to farnesane which can be blended with conventional jet-fuel (Amyris, Inc.) [23]. Farnesane has already been approved as a jet blendstock and meets ASTM standards. The technology is sometimes also referred to as DSHC (direct sugars to hydrocarbons). This technology is expensive and the intermediate is a far more valuable chemical than the final fuel. Generally, the biomass is pretreated, enzymatically hydrolyzed to form solubilized C_5 and C_6 sugars which are then separated and concentrated to convert to the desired hydrocarbons. This product can then be oligomerized and hydrotreated to produce jet fuel [24]. LS9 is another company that is trying to produce fatty acids aerobically using *E.Coli* to produce fatty alcohols, esters and alkanes [25].

2.4 Alcohol-to-Jet (ATJ) Fuel

This route also involves sugar (e.g., from sugar cane, sugar beet) fermentation but to alcohols like ethanol or butanol which can be upgraded to bio-jet (Fig. 3). Other feedstocks include hydrolyzed grain starch from wheat or corn and hydrolyzed polysaccharides from lignocellulosic biomass [26]. These alcohols, because of their high volatility, low flash point and low energy density, cannot be used directly as jet fuel [27]. The upgrading of the alcohols involves dehydration, oligomerization, distillation and hydrogenation steps. A number of companies are working on commercializing this technology. For example, Terrabon was awarded a $9.6 million DARPA grant but filed for bankruptcy in 2012. Swedish Biofuels and Gevo are companies which use this technology to produce bio-jet fuel. Gevo converts isobutanol to jet fuel and received ASTM certification in 2015 [28]. Last year, ArcelorMittal partnered with LanzaTech and Primetals Technologies to invest 87 million Euros in an ethanol fermentation pilot plant in Ghent, Belgium. Gases from the steel plant will be converted to ethanol which could then be further converted to jet fuel [29]. Virent's BioForming [30] technology is another example where biomass is first converted to solubilized sugars through biomass pretreatment and enzymatic hydrolysis, then the carbohydrates are converted to alcohols in the presence of hydrogen. The alcohols are sent to an aqueous phase reforming reactor where they react with water over a catalyst at 450–575 K and 10–90 bar to produce H_2, alkenes and oxygenates. The latter can be catalytically converted to jet-fuel through acid or aldol condensation and (hydro)-dehydration.

2.5 Hydrogenated Pyrolysis Oils (HPO) Fuel

Biomass is first pyrolysed to form biocrude which is highly oxygenated and therefore has to be upgraded using hydrogen and a catalyst to produce stable oil. This technology is still at the developmental level. UOP Honeywell and Ensyn Corp (Ensyn/Envergent technologies) have a licensed RTP (rapid thermal processing) technology and have started the construction of facility which will convert 65,000 dry metric tons of forest residue to 10 million gallons of renewable fuels per year (about 34,000 tons/year). It is projected to be operational in late 2017 [31]. BTG (Biomass Technology Group) Bioliquids from Netherlands uses non-food biomass to produce oil through fast pyrolysis and delivers commercial scale pyrolysis plants to customers (2 tons/h plant in Malaysia, 5 tonnes/h plant in Netherlands-Empyro BV). The technology mixes dried biomass particles with hot sand (heat carrier) in a rotating cone reactor to form oil vapors that are cooled to get the bio-oil. In 2016, Technip (France) signed an agreement with BTG to use BTL's fast pyrolysis technology to find commercial uses of the renewable oil [32].

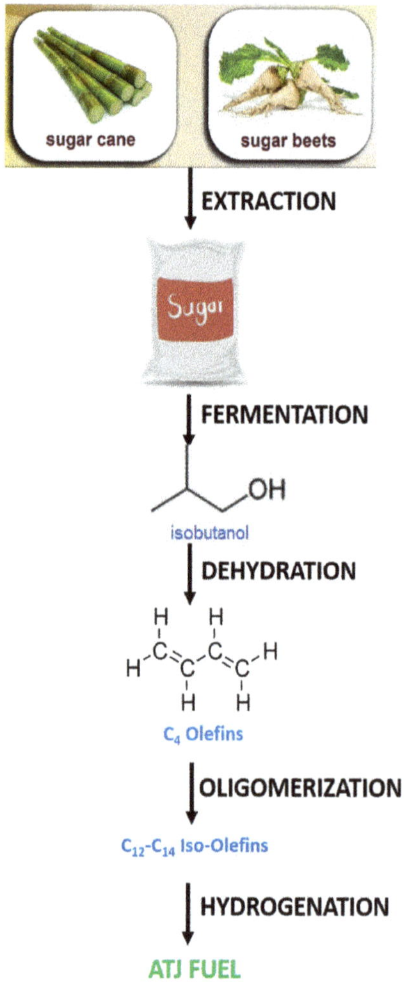

Fig. 3 Alcohol-to-Jet (ATJ) fuel process

2.6 *Future Perspective*

Some of the major challenges for using biofuels as an aviation fuel include (a) decreasing production costs to make the fuel more economically viable and competitive, (b) expanding supply potential by investing in feedstock production and conversion technology/facilities, and (c) ensuring efficiency over the entire supply chain [12]. Even though aviation biofuels are not expected to be as competitive as the fossil jet fuels, especially in the short-term, it is expected that legislation and public and political pressure to address environmental impacts might be the main drivers for this industry to use biofuels at least as a blend. It is believed that compared to the other transportation sectors, aviation is less complex and at a smaller scale, is more suitable for biofuel implementation. The costs of the feedstock together with supply and sustainability concerns, make it hard to scale up production of these biofuels to meet demand. For example, in May 2017, palm oil price was $1.9 per gallon while jet fuel was only $1.42 per gallon. It is believed that the HEFA/HVO fuel pathway is the main technology that can be used to establish the foundation for future and large-scale supply chains. More advanced technologies will have to wait at least another 5–10 years for maturity because of the different feedstock (biomass, wastes, etc.) being tested. Thermochemical routes will most likely be the front runner in terms of future technology because of the valuable intermediate products which could also be derived from these processes.

The competition between the renewable ground transportation fuel (HEPA-diesel) and HEPA for jets will be a challenge and proper policies need to be established to encourage the production of HEFA for jet fuel. However, the international nature of aviation complicates the policy establishment and its implementation. It is to be noted that blending conventional fossil fuel with low percentages of bio-diesel is currently in the ASTM certification process which could increase the adoption in aviation [23]. The Commercial Aviation Alternative Fuels Initiative's fuel readiness level (FRL) scale rates ASTM-certified fuels at FRL7 or higher with FRL9 being the maximum (full scale plant is operational). The FRL approach provides a descriptive hierarchy showing the technology's progress towards commercialization and takes into account the risks associated with fuel development [33]. Based on this approach, a recent study concluded that HEFA was at FRL9, FT at FRL7–8, SIP at FRL5–7, ATJ at FRL4–6 and HPO at FRL6. Among the certified fuels, HEFA is the most highly developed and has been used in most tests. Even though FT seems to be technologically ready, the economic feasibility is still not at the required level for it to be commercial. It is clear from literature that feedstock seems to be the most important parameter when looking at both the technical and commercial feasibility of the process. Another big challenge facing biofuel use in aviation is the high quality standard requirements and therefore testing the biofuels is crucial.

3 Biofuels in Ground Transportation

Currently a large volume of biofuel is used in ground transportation mostly in the form of bioethanol and biodiesel. At present, 10% bioethanol blends (E10) are already dispensed in fuel stations around the world with more than 50 countries having some mandated or targeted blends. For instance, the U.S. has a mandate of 136 billion liters of biofuels, out of which 60 should be cellulosic ethanol, by the year 2022 [34]. Biodiesel is also being blended with traditional fuels: 5% (BD5) in France to 20% (BD20) in the U.S [35]. In 2016, the U.S. produced an estimated 15.3 billion gallons of fuel ethanol and 1.6 billion gallons of biodiesel [36]. Biofuel technologies are often categorized as conventional and advanced, with most of the conventional technologies being commercially available while most of the advanced technologies are in the demonstration stage or moving towards that stage. As of Fall 2015, 67 biorefineries worldwide produced second generation ethanol, out of which 20 (with 9 operating on a commercial scale) were in North America [37]. There are a number of biofuel pathways that have been explored over the years as discussed below.

3.1 *Bioethanol*

Traditionally, bioethanol was obtained by glucose fermentation by yeasts to convert the glucose into ethanol and carbon dioxide as follows:

$$C_6H_{12}O_6 \rightarrow 2\ CH_3CH_2OH + 2\ CO_2$$

The conventional (i.e., first generation) feedstocks are sugars from sugar crops such as sugarcane, sugar beet, sweet sorghum and starch crops such as corn and wheat which would require an additional hydrolysis step to convert the starch to glucose. This technology is mature and recent improvements include optimization of energy integration and valorizing by-products. For the starch processes, for example, the economic efficiency depends a lot on the value of the co-products such as dried distiller's grains with solubles (DDGS) and fructose. Because of the concern of diverting food supply to fuel, other non-food feedstock (second generation) needed to be used.

Advanced cellulosic ethanol is produced through biochemical conversion of mainly the cellulose and hemicellulose components of biomass into fermentable sugars which can then be converted to ethanol [38]. Some of the agricultural and forest residues that have been used for this technology are corn stover, switchgrass, miscanthus, wood chips, and municipal solid wastes. However, the conversion of lignocellulosics to ethanol is more challenging than conversion from sugars or starch. This new technology consists of three main steps: (i) feedstock pre-treatment including harvesting, size reduction, fractionation, and removal of the lignin, (ii) enzymatic hydrolysis which uses enzymes to depolymerize the lignocellulosic biomass to sugars (i.e., glucose and xylose or C_6 and C_5, respectively) and finally, (iii) fermentation of the sugars to ethanol. Some of these sub-processes are developed but may not be fully commercial yet due to the enzyme cost. The key challenges in the production of second-generation ethanol are due to the low accessibility of cellulose and hemicellulose caused by tightly bound lignocellulosic structure, low activity of enzymes, and difficulty in fermenting C_5 sugars to ethanol. Research is still being conducted to valorize lignin as an energy carrier or material feedstock [35]. Also, improvements in micro-organisms and enzymes, the use of C_5 sugars for fermentation or value added products also need to be investigated. Commercial production of second-generation biofuels was started with Dupont opening one of the world's largest cellulosic ethanol plant in Nevada (Iowa) in 2015. A number of other companies soon followed but some of them like Abengoa have suspended production. As of July 2017 [39], 7 cellulosic ethanol plants were into operation (Table 1) while there are some 200 plants that operate on the sugar/starch platform.

Table 1 Cellulosic ethanol plants in operation as of July 2017 in the United States

Plant	State	Feedstock	Capacity, millions gallons/year
American Process Inc., Thomaston Biorefinery	GA	Other	0.30
DuPont Cellulosic Ethanol LLC, Nevada	IA	Crop residue	30.0
Fiberight Demonstration Plant	VA	Waste stream	0.50
ICM, Inc. Pilot Integrated Cellulosic Biorefinery	MO	Biomass crops	0.32
Poet—DSM Advanced Biofuels, LLC, Project Liberty	IA	Crop residue	20.0
Quad County Cellulosic Ethanol Plant	IA	Corn	2.00
ZeaChem Inc., Demonstration Plant	OR	Biomass crops	0.25

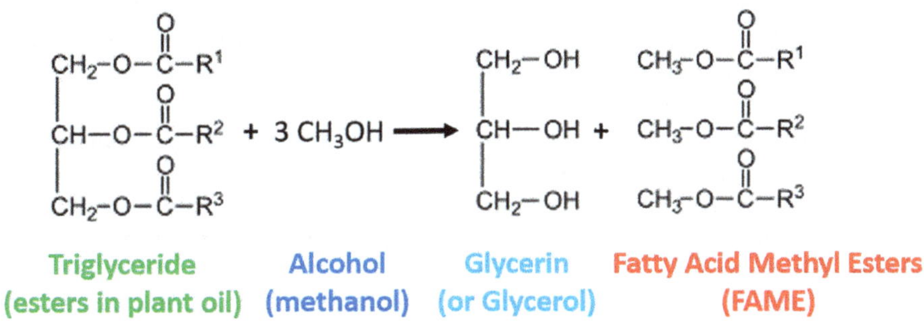

Fig. 4 Biodiesel production from triglycerides

Research in developing cost-effective technologies are emphasizing on feedstock development (genetic modification), pretreatment technology (lowering temperature and pressure), enzyme cost reduction, co-fermentation of glucose from cellulose (a mature technology) with xylose from hemicellulose (a complex technology) [40].

3.2 Biodiesel

Traditional biodiesel is derived from vegetable oils (e.g., soybean, canola, oil palm, sunflower) or animal fats. The feedstock, mostly triglycerides, is reacted with an alcohol (usually methanol or ethanol) to form fatty acid methyl esters (FAME) and glycerin (Fig. 4). This reaction can be catalyzed by alkalis (NaOH), acids (H_2SO_4), or enzymes (lipase). The biodiesel is usually separated from the glycerin through settling, filtration and decantation.

In general, the biodiesel can be used on its own or blended with diesel. Availability of land for crop production is one of the main limitations of biodiesel production. It is estimated that to replace half of the transportation fuel in the U.S. with biodiesel would require 24% of the existing crop land [41]. Advanced biodiesel include hydrotreated vegetable oil (HVO) and biomass-to-liquids (BtL) diesel or Fischer-Tropsch (FT) diesel. HVO is produced by hydrogenation of the vegetable oils or animal fats [42]. BtL is produced via a two-step process where biomass is first gasified to H_2 and CO-rich syngas which is conditioned and converted to hydrocarbon liquids via Fisher-Tropsch synthesis. Research however is still being conducted to improve the catalyst life and robustness and to reduce cost associated with syngas clean-up.

Attention has been given to another feedstock, (micro)algae, termed as the third generation feedstock, which can be grown on a non-arable land in different types of waters (saline, brackish) and has great potential in terms of productivity (high growth rates and high oil content). It has also been estimated that annual oil production from high-oil microalgal species (*Botryococcus brauni, cylindrotheca sp., isochrysis sp., nannochloropsis sp., schizochytrium sp.*) can reach 58,000–137,000 L (15,300–36,200 gallons) per hectare [41]. High-quality diesel and jet fuel analogues are expected to be produced. Usually, algae is grown in either open ponds or closed photobioreactors at around 20–30 °C under autotrophic or heterotrophic conditions. However, large-scale algae cultivation and the oil extraction are very expensive. In order to improve the economic feasibility of biodiesel from algae, research and development has to concentrate on developing high-yield microalgae, algae with better tolerance to high/low temperatures and high oxygen concentration and better growth and harvesting systems.

3.3 *Biofuel from Thermochemical Processes*

Liquefaction, pyrolysis, and gasification are all thermochemical processes that can convert biomass to gas and liquid products. However, all of these technologies produce intermediates that need to be upgraded or reacted to form the final fuel. For example, (hydrothermal) liquefaction produces a heavy viscous biocrude. Usually, temperatures of around 350–370 °C and pressures around 50–200 bar are used whereby the biomass is depolymerized into unstable and reactive smaller molecules. These molecules contain high amount of oxygen and are upgraded to bio-oil through catalytic hydrogenation [43]. In pyrolysis, a liquid is produced by breaking down biomass in the absence of oxygen. Slow pyrolysis operates at 300–700 °C with 30–200 s residence time while fast pyrolysis operates at 400–700 °C with 1–5 s residence time [44]. The formed bio-oil is further hydrotreated (where oxygen is removed as H_2O or CO_2) and hydrocracked to produce naphtha and diesel range fuels. Research is still being conducted in trying to combine the pyrolysis and hydrotreating steps like the one performed using a zeolite catalyst to increase the yield [45]. Gasification uses 800–900 °C to convert biomass to CO, CO_2, CH_4 and H_2 along with some tar. The synthetic gas formed is conditioned and cleaned before upgrading to liquid fuels via FT Synthesis, where H_2 and CO are combined using Fe/Co catalyst to form suitable long chain hydrocarbons as gasoline. It is to be noted that thermochemical processes can also be combined with biochemical processes by using a biocatalyst to convert the syngas to synthetic fuel, bioethanol.

3.4 Future Perspective

Second generation biofuels have a potential to complement and replace fossil fuels because of their net zero carbon emissions and the abundance of the feedstock. Compared to gasoline, ethanol has a number of advantages: higher octane number, flammability limits, flame speeds, heat of vaporization but has lower energy density, flame luminosity, vapor pressure and is corrosive and toxic to ecosystems. Biodiesel properties are not far from that of conventional diesel. For second and third generation biofuels to really succeed there needs to be strong policy support and investment in research and development to allow for the biofuels to reach their full potential. Also valorization of byproducts is important for economic sustainability; currently, only cellulose and hemicellulose are used in ethanol production and only lipids in biodiesel production.

4 Summary and Conclusions

It is estimated that by 2050, 90% of the world population will be living in developing countries [46]. The potential of biomass to provide a cost-effective and sustainable source of energy is high. Even though research has and is still being done on how to convert biomass to biofuels, there are many challenges (technical and economic) that need to be addressed to make the fuels commercial and market-ready. By 2015, some 2000 flights had used bio jet fuel and some airlines like KLM and Boeing have made commitments to use even more. As for ground transportation, blending bioethanol and biodiesel with conventional fossil fuels is the most common way to use alternative fuels. Research is still being carried out to convert lignocellulosic material to drop-in fuels but high oxygen content of the feedstock and economies of scale are still hurdles that need to be overcome. In the future, it is expected that electricity production will partly be supplied via direct combustion of wastes and residues, whereas bioethanol and biodiesel will be used as liquid fuels for transportation. Energy crops could contribute to combined heat and power production. This means that the share of biofuels in the aviation and transportation industry is expected to grow in the future. The reduction in biofuel costs by enhancing the economics of the process (through value-added co-products) will bring about development in the industrialized but most importantly in the rural areas. The promise of advanced biofuels is that is it clean, carbon neutral, renewable but also benefits the local communities/economies.

References

1. B. Metz, O.R. Davidson, P.R. Bosch, R. dave, L.A. Meyer (ed.), *IPCC Climate Change 2007: Mitigation, Contribution of Working Group 3 to the Fourth Assessment Report of the Intergovernmental Panel on Climate Change* (Cambridge University Press, Cambridge, United Kingdom and New York, USA, 2007)
2. R.B. Gupta, A. Demirbas, *Gasoline, Diesel, and Ethanol Biofuels from Grasses and Plants* (Cambridge University Press, 2010)
3. ATAG Aviation: Benefits beyond Borders. http://aviationbenefits.org/media/149668/abbb2016_full_a4_web.pdf
4. (a) D. Lee, L. Lim, B. Owen, Bridging the aviation CO_2 emissions gap: why emissions trading is needed, 22 (2015). Accessed May 2013; (b) D. Kennedy, B. Combes, O. Bellamy, Meeting the UK Aviation Target—Options for Reducing Emissions to 2050. International Civil Aviation Organization (2010)
5. ICAO (International Civil Aviation Organization) Environmental Report 2016. https://www.icao.int/environmental-protection/Documents/ICAO%20Environmental%20Report%202016.pdf
6. ATAG, The Right Flightpath to Reduce Aviation Emissions (2010)
7. W.-C. Wang, L. Tao, J. Markham, Y. Zhang, E. Tan, L. Batan, E. Warner, M. Biddy, *Review of Biojet Fuel Conversion Technologies* (NREL (National Renewable Energy Laboratory), Golden, CO, United States, 2016)
8. European Commission, Roadmap to a Single European Transport Area—Towards a competitive and resource efficient transport system. https://ec.europa.eu/transport/themes/strategies/2011_white_paper_en
9. N. Nikitevich, A. Romushkin, Biofuels in aviation. Актуальные проблемы авиации и космонавтики **2**(12) (2016)
10. IATA Report on Alternative Fuels. http://www.iata.org/publications/Documents/2015-report-alternative-fuels.pdf
11. KLMMediarelations KLM to operate biofuel flights out of Los Angeles. http://news.klm.com/klm-to-operate-biofuel-flights-out-of-los-angeles. Accessed 16 May 2017
12. M. Kousoulidou, L. Lonza, Biofuels in aviation: fuel demand and CO_2 emissions evolution in Europe toward 2030. Transp. Res. Part D: Transp. Environ. **46**, 166–181 (2016)
13. (a) R. Arvidsson, S. Persson, M. Fröling, M. Svanström, Life cycle assessment of hydrotreated vegetable oil from rape, oil palm and Jatropha. J. Clean. Product. **19**(2), 129–137 (2011); (b) M. Pearlson, C. Wollersheim, J. Hileman, A techno-economic review of hydroprocessed renewable esters and fatty acids for jet fuel production. Biofuels Bioproducts Biorefining **7**(1), 89–96 (2013)
14. M. Lapuerta, M. Villajos, J.R. Agudelo, A.L. Boehman, Key properties and blending strategies of hydrotreated vegetable oil as biofuel for diesel engines. Fuel Process. Technol. **92**(12), 2406–2411 (2011)
15. R. Mawhood, E. Gazis, S. de Jong, R. Hoefnagels, R. Slade, Production pathways for renewable jet fuel: a review of commercialization status and future prospects. Biofuels Bioproducts Biorefining **10**(4), 462–484 (2016)
16. S. De Jong, R. Hoefnagels, A. Faaij, R. Slade, R. Mawhood, M. Junginger, A techno-economic analysis of short-term production strategies for renewable jet fuels–the path toward fossil fuel price parity. Biofuels Bioproducts Biorefining **9**(6), 778–800 (2015). (Advance Online Publication)
17. B. Saynor, A. Bauen, M. Leach, The potential for renewable energy sources in aviation. Report commissioned for the DTI, Imperial College Centre for Energy Policy and Technology (2003)
18. J. Sandquist, B. Guell, Overview of biofuels for aviation. Chem. Eng. Trans. **29** (2012)
19. (a) K. Maniatis, M. Weitz, A. Zschocke, Million tons per year: a performing biofuels supply chain for EU aviation Technical Paper of European Commission/DG ENER (2011); (b) P. Morgan, P. Roets, In The Synthetic Jet Fuel Journey, 20th World Petroleum Congress, World Petroleum Congress (2011)

20. (a) V. Balan, D. Chiaramonti, S. Kumar, Review of US and EU initiatives toward development, demonstration, and commercialization of lignocellulosic biofuels. Biofuels Bioproducts Biorefining **7**(6), 732–759 (2013); (b) GreenAir Online, Low oil price and current lack of government support stalls British Airways' waste to jet biofuel project. http://www.greenaironline.com/news.php?viewStory=2163
21. Total BioTfueL: Developing Second-generation Biofuels. http://www.total.com/en/energy-expertise/projects/bioenergies/biotfuel-converting-plant-wastes-into-fuel
22. Fulcrum Bioenergy. http://fulcrum-bioenergy.com/
23. IRENA Biofuels for Aviation-Technology Brief. https://www.irena.org/DocumentDownloads/Publications/IRENA_Biofuels_for_Aviation_2017.pdf
24. R. Davis, L. Tao, E. Tan, M. Biddy, G. Beckham, C. Scarlata, J. Jacobson, K. Cafferty, J. Ross, J. Lukas, *Process Design and Economics for the Conversion of Lignocellulosic Biomass to Hydrocarbons: Dilute-acid and Enzymatic Deconstruction of Biomass to Sugars and Biological Conversion of Sugars to Hydrocarbons* (National Renewable Energy Laboratory (NREL), Golden, CO, 2013)
25. R. Mawhood, E. Gazis, R. Hoefnagels, S. De Jong, R. Slade, Technological and commercial maturity of aviation biofuels: emerging options to produce jet from lignocellulosic biomass (2015)
26. C.A. Cardona, Ó.J. Sánchez, Fuel ethanol production: process design trends and integration opportunities. Biores. Technol. **98**(12), 2415–2457 (2007)
27. J.I. Hileman, D.S. Ortiz, J.T. Bartis, H.M. Wong, P.E. Donohoo, M.A. Weiss, I.A. Waitz, *Near-term Feasibility of Alternative Jet Fuels* (RAND Corporation and Massachusetts Institute of Technology, Santa Monica, CA, USA, 2009)
28. Gevo Gevo's Alcohol to Jet Fuel Meets Approved ASTM Standard. http://ir.gevo.com/phoenix.zhtml?c=238618&p=RssLanding&cat=news&id=2151133
29. ArcelorMittal, LanzaTech, and Primetals Technologies announce partnership to construct breakthrough €87 million biofuel production facility. http://www.lanzatech.com/arcelormittal-lanzatech-primetals-technologies-announce-partnership-construct-breakthrough-e87m-biofuel-production-facility/
30. P. Blommer, R. Cortright, *Production of Conventional Liquid Fuels from Sugars* (Virent Energy Systems. Inc., Madison, WI, Online Whitepaper, 2008)
31. Envergent Honeywell's Envergent RTP® Technology to be used in new renewable fuels facility in Quebec. https://www.envergenttech.com/?press_release=honeywells-envergent-rtp-technology-to-be-used-in-new-renewable-fuels-facility-in-quebec
32. J. Lane, Technip inks pact with BTG Bioliquids for biomass-to-oil. http://www.biofuelsdigest.com/bdigest/2016/08/28/technip-inks-pact-with-btg-bioliquids-for-biomass-to-oil/
33. (a) U. Arup, E4tech. (2013), Advanced Biofuel Feedstocks–An Assessment of Sustainability. UK: ARUP URS & E4tech; (b) J.J. Steiner, K.C. Lewis, H.S. Baumes, N.L. Brown, A feedstock readiness level tool to complement the aviation industry fuel readiness level tool. BioEnergy Res. **5**(2), 492–503 (2012)
34. H. Ramsurn, R.B. Gupta, Production of biocrude from biomass by acidic subcritical water followed by alkaline supercritical water two-step liquefaction. Energy Fuels **26**(4), 2365–2375 (2012)
35. W.-R. Chang, J.-J. Hwang, W. Wu, Environmental impact and sustainability study on biofuels for transportation applications. Renew. Sustain. Energy Rev. **67**, 277–288 (2017)
36. Energy Review, U.S. Energy Infiormation Administration. https://www.eia.gov/totalenergy/data/browser/?tbl=T10.03#/?f=M
37. Q. Nguyen, J. Bowyer, J. Howe, S. Bratkovich, H. Groot, E. Pepke, K. Fernholz, Global production of second generation biofuels: trends and influences (2017)
38. D.L. Klass, *Biomass for Renewable Energy, Fuels, and Chemicals* (Academic press, 1998)
39. Ethanol producer Magazine, U.S. Ethanol Plants. http://www.ethanolproducer.com/plants/listplants/US/Operational/Sugar-Starch
40. J.J. Cheng, G.R. Timilsina, Status and barriers of advanced biofuel technologies: a review. Renew. Energy **36**(12), 3541–3549 (2011)

41. N.E. Bassam, *Energy Plant Species: Their Use and Impact on Environment and Development* (Routledge, 2013)
42. D. Bacovsky, M. Dallos, M. Wörgetter, I.B. Task, Status of 2nd generation biofuels demonstration facilities in June 2010. IEA Bioenergy Task 39: Commercializing 1st and 2nd Generation Liquid Biofuels from Biomass **39**, 1–126 (2010)
43. H. Ramsurn, R.B. Gupta, Deoxy-liquefaction of switchgrass in supercritical water with calcium formate as an in-situ hydrogen donor. Biores. Technol. **143**, 575–583 (2013)
44. H. Suopajärvi, E. Pongrácz, T. Fabritius, The potential of using biomass-based reducing agents in the blast furnace: a review of thermochemical conversion technologies and assessments related to sustainability. Renew. Sustain. Energy Rev. **25**, 511–528 (2013)
45. N. Thegarid, G. Fogassy, Y. Schuurman, C. Mirodatos, S. Stefanidis, E. Iliopoulou, K. Kalogiannis, A. Lappas, Second-generation biofuels by co-processing catalytic pyrolysis oil in FCC units. Appl. Catal. B **145**, 161–166 (2014)
46. U. Nations, World Urbanization Prospects: The 2014 Revision, Highlights. Department of Economic and Social Affairs. Population Division, United Nations (2014)

Innovative Bio-char Briquetting from Corn Residue Using Torrefaction Process

Krongkaew Laohalidanond and Somrat Kerdsuwan

Abstract Each year, millions of tons of corn are harvested from the farm fields. Their residues, e.g. stems/leaves and husks, are left in the field, while the empty cobs are processed at the mills and the byproduct is inefficiently used as low-grade fuel. The stems/leaves and husks are not used as fuel because of their poor fuel quality and because they are difficult to handle. Innovative bio-char briquetting from corn residues to improve their fuel properties has been proposed using the torrefaction process. The temperature and retention time for this process affects properties of the empty cobs, stems/leaves, and husks from the residual corn, which helps to improve their heating value and changes the volatile matter and fixed carbon proportion in the fuel. All of the torrefied corn residues is hydrophobic, which has less ability to absorb water due to the change in pore structure. The biochar from the corn residual can be used as premium feedstock for heating purposes with high heating value and low smoke.

Keywords Torrefaction process · Biochar · Fuel property · Corn residue

1 Introduction

As the key driver for the development of the industrial and energy sector, fossil fuel plays an important role as a primary energy source in many countries. In 2014, the share of oil, coal, and natural gas consumption in the world's total primary energy consumption was reported to be 32.65%, 30.04%, and 23.67%, respectively, and the proportion of this fossil fuel in the world's primary energy consumption in 2015 remained almost unchanged. In contrast to fossil fuel, the utilization of renewable energy was limited to only approximately 2.5% of the world's total primary energy consumption [1].

K. Laohalidanond · S. Kerdsuwan (✉)
Department of Mechanical and Aerospace Engineering, King Mongkut's University of Technology North Bangkok, Bangkok, Thailand
e-mail: somrat_k@yahoo.com

A. K. Runchal et al. (eds.), *Energy for Propulsion*, Green Energy and Technology,
https://doi.org/10.1007/978-981-10-7473-8_20

Relying significantly on fossil fuel, which has depleted continuously, however, has resulted in the security and sustainability of certain countries, especially the countries whose primary energy sources are deficient. Hence, the world energy trend should switch to greater utilization of renewable energy rather than fossil fuel to solve the problem of the depletion of fossil fuel [2]. Not only the depletion of fossil fuel, but the switch to renewable energy will also reduce the negative impact on the environment, especially in terms of the global warming effect.

Thailand today is facing an agricultural waste and limited non-renewable energy resource problem because of continuous increase in population and economic and industrial growth. To be able to meet these increasing demands, Thailand has to count on various energy imports being used in manufacturing process as commercial energy for domestic usage [3]. This causes the loss of a great deal of money from the country. Therefore, there is a high necessity to the new paradigm shift to use more renewable energy for energy security, and the improvement in the economy and ecology. The Alternative Energy Development Plan (AEDP2015) stated that the share of alternative energy for power production is targeted at 20% in 2036 [4]. Within this, 20% of biomass from agricultural products and forestry is promoted to be the main renewable energy source. Biomass has been considered as substitute renewable fuel because of its availability and carbon neutral balance [5]. In the northern and middle areas of Thailand, there are many corn farms that have a good deal of agricultural waste, such as empty cobs, stems/leaves, and husks. This waste can be used as a renewable energy in the form of thermal energy. However, using it in producing thermal energy has many disadvantages, for example, high moisture, low heat capacity, and low energy density, thus causing low thermal energy efficiency. These disadvantages can be minimized by upgrading fuel through the torrefaction process and densification process before using this waste as feedstock in the thermal conversion process.

Torrefaction has been considered as a mild pyrolysis, which improves biomass characteristics through thermochemical treatment at a temperature range of 250–350 °C, carried out under atmospheric pressure and in an inert atmosphere for 30–90 min [6–9]. The mechanism of the torrefaction process can be divided into 5 regimes, as illustrated in Fig. 1.

In regime A, moisture containing in biomass is carried out without any chemical reactions. Consequently, the chemical composition of the biomass is still unchanged. At the higher temperature of 120–150 °C in regime B, the lignin structure or the fiber in the biomass is softened and begins to decompose. The chemical bonds in the biomass are completely decomposed in regime C, called the reactive drying process, with a temperature ranging from 150 to 200 °C. In this regime, long-chain hydrocarbon compounds are destroyed into short-chain polymers. Hemicellulose is first decomposed at the temperature of 200–250 °C in regime D and cellulose is finally decomposed and devolatilizes as volatile matter in Regime E at the temperature of 250–350 °C.

As described above, during the torrefaction process, the heat will remove some of the volatile matters (H_2O and CO_2) from the biomass, which highly changes the biomass properties and leads to less moisture, and less hydrogen and oxygen in

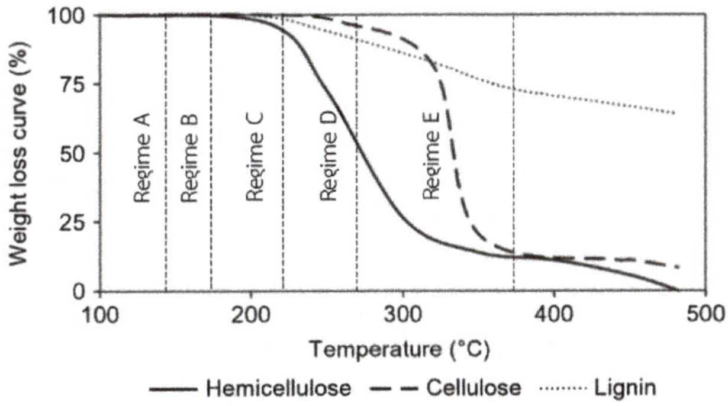

Fig. 1 Mechanism of the torrefaction process (Adapted from [8])

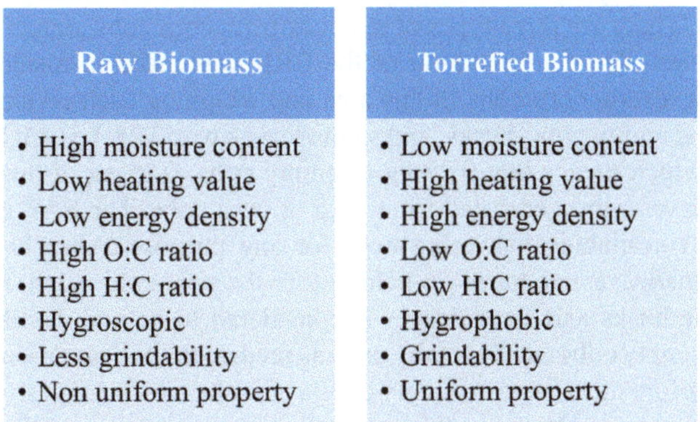

Fig. 2 Comparison of the fuel properties of torrefied biomass with raw biomass (Adapted from [3])

the biomass then the fixed carbon increases. This gives the biomass higher heating value and higher energy density, and it can be used as upgraded-fuel for producing thermal energy, vapor, or electricity with high efficiency. Figure 2 compares the fuel properties of torrefied biomass to raw biomass.

This chapter discusses the improvement of the fuel properties of corn residue through the torrefaction and densification process, where a feasibility study was conducted using briquettes of torrefied products in the gasification process. The gasification behaviors using briquettes from raw residue and torrefied residue are compared.

2 Corn Residue Potential

Corn cob is considered as an important economic crop in Thailand. Almost 5 million tons of cobs were cultivated throughout the country in three consecutive crop years from 2012 to 2015. More than two-thirds of them were produced in the northern Thailand, as quantitatively listed in Table 1.

In the northern part of Thailand, Petchaboon is a top-ranked province in terms of corn cob plantation. In the last 5 years from 2010 to 2014, more than 600,000 tons of corn cobs were annually cultivated there [10]. Hence, Petchaboon was selected as the focus area for this study.

In order to investigate the corn residue potential in the focus area, the Crop-to-Residue Ratio (CRR) and Surplus-Availability-Factor (SAF) were first determined through on-site surveying. The CRR describes the amount of residue generated from 1 kg of a crop. There are three types of residues from corn cobs: empty cobs, husks, and stems/leaves. The determination of the CRR was made in a selected area of 3×3 m by harvesting corn cobs in this area and weighing each residue generated. The CRR of the empty cobs, husks, and stems/leaves was 0.28:1, 0.15:1, and 1.28:1, respectively, which means that 0.28 kg of empty cobs, 0.15 kg of husks, and 1.28 kg of stems/leaves were generated from 1 kg of corn cobs. The SAF is the amount of residue still available after being used for any purpose. After harvesting, the husks and stems/leaves are burnt on-field before the next cultivation, and therefore the SAF of the husks and stems/leaves is considered to be one. Unlike husks and stems/leaves, empty cobs can be totally used as feedstock for heat production in corn mills, and therefore there were no empty cobs available, which suggests that the SAF of empty cobs is zero. However, this study took empty cobs into account because their utilization is ineffective way when used with old-fashioned boilers due to their low fuel properties. Figure 3 shows the potential of the corn residue in the focus area in 2014.

3 Materials

The corn residue used as feedstock in this study was empty cobs, husks, and stems/leaves generated from corn plantation in central Thailand. Empty cobs were collected from corn mill after corn processing, while the husks and stems/leaves were collected in the field after harvesting. Figure 4 presents the feedstock from the corn residue.

Feedstock was prepared in order to investigate its fuel properties, e.g. proximate and ultimate analysis, as well as its heating value. The feedstock was oven-dried in a DGH-9070b oven at 105 °C for 24 h for the determination of moisture content. Then, the feedstock was fine-ground for proximate and ultimate analysis, which was conducted according to the ASTM standard. The determination of the heating value was carried out in an adiabatic bomb calorimeter Gallenkamp. The properties of the

Table 1 Plantation of corn cob in Thailand

Region	Cultivated area (rai)			Harvested area (rai)			Crop (ton)		
	2012	2013	2014	2012	2013	2014	2012	2013	2014
Thailand	7,529,151	7,426,514	7,292,697	7,153,788	7,214,650	7,087,021	4,947,530	4,876,180	4,804,670
North	5,078,643	5,048,729	4,958,330	4,891,058	4,925,085	4,834,629	3,355,902	3,332,002	3,289,539
North-East	1,736,312	1,639,855	1,610,090	1,602,856	1,600,421	1,567,976	1,121,183	1,070,489	1,050,845
Central	714,196	737,930	724,277	659,874	689,144	684,416	470,445	473,689	464,286

Note 1 ha = 6.25 rai

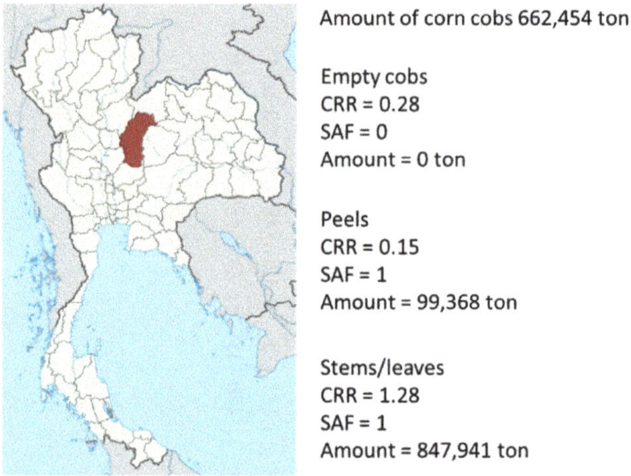

Fig. 3 Potential of corn residue in Petchaboon in 2017

Fig. 4 Corn residue **a** empty cobs **b** stems/leaves and **c** husks

corn residue used in this study compared to conventional fuel (lignite) are shown in Table 2.

In Table 2, it is clearly seen that the corn residue is a low-grade fuel compared to conventional fossil fuel because it has considerably high moisture content and volatile matter, ranging from 21 to 61 %wt. and from 78 to 88 %wt., respectively, whereas the fix carbon and heating value are much lower than conventional fossil fuel, lying between 9–16 %wt. and 16–18 MJ/kg, respectively.

Therefore, it is necessary to improve the fuel quality of corn residue using the torrefaction and briquetting process. After the improvement of the corn residue, the utilization of the corn residue with and without the torrefaction process in the gasification process will be conducted. The details of each process, including the experimental set up, procedure, as well as the results and discussion, will be separately described in the next sections.

Table 2 Properties of corn residue used in this study compared to conventional fuel (lignite)

No.	Properties	Empty cobs	Stems/leaves	Husks	Lignite [11]	Standard
Moisture content (%wt, as received)						
1	Moisture	49.88	61.71	21.52	28.84	ASTM D3173-87
Proximate analysis (%wt, dry basis)						
1	Volatile matter	81.70	78.22	87.99	40.74	ASTM D3175-89a
2	Fixed carbon	16.13	16.01	8.71	43.23	ASTM D3174-89
3	Ash	2.17	5.77	3.30	16.03	ASTM D3172-89
Ultimate analysis (%wt, dry basis)						
1	Carbon	47.04	45.19	46.17	58.35	ASTM D3178-89
2	Hydrogen	6.46	5.85	6.69	3.72	ASTM D3178-89
3	Nitrogen	0.34	0.81	0.53	0.85	ASTM D3178-89
4	Sulfur	0.07	0.13	0.10	2.36	ASTM D4239-85c
5	Chlorine	0.04	0.18	0.31	n.a.	ASTM D2361-91
6	Oxygen	43.88	42.07	42.89	18.69	By Difference
Higher heating value (kJ/kg, dry basis)						
1	HHV	17,734	15,570	16,010	21,984	ASTM D3286-91

4 Torrefaction Process

The target of the torrefaction process is to drive off the moisture content and the volatile matter from the corn residue, and as a consequence, the amount of the fix carbon and the heating value will increase.

4.1 Experimental Setup and Procedure

The torrefaction process was carried out in a batch-type electric muffle oven with an inside dimension of 480 mm × 670 mm × 800 mm, equipped with a 12 kW, 380 V electric wire. The oven had a multi-layer wall made of steel, ceramic fiber, and refractory brick. A stainless steel box with an inside dimension of 350 mm × 350 mm × 300 mm was used as the torrefaction reactor.

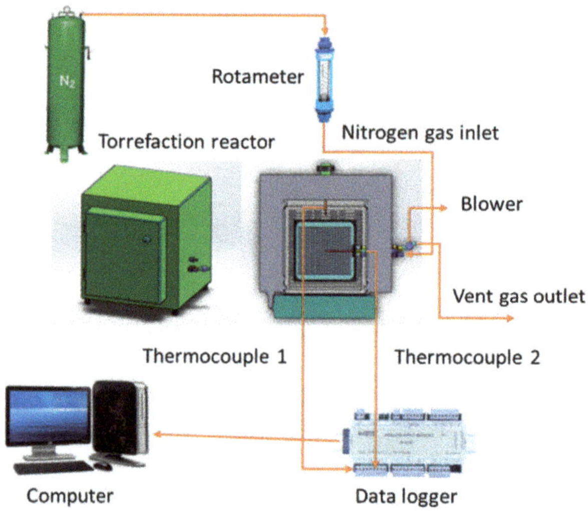

Fig. 5 Experimental setup for the torrefaction process

Two K-type thermocouples were installed at the back of the oven and inside the torrefaction reactor and were connected to the temperature control unit. There were two gas passages for the nitrogen gas inlet and the vent gas outlet. Figure 5 illustrates the experimental setup for the torrefaction process.

In each torrefaction experiment, 200–250 g of corn residue were introduced into the stainless steel box and placed in the torrefaction oven. The torrefaction temperature was adjusted to 250, 300, and 350 °C, while the retention time was set at 30, 60 and 90 min. After the torrefaction process, the torrefied product was taken from the oven and weighed in order to determine the mass loss, and after that the volatile matter, the fix carbon, and the higher heating value were investigated.

4.2 Results and Discussion

4.2.1 Mass Loss

Figure 6 shows the mass loss from the corn residue after the torrefaction process at different temperatures and retention times. It can be clearly seen that the mass loss during the torrefaction process relies on the torrefaction temperature and the retention time for all residue. The increase in the torrefaction temperature and retention time leads to the increase in the mass loss of the residue. Anyways, the torrefaction temperature plays a much more important role in terms of mass loss compared to retention time. The residue lost an extreme amount of mass when the temperature increased from 250 to 300 °C but the further increase in temperature did not have

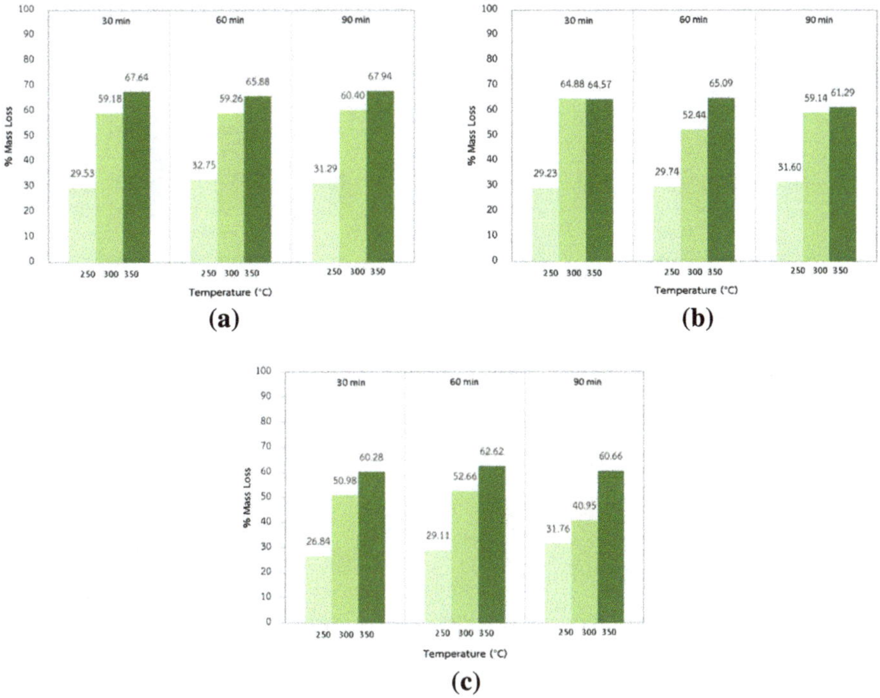

Fig. 6 Mass loss during the torrefaction process of **a** empty cobs, **b** husks, and **c** stems/leaves

a considerable effect on the mass loss. The mass loss from the torrefaction process was between 26.84 and 67.94 %wt.

4.2.2 Proximate Composition

Figures 7, 8 and 9 demonstrate the influence of the torrefaction temperature and retention time on the amount of volatile matter and fix carbon in the torrefied empty cobs, husks, and stems/leaves compared to raw residue.

Considering both the volatile matter and the fix carbon of all residue, it can be observed from Fig. 7, 8 and 9 that the amount of the volatile matter and the fix carbon in the torrefied residue depends enormously on the torrefaction temperature but slightly on the retention time. The increase in the torrefaction temperature leads to a decrease in the volatile matter and an increase in the fix carbon in the torrefied residue. By increasing the torrefaction temperature from 250 to 300 °C, the volatile matter decreased from 75.44–83.43 %wt. to 47.27–53.36 %wt., while the fix carbon content increased from 12.06–21.60 %wt. to 37.87–47.80 %wt. However, at a higher temperature, the volatile matter and ash content did not change remarkably. This

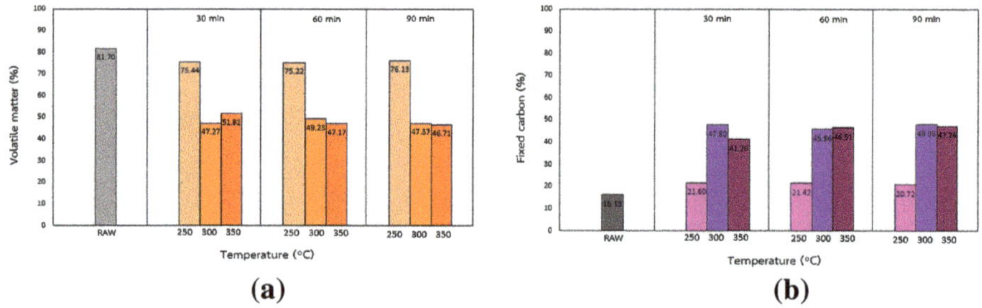

Fig. 7 **a** Volatile matter and **b** fix carbon in raw and torrefied empty cobs

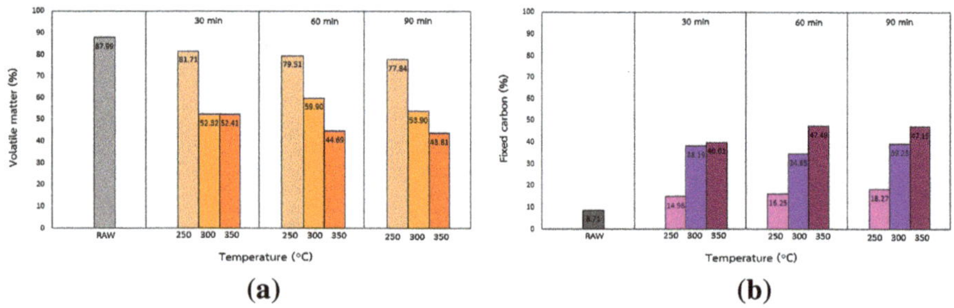

Fig. 8 **a** Volatile matter and **b** fix carbon in raw and torrefied husks

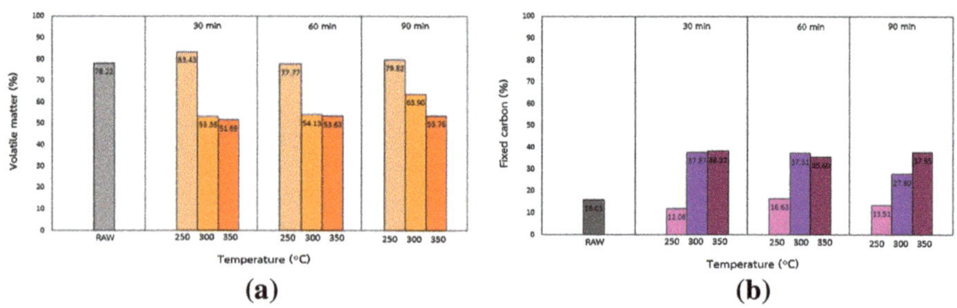

Fig. 9 **a** Volatile matter and **b** fix carbon in raw and torrefied stems/leaves

means that the volatile matter was driven from the residue in the torrefaction process. This can be the reason for the mass loss in the torrefied residue.

4.2.3 Heating Value

Figure 10 shows the dependency of the heating value on the torrefaction temperature and retention time for the torrefied residue.

According to Fig. 10, the trend of the heating value versus the torrefaction temperature and the retention time was as same as for the volatile matter but was in

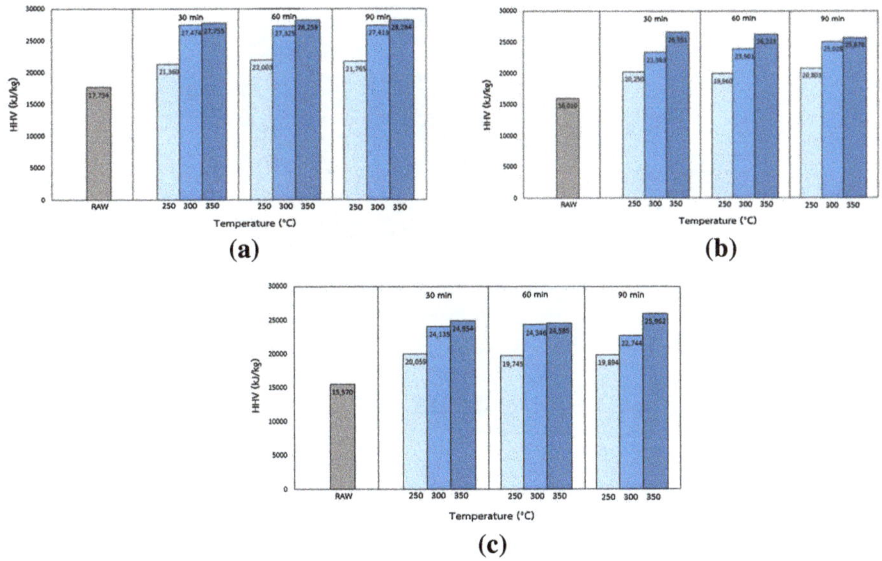

Fig. 10 Heating value of the torrefied corn residue **a** empty cobs, **b** husks, and **c** stems/leaves

contrast with the fix carbon. The heating value of the torrefied residue increased with the increased temperature, especially from 250 to 300 °C.

According to the results of the analysis of the torrefaction process, it can be concluded that for all residue the torrefaction temperature is important for the quality of torrefaction products, while the variation in retention time does not create a huge difference in the quality of torrefaction products. Hence, the retention time of 30 min was selected as the suitable value for the torrefaction process for corn residue. By increasing the torrefaction temperature to 300 °C at the retention time of 30 min, the volatile matter decreased 31.78–42.14 %wt. from the raw residue, the fix carbon increased 3–4.5 times from the value of the raw residue, and the heating value increased 45.93–55% from the raw residue. However, a further increase in the torrefaction temperature did not result in a significant change in the quality of the torrefied residue. Therefore, the torrefaction temperature of 300 °C was considered to be the proper value for the torrefaction process for the corn residue.

5 Briquetting Process

In practical terms, the agricultural residue has to be prepared into dense form or briquettes before being used as fuel. This step facilitates the transportation and storage of bio-fuel. In this study, raw residue and torrefied residue with high fuel quality

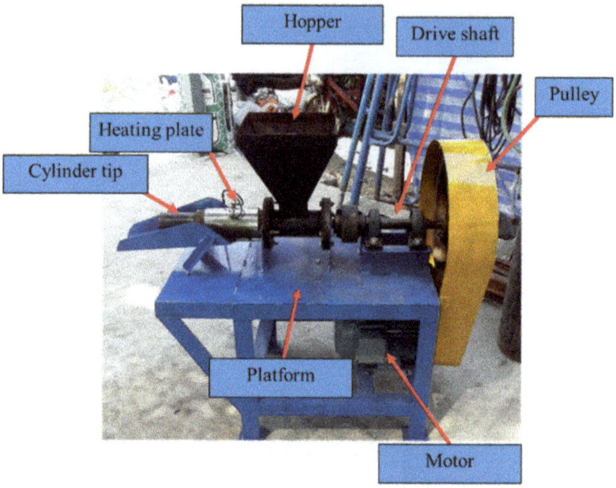

Fig. 11 Screw press briquetting machine

were mixed with binders and briquetted in a briquetting machine in order to produce briquette fuel.

5.1 Experimental Setup and Procedure

Figure 11 shows the screw-press briquetting machine, which was driven by a 3 HP 380 V motor. The dimension of the briquetting machine was 500 mm × 900 mm. A heating plate with a temperature control unit was installed around the cylinder to enable the briquetting of the corn residue.

Since the results from the torrefaction process revealed that the temperature of 300 °C and the retention time of 30 min were the proper operating condition, and only the torrefaction products from this condition were further processed at this step. As raw and torrefied empty cobs have a cylindrical shape, the briquetting process is not necessary. Unlike empty cobs, raw and torrefied husks, and stems/leaves, are amorphous, and therefore the briquetting process is essential. After removing the moisture content from the raw corn residue by heating it at a temperature of 105 °C for 24 h, the mixed raw and torrefied residue (stems/leaves: empty cobs: husks = 7: 2: 1) was crushed into a smaller size at around 0.5–1 mm using a cutter mill and this was mixed with the binder (water: tapioca flour with the ratio of 10: 1 by weight) at the proportion of residue: binder as 1: 0.7 by weight. Finally the prepared mixed residue was compressed with the briquetting machine and cut into 50–60 mm in length with a diameter of about 25 mm. Figure 12 illustrates the briquetting process of (a) empty cobs and (b) mix residue, respectively.

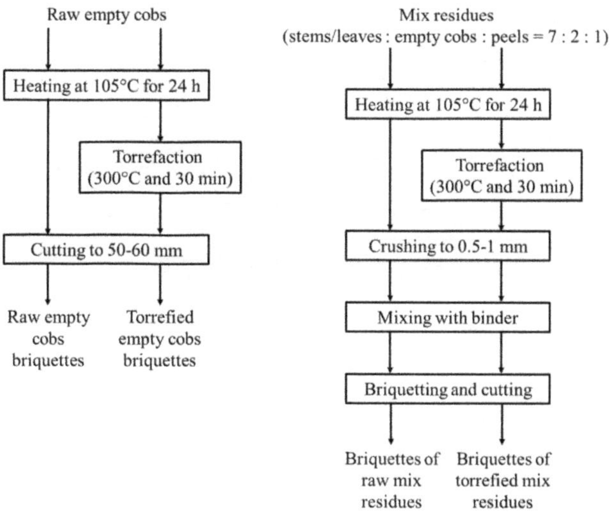

Fig. 12 Procedure for the production of briquettes from corn residue **a** empty cobs and **b** mixed residue

The fuel properties to be compared in this sub-process were density and water absorbability. The density of the fuel briquettes indicated the transportability of the fuel and water absorbability can be used to predict how long the fuel can be stored without degrading its properties.

There are two types of density—apparent density and bulk density—that were to be determined. The first one is the density calculated according to the mass per unit volume of the briquettes, while the second one is the density measured by the weight of the briquettes in a container whose volume was given. The formulas for the calculation of apparent density and bulk density is given in Eq. (1) and Eq. (2), respectively.

$$\rho_{\mathrm{app}} = \frac{m_{briquette}}{V_{briquette}} \tag{1}$$

$$\rho_{\mathrm{bulk}} = \frac{m_{briquettesincontainer}}{V_{container}} \tag{2}$$

In this study, the water absorbability was defined as the ability of the briquettes to adsorb water after a long period. The water absorbability was determined by placing the briquettes in a closed container in which the relative humidity was maintained at 76% by being filled with water and sodium chlorine solution. The briquettes were placed in a relative humidity-controlled container for 6, 12, 24, 48 and 96 h and the briquettes were weighed after each period. Equation (3) was used to calculate the water absorbability of the briquettes.

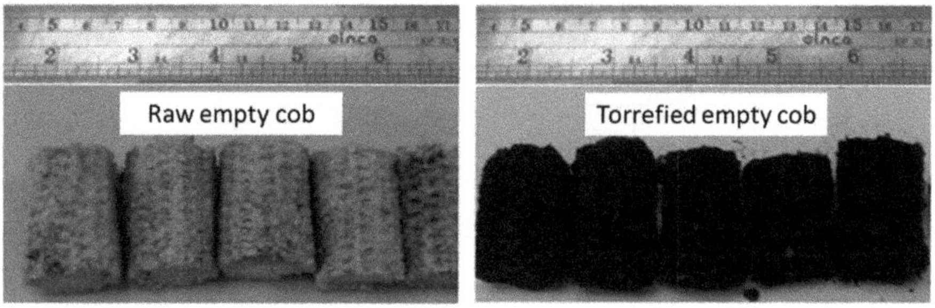

Fig. 13 Briquettes of empty cobs

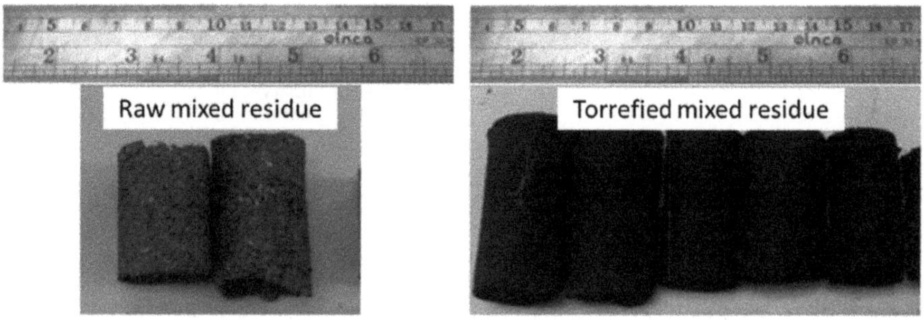

Fig. 14 Briquettes of mixed residue

$$\alpha_w = \frac{m_t - m_0}{m_0} \tag{3}$$

where

α_w is the water adsorbability
m_t is the mass of the briquettes at a given time
m_0 is the mass of the briquettes at the initial time.

5.2 *Results and Discussion*

The briquettes of empty cobs and mixed corn residue with and without using the torrefaction process are shown in Fig. 13 and Fig. 14, respectively. The briquettes have a cylindrical form with a diameter of around 25 mm and a length of 50–60 mm.

Table 3 Apparent and bulk density

Type of briquettes	Apparent density (kg/m^3)	Bulk density (kg/m^3)
Raw empty cobs	249.34	94.14
Torrefied empty cobs	162.67	78.45
Raw mixed residue	286.08	135.17
Torrefied mixed residue	680.98	271.55

5.2.1 Density

Table 3 lists the apparent and bulk density of the briquettes. It can be seen that the apparent and bulk density of the briquettes from the empty cobs considerably differs from the apparent and bulk density of the briquettes from the mixed residue.

In the case of empty cobs, the torrefaction process had a negative effect on both apparent and bulk density. Both values for the torrefied empty cobs, which were measured at 162.67 kg/m^3 and 78.45 kg/m^3, respectively, were less than the values of the raw empty cobs. On the other hand, the briquettes from the torrefied mixed residue had an apparent density of 680.98 kg/m^3 and a bulk density of 271.55 kg/m^3, which was much higher than the briquettes from the raw mixed residue.

5.2.2 Water Absorbability

After placing the different types of briquettes in the moisture-controlled container for the specified time, as shown in Fig. 15, it was found that during the first 48 h all of the briquettes were able to adsorb water and after 48 h the briquettes could absorb only a small amount of water or could not absorb water at all, as shown in Fig. 16. This implies that after 48 h all of the briquettes were almost saturated with water in a porous structure.

Comparing the water absorbability of the raw and torrified residue, it can be seen in Fig. 16 that the torrefied residue from both the empty cobs and mixed residue could absorb less water than the raw residue. The maximum amount of water absorbed by the torrefied empty cobs and torrefied mixed residue was approximately 5 %wt. and 9 %wt., respectively, after 48 h, whilst the raw empty cobs and the torrefied mixed residue could absorb the highest water amount of 11 %wt. and 15 %wt., respectively. This shows the hydrophobic characteristic of corn residue after the torrefaction process, which could help to store the biochar for a longer time without absorbtion of moisture.

Fig. 15 Different briquettes in a closed moisture-controlled container

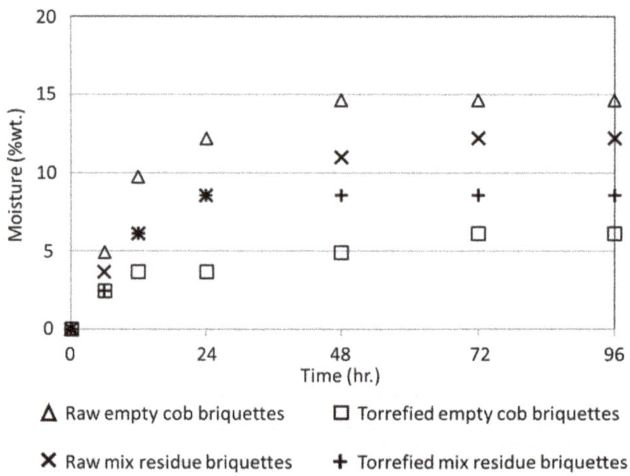

Fig. 16 Water absorbability of different briquettes

6 Gasification Process

The aim of performing a gasification test is to investigate and compare the gasification behavior of raw residue and torrefied residue in terms of temperature distribution, producer gas composition, producer gas heating value and cold gas efficiency.

6.1 Experimental Setup and Procedure

The lab scale gasification system used in this study consisted of 4 main parts: a 10 kg/h batch type downdraft gasifier, an air supply unit, a measuring unit, and a gas sampling unit, as illustrated in Fig. 17.

Figure 18 shows the gasifier which is separated into 4 zones: a hopper, a pyrolysis chamber, a reaction chamber, and an ash chamber. The reaction chamber can be divided into an oxidation zone and a reduction zone. The gasifier can sustain temperatures up to 1,200 °C. The air supply unit introduces the gasification air into the gasifier. Air flows into the tank through a 0.75 HP-blower and then preheated in the air preheater. Preheated air is introduced into the gasifier in the reaction chamber. The

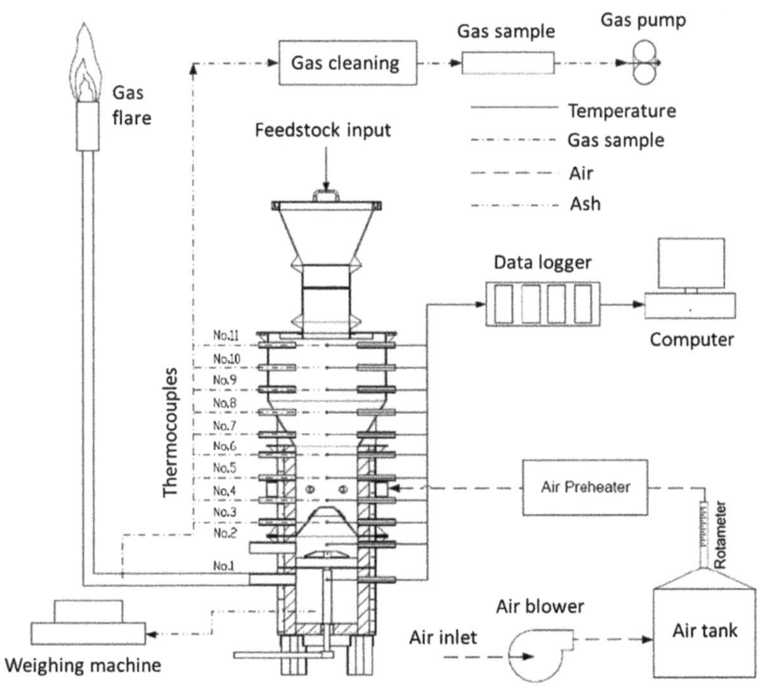

Fig. 17 A 10 kg/hr lab scale gasification system

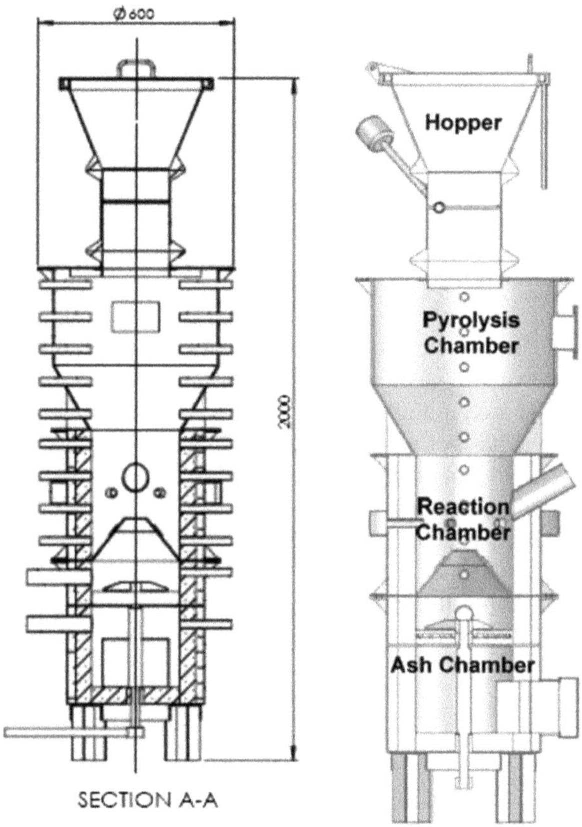

Fig. 18 A downdraft gasifier

Table 4 Number of experiments and operating conditions

No.	Type of feedstock briquettes	Air flow rate (L/min)	Equivalent ratio (-)
1	Raw empty cob	250	0.5
2	Torrified empty cob	250	0.5
3	Torrified empty cob	300	0.6
4	Torrified empty cob	350	0.7

air flow rate is controlled by the air rotameter. In this study, the gasification process was performed at the air flow rates of 250, 300, and 350 L/min, which corresponds to the equivalent ratio of 0.5–0.7. During the experiment, the temperature was measured and recorded by 11 k-type thermocouples installed in different heights of the gasifier (Fig. 17). The producer gas that occurred was taken as the gas sample using a gas sampling unit comprising a vacuum pump and condensing bottles. The number of experiments and the operating conditions were listed in Table 4.

The concentration of producer gas composition, including H_2, CO, CO_2, CH_4, O_2 and N_2, were measured using Shinaszu Gas Chromatography (GC) model GC-2014. The producer gas heating value (HV_{gas}), the feedstock consumption rate (FCR), and the cold gas efficiency (η_{th}) were determined using Eq. (4), Eq. (5), and Eq. (6), respectively.

$$HV_{gas} = \frac{[(30 \times \%CO) + (25.7 \times \%H_2) + (85.4 \times \%CH_4)] \times 4.2}{1000} \quad (4)$$

where %CO, $\%H_2$, and $\%CH_4$ are the concentration of CO, H_2, and CH_4 in the producer gas in % Vol., respectively

$$FCR = \frac{Weight\ of\ fuel\ used}{Operating\ time} \quad (5)$$

$$\eta_{th} = \frac{Q_{gas} \times HV_{gas}}{FCR \times HV_{fuel}} \quad (6)$$

where

Q_{gas} is the producer gas flow rate (m^3/min)
HV_{gas} is the producer gas heating value (kJ/m^3)
HV_{fuel} is the heating value of the feedstock (kJ/kg)
FCR is the fuel consumption rate (kJ/min).

6.2 *Results and Discussion*

6.2.1 Temperature Distribution

Figure 19 shows the temperature distribution of the different air flow rates for the raw empty cobs and the torrefied empty cobs.

As observed in Fig. 19, the temperature distribution of torrefied empty cobs did not enormously fluctuate with reaction time, whereas the temperature distribution of the raw empty cobs gasification highly fluctuated. Focusing on the effect of the air flow rate, it was clearly seen that the increase in the air flow rate introduced in the gasifier resulted in an increase in the temperature of the combustion zone and a wider temperature range in the combustion zone. This provided a greater heat supply for the pyrolysis and reduction zones, and consequently a higher temperature in the reduction zone [12]. Table 5 shows the temperature in each reaction zone at different air flow rates.

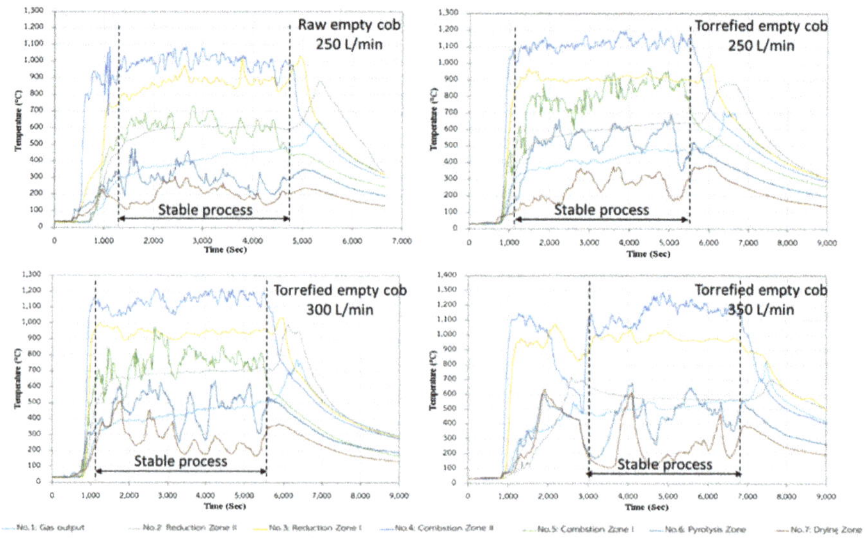

Fig. 19 Temperature distribution at different air flow rates

Table 5 Temperature in each reaction zone

No.	Feedstock	Air flow rate (L/min)	Temperature (°C)			
			Drying zone	Pyrolysis zone	Combustion zone	Reduction zone
1	Raw empty cobs	250	133–189	251–484	716–910	710–815
2	Torrefied empty cobs	250	137–192	456–667	939–1,089	720–794
3	Torrefied empty cobs	300	143–193	256–647	847–1,099	775–839
4	Torrefied empty cobs	350	161–195	327–683	1,126–1,292	764–808

6.2.2 Producer Gas Composition

Table 6 presents the producer gas composition obtained from gasification of various feedstock at different air flow rates. CO was the main producer gas composition, but there was also trace amount of H_2 and CH_4.

Table 6 indicated that the feedstock preparation affected the producer gas composition. The producer gas obtained from the torrefied empty cobs contained a higher amount of combustible gases, e.g. H_2, CO, and CH_4, than that derived from the raw empty cobs for all air flow rates, with a CO content of 18.60–27.79 %Vol, an H_2 content of 3.63–9.37 %Vol., and a CH_4 content of 0.17–0.33 %Vol. Comparing the producer gas composition yielded from different air flow rates, it was remarkably seen that the air flow rate played an important role on the amount of CO in the producer gas. The CO content increased from 18.60 %Vol. at the air flow rate of

Table 6 Producer gas composition

No.	Feedstock	Air flow rate (L/min)	Producer gas composition (%Vol.)					
			CO	H_2	CH_4	CO_2	O_2	N_2
1	Raw empty cobs	250	5.72	0.97	0.03	11.78	5.35	76.20
2	Torrefied empty cobs	250	18.60	3.63	0.31	6.07	4.35	67.05
3	Torrefied empty cobs	300	22.35	9.37	0.33	6.95	0.52	60.48
4	Torrefied empty cobs	350	27.79	6.39	0.17	3.61	1.23	60.81

Table 7 Heating value, feedstock consumption rate, and cold gas efficiency

No.	Feedstock	Air flow rate (L/min)	Parameters		
			FCR (kg/h)	HV (MJ/Nm^3)	η_{th} (%)
1	Raw empty cobs	250	6	0.84	31.34
2	Torrefied empty cobs	250	3.47	2.84	55.61
3	Torrefied empty cobs	300	4.64	3.95	86.90
4	Torrefied empty cobs	350	5.71	4.25	80.70

250 L/min to 27.79 %Vol. at the air flow rate of 350 L/min. Unlike the CO content, the air flow rate did not significantly affect the content of the H_2 or CH_4 in the producer gas. The maximum yield of H_2 and CH_4 in the producer gas can be achieved at the air flow rate of 300 L/min [12].

6.2.3 Heating Value, Feedstock Consumption Rate, and Cold Gas Efficiency

Because the feedstock preparation and air flow rate have an influence on the gasification process and producer gas composition, they also has the effect on heating values of producer gas, feedstock consumption rate, and cold gas efficiency, as listed in Table 7.

From Table 7, the consumption rate of the raw empty cobs was greater than that of the torrefied empty cobs as a result of the higher volatile matter and lower fixed carbon [12]. For the same feedstock but different air flow rates, it can be seen that the increase in the air flow rate provided an increase in the heating value as a consequence of the higher amount of combustible gas in the producer gas, and consequently, an increase in cold gas efficiency. The maximum heating value of the producer gas of 4.25 MJ/Nm^3 can be reached when using torrefied empty cobs as feedstock with the air flow rate of 350 L/min, and the highest cold gas efficiency of 86.9% can be gained from the gasification of torrefied empty cobs at the air flow rate of 300 L/min.

7 Conclusion

Corn residue, including corn cobs, corn stems/leaves and corn husks, has been studied in order to improve its fuel quality through the torrefaction process. It was found that the fuel obtained from this process can improve its heating value and can be used as high quality green fuel or biochar. The important parameter for the torrefaction process is the reaction temperature, where the retention time exerts less impact. The torrefaction temperature played a major role in affecting the heating value more than the retention time. The torreaction process also helped to improve the hydrophobic properties of the fuel, which would help in storage and keeping the fuel for use for a longer time without any absorption of water or moisture due to the closing of the open surface pores of the fuel. It can be concluded that the residual after harvesting corn in the field, including empty cobs, stems/leaves and husks, can be upgraded to be a premium fuel (biochar) through the torreaction process with a torrefaction temperature of 300 °C and a retention time of 30 min. The biochar from the corn residual can be used as premium feedstock for heating purposes with a high heating value and low smoke due to the higher fraction of fixed carbon and lower volatile matter.

Acknowledgements The authors would like to thanks EGAT-NSTDA Research and Development Co-funding Program for the financial support under the contract number of FOA-CO-2558-1160-TH. Additionally, this study was performed under the support of Department of Mechanical and Aerospace Engineering as well the Science and Technology Research Institute, King Mongkut's University of Technology North Bangkok, Thailand.

References

1. BP, *BP Statistical Review of World Energy* (2016)
2. BP, *BP Energy Outlook* (2013)
3. Energy Policy and Planning Office, Ministry of Energy, Thailand, Energy-Information (2016)
4. Department of Alternative Energy Development and Efficiency, The Alternative Energy Development Plan (2015)
5. International Energy Agency (IEA), *World Energy outlook 2016* (IEA/OECD, Paris, 2016)
6. W.-H. Chen, J. Peng, X.T. Bi, *A State of the Art Review of Biomass Torrefaction, Densification and Application* (Department of Nation Cheng Kung University, Taiwan and University of British Columbia, Canada, 2015)
7. D.R. Nhuchhen, P. Basu, B. Acharya, *A Comprehensive Review on Biomass Torrefaction* (IBIMA Publishing, 30 April 2014)
8. P. Basu, *Biomass Gasification, Pyrolysis and Torrefaction* (Elsevier Inc, 2013)
9. M.-F. Li, X. Li, J. Bian, J.-K. Xu, S. Yang, R.-C. Sun, Influence of temperature on bamboo torrefaction under carbon dioxide atmosphere. Ind. Crops Prod. **76**, 149–157 (2015)
10. Office of Agricultural Economics, Thailand. http://www.oae.go.th/public_stat.html
11. B.W. Farnum, S.A. Farnum, E.F. Bitzan, *Analysis of lignite liquefaction products processed with syngas and hydrogen sulfide* (Grand Forks Energy Technology Center)
12. K. Laohalidanond, N. Kongkaew, S. Kerdsuwan, Gasification behavior study of torrefied empty corn cobs. Energy Procedia **38**, 175–180 (2017)

Lightning Source UK Ltd.
Milton Keynes UK
UKHW02n1228100718
325496UK00002B/50/P

9 789811 074721